Interacting Electrons in Reduced Dimensions

NATO ASI Series

Advanced Science Institutes Series

A series presenting the results of activities sponsored by the NATO Science Committee, which aims at the dissemination of advanced scientific and technological knowledge, with a view to strengthening links between scientific communities.

The series is published by an international board of publishers in conjunction with the NATO Scientific Affairs Division

A	**Life Sciences**	Plenum Publishing Corporation
B	**Physics**	New York and London
C	**Mathematical and Physical Sciences**	Kluwer Academic Publishers Dordrecht, Boston, and London
D	**Behavioral and Social Sciences**	
E	**Applied Sciences**	
F	**Computer and Systems Sciences**	Springer-Verlag
G	**Ecological Sciences**	Berlin, Heidelberg, New York, London,
H	**Cell Biology**	Paris, and Tokyo

Recent Volumes in this Series

Volume 208—Measures of Complexity and Chaos
edited by Neal B. Abraham, Alfonso M. Albano, Anthony Passamante, and Paul E. Rapp

Volume 209—New Aspects of Nuclear Dynamics
edited by J. H. Koch and P. K. A. de Witt Huberts

Volume 210—Crystal Growth in Science and Technology
edited by H. Arend and J. Hulliger

Volume 211—Phase Transitions in Soft Condensed Matter
edited by Tormod Riste and David Sherrington

Volume 212—Atoms in Strong Fields
edited by Cleanthes A. Nicolaides, Charles W. Clark, and Munir H. Nayfeh

Volume 213—Interacting Electrons in Reduced Dimensions
edited by Dionys Baeriswyl and David K. Campbell

Volume 214—Science and Engineering of One- and Zero-Dimensional Semiconductors
edited by Steven P. Beaumont and Clivia M. Sotomayor Torres

Volume 215—Soil Colloids and their Associations in Aggregates
edited by Marcel F. De Boodt, Michael H. B. Hayes, and Adrien Herbillon

Series B: Physics

Interacting Electrons in Reduced Dimensions

Edited by

Dionys Baeriswyl

ETH
Zürich, Switzerland

and

David K. Campbell

Los Alamos National Laboratory
Los Alamos, New Mexico

Plenum Press
New York and London
Published in cooperation with NATO Scientific Affairs Division

Proceedings of a NATO Advanced Reseach Workshop (SDL Panel)
on Interacting Electrons in Reduced Dimensions,
held October 3-7, 1988,
in Turin, Italy

Library of Congress Cataloging-in-Publication Data

NATO Advanced Research Workshop (SDL Panel) on Interacting Electrons in Reduced Dimensions (1988 : Turin, Italy)
Interacting electrons in reduced dimensions / edited by Dionys Baeriswyl and David K. Campbell.
p. cm. -- (NATO ASI series. Series B, Physics ; vol. 213)
"Proceedings of a NATO Advanced Research Workshop (SDL Panel) on Interacting Electrons in Reduced Dimensions, held October 3-7, 1988, in Turin, Italy"--T.p. verso.
Published in cooperation with NATO Scientific Affairs Division.
Includes bibliographical references.
ISBN-13: 978-1-4612-7869-6 e-ISBN-13: 978-1-4613-0565-1
DOI: 10.1007/978-1-4613-0565-1
1. Solid state physics--Congresses. 2. Hubbard model--Congresses. I. Baeriswyl, D. (Dionys), 1944- . II. Campbell, David K. III. North Atlantic Treaty Organization. Scientific Affairs Division. IV. Title. V. Series: NATO ASI series. Series B, Physics ; v. 213.
QC176.A1N29 1988
530.4'1--dc20 89-48931
CIP

A Division of Plenum Publishing Corporation
233 Spring Street, New York, N.Y. 10013

SPECIAL PROGRAM ON CONDENSED SYSTEMS OF LOW DIMENSIONALITY

This book contains the proceedings of a NATO Advanced Research Workshop held within the program of activities of the NATO Special Program on Condensed Systems of Low Dimensionality, running from 1983 to 1988 as part of the activities of the NATO Science Committee.

Other books previously published as a result of the activities of the Special Program are:

SPECIAL PROGRAM ON CONDENSED SYSTEMS OF LOW DIMENSIONALITY

Volume 199 SCIENCE AND TECHNOLOGY OF FAST ION CONDUCTORS
edited by Harry L. Tuller and Minko Balkanski

Volume 200 GROWTH AND OPTICAL PROPERTIES OF WIDE-GAP II–VI LOW-DIMENSIONAL SEMICONDUCTORS
edited by T. C. McGill, C. M. Sotomayor Torres, and W. Gebhardt

Volume 202 POINT AND EXTENDED DEFECTS IN SEMICONDUCTORS
edited by G. Benedek, A. Cavallini, and W. Schröter

Volume 203 EVALUATION OF ADVANCED SEMICONDUCTOR MATERIALS BY ELECTRON MICROSCOPY
edited by David Cherns

Volume 206 SPECTROSCOPY OF SEMICONDUCTOR MICROSTRUCTURES
edited by Gerhard Fasol, Annalisa Fasolino, and Paolo Lugli

PREFACE

As its name suggests, the 1988 workshop on "Interacting Electrons in Reduced Dimensions" focused on the wide variety of physical effects that are associated with (possibly strongly) correlated electrons interacting in quasi-one- and quasi-two-dimensional materials. Among the phenomena discussed were superconductivity, magnetic ordering, the metal-insulator transition, localization, the fractional Quantum Hall effect (QHE), Peierls and spin-Peierls transitions, conductance fluctuations and sliding charge-density (CDW) and spin-density (SDW) waves. That these effects appear most pronounced in systems of reduced dimensionality was amply demonstrated at the meeting. Indeed, when concrete illustrations were presented, they typically involved chain-like materials such as conjugated polymers, inorganic CDW systems and organic conductors, or layered materials such as high-temperature copper-oxide superconductors, certain of the organic superconductors, and the QHE samples, or devices where the electrons are confined to a restricted region of sample, e.g., the depletion layer of a MOSFET.

To enable this broad subject to be covered in thirty-five lectures (and about half as many posters), the workshop was deliberately focused on theoretical models for these phenomena and on methods for describing as faithfully as possible the "true" behavior of these models. This latter emphasis was especially important, since the inherently many-body nature of problems involving interacting electrons renders conventional effective single-particle/mean-field methods (e.g., Hartree-Fock or the local-density approximation in density-functional theory) highly suspect. Again, this is particularly true in reduced dimensions, where strong quantum fluctuations can invalidate mean-field results. By convening a group of leading experts in theoretical methods ranging from rigorous analytical approaches ("bosonization" and Bethe *Ansatz*) for infinite systems through numerical algorithms, both "exact" for small systems (valence bond and Lanczos methods) and "stochastic" for intermediate systems (Quantum Monte Carlo (QMC)) to renormalization group and variational methods, the workshop was able to survey critically essentially the entire spectrum of relevant theoretical methods. As a result, the limitations of the various approaches were clearly exposed and a number of critical issues for future study were identified.

Both the formal presentations, recorded in these proceedings, and the informal discussions reflected the excitement that the confluence of analytic and computational approaches and the ever-increasing number of novel materials have brought to this topic. Among the particularly exciting new results were advances in QMC methodology, stimulated by work of the Trieste group, which offer substantial promise for fermionic QMC calculations in two (and higher !) dimensions. In another important area for numerical calculations – relating results for (small) finite-size clusters to the behavior of infinite systems – two new developments were presented, first the concept of "effective dispersion relations" and second the idea of "phase randomization/boundary condition averaging." Considerable discussion was devoted to holes in antiferromagnets, especially in the framework of the two-dimensional t-J model and its variants. Unfortunately (but unsurprisingly !), the intricate issue of superconductivity in these models was not answered in a definitive way. Similarly, no generally accepted "recipe" for describing materials as chemically and structurally complex as the high-temperature superconductors within the framework of simple model Hamiltonians was given. However, recent developments in *"ab initio"* methods (e.g. improvements of the local density approximation) may open the way for a semi-quantitative determination from the actual materials of certain crucial parameters in the models. We hope that the success of the conference, as measured by the level of interaction, discussion, and excitement at the meeting, is apparent to the readers of these proceedings.

The success of the conference owed much to the ambience of the Institute for Scientific Interchange (ISI) in Torino, Italy, and to the superb efforts of the staff, Tiziana Bertoletti, Carmen Novella, and Marinella Prato. As organizers we are also grateful to the President of the ISI, Prof. Tullio Regge, and its Director, Prof. Mario Rasetti, for their intellectual and financial support and are pleased to thank the NATO Office of Scientific Affairs for its support of the meeting under the NATO Advanced Research Workshop program.

Dionys Baeriswyl

David Campbell

CONTENTS

BRIEF HISTORY OF A ONE-DIMENSIONAL GROUND STATE WAVE FUNCTION

Bill Sutherland

Department of Physics
University of Utah
Salt Lake City, UT 84112 USA

THE WAVEFUNCTION

We present a brief discussion of a class of one-dimensional ground state wavefunctions. First, we review older results, then summarize very recent results of Haldane and Shastry, follow with a reappreciation of a paper of Gaudin, and finish with new results of our own.

The one-dimensional M-particle wave functions[1,2] we consider are of product form, and depend upon a parameter λ in the following way:

$$\Psi_\lambda(x_1 \ldots x_M) = \prod_{1=j<k}^{M} \psi^\lambda(x_k - x_j) = [\Psi_1]^\lambda .$$

We consider three explicit forms for the function $\psi(x)$:

$$\begin{aligned} \psi(x) &= x, && \text{algebraic;} \\ &= \sin(\pi x/L), && \text{trigonometric;} \\ &= \theta_1(\pi x/L \,|\, it), && \text{elliptic.} \end{aligned}$$

These three forms constitute a single family in that the trigonometric case is the limit of the elliptic case as $t \to \infty$, and the algebraic case is the limit of the trigonometric case as $L \to \infty$.

Interacting Electrons in Reduced Dimensions
Edited by D. Baeriswyl and D.K. Campbell
Plenum Press, New York

We note that since $\psi(0) = 0$, and the system is one-dimensional, we may use this wavefunction for a single ordering of particles, while for any other ordering, we multiply the absolute value by ±1 as the statistics demand. Thus for λ odd (even), the wavefunction is analytic in the particle coordinates for fermions (bosons). Diagonal elements of the reduced density matrices will be independent of statistics while off-diagonal elements will not. Since the wavefunction is positive for an ordering $x_1<...<x_M$, this is the groundstate wavefunction.

Why study this particular wavefunction? We might answer this question by: (a) physical motivation; (b) computational possibilities; (c) exact solubility. We have little to say here about physical motivation, except that similar forms have proven useful in the study of random Hamiltonian models for nuclear physics, the quantized Hall effect, and high T_c superconductivity.

As for computability, this rests on the following alternate forms for the wavefunction:

$$\prod_{1=j<k}^{M} (x_k - x_j) \cong \det_{jk} [x_j^{k-1}];$$

$$\prod_{1=j<k}^{M} \sin[\pi(x_k - x_j)/L \cong \det_{jk} [\exp\{i\pi x_j (M+1-2k)/L\}];$$

$$\prod_{1=j<k}^{M} \theta_1[\pi(x_k - x_j)/L \mid it] \cong \int_0^{1/M} d\gamma \det_{jk} [\theta_3\{\pi(x_j/L-k/M+\gamma) \mid it/M\}].$$

Here the symbol ≅ means up to a multiplicative normalization constant. The first two forms are the well-known expression for a van der Monde determinant, while the third is much less familiar. The content of this form, of course, is that the wavefunction for $\lambda=1$ is (nearly) a Slater determinant of one-particle orbitals. The "nearly" is because the orbitals are localized about the sites of a discrete lattice, and thus the Slater determinant is smeared to recover translational invariance.

Also, as will often prove useful, the probability density can also be viewed as a Boltzmann weight for a corresponding classical system at temperature T interacting by a pair potential U(x), through

$$\Psi_\lambda^* \Psi_\lambda = \exp\left[-\beta \sum_{1=j<k}^{M} U(x_k - x_j) \right],$$

where $\beta = 2\lambda = 1/T$ and $U(x) = -\ln[\psi(x)]$. Examining the functional form for $\psi(x)$, we see that this classical thermodynamic system is much like a one-component plasma, in that the interaction is very long-ranged.

THE HAMILTONIAN

To give the third reason why we study this particular wavefunction, we claim that it is the exact ground state wavefunction for a reasonable Hamiltonian:

$$H = -\sum_{j=1}^{M} \frac{\partial^2}{\partial x_j^2} + \sum_{1=j<k}^{M} V(x_k - x_j),$$

with pair potential V(x). To verify, we must satisfy the following form of the Schrodinger equation, obtained from the usual form by simply dividing by Ψ_λ, permissible since this is the ground state wavefunction, and hence is non-zero:

$$\frac{1}{\Psi_\lambda} \sum_{j=1}^{M} \frac{\partial^2 \Psi_\lambda}{\partial x_j^2} = \sum_{1=j<k}^{M} V(x_k - x_j) - E_0.$$

Substitution of the product form for the wavefunction gives a functional form for $\phi(x) = \psi'(x)/\psi(x)$, the logarithmic derivative of $\psi(x)$:

$$\phi(x)\,\phi(x+y) + \phi(y)\,\phi(x+y) - \phi(x)\,\phi(y) = v(x) + v(y) + v(x+y).$$

Essentially, this equation is the requirement that the 3-particle cross terms must be equivalent to a sum of pair potentials.

One may easily verify that our forms for $\psi(x)$ do satisfy the equation; the functional equation is equivalent to standard addition formulas for the appropriate trigonometric or elliptic function. In fact, Calogero has shown that the elliptic case is the most general solution to the functional equation.

The corresponding pair potentials are:

$$V(x) = 2\lambda(\lambda-1)/x^2, \quad \text{algebraic;}$$

$$= \frac{2\lambda(\lambda-1)\pi^2}{L^2 \sin^2(\pi x/L)}, \quad \text{trigonometric.}$$

The algebraic case is a scattering state; it can be made normalizable by a harmonic attraction. But the trigonometric case is the periodic version of the inverse square algebraic case, and is normalizable. It is this trigonometric case that we will study almost exclusively in the rest of this talk. We do not show the explicit form of the potential for the elliptic case, but it is best interpreted as a one-dimensional one-component plasma in a jellium background. It can be shown -- at least for $\lambda=1$ -- to crystallize into a Wigner crystal, with long-ranged crystalline order.

NORMALIZATION AND CORRELATIONS

The distributions

$$\Psi_\lambda{}^*\Psi_\lambda = \prod_{1=j<k}^{M} | e^{2\pi i x_k/L} - e^{2\pi i x_j/L} |^{2\lambda} ,$$

have been much studied by Dyson, Mehta, Gaudin, and others in the theory of random matrices[3]. In this theory, matrices are chosen from an appropriate ensemble, and the following values of λ are singled out as special: $\lambda = 1/2$, orthogonal; $\lambda = 1$, unitary; $\lambda = 2$, symplectic. Often quantities such as correlations can only be evaluated at these special values; remember that $\lambda=1$ is the rather trivial case of free fermions. Let us add to the special values of λ, the trivial free boson case of $\lambda=0$. We now quote the known results.

Normalization

The normalization constant for the wavefunction is known for all $\lambda \geq 0$. With the exact form as given in the equation immediately above, we find

$$Z(2\lambda) = \frac{1}{M!} \prod_{j=1}^{M} \int_0^L dx_j \, \Psi_\lambda{}^*\Psi_\lambda = \frac{L^M \Gamma(1+\lambda M)}{M! \Gamma(1+\lambda)^M} .$$

We simply want to point out that Z has no singularities as a function of λ for λ real and positive.

Correlations - Diagonal

For the special values of λ -- and only for these four values -- the diagonal elements of all m-particle density matrices can be explicitly evaluated. The results are independent of statistics. If we write the pair correlation function G(r) as

$$g(r) = \delta(r) + \rho + g(r),$$

where $\rho = M/L$ is the particle density, then the asymptotic result for g(r) is

$$g(r) \to \frac{a}{r^2} + \frac{b \cos(2\pi\rho r)}{r^{2/\lambda}} , \text{ as } r \to \infty.$$

Also one finds that the perfect screening condition

$$\int_{-\infty}^{+\infty} dr \, g(r) = -1$$

is satisfied for all λ. The asymptotic form for g(r) has been established only for the four special values of λ, but they are conjectured or argued to hold for all λ. For the elliptic function case, as we remarked, g(r) exhibits long-range crystalline order.

Correlations - Off-Diagonal

For the off-diagonal elements of the density matrices, statistics are important, and much less is known. The case of free bosons with $\lambda=0$, and free fermions with $\lambda=1$, are of course trivial. The case of bosons when $\lambda=1$ has long been studied; it goes by the name of hard-core bosons and is important for the Ising model near T_c. If we study the momentum distribution n(k), equal to the Fourier transform of the one-particle density matrix, Lenard long ago showed

$$n(k) \rightarrow \frac{a}{|k|^{1/2}} \quad , \text{ as } k \rightarrow 0, \text{ for } \lambda = 1, \text{bosons.}$$

A further result is an exact evaluation of n(k) for $\lambda=2$, bosons. One finds

$$n(k) \rightarrow -a \ln(|k|) \quad , \text{ as } k \rightarrow 0, \text{ for } \lambda = 2, \text{bosons.}$$

And these are all the known results for the correlations.

DISCRETE CASE - RESULTS OF HALDANE AND SHASTRY

Can we use these "nice" wavefunctions for corresponding discrete lattice problems? This is the question asked and answered by Haldane, and by Shastry, in two recent papers[4,5]. We here outline their methods. First, let us replace the length L by a lattice of N sites. Then the coordinates x_j of the particles take integer values $x_j=1,2,...,N$. Let us return to the basic Schrodinger equation, and ask: Can we find a one-body operator J_j for the lattice system, which does exactly what $-\partial^2/\partial x_j^2$ does for the continuum system. Clearly in a basis $\varphi_k(x) = e^{ikx}$, $-\pi<k\leq\pi$, we must have

$$J\,\varphi_k(x) = k^2\,\varphi_k(x),$$

which uniquely fixes J. However, the Leibniz rule does not work, for

$$J\,[\varphi_k(x)\,\varphi_{k'}(x)\,] = J\,[\,\varphi_{k+k'}(x)\,] = (k+k')^2\,\varphi_{k+k'}(x)$$

only when $|k+k'|\leq\pi$.

The only hope then is to keep the momenta of the wavefunction entirely within the first Brillouin zone. Let us examine Ψ_1. For a given x_j, the largest momentum k is $k=\pi(M-1)/N$. Thus, we will have a solution on the lattice for fermions if λ = odd integer, and for bosons if λ = even integer. However, in both cases we must keep k within the first Brillouin zone, and so $k\lambda\leq\pi$, or $\rho\lambda\leq1$. Exploiting particle-hole symmetry allows us to also obtain the wavefunction for holes if $(1-\rho)\lambda\leq1$. Then everything will work much as before. The figure (2) in the last section shows these regions where we have a solution.

NORMALIZATION FOR DISCRETE CASE -- RESULTS OF GAUDIN

We have established that for the discrete problem we can find a "nice" Hamiltonian H_λ, whose ground state is given by the "nice" wavefunction Ψ_λ, only for certain discrete values of λ, and certain densities. At this point, we give up on the Hamiltonian and keep the wavefunction Ψ_λ as physical.

What then shall we calculate? Let us calculate the normalization $\Psi_\lambda^+\Psi_\lambda = Z(2\lambda)$. The motivation is as follows: We have seen that this normalization can be interpreted as a partition function. Then the singularities of the normalization as a function of λ are equivalent to singularities of the partition function as a function of temperature. As we know, such singularities or critical temperatures usually indicate the boundary between order and disorder. But order in the thermodynamic system is equivalent to order in the ground state of the quantum system. Thus, singularities of the normalization as a function of coupling constant λ, indicate the on-set of long-ranged order in the ground state.

At this point, we recall the beautiful and important paper of Gaudin[6] in 1973. In this paper, Gaudin considered the partition function $Z(\beta,\rho)$ as a function of inverse temperature β and density ρ, for the discrete system determined by the normalization of the wavefunction exactly as we have defined it. He establishes the following: (a) an exact calculation of $Z(2\lambda,\rho)$ for λ integer, and $\lambda\rho\leq 1$; (b) an exact calculation of $Z(2\lambda,\rho)$ for $\lambda=1/2$, and all ρ; (c) a singularity as a function of λ occurs at λ_c, where $1/2\leq\lambda\leq 1$. Thus, the partition function is evaluated for all four of the special values of λ, for all ρ. We note that at $\lambda=2$, there is a singularity in ρ at $\rho=1/2$, where the third derivative of Z with respect to ρ is discontinuous. The result (a) evaluates Z for all the wavefunctions of the previous section, which remain within the first Brillouin zone. However, fact (b) goes beyond. We add these exactly known results to the figure (2) of the next section.

STRONG COUPLING FOR THE DISCRETE CASE

We find it very tantalizing to have so many widely scattered exact results. For this reason, we now analyze[7] the strong-coupling, or low temperature limit when $\lambda\to\infty$. We then simply want to minimize the potential energy U of the classical lattice gas. In terms of an Ising spin picture, where spin up represents no particle on a lattice site and spin down represents a particle on a lattice site, the potential energy is

$$U = \frac{1}{4}\left\{ \sum_{1=j<k}^{N} \sigma_j \sigma_k u_{jk} + \frac{N \ln N}{2} \right\}.$$

Here the pair potential $u_{jk} = u(|j-k|)$ is

$$u(r) = -\ln|2\sin(\pi r/N)|.$$

We explicitly exhibit the part of the energy which comes from the incommensurability of the particle spacing with the lattice spacing, by writing

$$U = M \ln N/2 - M \ln M/2 + M\,\varepsilon(\rho).$$

The first term is the interaction of the particles with the uniform neutralizing background; the second term is the interaction energy of the particles in the continuum case. Thus $\varepsilon(\rho)$ is the increase in energy per particle, due to the lattice.

Now it is a theorem first proven by Hubbard[8] and by Pokrovsky[9], that the configuration of particles which minimizes such a repulsive potential concave upwards, is $x_j = [j/\rho]$. Here the brackets means "take the integer part". This formula holds for ρ both rational or irrational. In figure (1), we show $\varepsilon(\rho)$ for ρ up to half filling, There are logarithmic cusps at every rational ρ.

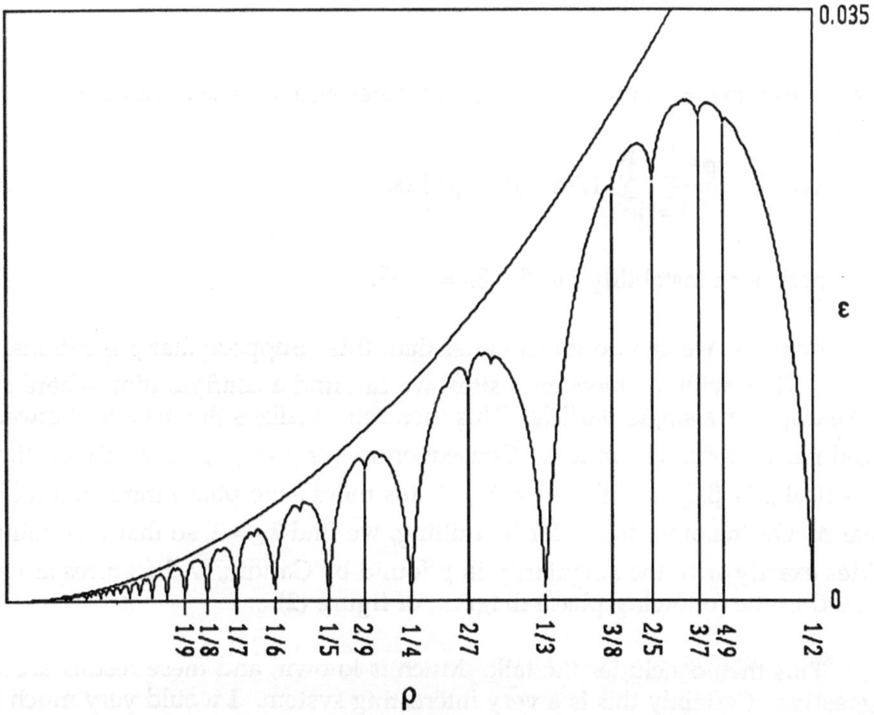

Fig. 1 The increase in the energy per particle, due to the lattice, is shown as a function of the particle density.

It thus appears that there might be ordered phases at every rational ρ. We now argue that this is likely the case[10]. The following brief argument can be made much more firm and convincing, although it falls short of a rigorous proof.

Let us consider a small deviation from the minimum energy configuration $x_j^o=[j/\rho]$, so $x_j = x_j^o + \delta x_j$. Then the energy is

$$U(\rho) \cong U_0(\rho) + \frac{1}{2} \sum_{1=j<k}^{N} u_0''(k-j)\,(\delta x_k - \delta x_j)^2,$$

where

$$u_0''(r) = \frac{\pi^2}{N^2 \sin^2(\pi r/N)} \cong \frac{1}{r^2}\,.$$

Finally with $x_j^o=[j/\rho]$,

$$\Delta U(\rho) \cong \frac{\rho^2}{2} \sum_{1=j<k}^{N} \frac{(\delta x_k - \delta x_j)^2}{(k-j)^2}.$$

If we choose $\delta x_j = 1, j = 1,\ldots,K$ and zero otherwise, we then evaluate

$$\Delta U(\rho) \cong \frac{\rho^2}{2} \sum_{k=1}^{K} \sum_{j\neq 1}^{K} 1/(k-j)^2 \cong \rho^2 \ln K.$$

Thus, there is an instability for $\beta \leq \beta_c = 1/\rho^2$.

Actually, we can do much better than this. Suppose that ρ is rational so that $\rho=m/n$. Then within a block of n sites, we can find a configuration where all δx_j are 0 except for a single particle. This then renormalizes the distance between displaced particles from n/m to n. Consequently, for $\rho=m/n$, the revised estimate for the critical β is $\beta_c(\rho=m/n) = n^2 = 2\lambda_c$. Thus there is no phase transition for irrational ρ. On the other hand, for half filling, we find $\lambda_c = 2$, so that this point coincides exactly with the singularity in ρ found by Gaudin. All features are summarized in the following phase diagram of figure (2).

This then concludes the talk. Much is known, and these results are very suggestive. Certainly this is a very interesting system. I would very much like to know more about it, and I hope that someone will be sufficiently clever to fill in the rest of the picture.

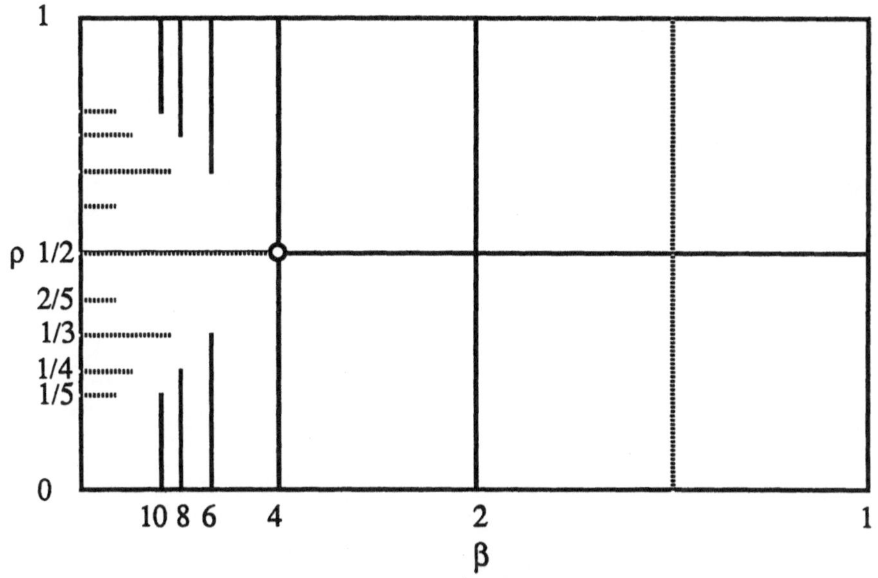

Fig. 2 The density - temperature phase diagram is shown. Solid lines are exactly known results; dashed horizontal lines are conjectured phases; the vertical dashed line is only to suggest the singularity between β=1 and β=2; the circle indicates a singularity. Finite resolution prohibits more detail.

REFERENCES

1. F. Calogero, 1983, Integrable Dynamical Systems and Related Mathematical Results, in: "Nonlinear Phenomena," Springer Lecture Notes in Physics, vol. 189, Springer-Verlag, Berlin.
2. B. Sutherland, 1985, Introduction to Bethe's ansatz, in: "Exactly Solvable Problems in Condensed Matter and Relativistic Field Theory," Springer Lecture Notes in Physics, vol. 242, B.S.Shastry, S.S.Jha and V.Singh, ed., Springer-Verlag, Berlin.
3. M. L. Mehta, "Random Matrices," Academic Press, New York (1967).
4. F. D. M. Haldane, Phys. Rev. Lett. 60:635 (1988).
5. B. S. Shastry, Phys. Rev. Lett. 60:639 (1988).
6. M. Gaudin, Jour. de Physique 34:511 (1973).
7. B. Sutherland, to be published, Phys. Rev.
8. J. Hubbard, Phys. Rev. B 17: 494 (1978).
9. V. L. Pokrovsky and G, V. Uimin, J. Phys. C 11:3535 (1978).
10. D. J. Thouless, Phys. Rev. 187:732 (1969).

TWO THEOREMS ON THE HUBBARD MODEL

Elliott H. Lieb

Departments of Physics and Mathematics
Princeton University, P.O. Box 708
Princeton, NJ 08544

ABSTRACT

In the attractive Hubbard model (and some extended versions of it) the ground state is proved to have spin angular momentum $S = 0$ for every (even) electron filling. In the repulsive case, and with a bipartite lattice and a half filled band, the ground state has $S = \frac{1}{2}||B| - |A||$, where $|B|$ (resp. $|A|$) is the number of sites in the B (resp. A) sublattice. In both cases the ground state is unique. These theorems hold for all values of U, the attraction or repulsion parameter. The second theorem confirms an old, unproved conjecture in the $|B| = |A|$ case; the generalization given here yields, with $|B| \neq |A|$, the first provable example of itinerant electron ferromagnetism. Since topology is irrelevant for the proofs, the theorems hold in all dimensions without even the necessity of a periodic lattice structure.

The importance of the Hubbard model of itinerant electrons is increasingly being appreciated. Because of the model's subtlety, rigorous results and exact solutions are clearly useful bench marks, but these are rare. Two theorems about the ground states are stated and proved here. Parts of them are resolutions of old conjectures while other parts are new. In particular the assertion that certain versions of the model show ferromagnetic behavior for the half-filled band is, I believe, new and yields the first provable example of itinerant electron ferromagnetism with finite forces and without ad hoc assumptions.

After some preliminary definitions the two theorems are stated. Each is followed by some remarks about their significance. Finally, the proofs are given. The proofs utilize a new kind of reflection positivity which does not involve the usual spatial reflections but rather reflections in spin-space. In fact spatial symmetry plays no role whatsoever and therefore the theorems apply in the widest generality to any collection of sites; all dimensions and topologies are included. The Hubbard model is not known to satisfy any kind of spatial reflection positivity or infrared bounds and this unfortunate fact has prevented the application of the usual proof techniques[1,2,3] for establishing the existence of long range order in periodic lattices.

Interacting Electrons in Reduced Dimensions
Edited by D. Baeriswyl and D.K. Campbell
Plenum Press, New York

The Hubbard model on a finite lattice Λ is defined by the Hamiltonian

$$H = \sum_{\sigma} \sum_{x,y\in\Lambda} t_{xy} c_{x\sigma}^{\dagger} c_{y\sigma} + \sum_{x\in\Lambda} U_x n_{x\uparrow} n_{x\downarrow} \tag{1}$$

with the following notation. The operators $c_{x\uparrow}$ and $c_{x\downarrow}$ and their adjoints $c_{x\sigma}^{\dagger}$ satisfy the usual fermion anticommutation relations $\{c_{x\sigma}^{\dagger}, c_{y\tau}\} = \delta_{xy}\delta_{\sigma\tau}$ and $\{c_{x\sigma}, c_{y\tau}\} = 0$. The hopping matrix elements t_{xy} are required to be real and satisfy $t_{xy} = t_{yx}$, but no other a-priori assumption is made about them (e.g. the condition $t_{xx} = 0$ is not assumed, which means that an x-dependent single particle potential $\sum_{x\sigma} t_{xx} n_{x\sigma}$ is allowed). The reality of the t_{xy}'s is consistent with their interpretation as overlap matrix elements of real operators in real, localized orbitals. The number operators are $n_{x\sigma} = c_{x\sigma}^{\dagger} c_{x\sigma}$, while U_x is the on-site energy which, for Theorem 1, is allowed to depend on the site x. The word "lattice" is certainly a misnomer because no particular topology (i.e. periodicity or dimensionality) is assumed; the generality assumed here is that Λ is merely a collection of sites. The number of these sites is denoted by $|\Lambda|$. There is said to be a *bond* between sites x and y if $t_{xy} \neq 0$, and Λ is said to be *connected* if there is a connected path of bonds between every pair of sites. Obviously it is no loss in generality to assume Λ is connected, and this will always be done here. Finally, Λ is said to be *bipartite* if the sites of Λ can be divided into two disjoint sets A and B such that $t_{xy} = 0$ whenever $x \in A$ and $y \in A$ or $x \in B$ and $y \in B$. The symbols $|A|$ and $|B|$ denote the number of sites in A and B in this case, whence $|\Lambda| = |A| + |B|$. The number of electrons is denoted by N; necessarily $N \leq 2|\Lambda|$.

The aim here is to study the ground state (or states if there is more than one) of H for a given N. Of central importance is the total spin S which is a conserved quantity. The spin operators are the quadratic operators

$$S^z = \frac{1}{2} \sum_{x\in\Lambda} (n_{x\uparrow} - n_{x\downarrow}), \quad S^+ = (S^-)^\dagger = \sum_{x\in\Lambda} c_{x\uparrow}^{\dagger} c_{x\downarrow}. \tag{2}$$

and $(S_{op})^2 = (S^z)^2 + \frac{1}{2}S^+S^- + \frac{1}{2}S^-S^+$, with eigenvalues $S(S+1)$.

Theorem 1 (Attractive case). Assume $U_x \leq 0$ for every x (but U_x is not necessarily constant) and that N is even. No extra assumption about Λ or the t_{xy}'s is made. Then

(a). Among the ground states of H there is one with spin $S = 0$.

(b). If $U_x < 0$ for every x, the ground state is unique (and hence has $S = 0$).

Remarks: (1). The theorem is "obvious" if all the U_x's are very large, for then the ground state consists of paired electrons on $N/2$ sites of Λ. In the other extreme that each $U_x = 0$ Theorem 1 is also obvious because one just fills the lowest $N/2$ levels of the $|\Lambda| \times |\Lambda|$ Hermitian hopping matrix $T = \{t_{xy}\}$.

(2). Theorem 1 can be considerably generalized to what is called an *extended Hubbard model*, and more. We can add any *real* operator, M, to H provided M satisfies the following two conditions. (A real operator is a polynomial in the $c_{x\sigma}$'s and $c_{x\sigma}^{\dagger}$'s with real coefficients.)

(i) M commutes with the spin operators S^z and $S^{\pm}$ and conserves both spin-up and spin-down particle numbers; (ii) M can be written as $M^{\uparrow} + M^{\downarrow} - M^{\uparrow\downarrow}$. Here $M^{\uparrow}$ (resp. $M^{\downarrow}$) is real, Hermitian, involves only spin-up (resp. spin-down) operators, and $M^{\uparrow}$ is identical to $M^{\downarrow}$ when the spins are flipped (i.e. $c_{x\uparrow}$ and $c_{x\downarrow}$ are interchanged). The up-down interaction, $M^{\uparrow\downarrow}$ can be written as a sum of terms of the form (in which μ merely denotes a summation index) $M^{\uparrow\downarrow} = \sum_{\mu} V_{\mu}^{\uparrow} (V_{\mu}^{\downarrow})^{\dagger}$ in which each V_{μ} operator is real (but not necessarily Hermitian) and involves only operators for one kind of electron. Again, $V_{\mu}^{\uparrow}$ must be the spin reflection of $V_{\mu}^{\downarrow}$ for each μ. The necessary changes in the proof are straightforward (see ref. 4). It is also easy to extend the proof to some multi-band Hubbard models; the details are left to the reader.

Theorem 2 (Repulsive Case). Asume $U_x = U =$ positive constant, independent of x. Assume $|\Lambda|$ is even, Λ is bipartite and $|B| \geq |A|$. No other assumption about Λ or the t_{xy}'s is made. Let $N = |\Lambda|$ (half-filled band). Then the ground state of H is unique (apart from the trivial $(2S+1)$-fold degeneracy) and has spin $S = \frac{1}{2}(|B|-|A|)$.

Remarks: (3). Theorem 2 is considerably more subtle than Theorem 1. The assumptions are more stringent. The theorem has long been assumed to be true in the $|B| = |A|$ case ("the half-filled band has spin zero"), but its proof has been elusive.

(4). The fact that $2S = |B| - |A|$ should be no surprise. In the limit $U = 0$ we fill the levels of the matrix $T = \{t_{xy}\}$ and one might hastily conclude that S must be zero in this limit. If so, Theorem 2 would be contradicted by a continuity argument with respect to U. However, the rank of T is at most $2|A|$ and so T has at least $|\Lambda| - 2|A| = |B| - |A|$ zero eigenvalues. The remaining eigenvalues of T come in plus-minus pairs, so T has at most $|A|$ negative eigenvalues. To achieve $2S = |B| - |A|$ we fill the negative levels twice with opposite spins and place the remaining electrons in the zero levels with a common spin, say spin-up. Thus the ground state is degenerate when $U = 0$, but $S = \frac{1}{2}(|B|-|A|)$ is among them. Therefore there is no contradiction with the continuity argument mentioned above. If, on the other hand, U is very large we know[5] from second order perturbation theory that H is effectively an isotropic spin $\frac{1}{2}$ Heisenberg antiferromagnet with Hamiltonian $h = (2/U)\sum_{x,y} {t_{xy}}^2 \left[\S_x \cdot \S_y - \frac{1}{4}\right]$. For such models it is also known[6] that the ground state is unique and has $2S = |B| - |A|$.

(5). It is easy to construct many regular, periodic lattices in every dimension greater than one with $|B| \neq |A|$. A classic (high T_c superconductor?) two-dimensional example is to start with a square lattice and then intercalate one site in the middle of each bond. The original vertices of the squares are A sites (copper) and the intercalated sites are B sites (oxygen). The half-filled band has 3 electrons per unit cell and Theorem 2 says that then the total net magnetization is $S = \frac{1}{2}(2-1) = \frac{1}{2}$ times the number of unit cells. I thank D.C. Mattis for this example.

Whether or not this example is physically realizable is less important than the fact that Theorem 2 applied to a periodic lattice with $|B| > |A|$ yields, for the first time, a natural, provable example of an *itinerant electron model of ferromagnetism.* In one-dimension $|B| = |A|$ by definition and therefore $S = 0$; this conclusion coincides with the known result[7] that S is always zero in one dimension with nearest neighbor nonpositive hopping $t_{xy} \leq 0$ and for *any many-body potential.* There is also the example of Thouless and Nagaoka[8,9,10] with $N = |\Lambda| - 1$, $U = \infty$ and $2S = |\Lambda| - 1$, but it is somewhat artificial.

(6). Theorem 2 also has some extensions similar to some of those described in Remark (2) above; hole-particle symmetry is required.

Proof of Theorem 1. S^2 and S^z are conserved and I work in the $S^z = 0$ subspace since all competitors have a representative there. Then there are $n = \frac{1}{2}N$ electrons of each type, spin-up and spin-down. Let $\{\psi^\alpha\}$ be any orthonormal basis for *one species* of n spinless fermions; there are $m = \binom{|\Lambda|}{n}$ of these and I require that they be *real* (i.e. each ψ^α is a *real,* homogeneous polynomial of order n in the $c_x^\dagger$'s acting on the vacuum). A ground state ψ can then be written as $\psi = \sum_{\alpha,\beta} W_{\alpha\beta} \psi_\uparrow^\alpha \otimes \psi_\downarrow^\beta$ with $W_{\alpha\beta}$ as coefficients. This $W_{\alpha\beta}$ is here viewed as a $m \times m$ matrix. Because all operators and basis vectors are real, it is obvious that if $W_{\alpha\beta}$ corresponds to a ground state then so does $W_{\beta\alpha}^* = (W^\dagger)_{\alpha\beta}$, and hence (by linearity) so does $W + W^\dagger$ and $i(W - W^\dagger)$. Thus for convenience we may henceforth assume $W = W^\dagger$ (but $W_{\alpha\beta}$ is not assumed to be real). The norm of ψ squared is $\langle\psi|\psi\rangle = \sum_{\alpha,\beta} |W_{\alpha\beta}|^2 = TrW^2$ and I assume this is unity. The hopping energy part of $\langle\psi|H|\psi\rangle$ is easily found to be $2TrKW^2$ where

$K_{\alpha\beta} = \left\langle \psi^\beta \middle| \sum_{x,y} t_{xy} c_x^\dagger c_y \middle| \psi^\alpha \right\rangle$. Clearly K is real and symmetric since each t_{xy} is real. The on-site energy is given by $-\sum_x U_x Tr(WL_xWL_x)$ with $(L_x)_{\alpha\beta} = \langle \psi^\beta | n_x | \psi^\alpha \rangle$, which is also real and symmetric. The total energy is then

$$E(W) = \langle \psi | H | \psi \rangle = 2TrKW^2 + \sum_x U_x Tr(WL_xWL_x). \tag{3}$$

and the equation for W, corresponding to the eigenvalue equation $H\psi = e\psi$ is

$$KW + WK + \sum_x U_x L_x W L_x = eW. \tag{4}$$

Now consider the positive semidefinite matrix $|W|$ defined by $|W|^2 = W^2$. Obviously, $TrW^2 = Tr|W|^2$. Moreover in an orthonormal basis (not to be confused with the ψ^α basis for the electrons) in which the Hermitian $m \times m$ matrix W is diagonal, with diagonal elements w_i the Hermitian matrix $|W|$ is also diagonal with elements $|w_i|$. In this diagonal basis I compute $TrWL_xWL_x = \sum_{i,j} w_i w_j |(L_x)_{ij}|^2 \leq \sum_{i,j} |w_i||w_j||(L_x)_{ij}|^2 = Tr|W|L_x|W|L_x$. Since $U_x \leq 0$, I conclude that $E(W) \geq E(|W|)$ and therefore that *among the ground states there is one satisfying* $W = |W| \geq 0$. This is the "spin-space reflection positivity" mentioned in the second paragraph. (This part of the proof is an adaptation of that given in ref. 4.) I choose this positive (possibly semidefinite) matrix W and now make the choice that the ψ^α's are the natural x-space basis for the electrons, i.e. α denotes n points in Λ and $\psi^\alpha = \Pi_{x\in\alpha} c_x^\dagger |0\rangle$. (Some arbitrary convention for the sign of the ψ^α's can be made here.) Since $W \geq 0$, it follows that $W_{\alpha\alpha} > 0$ for at least one α, for otherwise W vanishes identically. However, the vector $\phi^\alpha = \psi_\uparrow^\alpha \otimes \psi_\downarrow^\alpha$ satisfies $(S_{op})^2 \phi^\alpha = 0$ and therefore this ground state ψ has a nonzero projection onto the eigenspace of $(S_{op})^2$ in which $S = 0$. This would be impossible if conclusion (a) were false and thus conclusion (a) must be true.

To prove conclusion (b) I shall prove that necessarily a Hermitean W satisfies $W = |W|$ or $W = -|W|$ for every ground state when $U_x < 0$ for every x. This will prove conclusion (b) for the following reason. If there were two normalized, Hermitian ground state W's, say W^1 and W^2 with $W^1 \neq \pm W^2$ then for every real constant d, the Hermitian matrix $W^1 + dW^2 \equiv W_d$ is not zero and defines (after normalization) a ground state, by virtue of the linearity of the eigenvalue eq. (4). It is easy to verify that there must be a d for which W_d is neither positive nor negative semidefinite, and this contradicts the assertion that the ground state W_d satisfies $W_d = \pm|W_d|$.

With W given then, consider the Hermitian, positive semidefinite matrix $R \equiv |W| - W$ which is also a multiple of a ground state and satisfies (4). If Q denotes the kernel of R, i.e. $Q = \{\text{vectors } V \text{ such that } RV = 0\}$, then the assertion $W = \pm|W|$ is implied by the following statement which I shall prove: Q is either just the zero vector or else every vector is in Q. Let V be in Q and take the expectation of (4) in this state V, i.e. $\langle V | KR + RK + \sum_x U_x L_x R L_x | V \rangle = e\langle V | R | V \rangle$. Since $RV = 0$ and, for all x, $U_x < 0$ and $\langle V | L_x R L_x | V \rangle \geq 0$ by the positive semidefiniteness of R, I conclude that $\langle V | L_x R L_x | V \rangle = 0$ for all x. Since R is positive semidefinite, I conclude that $RL_xV = 0$. Thus each L_x maps Q into Q. Now let the matrices in (4) act on V (without taking expectation values). Since $RL_xV = 0$ and $RV = 0$, I conclude that $RKV = 0$. Thus K also maps Q into Q. As before, let α denote a collection of n points in Λ and define $L^\alpha = \Pi_{x\in\alpha} L_x$, which is the projector onto the basis vector μ^α in $\mathbf{C}^m$ (with components $(\mu^\alpha)_\gamma = \delta_{\alpha\gamma}$) and which has matrix elements $(L^\alpha)_{\gamma\delta} = \delta_{\alpha\gamma}\delta_{\gamma\delta}$. Note that the L_x's commute with each other and so the ordering of the L_x's is unimportant in the definition of L^α. Each L^α maps Q into Q because each L_x does. Since Λ is connected by T, it is easy to see that the α's are connected by K, i.e. for all α and β there are indices $\gamma_1, \gamma_2, \ldots, \gamma_p$ for some integer p such that the ordinary (not matrix) product $G_{\beta\alpha} \equiv K_{\beta\gamma_1} K_{\gamma_1\gamma_2} \cdots K_{\gamma_p\alpha}$ is not zero. If Q is not just the zero vector and $V \neq 0$ is in Q then $L^\alpha V \neq 0$ for some fixed choice of α (because $\sum_\alpha L^\alpha$ is the identity). Then $L^\alpha V = z\mu^\alpha$ for some nonzero constant z and with μ^α being the aforementioned basis vector. But then the vector $F \equiv L^\beta K L^{\gamma_1} K \cdots L^{\gamma_p} K L^\alpha V$ is in Q and, in fact, $F = zG_{\beta\alpha}\mu^\beta$. In short, Q contains a complete set of vectors (i.e. every μ^β) because $G_{\beta\alpha}$ is nonzero for every β by virtue of the connectivity. Thus every vector is in Q since Q is a linear space, which implies that $W = \pm|W|$ and which, in turn, implies uniqueness of the ground state. QED.

Proof of Theorem 2. First make the conventional unitary hole-particle transformation for the spin-up electrons followed by a sign change on the B sublattice, i.e. $c_{x\uparrow} \to \varepsilon(x)c^\dagger_{x\uparrow}$ and $c^\dagger_{x\uparrow} \to \varepsilon(x)c_{x\uparrow}$ with $\varepsilon(x) = 1$ for $x \in A, \varepsilon(x) = -1$ for $x \in B$. The spin-down electrons are unaltered, i.e. $c_{x\downarrow} \to c_{x\downarrow}$. Then $n_{x\uparrow} \to 1 - n_{x\uparrow}$ and the transformed H is $\widetilde{H} + UN_\downarrow$ with

$$\widetilde{H} = \sum_\sigma \sum_{x,y\in\Lambda} t_{xy} c^\dagger_{x\sigma} c_{y\sigma} - U \sum_{x\in\Lambda} n_{x\uparrow} n_{x\downarrow} \tag{5}$$

and with $N_\sigma = \sum_x n_{x\sigma}$. The original number operators N_σ transform as $N_\downarrow \to \widetilde{N}_\downarrow = N_\downarrow$ and $N_\uparrow \to \widetilde{N}_\uparrow = |\Lambda| - N_\uparrow$. The original condition $N = |\Lambda|$ becomes $N_\uparrow = N_\downarrow$. The spin operators (2) become the ***pseudo-spin operators***

$$\widetilde{S}^z = \frac{1}{2}(|\Lambda| - N_\uparrow - N_\downarrow), \quad \widetilde{S}^+ = \sum_{x\in\Lambda} \varepsilon(x) c_{x\uparrow} c_{x\downarrow} \tag{6}$$

and $S^2 \to (\widetilde{S}^z)^2 + \frac{1}{2}\widetilde{S}^+\widetilde{S}^- + \frac{1}{2}\widetilde{S}^-\widetilde{S}^+ \equiv (\widetilde{S})^2$. The $\widetilde{S}$ operators commute with $\widetilde{H}$, but so do the spin operators S given in (2). The S operators are the transforms of the pseudo-spin operators in the original variables and are of no special physical interest. The $\widetilde{S}$ operators are the ones of interest as far as $\widetilde{H}$ is concerned.

As before I can work in the $\widetilde{N}_\uparrow = \widetilde{N}_\downarrow = |\Lambda|/2$ subspace, which implies $N_\uparrow = N_\downarrow = |\Lambda|/2$. The uniqueness part of Theorem 2 is then a consequence of conclusion (b) of Theorem 1, which also states that the unique ground state, $\widetilde{\psi}$, of $\widetilde{H}$ has $S = 0$. This last fact is of secondary importance. The real problem is to prove that $2\widetilde{S} = |B| - |A|$. The shortest proof is to return to H and the $S^z = 0$ subspace (in the ***original variables***). For each $U > 0$ the ground state, $\psi(U)$, is nondegenerate as has been shown and I want to prove that $2S = |B| - |A|$. The nondegeneracy of the ground state for *all* $U > 0$ implies that the S of this unique ground state must be independent of U, for otherwise continuity in U would imply a degeneracy for some value of $U > 0$. However, when U is very large H, as stated before, is equivalent (to leading order in U) to h, the Heisenberg antiferromagnetic Hamiltonian defined in Remark (4). As stated there, h *also* has a unique ground state[6] (for $S^z = 0$) and this state has $2S = |B| - |A|$. The uniqueness property of h is crucial for it implies a gap (however small it may be) in the spectrum of h. Thus, for large enough U the S value of the ground state of h is identical to that of H. QED.

ACKNOWLEDGEMENTS

The partial support of the U.S. National Science Foundation grant PHY85-15288 A02 is gratefully acknowledged.

REFERENCES

1. F.J. Dyson, E.H. Lieb and B. Simon, J. Stat. Phys. **18**, 335 (1978).
2. J. Fröhlich, R. Israel, E.H. Lieb and B. Simon, Commun. Math. Phys. **62**, 1 (1978).
3. J. Fröhlich, R. Israel, E.H. Lieb and B. Simon, J. Stat. Phys. **22**, 297 (1980). The proof of Theorem 5.5 on p. 334 is incorrect.
4. T. Kennedy, E.H. Lieb and S. Shastry, J. Stat. Phys. **53**, no. 5/6 December, 1988.
5. P.W. Anderson, Phys. Rev. **115**, 2 (1959).
6. E.H. Lieb and D.C. Mattis, J. Math. Phys. **3**, 749 (1962). This paper uses a Perron-Frobenius argument to prove that in a certain basis the ground state ψ for the connected case is a positive vector and it is unique. By comparing this ψ with the ψ for a simple soluble model (which is also positive) in the $2S^z = |B| - |A|$ subspace we concluded $2S \leq |B| - |A|$. However, if the comparison is made in the $S^z = 0$ subspace, the same argument shows $2S = |B| - |A|$. This simple remark was overlooked in the 1962 paper, but it is crucial for the present work.
7. E.H. Lieb and D.C. Mattis, Phys. Rev. **125**, 164 (1962).
8. Y. Nagaoka, Phys. Rev. **147**, 392 (1966).
9. D.J. Thouless, Proc. Phys. Soc. (London) **86**, 893 (1965).
10. E.H. Lieb, in *Phase Transitions, Proceedings of the Fourteenth Solvay Conference*, (Wiley, Interscience, New York, 1971), pp. 45-63.

A MODEL OF CRYSTALLIZATION: A VARIATION ON THE HUBBARD MODEL

Elliott H. Lieb

Departments of Mathematics and Physics
Princeton University
Jadwin Hall, P.O. Box 708
Princeton, NJ 08544

ABSTRACT

A quantum mechanical lattice model of fermionic electrons interacting with infinitely massive nuclei is considered. (It can be viewed as a modified Hubbard model in which the spin-up electrons are not allowed to hop.) The electron-nucleus potential is "on-site" only. Neither this potential alone nor the kinetic energy alone can produce long range order. Thus, if long range order exists in this model it must come from an exchange mechanism. N, the electron plus nucleus number, is taken to be less than or equal to the number of lattice sites. We prove the following: (i) For all dimensions, d, the ground state has long range order; in fact it is a perfect crystal with spacing $\sqrt{2}$ times the lattice spacing. A gap in the ground state energy always exists at the half-filled band point (N = number of lattice sites). (ii) For small, positive temperature, T, the ordering persists when $d \geq 2$. If T is large there is no long range order and there is exponential clustering of all correlation functions.

INTRODUCTION

This lecture concerns joint work with Tom Kennedy [4], first announced in [10]. It also contains some related results on the Peierls instability, which were reported in [15] and whose details appear in [16].

Much attention has been paid to the question of proving the existence of long range order in model statistical mechanical systems in which the basic atomic constituents interact with short range forces. An important example is a lattice spin system in which the spin at each site represents the localized spin of an atom located at that site and where the short range, pairwise interaction (Ising or Heisenberg) reputedly comes from an interatomic exchange energy. Another problem - so far unsolved - is the existence of periodic crystals which are supposed to come from short range (e.g. Lennard- Jones) interatomic potentials.

Interacting Electrons in Reduced Dimensions
Edited by D. Baeriswyl and D.K. Campbell
Plenum Press, New York

In the real world, however, these interactions are not given a-priori; it is ultimately itinerant electrons and their correlations that give rise to the long range ordering. In other words, a deep unsolved problem is to derive magnetism or crystallization from the Schrödinger equation - or some caricature of it. The construction of a simple model based on itinerant electrons, and the rigorous derivation of ordering from it, is a challenge for mathematical physicists.

A lattice model of itinerant electrons that is believed to display ferro and antiferromagnetism - if it could be solved - is the Hubbard model [1-3, 18, 19]. We are also unable to solve it, but we have succeeded in proving that a simplified version of it does display crystallization. It is a toy model but it is, to our knowledge, the first example of this genre. Roughly, it has the same relation to the Hubbard model as the Ising model has to the quantum Heisenberg model. Here we shall give a brief report of our results, the full details of which appear elsewhere [4].

The Hubbard model, which is the motivation for our model, is defined by the second-quantized Hamiltonian

$$H^H = \sum_{\sigma} \sum_{x,y\in\Lambda} t_{xy} c^{\dagger}_{x\sigma} c_{y\sigma} + 2U \sum_{x\in\Lambda} n_{x\uparrow} n_{x\downarrow} \tag{1}$$

with the following notation: $\sigma = \pm 1$ denotes the 2 spin states of the electrons; Λ is a finite lattice; $c_{x\sigma}$ is a fermion annihilation operator for a spin σ electron at $x \in \Lambda$; $n_{x\sigma} = c^{\dagger}_{x\sigma} c_{x\sigma}$ is the number operator for spin σ at x. Electrons interact only at the same site with an energy $2U$, and $t_{xy} = t^*_{yx}$ is the hopping amplitude from x to y.

The crucial assumption will be made that Λ is the union of two sublattices $A \cup B$ such that $t_{xy} = 0$ unless $x \in A, y \in B$ or $x \in B, y \in A$. The number of sites in Λ, A, and B are denoted by $|\Lambda|, |A|$ and $|B|$. The two sublattices need not be isomorphic. Thus, for example, a face-centered cubic lattice is allowed with A=face centers and B=cube corners. Λ is said to be *connected* if every $x, y \in \Lambda$ can be joined by a "path" through nonzero t's.

In our model we assume that one kind of electron (say $\sigma = -1$) does not hop. One can say that these electrons are infinitely massive. The Hamiltonian is then

$$H = \sum_{x,y\in\Lambda} t_{xy} c^{\dagger}_x c_y + 2U \sum_{x\in\Lambda} n_x W(x) \tag{2}$$

with $n_x = c^{\dagger}_x c_x$ (the subscript σ is omitted since the dynamic electrons have $\sigma = +1$) and with $W(x) = +1$ if a fixed electron ($\sigma = -1$) is at x and $W(x) = 0$ otherwise. It will be recognized immediately that (2) is just a fancy way to say that the movable electrons are independent, with a single particle Hamiltonian

$$\tilde{h} = T + V, \tag{3}$$

with T being the $|\Lambda|$-square matrix t_{xy} and $V_{xy} = 2UW(x)\delta_{xy}$. It is convenient to write $\tilde{h} = h + U$ with

$$h = T + US \tag{4}$$

with $S_{xy} = s_x \delta_{xy}$, and $s_x = 1$ (resp. -1) if x is occupied (resp. unoccupied). The $\{s_x\}$ are like Ising spins.

We shall henceforth call the movable particles "electrons" and the fixed particles "nuclei". This terminology is most appropriate if $U < 0$, for then H does represent a lattice system of electrons and nuclei in which all Coulomb interactions except the on-site electron-nucleus attraction and the on-site infinite nuclear repulsion are regarded as "screened out". This conforms with the spirit of the original Hubbard model. The electron number, the nuclear number and the total particle number, all of which commute with H, are, respectively

$$N_e = \sum_{x\in\Lambda} n_x, \quad N_n = \frac{1}{2}\sum_{x\in\Lambda}[s_x + 1] = \sum_{x\in\Lambda} W(x), \quad \mathcal{N} = N_e + N_n. \tag{5}$$

It is to be emphasized that W does *not* represent a disordered potential. We take the "annealed", not the "quenched" system. The ground state for fixed N_e and N_n is defined by taking the ground state of H (with respect to the electrons) for each W and then minimizing the result with respect to the location of the nuclei. The ground state energy will be denoted by $E(N_e, N_n)$. Likewise, for positive temperature, we take $Tre^{-\beta H}$ with respect to *both* the electron variables and the nuclear locations.

Since $t_{xy} = 0$ for x, y on the same sublattice, the spectra of H and H^H are invariant under $t_{xy} \to -t_{xy}$ (all x, y). There is also a hole-particle symmetry. If $c_x^\dagger \to c_x, c_x \to c_x^\dagger, n_x \to 1 - n_x, t_{xy} \to -t_{xy}$, then $H(U) \to H(-U) + 2UN_n$. If $W(x) \to 1 - W(x)$, then $H(U) \to H(-U) + 2UN_e$. A similar symmetry holds for H^H. Thus, the $U > 0$ and $U < 0$ cases are similar — from the mathematical point of view.

Our results are of two kinds. The first concerns the ground state which we prove always has perfect crystalline ordering and an energy gap (defined later). The second concerns the positive temperature ($1/kT = \beta < \infty$) grand canonical state. For large β and dimension $d \geq 2$, the long range order persists. For small β it disappears and there is exponential clustering of the nuclear correlation functions. Some of our results were discovered independently, and at the same time, by Brandt and Schmidt [17] who used completely different methods.

THE GROUND STATE

Theorem *1*: (a) *Let $U < 0$. Under the condition $\mathcal{N} \equiv N_e + N_n \leq 2|A|$, the ground state (i.e. we minimize $E(N_e, N_n)$ over the set $N_e + N_n \leq 2|A|$) occurs for $N_e = N_n = |A|$ and a minimizing nuclear configuration is $W(x) = W_A(x) \equiv 1(x \in A), 0(x \notin A)$. Under the condition $\mathcal{N} \leq 2|B|$, the ground state is $N_e = N_n = |B|$ and the B sublattice is occupied ($W = W_B$). If Λ is connected, these are the only groundstates, i.e. if $|A| > |B|$ the ground state is unique; if $|A| = |B|$ it is doubly degenerate. No assumption is made about the sign or magnitude or periodicity of the t_{xy} other than $t_{xy} = 0$ for $x, y \in A$ or $x, y \in B$.*

(b) *Let $U > 0$. Under the condition $\mathcal{N} \geq |A| + |B|$, there are two ground states: $N_e = |A|, N_n = |B|, W = W_B$ and $N_e = |B|, N_n = |A|, W = W_A$. If Λ is connected, these are the only ground states.*

The condition $\mathcal{N} = |\Lambda|$ is called the *half-filled* band. If $|A| = |B| = |\Lambda|/2$, the crystal occurs at the half-filled band. If Λ is a cubic lattice, for example, this means that the ground state is a cubic lattice of period $\sqrt{2}$ oriented at 45° with respect to Λ.

Theorem 1 relies heavily on the fact that the electrons are fermions. The ground state would be completely different if they were bosons. For bosons and for Λ a cubic lattice, the nuclei would all be clumped together in the ground state instead of being spread out into a crystal. By using rearrangement inequalities it is possible to describe this clumping quantitatively.

Next, we define the *energy gap*. Actually two different definitions are of interest. First, let

$$E(\mathcal{N}) \equiv \min\{E(N_e, N_n) | N_e + N_n = \mathcal{N}\}. \tag{6}$$

The *chemical potential* is defined by

$$\mu(\mathcal{N}) \equiv E(\mathcal{N}+1) - E(\mathcal{N}). \tag{7}$$

We say there is a *gap of the first kind* at $\mathcal{N}$ if

$$\mu(\mathcal{N}) - \mu(\mathcal{N}-1) \geq \varepsilon_1 > 0 \tag{8}$$

with ε_1 being independent of the size of the system. We say there is a *gap of the second kind* at N_e, N_n if

$$E(N_e+1, N_n) + E(N_e-1, N_n) - 2E(N_e, N_n) \geq \varepsilon_2 > 0. \tag{9}$$

In other words, the nuclear number is fixed in the second definition.

A gap is one indication that the system is an insulator, for it implies that it costs more energy to put a particle into the system than is gained by removing one. The first kind of gap is relevant if one views our model as an approximation to the Hubbard model; the second is relevant from the "electrons and nuclei" point of view.

Theorem 2: *Assume that Λ is not only connected but that every $x, y \in \Lambda$ can be connected by a chain with $|t_{ab}| \geq \delta$ for every a, b on the chain. Also, assume that $\|T\| \leq \tau$. The following energy gaps exist with $\varepsilon_2 \geq \varepsilon_1 > 0$ and depending only on δ, τ and U:*

$\underline{U < 0}$: First kind at $\mathcal{N} = 2|A|$ and at $\mathcal{N} = 2|B|$. Second kind at $N_e = |A|, N_n = |A|$ and at $N_e = |B|, N_n = |B|$.

$\underline{U > 0}$: First kind at $\mathcal{N} = |A| + |B|$. Second kind at $N_e = |A|, N_n = |B|$ and at $N_e = |B|, N_n = |A|$.

In order to give the flavor of our methods, the proof of Theorem 1 will be given here. The proof of Theorem 2 is more complicated.

Proof of Theorem 1: Let $\lambda_1 \leq \lambda_2 \leq \ldots$ be the eigenvalues of h in (4). They depend on the nuclei. For N_e electrons the ground state energy, E, of H satisfies

$$E - UN_e = \sum_{j=1}^{N_e} \lambda_j \geq \sum_{\lambda_j < 0} \lambda_j = \frac{1}{2}[Trh - Tr|h|]. \tag{10}$$

But $Trh = U\sum s_x = 2UN_n - U|\Lambda|$ and $|h| = \{T^2 + U^2 + UJ\}^{1/2}$ with $J_{xy} = t_{xy}[s_x + s_y]$. Since the function $0 < x \to x^{1/2}$ is concave, $f(y) = Tr\{T^2 + U^2 + yUJ\}^{1/2}$ is concave in $y \in [-1,1]$. But $f(-1) = f(1)$ (since spec(h) is invariant under $T \to -T$), so $f(1) \leq f(0)$, with equality if and only if $J \equiv 0$. Thus

$$E \geq U\mathcal{N} - \frac{1}{2}U|\Lambda| - \frac{1}{2}Tr(T^2 + U^2)^{1/2}. \tag{11}$$

If Λ is connected, the only ways to have $J \equiv 0$ are either $W = W_A$ or W_B. Consider $U < 0$ and $\mathcal{N} \leq 2|A|$, whence, from (11),

$$E \geq U(2|A| - \frac{1}{2}|\Lambda|) - \frac{1}{2}Tr(T^2 + U^2)^{1/2}. \tag{12}$$

If $W = W_A$ then, as is easily seen, h has precisely $|A|$ negative and $|B|$ positive eigenvalues. Thus, if $W = W_A$ and $N_e = |A|$, then (12) is an equality. The other cases are similar. □

GRAND CANONICAL ENSEMBLE

First, we define the partition function Ξ. A nuclear configuration is denoted by $S = \{s_x\}, s_x = \pm 1$, and the λ_j are the eigenvalues of h in (4). If μ_n, μ_e are the nuclear and electronic chemical potentials,

$$\Xi = \sum_S \exp[\frac{1}{2}\beta\mu_n(\sum_x s_x + |\Lambda|)] \prod_{j=1}^{|\Lambda|}\{1 + \exp[-\beta(\lambda_j + U - \mu_e)]\}. \tag{13}$$

The product in (13) is just the well known Fermi-Dirac grand canonical partition function for the electrons. We want to choose μ_e, μ_n so that $\langle N_e \rangle = \frac{1}{2}|\Lambda|$ and $\langle N_n \rangle = \frac{1}{2}|\Lambda|$, or $\langle \sum s_x \rangle = 0$. From the fact that if $T \to -T$, spec(h) $\to$ spec(h), one has that when $S \to -S$, spec(h) $\to$ $-$spec(h). It is then easy to see that the desired chemical potentials are $\mu_e = \mu_n = U$. Since $\sum \lambda_j = U \sum s_x$, (13) becomes in this case (after dropping an irrelevant factor $2^{|\Lambda|}e^{\beta U|\Lambda|/2}$)

$$\Xi = \sum_S \exp[-\beta F(S)] \tag{14}$$

with

$$\begin{aligned} -\beta F(S) &= \sum_{j=1}^{|\Lambda|} \ell n \cosh(\frac{1}{2}\beta\lambda_j) = Tr\ell n \cosh[\frac{1}{2}\beta h] \\ &= Tr\ell n \cosh[\frac{1}{2}\beta(T^2 + U^2 + UJ)^{1/2}]. \end{aligned} \tag{15}$$

Thus, (14) is like an Ising model partition function but with a complicated, temperature dependent "spin-spin" interaction, $F(S)$, given by (15) in terms of the eigenvalues of h. With respect to this "spin measure" we can talk about the presence or absence of long range nuclear order in the thermodynamic limit. In order to discuss this limit we henceforth restrict ourselves to a translation invariant nearest neighbor hopping on a cubic lattice in d dimensions.

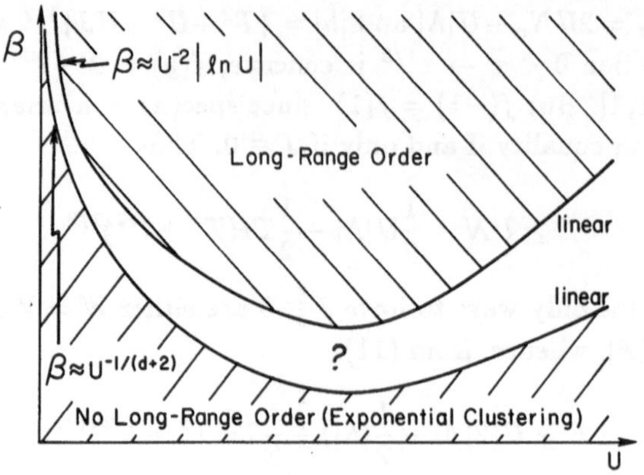

Fig. 1

What we are able to prove is summarized in the schematic figure above and in

Theorem 3: *For all U and sufficiently large β there is long range order for $d \geq 2$ (the same kind as in the ground state). For all U and sufficiently small β there is none; indeed there is exponential decay of all nuclear correlation functions.*

Presumably there is no intermediate phase, but we cannot prove this. For large U, β_c is clearly linear in U. For small U, we have the bound on the lower $\beta_c \sim U^{-1/(2+d)}$ and the bound on the upper $\beta_c \sim |\ell n U| U^{-2}$. Our guess is that the true state of affairs is $\beta_c \sim U^{-2}$.

For large β we use a Peierls argument; for small β we use Dobrushin's uniqueness theorem. A sketch of our proof - omitting many important details - is the following. For simplicity, we here consider only large $U > 0$.

Define, for $x > 0$,

$$P(x) = \ell n \cosh x^{1/2}. \tag{16}$$

[We note in passing that P is concave, and we see from the last expression in (15), using the proof in Theorem 1, that $F(S)$ has its minima at precisely the same values of W (or S) as in Theorem 1.] P is a Pick (or Herglotz) function with the representation

$$P'(x) = \frac{1}{2} x^{-1/2} \tanh x^{1/2} = \sum_{k=0}^{\infty} [(k + \frac{1}{2})^2 \pi^2 + x]^{-1}. \tag{17}$$

We are interested in $x = \frac{1}{4}\beta^2(T^2 + U^2 + UJ)$ with $J_{xy} = t_{xy}(s_x + s_y)$.

Long Range Order (large β): Choose S, and then define antiferromagnetic contours in the usual way. If γ is a connected contour component and $\Gamma + \gamma$ is the whole contour, we want to prove that $\Delta \equiv F(\Gamma + \gamma) - F(\Gamma) \geq C|\gamma|$ for a suitable constant $C = C(U)$.

Obviously, $J_{\Gamma+\gamma} = J_\Gamma + J_\gamma$. To remove γ, we change s_x to $-s_x$ inside γ. For $0 \leq t \leq 1$, define $J(t) = J_\Gamma + tJ_\gamma$. Then, assuming for simplicity that Γ lies entirely outside γ, we have, by differentiating (15) and using (17), that

$$-\beta\Delta = \frac{1}{4}\beta^2 U \sum_{k=0}^{\infty} \int_o^1 dt Tr(G_k J_\gamma), \tag{18}$$

with

$$G_k = [(k+\frac{1}{2})^2\pi^2 + \frac{1}{4}\beta^2(T^2 + U^2 + UJ_\Gamma + tUJ_\gamma)]^{-1}. \tag{19}$$

Integrating by parts, this becomes

$$-\beta\Delta = -(\beta^4 U^2/16)\sum_k \int_o^1 dt(1-t) Tr G_k J_\gamma G_k J_\gamma + \ldots \tag{20}$$

[The ... terms in (20) come from $t = 0$ in the partial integration. They are small and easily bounded for large U, but they have to be treated more judiciously when U is small.] If A, B, D are matrices with $A \geq B \geq 0$, then $Tr AD^\dagger AD \geq Tr BD^\dagger BD$. Also, $A \geq B > 0 \Rightarrow 0 < A^{-1} \leq B^{-1}$. Moreover, $T^2 \leq (2d)^2$ and $UJ \leq T^2 + U^2$. Using this, we have the matrix inequality

$$G_k \geq \left[(k+\frac{1}{2})^2\pi^2 + \frac{1}{2}\beta^2\left((2d)^2 + U^2\right)\right]^{-1}. \tag{21}$$

Summing on k,

$$\Delta \geq \text{(const.) } \mathrm{U}^2\left((2\mathrm{d})^2 + \mathrm{U}^2\right)^{-3/2} \mathrm{Tr}(\mathrm{J}_\gamma)^2.$$

Clearly $Tr(J_\gamma)^2 =$ (const.)$|\gamma|$. Thus, for U large, $C(U) \geq$ (const.)U^{-1}, and thus long range order exists in $d \geq 2$ if β/U is large enough.

Absence of Long Range Order (small β): Dobrushin's uniqueness theorem [5-7], together with the modification in [8], gives the following criterion for exponential clustering. We have to bound the change in F when we change the spins at x and y (taking the worst case with respect to the other spins). Call this f_{xy}. The requirement is that for some $m > 0$ and all x,

$$\beta \sum_y f_{xy} \exp[m|x-y|] < 1. \tag{22}$$

By an argument similar to the preceding (large U)

$$\begin{aligned} f_{xy} &\cong (\beta^3 U^2/16) \sum_k T_r G_k J_x G_k J_y \\ &\cong (\beta^3 U^2/16) \sum_k |G_k(x,y)|^2. \end{aligned} \tag{23}$$

Here, $J_x = \delta_x T + T\delta_x$ and $G_k(x,y)$ is the x, y matrix element of (19).

To implement (22) we now require an *upper* bound on $G_k(x,y)$ that has exponential decay. For this purpose the Combes-Thomas argument is ideal. Let Q be the matrix with elements $Q_{xy} = \delta_{xy} e^{n \cdot x}$ and with $|n| = 1$. Then

$$QG_k^{-1}Q^{-1} = G_k^{-1} + R_k \equiv L_k. \tag{24}$$

The "remainder" R_k can be bounded for large U: $\|R_k\| \le C\beta^2 U$ for some constant C. Similar to (21), for large U

$$G_k^{-1} \ge [(k+\frac{1}{2})^2\pi^2 + \frac{1}{8}\beta^2U^2] \equiv \alpha_k. \tag{25}$$

Thus,

$$\|L_k^{-1}\| \le [\alpha_k - C\beta^2U]^{-1}. \tag{26}$$

Since $|(L_k^{-1})_{xy}| \le \|L_k^{-1}\|$, we have from (24), $|(QG_kQ^{-1})_{xy}| \le \|L_k^{-1}\|$, and thus

$$|(G_k)_{xy}| \le \exp[n\cdot(y-x)]\{\alpha_k - C\beta^2U\}^{-1}. \tag{27}$$

This holds for all n, so summing on k,

$$f_{xy} \le \text{ (const.) } U^{-1}e^{-2|x-y|}.$$

Hence, (22) holds with $m=1$ if β/U is small enough.

THE PEIERLS INSTABILITY

Another itinerant electron model in which translation invariance is partially broken was introduced by Peierls [11], see also [14]. Consider a linear chain of $2N$ atoms (arranged in a ring) and let $w_i (i = 1,\ldots,2N)$ denote the distance from atom i to atom $i+1$ (with $2N+1 \equiv 1$). If $d > 0$ is the equilibrium distance, the distortion energy associated with the w_i is taken to be

$$U(\{w\}) = \kappa \sum_{i=1}^{2N} (w_i - d)^2. \tag{28}$$

Here $\kappa > 0$ is a measure of stiffness.

Now suppose there are $2N$ (spin $\frac{1}{2}$) electrons ($\frac{1}{2}$ filled band) which can hop between nearest neighbor atoms, with an amplitude $t_i = t(w_i)$ to hop from atom i to $i+1$ and from $i+1$ to i. The function $t(w)$ will be explained shortly. The ground state energy of these $2N$ electrons is

$$K(\{w\}) = 2\sum_{j=1}^{N} \lambda_j(T) \tag{29}$$

where $\lambda_1 \le \lambda_2 \le \ldots \le \lambda_{2N}$ are the eigenvalues of the $2N \times 2N$ matrix T having matrix elements $T_{i,i+1} = T_{i+1,i} = t_i = t(w_i)$ and $T_{ij} = 0$ otherwise. The total Hamiltonian is

$$H(\{w\}) = K(\{w\}) + U(\{w\}) \tag{30}$$

and the problem is to determine the ground state energy

$$E = \min_{\{w\}} H(\{w\}) \tag{31}$$

and the configuration(s) $\{w\}$ that attains it.

The function $t(w)$ should be something like ae^{-cw}, but here we shall adopt the choice of Su, Schrieffer and Heeger [12]:

$$t(w) = b - c(w - d) \tag{32}$$

with $b, c > 0$.

One guess for a minimizing $\{w\}$ is $w_i = W = \text{const}$. However, Peierls observed that if we take $w_i = W + (-1)^i\delta$ (a *dimerized configuration*) the energy will be lowered. The reason is that when $\{w\}$ has period two a gap in the spectrum of T opens up at $k = \pm\pi/2$; since there are just enough electrons to fill the lower band, the K energy will be lowered by an amount (const.) $\delta^2 ln|\delta|$ for small δ. (This comes from the fact that the perturbation theory is singular at the band edges.) On the other hand, U depends on δ like $\kappa\delta^2$, so it is favorable to have $\delta \neq 0$. This breaking of translation invariance is called the Peierls instability. The interesting question, which was open for a long time, is this: Is the translation invariance merely broken from period one to period two, or does something more complicated happen? Conceivably the minimizing $\{w\}$ could have a longer period – or perhaps no periodicity at all.

Our answer to this question is that period two always holds.

Theorem 4: *For all b, c, d, κ, N there are precisely two minima for E*

$$w_i = W + (-1)^i\delta \text{ or } w_i = W - (-1)^i\delta, \tag{33}$$

with W and δ being constants depending on κ, b, c, N. Moreover, there is a gap, i.e. a jump in the chemical potential:

$$\Delta\mu \equiv [E(2N + 1) - E(2N)] - [E(2N) - E(2N - 1)] > \varepsilon, \tag{34}$$

with ε depending on κ, b, c but independent of N, d. (Here $E(2N + 1)$ is defined as in (30), (31) but with K replaced by $2\sum_{j=1}^{N} \lambda_j(T) + \lambda_{N+1}(T)$ and, for $E(2N - 1)$, K is replaced by $2\sum_{j=1}^{N} \lambda_j(T) - \lambda_N(T)$.)

As in the proof of Theorem 1, the proof of Theorem 4 uses the concavity of $Tr\sqrt{X}$ with respect to X, and with $X = T^2$. The crucial observation is that for the dimerized configuration (33), T^2 is translation invariant.

It turns out that if the U in (28) is modified, an "integrable system" is obtained. In terms of t, using (32), we have

$$(w_i - d)^2 = \alpha + \beta t_i + \gamma t_i^2 \tag{35}$$

with α, β, γ constants. The modification is

$$U(\{w\}) = \alpha + \sum_{j=1}^{2N} [-\beta \ln t_i + \gamma t_i^2] \tag{36}$$

with α, β, γ arbitrary constants (but (32) is unchanged). This system is presumably not very different from ours (but there is no proof). It was investigated in [13]. However, the solution in [13] covers all electron numbers *except* the half-filled band ($2N$ electrons), which is the most interesting case since it is only here that one expects to see the instability manifest itself in the form proposed by Peierls - namely the partial breaking of translation invariance into a period two invariance.

ACKNOWLEDGEMENTS

The support of the U.S. National Science Foundation, grant PHY-8515288, is gratefully acknowledged.

REFERENCES

1. M. C. Gutzwiller, Phys. Rev. Letters *10*, 159-162 (1963), and Phys. Rev. *134*, A923-941 (1964), and *137*, A1726-1735 (1965).
2. J. Hubbard, Proc. Roy. Soc. (London), Ser. A*276*, 238-257 (1963), and *277*, 237-259 (1964).
3. J. Kanamori, Prog. Theor. Phys. *30*, 275-289 (1963).
4. T. Kennedy and E. H. Lieb, An itinerant electron model with crystalline or magnetic long range order, Physica *138A*, 320-358 (1986).
5. R. L. Dobrushin, Theory Probab. and Its. Appl. *13*, 197-224 (1968).
6. L. Gross, Commun. Math. Phys. *68*, 9-27 (1979).
7. H. Föllmer, J. Funct. Anal. *46*, 387-395 (1982).
8. B. Simon, Commun. Math. Phys. *68*, 183-185 (1979).
9. J. M. Combes and L. Thomas, Commun. Math. Phys. *34*, 251-270 (1973).
10. T. Kennedy and E.H. Lieb in Proceedings of the Groningen conference on statistical mechanics and field theory, August, 1985, Springer Lecture Notes in Physics, T.C. Dorlas, N.M. Hugenholtz and M. Winnick eds., vol. *257*, 1-9 (1986).
11. R.E. Peierls, *Quantum Theory of Solids*, Clarendon Press, Oxford (1955), p. 108.
12. W.P. Su, J.R. Schrieffer and A.J. Heeger, Phys. Rev. Lett. *42*, 1698-1701 (1979).
13. S.A. Brazovskii, N.E. Dzyaloshinskii and I.M. Krichever, Sov. Phys. JETP *56*, 212-225 (1982).
14. H. Fröhlich, Proc. Roy. Soc. *A223*, 296-305 (1954).
15. T. Kennedy and E.H. Lieb, A model for crystallization: a variation on the Hubbard model, Physica *140A*, 240-250 (1986).
16. T. Kennedy and E.H. Lieb, Proof of the Peierls instability in one dimension, Phys. Rev. Lett. *59*, 1309-1312 (1987).
17. U. Brandt and R. Schmidt, Exact results for the distribution of the f-level ground state occupation in the spinless Falicov-Kimball model, Z. Phys. B. - Condensed Matter *63*, 45-53 (1986).
18. P.W. Anderson, New approach to the theory of superexchange interactions, Phys. Rev. *115*, 2-13 (1959).
19. J.H. Van Vleck, Rev. Mod. Phys. *25*, 220 (1953).

THE FERMI-LINEARIZED HUBBARD MODEL: DIMER GROUND STATE

Arianna Montorsi *and* Mario Rasetti
Dipartimento di Fisica and Unitá CISM, Politecnico di Torino , Italy

Allan I. Solomon
Faculty of Mathematics, The Open University, Milton Keynes, UK

Abstract

The zero temperature ground state for a Hubbard model is constructed, in the frame of fermionic linearization, for a two-site (dimer) cluster, in the form of a supercoherent state of the dynamical superalgebra of the model. It is shown how the phase space is characterized by the existence of regions in which the pairing order parameter is non-vanishing.

In a set of recent papers [1],[2] a mean field method was proposed by the authors to deal with systems of interacting fermions. Such a method is based on a procedure (referred to as *linearization*) whereby bilinear products of anticommuting operators are replaced by their linear combinations. The coefficients of the latter belong to the odd sector of a graded field of numbers $\mathcal{G}$ with the following properties : the *odd* elements anticommute among themselves as well as with the fermi operators, the *even* elements commute.

Typically the linearized product has the form

$$AB \simeq \theta_A B - \theta_B A - \theta_A \theta_B \quad ;$$

$$\{R, \theta_S\} = 0 = \{\theta_R, \theta_S\} \quad . \tag{1}$$

In several interesting instances the hamiltonian describing the system reduces, upon linearization, to an element of some ***superalgebra*** S, and its spectrum and coherent states can be obtained by standard algebraic techniques [3], which straightforwardly extend to superalgebras, due to the properties of the θ-numbers. In particular, a generalized Bogolubov transformation $\Phi_S : \mathcal{S} \to \mathcal{S}$ can be defined, which rotates the hamiltonian into its diagonal form (*i.e.* into an element of the *Cartan* subalgebra of $\mathcal{S}$); moreover supercoherent states can be constructed by the action of the unitary operator $\mathcal{U}_S = \exp(Z); Z \in \mathcal{S}$ which implements the automorphism Φ_S, over some cyclic vector $|\omega>$ (*e.g.* the vacuum state for the rotated hamiltonian). As the linearized hamiltonian, as well as the operator $\mathcal{U}_S$, have coefficients which are both c-numbers and elements of $\mathcal{G}$, the Hilbert space has to be enlarged to allow the supercoherent states to have coefficients in $\mathcal{G}$.

$\mathcal{G}$ is the direct sum of two disjoint sectors, $\mathcal{G} = \mathcal{G}_e \oplus \mathcal{G}_o$, the elements of $\mathcal{G}_o$ being products of *odd* numbers and those of $\mathcal{G}_e$ of *even* numbers of θ's. If the condition (1) were extended to the case $A = B$, then we should have $\theta_A{}^2 = 0$, and $\mathcal{G}$ would be a (*Grassmann-Banach*) algebra [4].

The elements of $\mathcal{G}_e$ commute then with those of $\mathcal{G}_o$, as well as with any product of fermi operators.

In the case when $\mathcal{G}$ is an algebra, one may define an inner product $\prec \theta_R|\theta_S \succ \in \mathcal{C}$ over the vector space (actually a superspace) associated with $\mathcal{G}$, such that $\prec \theta_A|\theta_B \succ = - \prec \theta_B|\theta_A \succ$, and $\prec \theta_A|\theta_A \succ = 0$. Such a product is not necessarily associative. In the following, in view of the commutation properties of the elements of $\mathcal{G}_e$, we shall introduce a further linearization which consists in identifying them with the corresponding inner products. This amounts to treating them as *non-nilpotent* c-numbers (the non- associativity of $\prec \bullet|\bullet \succ$ allows this choice). Besides we will assume that, for $g \in \mathcal{G}_e$, $g + \bar{g}$ and $g\bar{g}$ are real numbers.

A lattice problem especially suitable to a solution with the method outlined above (namely linearization, recognition of a dynamical superalgebra and diagonalization by a generalized Bogolubov rotation) is the Hubbard model [5], described by the many fermion hamiltonian

$$H = \sum_{\mathbf{i},\sigma} \varepsilon_{\mathbf{i}} n_{\mathbf{i},\sigma} + \sum_{\mathbf{i}} U_{\mathbf{i}} n_{\mathbf{i},\Uparrow} n_{\mathbf{i},\Downarrow} + \sum_{<\mathbf{i},\mathbf{j}>,\sigma} t_{\mathbf{i},\mathbf{j}} a^{\dagger}_{\mathbf{i},\sigma} a_{\mathbf{j},\sigma} \quad ;$$

$$\{a_{\mathbf{i},\sigma}, a^{\dagger}_{\mathbf{j},\sigma'}\} = \delta_{\mathbf{i},\mathbf{j}} \delta_{\sigma,\sigma'} \quad , \quad \{a_{\mathbf{i},\sigma}, a_{\mathbf{j},\sigma'}\} = 0 \quad , \quad n_{\mathbf{i},\sigma} = a^{\dagger}_{\mathbf{i},\sigma} a_{\mathbf{i},\sigma} \quad ; \tag{2}$$

where $\mathbf{i} \in \Lambda$ and $\sigma \in \{\Uparrow, \Downarrow\}$. Λ denotes the lattice; $\varepsilon_{\mathbf{i}}, U_{\mathbf{i}}$, and $t_{\mathbf{i},\mathbf{j}}$ are assumed to be real.

Adoption of the linearization scheme (1) for the *hopping* terms $a^{\dagger}_{\mathbf{i},\sigma} a_{\mathbf{j},\sigma}$ in (2) leads to [2] :

$$H_{red} = \sum_{\mathbf{i} \in \Lambda} H_{\mathbf{i}} \in \mathcal{S} = \bigoplus_{\mathbf{i} \in \Lambda} \mathcal{S}_{\mathbf{i}} \quad ;$$

$$H_{\mathbf{i}} = \varepsilon_{\mathbf{i}} N_{\mathbf{i}} + U_{\mathbf{i}} P_{\mathbf{i}} + \sqrt{2}\{\theta_{\mathbf{i}} A_{\mathbf{i}} - \bar{\theta}_{\mathbf{i}} A^{\dagger}_{\mathbf{i}}\} \quad , \tag{3}$$

$$N_{\mathbf{i}} = n_{\mathbf{i},\Uparrow} + n_{\mathbf{i},\Downarrow} \quad , \quad P_{\mathbf{i}} = n_{\mathbf{i},\Uparrow} n_{\mathbf{i},\Downarrow} \quad , \quad A_{\mathbf{i}} = \frac{1}{\sqrt{2}}(a_{\mathbf{i},\Uparrow} + a_{\mathbf{i},\Downarrow}) \quad ;$$

where, denoting by $< \bullet >$ the average in some selected equilibrium state,

$$\vartheta \equiv \theta_{\mathbf{i}} = \theta_{\mathbf{i},\sigma} = \sum_{\mathbf{j}=n.n.(\mathbf{i})} t_{\mathbf{i},\mathbf{j}} < a^{\dagger}_{\mathbf{j},\sigma} > \in \mathcal{G}_o \quad . \tag{4}$$

It is easily checked that $\mathcal{S}_{\mathbf{i}} \sim u(1|1) \oplus u(1|1)$. This allowed us to obtain – by algebraic implementation of the Bogolubov automorphism (ref.[2]) – the spectrum of H as well as some physically interesting properties of the system as functions also of the c-number $\vartheta\bar{\vartheta}$. The latter can be self- consistently evaluated once the averaging procedure $< \bullet >$ is fixed.

A further step towards a reliable approximation of H can be made by a *cluster* mean field reduction : one considers the lattice Λ as the sum of two disjoint sublattices, $\Lambda = \Lambda_O \cup \Lambda_E$, such that the sites of each have as nearest neighbours only sites of the other. The hamiltonian is then approximated as $H = \sum_{\gamma \in \Gamma} H_\gamma$, where Γ denotes any covering of Λ by *dimers* γ (*i.e.* two-sites clusters, each having a site in Λ_O and the other in Λ_E). H_γ is the sum of two single-site hamiltonians of the form of $H_{\mathbf{i}}$ in (3) [we shall denote by $\mathbf{i} = E$ and $\mathbf{i} = O$ the sites of γ belonging to Λ_E and Λ_O respectively] plus the interaction (hopping) between the two sites E and O, which is kept into account exactly. In other words,

$$H_\gamma = H_E + H_O + \{tA_O^\dagger A_E + \text{h.c.}\} \quad ; \tag{5}$$

Notice that in (5) the coupling between the two sublattices is considered with equal amplitude for both ***parallel*** and ***spin-flip*** hopping processes. However, diversity in this amplitude can be thought of as included in $\theta_{\mathbf{i}}\,; \mathbf{i} = E, O$ by the fermionic linearization procedure. Some preliminary results about this model were presented in ref. [2]. Here we shall construct the supercoherent states of H_γ – in particular its ground state – and the corresponding energy and pairing ($< A_O^\dagger A_E^\dagger >$) expectation values.

The first step is to recognize H_γ as an element of a superalgebra $\mathcal{S}_\gamma$. One can check that $\mathcal{S}_\gamma \equiv u(2|1) \oplus u(2|1) \oplus u(2|1) \oplus u(2|1)$. $u(2|1)$ is a superalgebra generated by 9 elements, five of which are in the bosonic sector $\mathcal{B}(u(2|1))$ [of these one is central and three are Cartan] and four are in the fermionic sector $\mathcal{F}(u(2|1))$:

$$\mathcal{B}(u(2|1)) = \{\mathbf{I}, A_E^\dagger A_E, A_O^\dagger A_O, A_E^\dagger A_O, A_O^\dagger A_E\} \quad ;$$

$$\mathcal{F}(u(2|1)) = \{A_E, A_O, A_E^\dagger, A_O^\dagger\} \quad . \tag{6}$$

The four orthogonal copies of $u(2|1)$ which generate the whole superalgebra $\mathcal{S}_\gamma$ are obtained by multiplying the odd and even sectors of $u(2|1)$, as given in (6), successively by $D_E D_O, (\mathbf{I} - D_E)(\mathbf{I} - D_O), D_E(\mathbf{I} - D_O), (\mathbf{I} - D_E)D_O$, where

$$D_\alpha = \frac{1}{2}(n_{\alpha,\Uparrow} + n_{\alpha,\Downarrow} + a_{\alpha,\Uparrow}^\dagger a_{\alpha,\Downarrow} + a_{\alpha,\Downarrow}^\dagger a_{\alpha,\Uparrow}) \quad ; \quad \alpha = E, O \quad , \tag{7}$$

are central elements of $\mathcal{S}_\gamma$. As $D_\alpha^2 = D_\alpha$, the D_α's have eigenvalues 1 and 0 , and the four copies of $u(2|1)$ generating $\mathcal{S}_\gamma$ are in fact labelled by the four possible different combinations of eigenvalues of D_E and D_O.

In the sequel, in order to evaluate supercoherent state expectation values, we shall need a faithful (matrix) representation of $\mathcal{S}_\gamma$; this will therefore be simply given by a four-block-diagonal matrix, each block of which is a matrix representation of

$u(2|1)$ acting in a sector of the Hilbert space characterized by two specified eigenvalues of D_E and D_O.

We have adopted for $u(2|1)$ the following four-dimensional representation :

$$A_E^{\dagger} A_E = \frac{1}{2} \mathbf{1}_2 \otimes (\mathbf{1}_2 - 2\sigma_z) \quad ; \quad A_O^{\dagger} A_O = \frac{1}{2} (\mathbf{1}_2 - 2\sigma_z) \otimes \mathbf{1}_2 \quad ;$$

$$A_E^{\dagger} A_O = \sigma_+ \otimes \sigma_- \quad ; \quad A_O^{\dagger} A_E = \sigma_- \otimes \sigma_+ \quad ; \quad \mathbf{I} = \mathbf{1}_2 \otimes \mathbf{1}_2 \quad ;$$

$$\theta_E \, A_E = \theta_E \, \mathbf{1}_2 \otimes \sigma_+ \quad ; \quad \theta_O \, A_O = \theta_O \, \sigma_+ \otimes 2\sigma_z \quad ;$$

$$A_E^{\dagger} \, \bar{\theta}_E = \mathbf{1}_2 \otimes \sigma_- \, \bar{\theta}_E \quad ; \quad A_O^{\dagger} \, \bar{\theta}_O = \sigma_- \otimes 2\sigma_z \, \bar{\theta}_O \quad . \tag{8}$$

where $\sigma_{\pm} = \sigma_x \pm i\sigma_y$ and $\sigma_x, \sigma_y, \sigma_z$ are the customary two-dimensional Pauli matrices, and $\mathbf{1}_2$ the 2×2 unit matrix.

We can now proceed to the construction of the ground state $|\psi_G>$ for H_γ. $|\psi_G>$ will be obtained as that particular supercoherent state $|\psi_S>$ of $\mathcal{S}_\gamma$ by which the expectation value of H_γ, namely the zero-temperature free energy, is minimized.

We define first the generic supercoherent state $|\psi_S>$ as :

$$|\psi_S> = \mathcal{U}_S |\omega>_\gamma \quad ; \quad \mathcal{U}_S = \exp(Z) \quad , \quad Z \in \mathcal{S}_\gamma \quad ; \tag{9}$$

where $|\omega>_\gamma$ is the highest weight vector for the superalgebra $\mathcal{S}_\gamma$. If Z is such that $\exp(\mathrm{ad}Z)(H_\gamma)$ is diagonal, then $|\omega>_\gamma$ coincides with the vacuum of H_γ .

We assume

$$|\omega>_\gamma = \frac{1}{2} \sum_{\alpha=1}^{4} |\omega_\gamma>^\alpha \quad ;$$

$$|\omega_\gamma>^\alpha = \frac{1}{\sqrt{2}} (\,|1_E 0_O>^\alpha - |0_E 1_O>^\alpha) \, ; \; {}_\gamma<\omega|\omega>_\gamma = 1 = {}^\alpha<\omega_\gamma|\omega_\gamma>^\alpha ; \tag{10}$$

where by $|i_E j_O>^\alpha$; $i, j \in \{0, 1\}$ we denote the state with total number of particles equal to i on site E , and to j on site O . $|\omega_\gamma>^\alpha$ is the selected maximal vector for the representation given by (8), in which it is represented by a 16-dimensional column vector, made of four 4-vectors . Three of the latter have all their components zero, the α-th is given by

$$|v> = \begin{pmatrix} 0 \\ 1 \\ 1 \\ 0 \end{pmatrix} \quad . \tag{11}$$

The operator Z in (9) is, in principle, an antihermitian linear combination of all the elements of $\mathcal{S}_\gamma$. However, since we are going to evaluate expectation values in $|\psi_S>$ (*i.e.*, for any operator $\mathcal{O}$, $<\mathcal{O}> \equiv <\psi_S|\mathcal{O}|\psi_S>$), we can – with no loss of generality – ignore all the operators belonging to the Cartan subalgebra of $\mathcal{B}(\mathcal{S}_\gamma)$. For analogous reasons we can use the following "weak" identity (indeed an equivalence, denoted by the symbol $\simeq$, implying equality *modulo* elements of the Cartan subalgebra of $\mathcal{S}_\gamma$) :

$$\exp Z \simeq \exp Z^{(b)} \exp Z^{(f)} \quad ; \quad \cdot \tag{12}$$

where

$$Z^{(b)} = p(A_E^{\dagger}A_O - A_O^{\dagger}A_E) \quad ; \quad Z^{(f)} = \sum_{\substack{i=E,O \\ \text{mod } 2}} (\zeta_i A_i + \zeta_i^{\dagger} A_i^{\dagger}) \quad ;$$

$$p = p_1 + p_2 D_E + p_3 D_O + p_4 D_E D_O \quad ; \quad \{p_\alpha | \alpha = 1, \ldots, 4\} \in \mathcal{G}_e \quad ;$$

$$\zeta_i = \beta_i + \gamma_i D_i + \delta_i D_{i+1} + \eta_i D_i D_{i+1} \quad ;$$

$$\{\pi_\alpha^i | \alpha = 1, \ldots, 4; i = E, O\} \equiv \{\beta_i, \gamma_i, \delta_i, \eta_i \,|\, i = E, O\} \in \mathcal{G}_o \quad . \tag{13}$$

Eqs. (9), (12), and (13) show that $|\psi_S>$ is indeed a supercoherent state, living, as expected, in an extended Hilbert space with coefficients in $\mathcal{G}$.

We are now finally able to write the expectation value of $H_\gamma - \mu N_\gamma$ (where μ denotes the chemical potential, and $N_\gamma = N_E + N_O$ is the total number of particles on dimer γ) in the supercoherent state (9) :

$$\begin{aligned} \mathcal{H} &\equiv <\psi_S | H_\gamma - \mu N_\gamma | \psi_S> \\ &= {}_\gamma\!<\omega| \exp\{\text{ad}Z^{(b)}[\exp(\text{ad}Z^{(f)})]\}(H_\gamma - \mu N_\gamma)|\omega>_\gamma \\ &= \mathcal{H}_\mathcal{B}(\{p_\alpha\}) + \mathcal{H}_\mathcal{F}(\{\pi_\alpha^i\}) \quad ; \end{aligned} \tag{14}$$

$\mathcal{H}$ as given by (14) turns out to be a real function of the twelve parameters $\{p_\alpha, \pi_\alpha^i\}$ (with $\alpha = 1, \ldots, 4$; $i = E, O$). Notice the distribution of the parameters $\{p_\alpha\}$ and $\{\pi_\alpha^i\}$ between the two factors $\mathcal{H}_\mathcal{B}$, which represents the expectation value of $H_\gamma - \mu N_\gamma$ for the non-interacting dimer, and $\mathcal{H}_\mathcal{F}$, which describes the action of the rest of the lattice on the dimer γ by the fermionic mean field.

As the parameters $\{\pi_\alpha^i\}$ are necessarily linear combinations of θ_E and θ_O and they appear in $\mathcal{H}_\mathcal{F}$ only in bilinear products of the form $\pi_\alpha^i \bar{\pi}_\beta^j$, the only three elements of $\mathcal{G}_e$ entering $\mathcal{H}_\mathcal{F}$ are $f_i = \theta_i \bar{\theta}_i$, $i = E, O$, and $g + \bar{g} = \theta_E \bar{\theta}_O + \theta_O \bar{\theta}_E$. These three last quantities are to be evaluated self-consistently (eqs. (4), for $\mathrm{i} \equiv i = E, O$), keeping the constraint

$$f_E f_O \equiv \theta_E \bar{\theta}_E \theta_O \bar{\theta}_O = -\theta_E \bar{\theta}_O \theta_O \bar{\theta}_E \equiv -g\bar{g} \quad . \tag{15}$$

into account. The self-consistency conditions are easily written in explicit form by computing on the one hand the average in the supercoherent state $|\psi_S>$ of the quantities $(\theta_i a_i - \bar{\theta}_i a_i^{\dagger})$ and on the other using eqs. (4) :

$$\begin{aligned} <(\theta_i a_i - \bar{\theta}_i a_i^{\dagger})> &= \frac{1}{\sqrt{2}\,t\,(\nu_{i+1} - 1)}(g + \bar{g}) \equiv \ell_i(g + \bar{g}) \\ &= \sqrt{2}\,\{\theta_i(\bar{\beta}_i + \frac{1}{2}(\bar{\gamma}_i + \bar{\delta}_i) + \frac{1}{4}\bar{\eta}_i) + \text{h.c.}\} \quad . \end{aligned} \tag{16}$$

ν_i denotes the number of nearest neighbours of site i.

If one performs now the reparametrization of the superspace of the supergroup associated with $\mathcal{S}_\gamma$ which amounts to replacing the set $\{\pi_\alpha^i\}$ by the new set $\{\hat{\pi}_\alpha^i\} \equiv \{k_i, h_i, m_i, r_i | i = E, O\} \in \mathcal{G}_e$, defined by

$$\begin{aligned} \beta_i = \theta_{i+1} k_i \quad &; \quad \gamma_i = \theta_{i+1} f_E h_i \quad ; \\ \delta_i = \theta_{i+1} f_O m_i \quad &; \quad \eta_i = \theta_{i+1}(g + \bar{g}) r_i \quad ; \end{aligned} \tag{17}$$

In view of the self-consistency equations (16), we fix the gauge for the θ-variables by assuming *e.g.* $k_E \equiv 1$ (we shall then simply write k for k_O); thus the set $\{\hat{\pi}^i_\alpha\}$ becomes $\{\Pi_\beta | \beta = 1, \ldots, 7\} \equiv \{k, \{h_i, m_i, r_i \,;\, i = E, O\}$.

Due to (17), $\mathcal{H}_\mathcal{F}$ becomes a function of $(f_E, f_O, g + \bar{g}, \{\Pi_\beta\})$ of the form

$$\mathcal{H}_\mathcal{F} = \mathcal{H}_1(f_E, f_O, \{\Pi_\beta\}) + \frac{1}{2}\frac{g+\bar{g}}{|g\bar{g}|^{\frac{1}{2}}}\mathcal{H}_2(f_E, f_O, \{\Pi_\beta\}) \quad ; \tag{18}$$

Notice that the factor $\frac{1}{2}(g + \bar{g})\,|g\bar{g}|^{-\frac{1}{2}}$ multiplying $\mathcal{H}_2$ in (18) is nothing but the cosine of the phase of g.

One can check, moreover, that the non-trivial solutions of (15) and (16) are simply given by :

$$\begin{aligned} f_O &= \frac{r_E(\ell_E - k) - r_O(\ell_O - 1)}{r_E m_O - r_O m_E} \equiv F_O(\{\Pi_\beta\}) \quad ; \\ f_E &= \frac{r_E(\ell_E - k) - r_O(\ell_O - 1)}{r_E h_O - r_O h_E} \equiv F_E(\{\Pi_\beta\}) \quad ; \end{aligned} \tag{19}$$

$$g+\bar{g} = \frac{m_O(\ell_O - 1) - m_E(\ell_E - k)}{r_E m_O - r_O m_E} + \frac{h_O(\ell_O - 1) - h_E(\ell_E - k)}{r_E h_O - r_O h_E} \equiv G(\{\Pi_\beta\}) \quad . \tag{20}$$

There exists also a set of trivial solutions $f_E = f_O = g = 0$, which we don't take into consideration here because they describe a dimer not interacting with the rest of the lattice (for the range of parameters for which the nonvanishing solutions (19) and (20) hold, these zero solutions are unstable; we expect them to become stable above the critical temperature [2]).

After insertion of the self-consistent solutions (19) and (20) , $\mathcal{H}_\mathcal{F}$ becomes a function of $\{\Pi_\beta\}$ only.

Upon denoting by $\mathcal{N} \equiv \langle N_\gamma \rangle = \mathcal{N}(\{p_\alpha, \Pi_\beta\})$ the dimer occupation number expectation value, the ground state free energy $\mathcal{E}$ is then obtained by eliminating μ between the two equations

$$\begin{aligned} \mathcal{N}(\{\tilde{p}_\alpha, \tilde{\Pi}_\beta\}) &= \mathcal{N}^{(0)}_\gamma \quad ; \\ \mathcal{H}(\{\tilde{p}_\alpha, \tilde{\Pi}_\beta\}) &= \mathcal{E} \quad ; \end{aligned} \tag{21}$$

where $\{\tilde{p}_\alpha, \tilde{\Pi}_\beta\}$ is the solution to the variational equation

$$\mathcal{H}(\{\tilde{p}_\alpha, \tilde{\Pi}_\beta\}) = \min_{\{p_\alpha, \Pi_\beta\}} \mathcal{H}(\{p_\alpha, \Pi_\beta\}) \quad ; \tag{22}$$

and $\mathcal{N}^{(0)}_\gamma$ is given (it is the total number of particles on the dimer γ).

The corresponding ground state is then given by

$$|\psi_G\rangle = \exp(\tilde{Z})\,|\omega\rangle_\gamma \quad ; \tag{23}$$

where of course $\tilde{Z}$ is the operator defined in (12) , (13) with the parameters $\{p_\alpha\}$ and $\{\pi^i_\alpha\}$ replaced respectively by $\{\bar{p}_\alpha\}$ and $\{\pi^i_\alpha(\{\Pi_\beta\})\}$. We are now able to compute the physical expectation values in $|\psi_G>$, in particular those giving rise to interesting *order parameters*, such as

$$< a^\dagger_E a^\dagger_O > = {}_\gamma< \omega|\tilde{\zeta}_E\,\tilde{\zeta}_O|\omega>_\gamma = \varphi(\{\tilde{\Pi}_\beta\})\,h \quad , \tag{24}$$

where $h \in \mathcal{G}_e$ is such that $h\bar{h} = f_E f_O$, and

$$\begin{aligned} < a^\dagger_E a_O + a^\dagger_O a_E > &= {}_\gamma< \omega|(a^\dagger_E a_O + a^\dagger_O a_E)|\omega>_\gamma + {}_\gamma< \omega|(\tilde{\zeta}_E\tilde{\zeta}^\dagger_O + \tilde{\zeta}_O\tilde{\zeta}^\dagger_E)|\omega>_\gamma \\ &= -\frac{1}{4} + \psi(\{\tilde{\Pi}_\beta\})\,(g+\bar{g}) \quad , \end{aligned} \tag{25}$$

$\varphi(\bullet)$ and $\psi(\bullet)$ in ((24) and (25) are known functions (that we don't report explicitly for lack of space) straightforwardly obtained from (10) and (13).

Notice that if one replaced the trivial solution in the order parameters (24) and (25) , the latter, which is also related to hopping inside the dimer, would not vanish; while the pairing expectation value (24) would become zero. On the other hand, for solutions different from the trivial one, both $< a^\dagger_E a_O >$ and $< a^\dagger_E a_O + a^\dagger_O a_E >$ are *non* zero. This implies that it is indeed the rest of the lattice which induces a pairing on the dimer.

The evaluation of the order parameters (24) , (25) as well as of the physical values of $f_E, f_O, g+\bar{g}$ – from which all other physically relevant quantities depend – is linked to the solution of eq. (22).

One can see from eq. (14) that the problem of minimizing $\mathcal{H}$ can be tackled by minimizing independently $\mathcal{H}_{\mathcal{B}}$ with respect to the set of parameters $\{p_\alpha\}$ characterizing the rotation in $\mathcal{B}(S_\gamma)$, and $\mathcal{H}_{\mathcal{F}}$ with respect to $\{\Pi_\beta\}$ which define the rotation in $\mathcal{F}(S_\gamma)$.

The first step is easily performed, and leads to

$$p_1 = \frac{1}{2}\arctan\big(\frac{\varepsilon_E - \varepsilon_O}{4t}\big) \quad ,$$

$$p_2 = \frac{1}{2}\arctan\big(\frac{\varepsilon_E - \varepsilon_O + U_E}{4t}\big) - p_1 \quad , \quad p_3 = \frac{1}{2}\arctan\big(\frac{\varepsilon_E - \varepsilon_O - U_O}{4t}\big) - p_1 \quad ,$$

$$p_4 = \frac{1}{2}\arctan\big(\frac{\varepsilon_E - \varepsilon_O + U_E - U_O}{4t}\big) - p_1 - p_2 - p_3 \quad . \tag{26}$$

(independent on μ). For example, in the simple symmetric case when $\varepsilon_E = \varepsilon_O \equiv \varepsilon$; $U_E = U_O \equiv U$, the contribution of $\mathcal{H}_{\mathcal{B}}$ to the ground state free energy would be $\mathcal{E}^B_G = 2(\varepsilon - \mu) + \frac{1}{2}U - 2t\,[1 + (\frac{1}{4}U)^2]^{-1}$, which is of course the total ground state free energy for the non-interacting dimer.

As for $\mathcal{H}_{\mathcal{F}}$, in practice it turns out to be more convenient, instead of approaching eq. (22) directly, to minimize it in its form (18) by taking into account the three constraints (19) and (20) explicitly. This requires the introduction of three Lagrange multipliers, say $\lambda_i\,; i = E, O,$ and σ.

The set of equations

$$\frac{\partial}{\partial x_j}\{\mathcal{H}_{\mathcal{F}} - \sum_{i=E,O} \lambda_i[f_i - F_i(\{\Pi_\beta\})] - \sigma[g + \bar{g} - G(\{\Pi_\beta\})]\} = 0 \quad ; \quad j = 1, \dots, 10 \quad ; \tag{27}$$

where $x_j \equiv \Pi_\beta$ for $j = 1, \dots, 7$ and $x_8 \equiv f_E$, $x_9 \equiv f_O$, $x_{10} \equiv (g + \bar{g})$, together with the three constraints, form a system of thirteen equations in the unknowns $\{x_j, \lambda_i\}, \sigma$.

Eleven of such equations, upon considering σ and $(g + \bar{g})$ as parameters, form a linear (and non-homogeneous) system in the remaining eleven unknowns. The solutions of the latter (that we obtained by the aid of the *MACSYMA* symbolic manipulation program), inserted in the two equations left, transform them in two coupled equations for σ and $(g + \bar{g})$, the first of degrees 6 in σ and 9 in $(g + \bar{g})$, the other of degrees 8 in σ and 10 in $(g + \bar{g})$. Both factorize a term which generates the trivial solution. In particular, the former, in the simple case considered above when the two sites E and O have equal parameters ε and U, becomes of degree 5 in σ, with a root equal to 0 and the remaining equation biquadratic. By substitution of its solutions into the $(g + \bar{g})$ equation, one obtains a trascendental equation which cannot but be handled numerically. The structure of the phase space resulting from numerical analysis will be given elsewhere.

References

[1] A. Montorsi, M. Rasetti and A.I. Solomon, *Phys. Rev. Lett.* **59**, 2243 (1987)
[2] A. Montorsi, M. Rasetti and A.I. Solomon, *Intl. J. Mod. Phys.* **3** (1989)
[3] M. Rasetti, *Intl. J. Theor. Phys.* **13**, 425 (1975)
[4] A. Jadczyk and K. Pilch, *Commun. Math. Phys.* **78**, 373 (1981)
[5] J. Hubbard, *Proc. Roy. Soc. London*, Ser. **A, 227**, 237 and **276**, 238 (1964)

SUPERCONDUCTIVITY OF ITINERANT ELECTRONS COUPLED TO SPIN CHAINS

R. Shankar

Sloane Physics Laboratory
Center for Theoretical Physics
Yale University
New Haven, Conn. 06520

INTRODUCTION

Let us take as our starting point the Cu-O layers. The question is this: if one dopes the system away from half filling is there a mechanism that can lead to superconductivity? We will also assume that the carriers reside at the oxygen sites and that the spins in the copper sites provide an antiferromagnetic background for the former. A possible mechanism for pairing is due to spin polarons. Consider just one carrier: if there is a coupling between the spins of the itinerant electrons (IE) and the fixed spins (say ferromagnetic), then the carriers will prefer a ferromagnetic domain (aligned with its spin), i.e. a polaron. The polaron size will be determined by a competition between loss of exchange energy and localization energy. If a second carrier is now introduced, it will reside in the same polaron (if the coulomb interaction is not too strong) and also have its spin parallel to the first. Thus the polaron can lead to pairing in the spin 1 channel.

All this is of course speculation. Unfortunately a reliable calculation is not possible in d=2 since there is no obvious small parameter in the problem. There are several options. One can either introduce a small parameter like 1/N or do mean field theory using 1/R or 1/E as the small parameter to inhibit fluctuations, R and E being the author's reputation or ego. Another option is to try for an exact solution. This seems possible in d=1. I will now describe a calculation Subir Sachdev[1] and I performed in a d=1 model that was solvable in the long distance limit. The calculation showed that superconductivity was indeed possible and that too in the parallel spin channel anticipated by the polaron model.

Interacting Electrons in Reduced Dimensions
Edited by D. Baeriswyl and D.K. Campbell
Plenum Press, New York

The Model Hamiltonian

The hamiltonian that was chosen to model the Cu-o physics was the following

$$H = \sum_{n} \vec{S}(n) \cdot \vec{S}(n+1) + t \sum_{n,\alpha=\pm,} C_{n+1}^{\alpha\dagger} C_{n}^{\alpha} + C_{n}^{\alpha\dagger} C_{n+1}^{\alpha} \tag{2.1}$$

$$+ m \sum_{n} s_z C^{\dagger} \sigma_z C + \text{Hubbard repulsion.}$$

The first term describes the antiferromagnetic interactions between the fixed spins (S=1/2), the second the hopping of the IE, the third the spin-spin coupling between the two species. The label α describes the spin of the IE. For technical reasons it was not possible to consider a rotationally invariant coupling between the species; however, this does not undermine the polaron mechanism. The Hubbard replusion includes on site (U) and nearest neighbor terms (V).

Although this hamiltonian was not solved exactly, its continuum limit (i.e. long distance physics) is totally known and in particular one can tell if superconductivity is possible or not. To be precise, since in d=1 only quasi-long range order is possible, the question is whether the Fourier transform of $\chi_P = \langle C_{\uparrow}(o)C_{\uparrow}(o)C_{\uparrow}^{\dagger}(n)C_{\uparrow}^{\dagger}(n)\rangle$ (where "P" reminds us the spins are parallel) behaves as $k^{-\alpha}$ with $\alpha > o$. We find that indeed it does if m in Eq. (1) exceeds a minimum value determined by the U and V terms. In contrast χ_A, for antiparallel spins, does not diverge. We also calculated CDW and SDW responses; these will not be discussed here.

We will now turn to the actual computations. I will mostly give results and not the proofs. Given that one is forced to make choices due to time and space constraints you will agree that this is a better alternative than giving a lot of proofs and no results!

Computational Details

Since it is not possible to present all the details of the calculation here, I will attempt a pedagogical treatment of two main ideas used: Jordan Wigner transformation and bosozization. These ideas were originally used in similar problems by Lieb and Mattis[2] and more recently and more along the lines we follow, by Luther and Peschel[3] and den Nijs[4]. Our contribution was to realize that the present problem could be attacked using these notions.

To illustrate how these ideas are used, let us throw out all terms but the $\vec{S}.\vec{S}$ term. Let us generalize it to an anisotropic model

$$H_{xxz} = \sum_{n} \frac{1}{2} \left[S_+(n)\ S_-(n+1) + S_-(n) S_+(n+1) \right]$$

$$+\Delta\ S_z(n)\ S_z(n+1) \tag{3.1}$$

and throw away the Δ term also, ending up with just H_{xx}:

$$H_{xx} = \frac{1}{2} \sum_{n} \left[S_+(n)\ S_-(n+1) + S_-\ (n)\ S_+\ (n+1) \right] \tag{3.2}$$

We have a translationally invariant, quadratic hamiltonian. Why don't we solve it by Fourier transformation? The reason is that the spins operators $S_{\pm}$ form a mixed algebra: they behave like fermions ψ and $\psi^{\dagger}$ at one site and commute between different sites. The Fourier cofficients will therefore not obey any simple algebra. The Jordan Wigner transformation trades the $s_{\pm}$'s for a global Fermi field:

$$\psi(n) = (-1)^n \left[\prod_{-\infty}^{n-1} 2S_z \right]\ S_+\ (n) \tag{3.3}$$

with $\psi^{\dagger}$ (n) defined as usual. One can check that ψ is a genuine one component Fermi field. But the real question is this: how does H_{xx} look in terms of ψ ? Does the nonlocal change of variables (3.3) make H_{xx} into a nonlocal function of ψ? The answer is no:

$$H_{xx} = -2\sum_{n} \left(\psi^{\dagger}_{n+1}\ \psi_n + \psi_n^{\dagger}\ \psi_{n+1} \right) \tag{3.4}$$

(Try adding a local term $\sum_{n} S_x$ to H_{xx} and see how it looks in the fermion language!). We now go to plane waves and solve (3.4):

$$H_{xx} = -\sum_{k=-\pi}^{\pi} (\cos k)\ C^{\dagger}(k)\ C(k) \tag{3.5}$$

The ground state has all states with $-\pi/2 \leq k \leq \pi/2$ filled, i.e., $k_F = \pi/2$ and we have a half filled band.

We now put back the Δ term in (3.1).

Now

$$S_z = \psi^\dagger \psi - \tfrac{1}{2} = :\psi^\dagger \psi: \qquad (3.6a)$$

where : :, means the vacuum expectation value of 1/2 (due to half filling) is removed from $\psi^\dagger \psi$.

We are now left with a local but interacting four-Fermi interaction

$$H_{XXZ} = H_{XX} + \Sigma_n \Delta :\psi^\dagger(n)\psi(n)::\psi^\dagger(n+1)\ \psi\ (n+1):. \qquad (3.6b)$$

In higher dimensions our only hope is perturbation theory; in d=1 we can go beyond, using bosonization. Briefly speaking, bosonization is a mapping of a fermion problem to a boson problem with the attractive feature that quartic fermi interaction in question will turn into a quadratic boson term. The details are as follows:

Fermion		Boson	
$H^F=\int (\psi^\dagger\ (\alpha.P)\ \psi)\ dx$		$H^B = \frac{1}{2}\int \left[\Pi^2 + (\partial\phi)^2\right]dx$	(3.7a)
$\langle O^F(x_1)\ O^F(x_2)\ldots\rangle_{0,\alpha}$	=	$\langle O^B(x_1)\ldots\rangle_{0,\alpha}$	(3.7b)
$O^F = \psi^\dagger\psi = \psi^\dagger{}_1\psi_1 + \psi_2{}^\dagger\psi_2$		$O^B = \frac{1}{\sqrt{\pi}}\ \partial\phi$	(3.7c)
$O^F = i\psi_1{}^\dagger\ \psi_2 - i\psi_2{}^\dagger\psi_1 = \bar{\psi}\psi$		$O^B = \frac{-1}{\pi\alpha}\ \cos\sqrt{4\pi}\ \phi$	(3.7d)

The above table means the following: $H^{F,B}$ describe massless F/B theories in d=1. $\langle\rangle_{0,\alpha}$ define vacuum averages of spectators O^F or O^B, with all momentum integrals cut-off by $e^{-\alpha|p|}$ in both cases. Eq. (3.7b) tells us that for each string of fermionic operators O^F there is a corresponding string of bosonic operators O^B with the same correlation functions. For example $\psi^\dagger\psi$ in one case corresponds to $(\partial\phi)/\sqrt{\pi}$ in the other ect. Using this dictionary one can now map interacting fermion threories into the corresponding bosonic theories since they are guaranteed to have the same perturbation series. Note in particular

Eqs. (3.7b) which allows to replace fermion bilinear $\psi^{\dagger}\psi$ with a boson linear term $\partial\phi$.

There is however, an obstacle to be crossed before (3.7) is used: it refers to a two component relativistic Dirac field $\psi=(\psi_1,\psi_2)$ while Eq. (3.6) describes a one component field with a nonrelativistic dispersion relation. Both problems are solved as follows. If one is interested in long distance physics one can focus on excitations near Ik_F and linearize the dispersion relation at these points: these give two species (the putative ψ_1 and ψ_2) with $E \propto \pm k$, k being the deviation from the Fermi surface:

$$H_{xx} = E_{sea} + \Sigma\, C_1^{\dagger} C_1 k - \Sigma\, C_2^{\dagger} C_2 k \tag{3.8}$$

In real space this amounts to

$$\psi(n) = \psi_1(n) e^{ik_F n} + \psi_2(n) e^{-ik_F n} \tag{3.9}$$

Going to continuum fields obtained by dividing ψ by $\sqrt{a}$, a being the lattice spacing, we get in continuum notation.

$$H_{xx} = \int dx\, \psi^{\dagger}(\alpha.p)\psi + \text{energy of filled states} \tag{3.10}$$

where α is the pauli matrix σ_z. Consider now the interaction term.

$$=:\psi^{\dagger}(n)\psi(n): = :\left[\psi_1^{\dagger}(n)e^{-ik_F n} + \psi_2^{\dagger}(n)e^{+ik_F n}\right]\left[\psi_1(n)e^{ik_F n} + \psi_2 e^{-k_F n}\right]:$$

$$= :\psi_1^{\dagger}\psi_1 + \psi_2^{\dagger}\psi_2: + \psi_1^{\dagger}\psi_2 e^{-21k_F n} + \psi_2^{\dagger}\psi_1 e^{2ik_F n} \tag{3.11}$$

If $:\psi^{\dagger}\psi:$ were the term to be added we will ignore the terms with the rapidly oscillating factors $e^{2ik_F n} = (-1)^n$ since they average to zero over a very short distance. However, we need $:\psi^{\dagger}(n)\psi(n)::\psi(n+1)\psi(n+1):$ In this product such factors can cancel. One must be careful to retain the notorious umklapp term with a factor $e^{4ik_F n}$ since this equals 1 in view of the fact $k_F=\pi/2$.

The result of all this is that after bosonization

$$H_{xxz} = \int \left[1/2 \left[\Pi^2 + \left(1+\frac{4\Delta}{\pi}\right)(\partial\varphi)^2\right] + A\cos\sqrt{16\pi}\,\phi \right] dx \qquad (3.12)$$

where $A \propto \Delta$ is the coefficient of the umklapp term. Rescaling $\phi \to \phi(1+4\Delta/\pi)^{1/4}$, $\Pi \to \Pi\,(1+4\Delta/\pi)^{-1/4}$, and rescaling H by a factor $(1+4\Delta/\pi)^{1/2}$, we get

$$H_{xxz} = \int \left[\frac{1}{2}\left(\pi^2 + (\partial\phi)^2\right) + \alpha_0 \cos\beta_0\phi\right] dx \qquad (3.13)$$

where $\alpha_0 \propto \Delta$ and $\beta_0 \sim \sqrt{16\pi}$ for small Δ.

The above Sine-Gordon model is well understood and is described by the famous Kosterlitz-Thouless[5] R G flows of Figure 1.

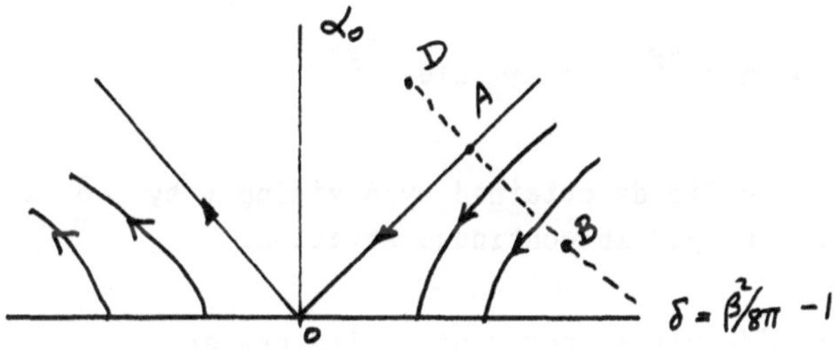

Fig. 1. R G flow into the $\alpha_0 - \beta_0$ plane. Points to the right of the separatrix OA flow into the massless fixed points on the α_0=o axis. On the dotted line, point A describes $\Delta = 1$, B is a point of $\Delta<1$, D of $\Delta>1$.

At the point O, it is easy to show correlation functions are isotropic, it follows that $\Delta = 1$ must be the point A on the separatrix. Recently Singh[6] et al. showed explicitly that on OA correlations are isotropic and decay as $(\ln r)^{1/2}/r$. It is clear from the figure that it is the umklapp term that finally makes the model massive for $\Delta>1$.

This completes the analysis of the Heisenberg magnet. Let us now reinstate the other terms in (2.1). If the reader has followed the previous arguments it will be clear that none of the terms make the model massive. Consider the hopping term of the itinerant electrons. They too will have a cosine despersion relation. However, a chemical potential (not shown in 2.1) will be used to fill a small fraction of the band (i.e; $k_F \ll \pi/2$) to represent a small concentration of carriers due to doping. Both $\uparrow$ and $\downarrow$ species will be first described by linearized Dirac fermions and then bosonized into fields $\phi_\uparrow$ and $\phi_\downarrow$, which will have a massless hamiltonian. The m term in Eq. (2.1), becomes

$$= :\psi^\dagger\psi: \left[\psi^\dagger_\uparrow \psi_\uparrow - \psi^\dagger_\downarrow \psi_\downarrow\right] \to \left[\frac{\partial\phi}{\sqrt{\pi}}\right] \frac{\partial\varphi_\uparrow - \partial\varphi_\downarrow}{\sqrt{\pi}} \tag{3.14}$$

Note that all terms of the form exp $2ik_Fn$ are dropped since they cannot be neutralized by other such terms. Thus $\psi^\dagger\psi \to \psi^\dagger_1 \psi_1 + \psi^\dagger_2\psi_2 \to \partial\phi/\sqrt{\pi}$ etc. (recall $k_F \neq \pi/2$ for itinerant electrons).

The Hubbard U and V terms proportional to $(\psi^\dagger\psi)^2$ are again bilinear in the boson fields. Thus everything is massless, as long as the umklapp terms at Δ - 1 is not rendered <u>relevant</u> due to the m, U and V terms. It is easily checked by calculation within the massless Gaussian theory described above that the m, U and V terms only make the umklepp term <u>more</u> irrelevant. Given the irrelevance of this possible interaction term, it is easy to calculate the exponents describing various susceptibilities like χ_P, χ_A etc. and to verify the claims made earlier i.e. that superconductivity for parallel spin pairing results for $m > m_c(U,V)$. The details may be found in Ref. 1.

CONCLUSION

We set out to test the viability of the polaron mechanism for pairing. We took a d=1 version of the Cu-O 2-band problem in which the itinerant spins were coupled to the background antiferromagnetic spins. Quasi-long range order in the parallel spin channel was shown to follow, confirming the soundness of the picture. Since order in d=2 is even stronger than in d=1, there is no reason to feel that the d=1 conclusions are not representative of the original problem.

ACKNOWLEDGEMENTS

I am pleased to thank the organizers of the workshop, particularly David Campbell for the kind invitation and hospitality.

REFERENCES

1. Subir Sachdev and R. Shankar; Phys. Rev. <u>B38</u>, 826, (1988).
2. E. H. Lieb and D. Matties; J. Math. Phys., <u>6</u>, 304, (1965).
3. A. Luther and I. Peschel, Phys. Rev. <u>B12</u>, 3908, (1975).
4. M. P. M. den Nijs, ibid, <u>23</u>, 6111, (1981).
5. J. M. Kosterlitz and D. J. Thouless, J. Phys. <u>C6</u>, 1181, (1973).
6. R. Singh, M. E. Fisher and R. Shankar to appear in Phys. Rev. B.

VALENCE BOND SPIN MODELS: GROUND STATES AND EXCITATIONS

Elliott H. Lieb

Departments of Physics and Mathematics
Princeton University, P.O. Box 708
Princeton, NJ 08544

This report is about quantum spin Hamiltonians which are

- Antiferromagnetic
- Isotropic in Spin Space
- Short Range.

An example is the Heisenberg Hamiltonian

$$H = \sum_{\langle i,j \rangle} \vec{S}_i \cdot \vec{S}_j \tag{1}$$

where $\langle i,j \rangle$ denotes nearest neighbors on a lattice. This work was done jointly with I. Affleck, T. Kennedy, H. Tasaki and S. Shastry[1,2,3,4,5]. We study ground states and gaps (if any) to excited states. Our goal is to find out what types of behavior can occur and to find rigorous examples of the various possiblities.

I. ONE DIMENSIONAL SYSTEMS

The Heisenberg Hamiltonian for spin $\frac{1}{2}$ is solvable by the Bethe ansatz. There are excitations with arbitrarily small energy (no gap), and correlation functions have a power law decay. For higher spin one might have expected the same but the important conjecture of Haldane[6] is that there is a gap for $S = 1, 2, 3, \ldots$ and no gap for $S = \frac{1}{2}, \frac{3}{2}, \frac{5}{2}, \ldots$. The second part of this was proved for the Heisenberg and $SU(n)$ models by Affleck and Lieb[3]. In addition to there being a gap for integral S one is also supposed to have the following

(i) Correlation functions have exponential decay for integral S and power law decay for half integral S

(ii) The infinite volume limit ground state is unique.

Why add (ii) to the list? The point is that there is the following model due to Majumdar and Ghosh[7] for $S = \frac{1}{2}$.

$$H = \sum_i \vec{S}_i \cdot \vec{S}_{i+1} + \frac{1}{2} \vec{S}_i \cdot \vec{S}_{i+2} \, . \tag{2}$$

Interacting Electrons in Reduced Dimensions
Edited by D. Baeriswyl and D.K. Campbell
Plenum Press, New York

After some considerable period of uncertainty, it was finally proved[1] that this model has

(a) Ultrashort range correlations.

(b) There is a gap.

(c) But there are exactly two ground states in the infinite volume limit (i.e. translational symmetry is broken).

A generalization of the nearest neighbor Heisenberg model for $S = 1$ is

$$H = \sum_i \vec{S}_i \cdot \vec{S}_{i+1} - \beta(\vec{S}_i \cdot \vec{S}_{i+1})^2. \tag{3}$$

This model is solvable by the Bethe ansatz when $\beta = +1$ and thus *presumably* has no gap and has power law decay. For some range of β, however, it is believed to be in the Haldane phase. Indeed, this is true for $\beta = -1/3$ as the following theorem[1] states

Theorem 1. *For $\beta = -1/3$, the Hamiltonian in (3) has a unique infinite volume ground state with exponentially decaying correlation functions and with a gap in the spectrum above the ground state.*

The proof of the gap (also for the Majumdar-Ghosh model) was simplified by Knabe[8].

The special feature of $\beta = -1/3$ is that

$$\vec{S}_i \cdot \vec{S}_{i+1} + \frac{1}{3}(\vec{S}_i \cdot \vec{S}_{i+1})^2 = 2P^{(2)}_{i,i+1} - \frac{2}{3} \tag{4}$$

where $P^{(2)}_{i,j}$ is the projector onto spin 2 for the total spin of the pair $S_i + S_j$. Such models defined by projectors (we shall come to other examples later) are called **Valence Bond Models** and their ground states are called **Valence Bond States (VBS)**.

The VBS ground state ψ has $P^{(2)}_{i,i+1}\psi = 0$ for all i. One cannot improve on this since $P^{(2)}_{i,i+1}$ is a positive operator. To construct the state, regard each spin 1 as a pair of spin $\frac{1}{2}$'s. First couple a pair of nearest neighbor spin $\frac{1}{2}$'s into a singlet state $\uparrow\downarrow - \downarrow\uparrow$. This state does not have $S = 1$ at each site but this can be remedied by symmetrizing the wave function at each site. The final result no longer has $S = 0$ for each nearest neighbor bond, but it does have the property that there is no $S = 2$ component (as required). The VBS state is not unique for a *finite* system because of the unpaired spin 1/2's on the boundary. The problem in proving Theorem 1 is to show that boundary effects do not matter in the infinite volume limit. Possibly boundary effects do play a role in sufficiently high dimension (see next section), in which case the VBS states would not be unique and would therefore have long range order when in the ground state.

Obviously this construction can be generalized – but only to integral S at each site. The state ψ is not a simple tensor product at each site and it has nontrivial correlations, but they decay exponentially.

II. HIGHER DIMENSIONAL SYSTEMS

Let us first review the results for the usual Heisenberg nearest neighbor model (1). Dyson, Lieb and Simon[9] showed the existence of long range Néel order (LRO) at positive temperature for dimension $d \geq 3$ and $S \geq 1$, and also for $S = 1/2$ but larger d. Jordão Neves and Fernando Perez[10] realized that the DLS method could be used to say something about the *ground state* for $d = 2$. They prove LRO for $d = 2$ and $S \geq 1$. This was improved by Kennedy, Lieb and Shastry[4] (using a new sum rule not contained in DLS) to LRO for $d \geq 3$ and $S = 1/2$. The only open case is $d = 2, S = 1/2$; there are some indications in ref. 4 that LRO also exists in this case as well. Kennedy, Lieb and Shastry[5] did show LRO for the anisotropic XY model for all $d \geq 2$ and all $S \geq 1/2$.

Now let us return to VBS models. Take the hexagonal (or honeycomb) lattice with $S = 3/2$ on each site and let

$$H = \sum_{\langle i,j \rangle} P^{(3)}_{i,j} \tag{5}$$

where $P^{(3)}$ is the projection onto $S = 3$ for a pair. The following was proved by Kennedy, Lieb and Tasaki[2].

Theorem 2. *This model (5) has a unique infinite volume ground state and in this ground state correlation functions decay exponentially.*

A gap has not yet been proved to exist, but we conjecture that it does. A similar theorem can "almost" be proved for the square lattice (with $S = 2$ and $P^{(4)}$) in the sense that we can show[2] that the ground state is, indeed, independent of boundary conditions for what is expected to be the worst case (all spins parallel on the boundary).

In the proof of Theorem 2 extensive use is made of a polynomial representation of the VBS states due to Arovas, Auerbach and Haldane[11]. The proof of uniqueness in Theorem 2 (i.e. showing that boundary spins do not matter) is complicated. Using this representation the problem is reduced to the analysis of a gas of "loops" on the lattice with an activity equal to $(1/3)^{L-1}$ where L is the length of a loop. There is also a hard core condition. We have to show that the usual polymer expansion converges (and hence that boundary effects play no role). It turns out that 1/3 is small enough to permit this, but not small enough to make the proof easy. A computer must be used to count a certain number of small loops.

REFERENCES

1. I. Affleck, T. Kennedy, E.H. Lieb and H. Tasaki, *Phys. Rev. Lett. 59*: 799 (1987) and *Commun. Math. Phys. 115*: 477 (1988).
2. T. Kennedy, E.H. Lieb and H. Tasaki, A two dimensional isotropic quantum antiferromagent with unique disordered ground state, *Jour. Stat. Phys.* (1988) (in press).
3. I. Affleck and E.H. Lieb, *Lett. Math. Phys. 12*: 57 (1986).

4. T. Kennedy, E.H. Lieb and S. Shastry, Existence of Néel order in some spin $\frac{1}{2}$ Heisenberg antiferromagnets, *J. Stat. Phys.* (1988) (in press).
5. T. Kennedy, E.H. Lieb and S. Shastry, The XY model has long range order for all spins and all dimensions greater than one, *Phys. Rev. Lett.* (1988) (in press).
6. F.D.M. Haldane, *Phys. Lett.* *93A*: 464 (1983); *Phys. Rev. Lett.* *50*: 1153 (1983); *J. Appl. Phys.* *57*: 3359 (1985).
7. C.K. Majumdar and D.K. Ghosh, *J. Math. Phys.* *10*: 1388 (1969).
8. S. Knabe, *J. Stat. Phys.* *52*: 627 (1988).
9. F.J. Dyson, E.H. Lieb and B. Simon, *J. Stat. Phys.* *18*: 335 (1978).
10. E. Jordão Neves and J. Fernando Perez, *Phys. Lett.* *114A*: 331 (1986).
11. D.P. Arovas, A. Auerbach and F.D.M. Haldane, *Phys. Rev. Lett.* *60*: 531 (1988).

NUMERICAL SIMULATION OF THE TWO DIMENSIONAL HUBBARD MODEL

S. Sorella and M. Parrinello

International School for Advanced Studies (SISSA)
Strada Costiera 11, I-34014, Trieste, Italy

Abstract

A new method for simulating strongly correlated fermionic systems, has been applied to the study of the ground state properties of the 2D Hubbard model at various fillings. Comparison has been made with exact diagonalizations in the 2×2 and 4×4 lattices where very good agreement has been verified.
Numerical results concerning spin correlations and momentum distribution have been obtained with this method for sizes up to 16×16 sites and several fillings. Away from half filling the 2D antiferromagnetic order is initially destroyed, albeit without any clear sign of a Fermi liquid behaviour. A metallic jump in the momentum distribution appears only very far from half filling.

The suggestion that the properties of the high T_c superconductors[1] can be related to that of two dimensional Hubbard model, has renewed considerable attention on this model. However in spite of its apparent simplicity the model has resisted several decades of efforts, with the noticeable exception of the 1D case for which exact analytical results are available[2]. In all the other cases the situation is still unclear and numerical results are expected to provide the necessary insight. Efforts in this direction have so far been hampered by two major difficulties[3−7] a) the numerical instabilities that standard fermion monte carlo algorithms incur in the low temperature regime b) the occurrence of non–positive definite weights which makes the MC estimation of physical properties rather difficult.

We present here some detail on a recently proposed new method[8,9] that completely solves the problem of numerical stability and considerably alleviates the negative sign problem. We then present some numerical results for the two dimensional case. Following Koonin et al.[10,11] we introduce a projected partition function

$$Q = \langle \Psi_T \mid e^{-\beta H} \mid \Psi_T \rangle, \tag{1}$$

where $\mid \Psi_T \rangle$ is a trial wavefunction non orthogonal to the ground state $\mid \Psi_0 \rangle$ and β can be thought of as an imaginary time. Since the imaginary time propagator $e^{-\beta H}$ for $\beta \to \infty$ projects from $\mid \Psi_T \rangle$ its component along $\mid \Psi_0 \rangle$ in terms of Q the ground

Interacting Electrons in Reduced Dimensions
Edited by D. Baeriswyl and D.K. Campbell
Plenum Press, New York

state energy is given by:

$$E_o \underset{\beta\to\infty}{=} -\frac{1}{\beta}\ln Q. \tag{2}$$

Expressions can be found for other ground state properties, by differentiating Eq.(2) with respect to appropriate external fields coupled to the quantities of interest.

We now apply our method to the 2D Hubbard model. The Hubbard model is described by the hamiltonian $H = -\sum_{\langle ij\rangle,\alpha} c^+_{i\alpha}c_{j\alpha} + U\sum_i c^+_{i\uparrow}c_{i\uparrow}c^+_{i\downarrow}c_{i\downarrow}$ where $\sum_{\langle ij\rangle}$ indicates nearest neighbours sum, the indices run over the M lattice sites, $c^+_{i\alpha}$ ($c_{i\alpha}$) are the usual creation (annihilation) operators at site i with spin α and the two body interaction is on site and repulsive ($U > 0$). Note, however, that our method can be extended to treat hamiltonians with two body interactions of fairly general range and sign.

In order to evaluate $e^{-\beta H} \mid \Psi_T\rangle$ we split the imaginary-time propagator $e^{-\beta H}$ into a product of P short-time propagators, and apply the Hubbard-Stratonovich (HS) transformation to each of them. Finally Eq.(1) reads :

$$Q \approx \int d\sigma e^{-\frac{1}{2}\sigma^2}\langle\Psi_T \mid U_\sigma \mid \Psi_T\rangle. \tag{3}$$

where where $d\sigma = \prod_{j=1}^P \prod_{r=1}^M \frac{d\sigma_r(j)}{\sqrt{2\pi}}$, $\sigma^2 = \sum_{r,j}\sigma_r(j)^2$, and U_σ is the (discretized) imaginary time propagator in the time dependent external magnetic field $\sigma_r(j)$:

$$U_\sigma \equiv U_\sigma(\beta,0) = e^{-\frac{U\beta N}{2}} \prod_{j=1}^{P} e^{-H_0\frac{\Delta\tau}{2}} e^{-\lambda\sum_r \sigma_r(j)m_r} e^{-H_0\frac{\Delta\tau}{2}}, \tag{4}$$

where H_0 is the one body part in the hamiltonian H, $\lambda = \sqrt{U\frac{\beta}{P}}$, $m_r = c^+_{r\uparrow}c_{r\uparrow} - c^+_{r\downarrow}c_{r\downarrow}$ is the magnetization operator at site r, and N is the total number of electrons. Eq.(3) becomes exact in the $P\to\infty$ limit. $U_\sigma(\beta,0)$ can be written as a product of operators that act separately on the different spin components: $U_\sigma = U^\uparrow_\sigma U^\downarrow_\sigma$.

The most convenient form for $\mid\Psi_T\rangle$ is to assume that it is a Slater determinant of single particle orbitals $|\phi^\alpha_i\rangle$ for each spin component. With this choice Q becomes: $Q = \int d\sigma\ e^{-\frac{\sigma^2}{2}} det A^\uparrow_\sigma det A^\downarrow_\sigma$ where A^α_σ are square matrices of components $(A^\alpha_\sigma)_{m,n} = \langle\phi^\alpha_m \mid U^\alpha_\sigma \mid \phi^\alpha_n\rangle$. These matrix elements involve only propagation of single particle orbitals and can be easily evaluated.

If we assume for the moment that $\langle\Psi_T \mid U_\sigma \mid \Psi_T\rangle$ is positive definite, Q takes the form of a classical partition function: $Q = \int d\sigma e^{-V_{eff}(\sigma)}$, where $V_{eff}(\sigma) = \frac{1}{2}\sigma^2 - \ln\left(det A^\uparrow_\sigma \times det A^\downarrow_\sigma\right)$. The sampling of this distribution can be done by standard means. We have used here a second order force based Langevin dynamics method[12,13] rather than the standard MC procedure given the highly non–local nature of the interaction which makes the MC update of a single degree of freedom rather costly.

In the large-β limit, a straightforward implementation of our approach undergoes numerical instabilities since the single particle orbitals $\mid\phi^\alpha_n\rangle$ tend to become parallel under the action of U_σ, albeit remaining linearly independent. The numerical calculation of a determinant having all the columns nearly parallel is ill-conditioned in finite precision arithmetics. Therefore for large enough β all the useful information is lost. This numerical instability can be successfully removed without any approximation if one takes the precaution of orthonormalizing during the propagation the single particle orbitals. This operation is permissible because a Slater determinant of non orthonormal orbitals can always be replaced by a determinant of orthogonalized orbitals, times a multiplicative constant. Since the orthogonalization procedure is rather costly, in practice one does not perform this operation at every imaginary time step but at the largest interval which is compatible with the stability of the calculation.

This procedure has been adopted for the first time by Sugiyama and Koonin[11]. These authors however justify the use of the orthogonalization procedure merely as a means of preserving antisymmetry. We believe this justification to be incorrect since antisymmetry is automatically guaranteed by the use of a Slater determinant $|\psi_T>$.

Once the algorithm has been set up one has to find for each operator of interest, O, appropriate estimators, namely functions $E_O(\sigma)$, such that

$$\langle \Psi_0 \mid O \mid \Psi_0 \rangle \underset{\beta \to \infty}{=} Q^{-1} \int d\sigma\, e^{-\frac{1}{2}\sigma^2}\, det(A^{\uparrow}) det(A^{\downarrow}) E_O(\sigma). \tag{5}$$

For instance the estimator of the density-matrix operator $c_r^+ c_{r'}$ is given by:

$$E_{c_r^+ c_{r'}} = \sum_{m,n,\alpha} \phi_m^{\alpha\rangle}(r,\beta/2) A^{-1}_{m,n} \psi_n^{\alpha\langle}(r',\beta/2) \tag{6}$$

where $|\phi_m^{\alpha\rangle}(\beta/2)\rangle = U_\sigma(\beta/2,0)|\phi_m^\alpha\rangle$, and $|\phi_m^{\alpha\langle}(\beta/2)\rangle = U_\sigma(\beta/2,\beta)|\phi_m^\alpha\rangle$.

So far we have assumed that $\langle \Psi_T \mid U_\sigma \mid \Psi_T \rangle$ is a positive definite quantity. This is in general not true. The standard procedure in this case would be to introduce the auxiliary partition function

$$Q_M = \int d\sigma\, e^{-\frac{1}{2}\sigma^2} |\langle \Psi_T \mid U_\sigma \mid \Psi_T \rangle| \tag{7}$$

and to relate averages in the Q_M ensemble denoted by $\langle\ \rangle_M$ to the needed averages in the Q ensemble by

$$\langle A \rangle = \frac{\langle A \times s \rangle_M}{\langle s \rangle_M}, \tag{8}$$

where $s = sgn\langle \Psi_T \mid U_\sigma \mid \Psi_T \rangle$. This method relies on the fact that $\langle s \rangle_M \neq 0$. However, very often $\langle s \rangle_M$ is very small and its variance very large. Thus estimates based on Eq.(8) are subject to a large numerical uncertainty and require long simulation runs in order to converge to a reliable value. We show below that, in many cases, for the calculation of ground state properties this is an unnecessarily complicated procedure.

To this end, we consider first the $\beta \to \infty$ limit of $\frac{Q}{Q_M} = \langle s \rangle_M$. It will be shown[14] that for $\beta \to \infty$ only two situations are possible:

$$a) \quad < S >_M \geq |< \psi_0 | \psi_T >|^2$$

$$b) \quad < S >_M = \frac{|< \psi_0 | \psi_T >|^2}{h_M(\beta)} \exp[-\Delta\beta] \tag{9}$$

where Δ is a constant and $h_M(\beta)$ is a bounded function of β. In case (a), which is the most favorable one, one has $Q_M = h_M(\beta)\exp[-\beta E_0]$. One can show that for the Hubbard model case (a) applies rigorously in 1D at arbitrary filling and in 2D at half filling. Thus the ground state energy can be obtained as: $E_0 =_{\beta\to\infty} -\frac{1}{\beta}\ln Q_M$.. The consequence of this is that Q_M can replace Q in the calculation of ground state properties. In case (b) although the origin of Δ is unknown and the value of Δ possibly depends on the particular form of the HS transformation we have experienced that Δ is either zero or very small as compared with E_0. In this case Q_M can still be usefully employed since one can calculate $E_0 - \Delta$ from

$$E_0 - \Delta = -\frac{1}{\beta}\ln Q_M \tag{10}$$

and extimate Δ from eq.[9].

The use of Q_M however presents some technical difficulties since the corresponding classical potential energy is singular along the nodal surface of $\langle\Psi_T \mid U_\sigma \mid \Psi_T\rangle$ and this breaks up the integration domain into disconnected regions separated by infinite potential barriers. This difficulty can be removed and the statistical quality of the results further improved by introducing a third partition function

$$Q_N = \int d\sigma e^{-\frac{1}{2}\sigma^2}\langle\phi_T \mid U_\sigma^+ U_\sigma \mid \phi_T\rangle^{\frac{1}{2}}. \tag{11}$$

It will be shown elsewhere[12] that Q_N satisfies the following inequalities:

$$Q_M(\beta) \le Q_N(\beta) \le D Q_M(2\beta)^{\frac{1}{2}}, \tag{12}$$

where D is the (finite) dimension of the Hilbert space. From eq.[10] one must have

$$E_0 - \Delta = -\frac{1}{\beta}\ln Q_N \tag{13}$$

where Δ is zero in case (a). Therefore also Q_N can be used to calculate ground state properties.

In order to verify the validity of our approach we have conducted extensive numerical simulations of 1D Hubbard model for which analytical[2] and numerical solution are available. A detailed discussion of the 1D results will be presented elsewhere.

In 2D no analytical results are available except for the 2×2 case which can be mapped into a 4 site 1D ring. However very recently A. Parola[15] has calculated exactly ground state properties up to sixteen electrons in a 4×4 Hubbard model. The comparison of this exact results with the ones obtained with the present algorithm are shown in Table 1. A very good agreement is found even in case (10 electrons) where possibly $\Delta \neq 0$.

We now present a selected set of results for different filling $\nu = \frac{N}{M}$ ground state properties. In Fig. 1 we show the momentum distribution $n(\mathbf{k})$ of the half-filled ($\nu = 1$) 16×16 lattice for $U = 0.5$ and $U = 4$. The Fermi surface is smeared out in both cases, although very little so for $U = 0.5$. Further evidence that the insulating character is present even at $U = 0.5$ is provided also by the direct evaluation of the single-particle

Table 1. Numerical results for the 2D 2×2 and 4×4 Hubbard model obtained with the present method as compared with the corresponding exact ones[15] displayed between brackets. Total and kinetic energy per site, staggered magnetic structure factor at (π, π) wavevector are shown in the table for different filling factors.

Size	# el	U	E/N	E_{kin}/N	$S_{magn.}(\pi,\pi)$
2×2	4	4	-1.406(5)	-1.880(6)	1.92(8)
			(-1.414)	(-1.867)	(1.93)
3×3	10	4	-0.700(1)	-1.626(6)	
			(-0.699)	(-1.626)	
4×4	10	4	-1.227(3)	-1.414(5)	0.73(1)
			(-1.224)	(-1.408)	(0.73)
4×4	10	8	-1.088(6)	-1.255(8)	0.76(2)
			(-1.094)	(-1.272)	(0.75)
4×4	16	4	-0.848(5)	-1.324(6)	3.4(2)
			(-0.8514)	(-1.3118)	(3.64)
4×4	14	4	-0.988(3)	-1.331(4)	1.60(7)
			(-0.9840)	(-1.3366)	(2.14)

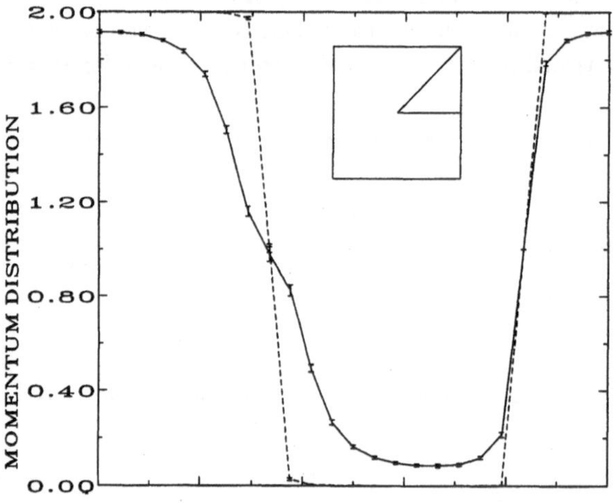

Figure 1. Momentum distribution of the half filled 2D 16×16 Hubbard model for $U = 0.5$ (dashed line) and $U = 4$ (continuous line). The imaginary time for this simulation was $\beta = 12$. The ground state energy per site was $E_0 = -1.491$ for $U = .5$ and $-.85(7)$ for $U = 4$. The corresponding HF values are -1.488 and $-.797$ The path in the Brillouin zone is shown in the inset. Note that the $Q\Gamma$ k-scale is compressed.

gap $\mu_+ - \mu_-$ (in the Lieb-Wu notation) which is 0.2 ± 0.05. It seems likely that the insulating antiferromagnetic behavior should continue all the way down to $U \to 0$.

In Fig. 2 we display the spin-spin correlation function $\zeta(\mathbf{r}_i - \mathbf{r}_j) = \frac{4}{3} < S(\mathbf{r}_i)S(\mathbf{r}_j) >$

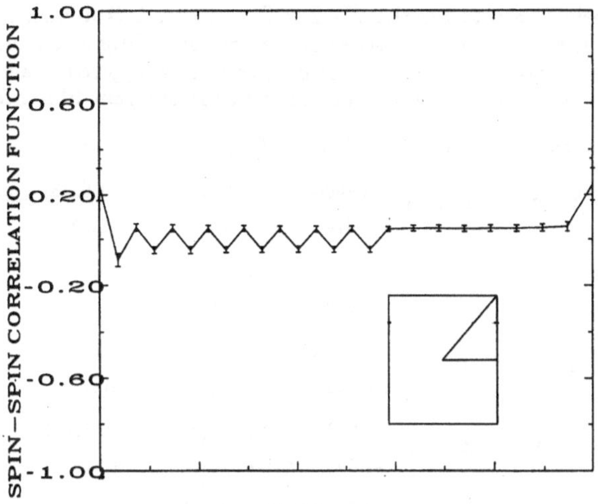

Figure 2. Spin-spin correlation function of a half-filled 2D 16×16 Hubbard model for $U = 4$. The path in the unit cell is shown in the inset.

for the same systems of Fig. 1, clearly exhibiting long-range antiferromagnetism. The staggered magnetization –which would be 1 for the ideal, fully aligned, AF state– is approximately 0.37 for $U = 4$ (HF value 0.7) and 0.14 for $U = 0.5$ (HF value 0.15). The antiferromagnetic order is likely to extend over all values of U. Therefore in 2D the metal-insulator transition at $\nu = 1$ seems to occur only for $U = 0$, similarly to the 1D case.

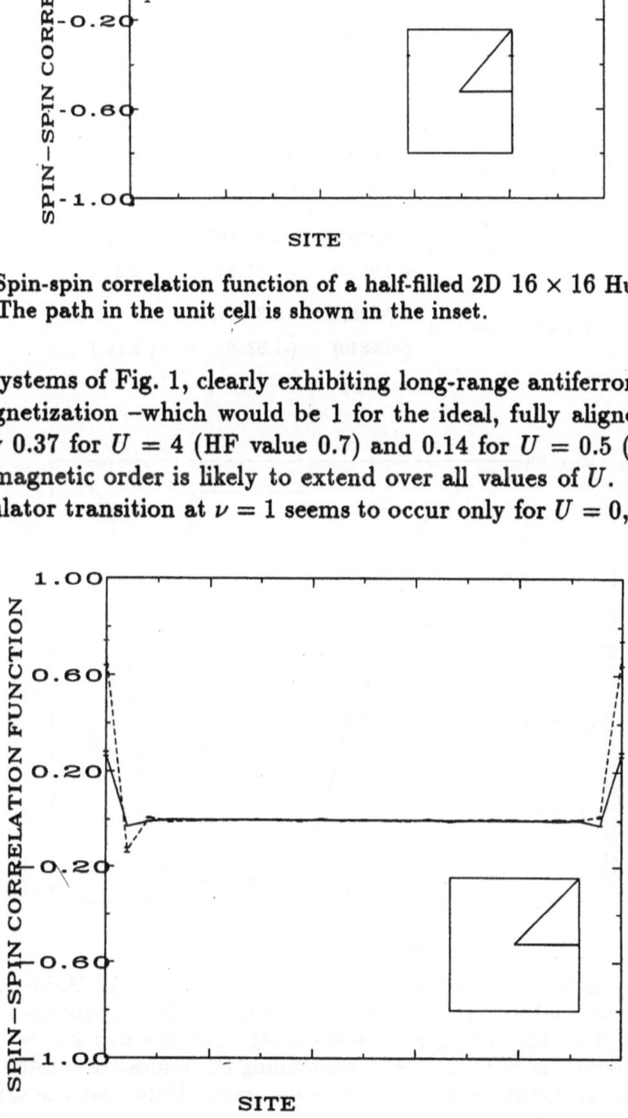

Figure 3. Spin-spin correlation function of a 2D 16×16 Hubbard model for $U = 4$, $\nu = \frac{226}{256}$ (continuous line) and $\nu = \frac{74}{256}$ (dashed line). The path in the unit cell is shown in the inset.

The 2D antiferromagnetic order is readily destroyed by doping. In Fig. 3 we display the spin-spin correlation function ζ in the 16×16 lattice at $U = 4$ for $\nu = \frac{226}{256}$ ($\delta = 12\%$) and for $\nu = \frac{74}{256}$ ($\delta = 71\%$). These results have been obtained using a paramagnetic trial wavefunction. The same results are also recovered starting from an

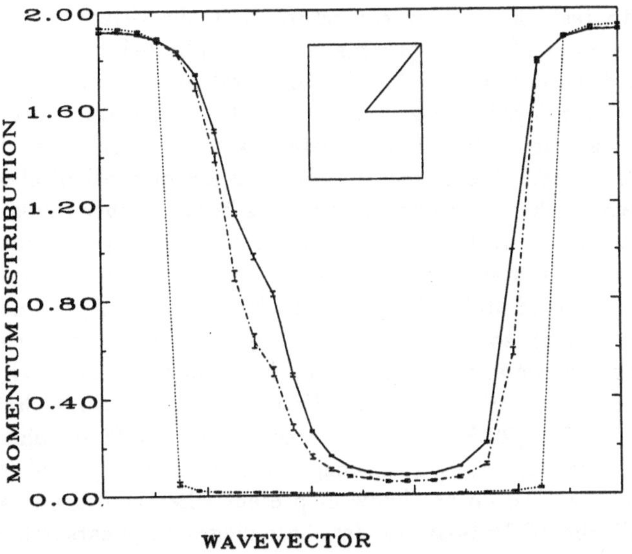

Figure 4. Momentum distribution of the 2D 16×16 Hubbard model with $U = 4$,for $\nu = \frac{256}{256}$ (dashed line), $\nu = \frac{226}{256}$ (dotted line), and $\nu = \frac{74}{256}$ (continuous line). The path in the Brillouin zone is shown in the inset. The imaginary time was $\beta = 12$ (half filled) and $\beta = 24$ (less than half filled).

antiferromagnetic wavefunction. However in this case a longer Langevin simulation is needed to reach equilibrium and melt the magnetic order.

The results of Fig. 3 confirm the suggestion by Hirsch that away from $\nu = 1$ the magnetic order is destroyed in the 2D Hubbard model. The question whether spin correlations in this system decay exponentially or as a power law is however difficult to answer within the present numerical accuracy. In fact, the corresponding spin-spin correlations for the $U = 0$ Fermi-liquid look quite similar.

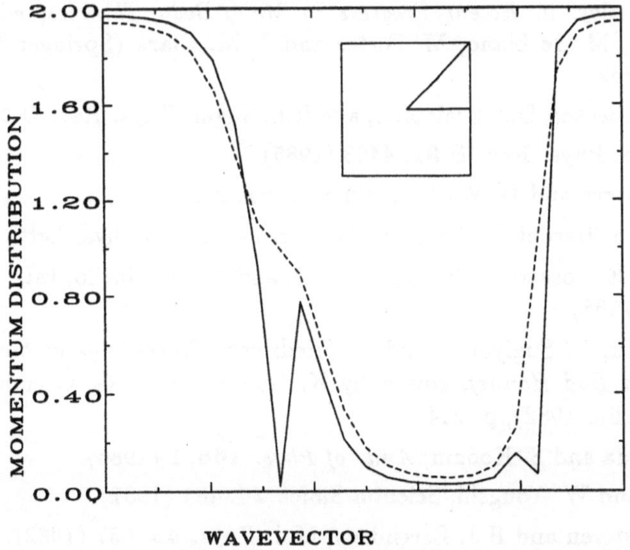

Figure 5. Momentum distribution of the Hartree-Fock AF solution of the 2D 16×16 Hubbard model with $U = 4$. The dashed line correspond to half filling, whereas the continuous line corresponds to $\nu = \frac{226}{256}$.

In order to test the further conjecture that the disappearance of magnetic order signals the onset of simple paramagnetic metallic behavior, we have measured the momentum distribution $n(\mathbf{k})$ for the three values of δ considered above. Our results, which are displayed in Fig. 4, show that the momentum distribution for the half-filled case is a smooth function consistent with the insulating character of the antiferromagnetic state. The momentum distribution of the half-filled AF Hartree-Fock state, shown by the dashed line of Fig. 5, is in fact rather similar to that of the true ground state of Fig. 4. The situation changes completely with moderate doping, $\nu = 88\%$. The new HF state displays a sharp dip corresponding to a Fermi surface (Fig.5). The true ground state $n(\mathbf{k})$ shown in Fig. 4, however, has no such dip, and roughly looks as though some kind of insulating character were preserved, in spite of the fact that magnetic order is destroyed. For heavy doping, finally ($\nu = 29\%$) a sharp Fermi surface is clearly recovered. We tentatively conclude that each of these three values of ν $(1, 88\%, 29\%)$ seems to correspond to a different phase, respectively an AF insulator, a yet unidentified non-magnetic non Fermi-liquid phase, and a Fermi liquid metallic phase. We do not know at present whether other phases might also appear. All in all, the 2D Hubbard model phase diagram appears richer than hitherto suspected.

ACKNOWLEDGEMENTS

We are deeply indebt to S. Baroni, R. Car, A. Parola and E. Tosatti for many useful discussions. This work was carried out as a part of the SISSA–CINECA collaborative project, under the sponsorship of the Italian Ministry of Education.

REFERENCES

[1] P.W. Anderson , Science **235**, 1196 (1987).

[2] E.H. Lieb and F.Y. Wu, Phys. Rev. Lett. **20**, 1445 (1968).

[3] M. H. Kalos, in *Monte Carlo Method in Quantum Physics*, edited by M. H. Kalos NATO ASI Series, (D. Reidel Publ. Co., Dordrecth, 1984) p.19.

[4] D.M. Ceperley, in *Recent Progress in Many-Body Theories*, edited by J. G. Zabolitzky, M. de Liano, M. Fortes and J. M. Clark (Springer Verlag, Berlin, 1981), p. 262.

[5] R. Blankenbecker, D.J. Scalapino, and R.L. Sugar, Phys. Rev. D **24** 2278 (1981).

[6] J.E. Hirsch, Phys. Rev. B **31**, 4403 (1985).

[7] I. Morgenstern and D. Wortz Z. Phys. B **61**, 219 (1985).

[8] S. Sorella, S. Baroni, R. Car, and M. Parrinello, Europhys. Lett. **8**, 663 (1989).

[9] S. Sorella, E. Tosatti, S. Baroni, R. Car, and M. Parrinello, Int. J. Mod. Phys. B **1**, 993 (1988).

[10] S.E. Koonin, G. Sugiyama, and H. Friedriech, *Proceedings of the International Symposium Bad Honnef*, edited by K. Goeke and P. G. Greinhard (Springer Verlag, Berlin, 1982), p. 214.

[11] G. Sugiyama and S. Koonin, *Ann. of Phys.* **168**, 1 (1986).

[12] G. Parisi and W. Yougshi, Scientia Sinica **24**, 483 (1981).

[13] W.F. Gunsteren and H.J. Berendsen, Mol. Phys. **45**, 637 (1982).

[14] S. Sorella, to be published.

[15] A. Parola to be published.

STABLE MATRIX-MULTIPLICATION ALGORITHMS FOR LOW-TEMPERATURE NUMERICAL SIMULATIONS OF FERMIONS

E. Y. Loh, Jr. and J. E. Gubernatis

Theoretical Division
Los Alamos National Laboratory
Los Alamos, NM 87545

R. T. Scalettar

Department of Physics
University of Illinois
Urbana, IL 61801

R. L. Sugar and S. R. White

Department of Physics
University of California
Santa Barbara, CA 93106

INTRODUCTION

In this note, we discuss the use of matrix factorizations to stabilize the numerical matrix multiplications and inversions needed to simulate systems of interacting fermions at low temperatures. While the essence of a specific stable numerical algorithm is presented, we mainly emphasize the concepts of stabilization. A detailed description of a ground state and finite temperature algorithm is given elsewhere.[1]

In order to perform a numerical simulation of fermion systems in more than one dimension, it is necessary to integrate out the fermion degrees of freedom. This ordinarily requires the use of a Hubbard-Stratonovich transformation, which can be either discrete[2] or continuous. This transformation reduces a system of self-interacting fermions to one in which the fermions interact only with the time-varying Hubbard-Stratonovich (HS) field. Since the problem now has no self-interactions, the fermion degrees of freedom are integrated away analytically[3], leaving one with a sum or integral over the HS fields which can be performed by Monte Carlo[3], hybrid-molecular dynamics,[4] or Langevin[5] methods.

Integrating out the fermion degrees of freedom leaves the Hubbard-Stratonvich fields weighted by

$$\det(\mathbf{1} + \mathbf{B}_\tau \cdots \mathbf{B}_{\Delta\tau}\mathbf{B}_\beta \cdots \mathbf{B}_{\tau+\Delta\tau})$$

where each NxN matrix $\mathbf{B}_\tau$ is a single-particle propagator on a lattice of N sites from imaginary-time $\tau - \Delta\tau$ to τ and the imaginary-time path integral has been discretized in units of the "Trotter" parameter $\Delta\tau$. The change in this quantity resulting in the change in one of the HS fields involves the equal-time Green's function

$$\mathbf{G}(\tau,\tau) = (\mathbf{1} + \mathbf{B}_\tau \cdots \mathbf{B}_{\Delta\tau}\mathbf{B}_\beta \cdots \mathbf{B}_{\tau+\Delta\tau})^{-1} \tag{1}$$

Given $\mathbf{G}(\tau,\tau)$, Blankenbecler *et al.*[3] give simple rules for updating the HS fields, for correcting $\mathbf{G}$ when the HS fields are changed, for quickly calculating $\mathbf{G}(\tau+\Delta\tau,\tau+\Delta\tau)$, and for measuring physical observables at imaginary-time τ. Unfortunately, as β becomes large, the matrices needed to produce $\mathbf{G}(\tau,\tau)$ become ill–conditioned so the Green's function cannot be computed reliably. The reason is that as many $\mathbf{B}$ matrices are multiplied together, high-energy states generate small-scale numerical features that get buried by the large-scale numerical features generated by the low-energy states. The information from the high-energy states is not lost; rather, one formally calculates Slater determinants in which it becomes necessary to take differences between numerically large matrix elements to pick up the numerically smaller information that comes from the high-energy features of the band. This is a hopelessly noisy procedure that is intolerable for fermionic systems since for a typical HS configuration, information about all energy scales is needed to accurately compute the fermion matrices. Conventional simulations ultimately fail simply because these important small-scale features cannot be pulled out of $\mathbf{B}_\tau \cdots \mathbf{B}_{\tau+\Delta\tau}$ using computers with finite precision. Recents efforts[6,7] at low-temperature stabilization use matrices of increased dimension, at a substantial cost of computer time and memory.

STABILIZATION METHODS

To protect important, small-scale features present in the product of the $\mathbf{B}$ matrices, we take our clues from basic texts on numerical analysis.[8] The *condition number* of a matrix is roughly the ratio of the largest singular value of the matrix to the smallest one and represents an upper bound to the amplification of errors in matrix multiplications. The realized amplification may, in fact, be much less. Multiplication of two diagonal matrices, for example, is very stable no matter how ill-conditioned the factors may be. Meanwhile, orthogonal matrices are also stable multipliers since their condition number is one: there is absolutely no variation in scales. Thus, orthogonal and diagonal matrices figure prominently in matrix factorization.

Our aim then is to decompose ill-conditioned matrices by representing them in the form $\mathbf{UDV}$, where the diagonal matrix $\mathbf{D}$ contains the diverging singular values and has the large condition number but where $\mathbf{U}$ and $\mathbf{V}$ are "sufficiently well-conditioned," a property which will be made more precise later. If we choose both $\mathbf{U}$ and $\mathbf{V}$ to be orthogonal, the resulting decomposition is the singular-value decomposition (SVD), which is known to be very stable. Unfortunately, SVD decompositions are relatively slow to perform on a computer, so in practice we use the modified Gram-Schmidt (MGS) factorization $\mathbf{UDV}$, where $\mathbf{U}$ is orthogonal and $\mathbf{D}$ is diagonal but $\mathbf{V}$ is unit triangular. We found the time form a MGS decomposition is in some cases up to 20 times less then that for SVD. Fortunately, we also found that the unit triangular matrices are "sufficiently well-conditioned" so we can multiply large numbers of them in our simulations without destroying stability.

To stably compute the product of many matrices, we decouple the various scales present throughout the calculation. To illustrate what we mean, we first imagine we have decomposed a partial product of the $\mathbf{B}$ matrices into the form $\mathbf{UDV}$. To multiply several more $\mathbf{B}$ matrices on the left, we then write

$$\mathbf{B}\cdots\mathbf{B}(\mathbf{UDV}) = (\mathbf{B}\cdots\mathbf{BUD})\mathbf{V} = (\mathbf{U'D'V'})\mathbf{V} = \mathbf{U'D'}(\mathbf{V'V}) \quad , \tag{2}$$

giving the decomposition of the next partial product. Again, the $\mathbf{V}$ matrices must be sufficiently well-conditioned that we can multiply many of them together stably;

we can do this for both orthogonal (for SVD) and unit-triangular (for MGS) $\mathbf{V}$ matrices. [9] The other assumption in (2) is that the decomposition of $\mathbf{B}\cdots\mathbf{BUD}$ into $\mathbf{U'D'V'}$ can be performed stably. This is in fact possible since, for a limited number of $\mathbf{B}$'s, the diverse scales in $\mathbf{B}\cdots\mathbf{BUD}$ are well separated in different columns.

The most obvious way of stabilizing a fermionic simulation is simply to replace the straight-forward computation of the Green's function by a computation which handles decomposed matrices. One may build up the decomposition for

$$\mathbf{B}_\tau\cdots\mathbf{B}_{\tau+\Delta\tau} = \mathbf{UDV}$$

by generating successive partial products in the manner described above. Then, to stablize the matrix inversion, we follow the steps

$$\begin{aligned}\mathbf{G} &= (\mathbf{1}+\mathbf{B}_\tau\cdots\mathbf{B}_{\tau+\Delta\tau})^{-1} = (\mathbf{1}+\mathbf{UDV})^{-1} = \mathbf{V}^{-1}(\mathbf{U}^{-1}\mathbf{V}^{-1}+\mathbf{D})^{-1}\mathbf{U}^{-1}\\ &= \mathbf{V}^{-1}(\mathbf{U'D'V'})^{-1}\mathbf{U}^{-1} = \mathbf{V}^{-1}\mathbf{V'}^{-1}\mathbf{D'}^{-1}\mathbf{U'}^{-1}\mathbf{U}^{-1} \quad . \end{aligned} \tag{3}$$

These steps isolate as long as possible the matrices which contain the bulk of the scale variations, and the ill-conditioned sum $(\mathbf{U}^{-1}\mathbf{V}^{-1}+\mathbf{D})$ is inverted by first decomposing and separating scales. Thus, we only ever need to invert orthogonal, diagonal, and triangular matrices, all of which are easy to invert. The factors in the final expression in (3) can be multiplied in any order.

It is useful to remark that although the effect of the HS transformation was to remove interactions among the fermions, they still interact with a time-dependent external field and that the time-dependence of this field reduces computational options. If the field were time-independent, we could rewrite (3) as

$$\mathbf{G} = (\mathbf{1}+\exp(-\beta\mathbf{H}))^{-1} = \mathbf{S}(\mathbf{1}+\mathbf{D})\mathbf{S}^{-1} \tag{4}$$

where $\mathbf{S}$ is the similarity transformation that diagonalizes the Hamiltonian $\mathbf{H}$. In (4), the elements of $(\mathbf{1}+\mathbf{D})^{-1}$ are $1/(1+\exp(-\beta(E_k-\mu)))$ in which terms, 1 and $\exp(-\beta(E_k-\mu))$, of very different numerical scales at low temperatures are combined, but the effects of the small eigenvalues of $\exp(-\beta\mathbf{H})$, the $\exp(-\beta(E_k-\mu)$, are naturally cut-off by their addition to 1 so they are unable to corrupt the inversion. What we are doing in (3), from one point of view, is the analog for non-symmetric matrices of what we did in (4) for symmetric matrices. The decomposition allows us not only to work with matrices whose elements are real numbers but also to focus our attention on the singular values, as opposed to the eigenvalues, of the matrices. In (3), where we have prolonged the combination of numerical scales as much as possible until the last step, $\mathbf{UV}^{-1}$ cuts off the scales in $\mathbf{D}$ in a manner analogous the role played by $\mathbf{1}$ in (4).

OTHER APPLICATIONS

An alternative to the great deal of recomputation involved in re-evaluating (3) for many different τ is to use computer memory to store partial products. Now, we imagine we have the decompositions of *two* partial products:

$$\mathbf{B}_\tau\cdots\mathbf{B}_{\Delta\tau} = \mathbf{U}_L\mathbf{D}_L\mathbf{V}_L$$

and

$$\mathbf{B}_\beta\cdots\mathbf{B}_{\tau+\Delta\tau} = \mathbf{U}_R\mathbf{D}_R\mathbf{V}_R \quad .$$

Then we can write

$$\begin{aligned}\mathbf{G} &= (\mathbf{1}+\mathbf{U}_L\mathbf{D}_L\mathbf{V}_L\mathbf{U}_R\mathbf{D}_R\mathbf{V}_R)^{-1}\\ &= \mathbf{V}_R{}^{-1}(\mathbf{U}_L{}^{-1}\mathbf{V}_R{}^{-1}+\mathbf{D}_L\mathbf{V}_L\mathbf{U}_R\mathbf{D}_R)^{-1}\mathbf{U}_L{}^{-1} \quad . \end{aligned} \tag{5}$$

Again, the inverse of the ill-conditioned sum can be stabilized by decomposing the sum and then inverting its individual pieces. Because we kept the diagonal matrices on the outsides of the terms, elements of different scales are added together only to "cut scales off", as before. To implement (5), one needs to compute the decompositions $\mathbf{U}_L\mathbf{D}_L\mathbf{V}_L$ and $\mathbf{U}_R\mathbf{D}_R\mathbf{V}_R$ of the partial products $\mathbf{B}_\tau \cdots \mathbf{B}_{\Delta\tau}$ and $\mathbf{B}_\beta \cdots \mathbf{B}_{\tau+\Delta\tau}$ from scratch only once, at the beginning of the program. Thereafter, moving from one imaginary time τ to a nearby time involves only the multiplication of a few additional factors of $\mathbf{B}$ to one partial product using (2) and the deletion of these factors from the other partial product by recalling a previously stored partial product. While the memory needs of this algorithm are considerable, they are affordable.

For calculating zero-temperature, equal-time correlation functions, Sorella, *et al.* [5] have shown many benefits of projecting the ground state of n electrons from a trial wavefunction $|\psi_0\rangle$ using $\exp(-\beta\mathbf{H})$,

$$\langle\psi_0|\exp(-\beta\mathbf{H})|\psi_0\rangle = \det \mathbf{M}^{\mathbf{T}}\mathbf{B}_\tau \cdots \mathbf{B}_{\Delta\tau}\mathbf{B}_\beta \cdots \mathbf{B}_{\tau+\Delta\tau}\mathbf{M} \tag{6}$$

where $\mathbf{M}$ is an $N \times n$ matrix that transforms $|\psi_0\rangle$ into the lattice site representation that is generally chosen for the $\mathbf{B}$ matrices. Hence the determinant is that of an $n \times n$ matrix which means both the computer time and memory requirements decrease quadratically with the density of particles n/N for a fixed β and N. In this projection framework, stabilization of (6) in the manner of (2) can be simplified in several respects and is equivalent to what Sorella, *et al.* arrive at from a different context.

Finally, while equal-time measurements can always be expressed in terms of averages of the equal-time Green's function $\mathbf{G}(\tau,\tau)$, calculations of susceptibilities and other time-dependent quantities require matrix elements of

$$\mathbf{G}(\tau+\tau',\tau) = \mathbf{B}_{\tau+\tau'} \cdots \mathbf{B}_{\tau+\Delta\tau}\mathbf{G}(\tau,\tau).$$

Again writing

$$\mathbf{B}_\tau \cdots \mathbf{B}_{\Delta\tau} = \mathbf{U}_L\mathbf{D}_L\mathbf{V}_L$$

and

$$\mathbf{B}_\beta \cdots \mathbf{B}_{\tau+\Delta\tau} = \mathbf{U}_R\mathbf{D}_R\mathbf{V}_R \quad .$$

we can calculate an unequal-time Green's function stably by writing

$$\begin{aligned} \mathbf{G}(\tau,0) &= \mathbf{U}_L\mathbf{D}_L\mathbf{V}_L(1+\mathbf{U}_L\mathbf{D}_L\mathbf{V}_L\mathbf{U}_R\mathbf{D}_R\mathbf{V}_R)^{-1} \\ &= \mathbf{V}_R{}^{-1}(\mathbf{D}_L{}^{-1}\mathbf{U}_L{}^{-1}\mathbf{V}_R{}^{-1} + \mathbf{V}_L\mathbf{U}_R\mathbf{D}_R)^{-1}\mathbf{V}_L \quad . \end{aligned} \tag{7}$$

Once again, we decompose the ill-conditioned sum before inverting it.

STABILITY AND ACCURACY TESTS

Various tests were performed to study the stability of these algorithms both for "toy problems" using random Hubbard-Stratonovich spins and in actual simulations. Tests were performed in 32-bit, 64-bit, and 128-bit floating point precision. In testing the Green's function calculated from (3), stability requires that elements of the Green's function and measurements from simulations are invariant with machine precision. We found that matrices could be propagated stably in imaginary-time up to some τ_0 before decomposition is needed; τ_0 depends strongly on model parameters, machine precision, and somewhat on the HS configuration, but it is insensitive to the discretization parameter $\Delta\tau$ and to the inverse temperature β. Typically, τ_0 ranges from 2 to 5 and sets the limit on β for an unstable calculation. Instability in calculations sets in suddenly as the imaginary-time τ_0 between decompositions is increased too much.

There is no difficulty in pushing simulations using (3) to extreme temperatures ($\beta \approx 100$) even with 32-bit floating-point arithmetic. At such low temperatures we find agreement between physical quantities calculated with a finite temperature based on (3) and a ground state algorithm based on (6). We can check the results from (6) against those obtained by exact-diagonalization for a small system and free-fermion results for special problems. For example, for 2×2 periodic Hubbard cluster with $t = 1$, attractive interaction $U = -8$, and discretization parameter $\Delta\tau = 0.1$, projecting the ground state from a trial wavefunction gives on-site double occupancy $< n_{l\uparrow}n_{l\downarrow} >= 0.4366$ and nearest-neighbor hopping $< c^{\dagger}_{l'}c_l >= 0.421$ at $\beta = 100$. The statistical uncertainity in the Monte Carlo results is in the last quoted digit in each case. At $\Delta\tau = 0.05$, we obtain $< n_{l\uparrow}n_{l\downarrow} >= 0.4303$ and $< c^{\dagger}_{l'}c_l >= 0.411$. After extrapolating these values to $\Delta\tau \to 0$, we find that they agree with results from exact diagonalizations (0.4282 and 0.408) to the stated precision, which, for any finite running time, is set by the statistical fluctuations. This performance on a 2×2 cluster is typical since stability is related primarily to β and Hamiltonian parameters that set the band edges but not to lattice size.

Preliminary tests using (5) and (7) suggest they help stablize the simulation numerics but prehaps are not as robust as (3). With respect to computational efficiency, our computing time scales no worse then $(\beta/\tau_0)^2$, while for several recently proposed methods [6,7] it scales as $(\beta/\tau_0)^3$. We will report elsewhere more extensive studies using these techniques.[1,10]

ACKNOWLEDGEMENTS

We would like to thank V. Faber, T. Manteuffel, M. Parrinello, D. Scalapino, and S.Sorella for helpful discussions. This work was supported in part by the Department of Energy grant DE-FG03-88ER45197, National Science Foundation grants PHY86-14185 and DMR 86-15454, and the University of San Diego Supercomputer Center. One of us (E.Y.L.) would also like to thank Los Alamos National Laboratory for supporting basic research at the Laboratory. S.R.W. gratefully acknowledges the support of IBM through a post-doctoral fellowship.

REFERENCES

1. S. R. White, D. J. Scalapino, R. L. Sugar, E. Y. Loh, Jr., J. E. Gubernatis, and R. T. Scalettar, "A Numerical Study of the Two-Dimensional Hubbard Model with Repulsive Interaction," preprint.

2. J. E. Hirsch, *Phys. Rev. B* **31**:4403 (1985).

3. R. Blankenbecler, D. J. Scalapino, and R. L. Sugar, *Phys. Rev. D* **24**:2278 (1981).

4. R.T. Scalettar, D.J. Scalapino, R.L. Sugar and D. Toussaint, *Phys. Rev. B* **36**:8632 (1987).

5. S. Sorella, E. Tosatti, S. Baroni, R. Car, and M. Parrinello, preprint and private discussions.

6. S. R. White, R. L. Sugar, and R. T. Scalettar, "An Algorithm for the Simulation of Many Electron Systems at Low Temperatures," preprint.

7. J. E. Hirsch, "Stable Monte Carlo Algorithm for Fermion Lattice Systems at Low Temperatures," preprint.

8. W.H.Press, B.P.Flannery, S.A.Teukolsky, and W.T. Vetterling, **Numerical Recipes: The Art of Scientific Computing**, (Cambridge University Press, Cambridge, 1988); J. R. Rice, **Numerical Methods, Software, and Analysis**, (McGraw-Hill, New York, 1983).

9. The SVD decomposition proceeds by multiplying on both sides of $\mathbf{B} \cdots \mathbf{BUD}$ by orthogonal matrices; MGS multiplies only on the right side to orthogonalize the columns. Multiplication on the left side is clearly stable since it only mixes elements within the same column — elements of the same scale. Only right multiplication is tricky. For MGS, there is no difficulty since orthogonalizing a column $\mathbf{v}$ with respect to some column $\mathbf{u}$ simply entails replacing $\mathbf{v}$ with $\mathbf{v} - \mathbf{u}(\mathbf{u}^T\mathbf{v})/(\mathbf{u}^T\mathbf{u})$. Thus, independent of the relative scales of $\mathbf{u}$ and $\mathbf{v}$, one always subtracts from $\mathbf{v}$ a term which is smaller by a factor of the direction cosine between $\mathbf{u}$ and $\mathbf{v}$. For SVD, some "pivoting" is required to make the decomposition stable.

10. E. Y. Loh, Jr., J. E. Gubernatis, R. T. Scalettar, R. L. Sugar, and S. R. White, "Matrix-Multiplication Methods for the Low-Temperature Simulation of Fermions," in preparation.

NUMERICAL SIMULATIONS OF THE TWO-DIMENSIONAL HUBBARD MODEL

I. Morgenstern

IBM Research Division
Zurich Research Laboratory
8803 Rüschlikon, Switzerland

Monte-Carlo simulations based on the Trotter-Suzuki transformation were performed. For intermediate-range repulsive interaction U, they show suppression of the nearest-neighbor hole correlation function, while next-nearest-neighbor correlations are enhanced.

1. INTRODUCTION

In the theoretical approach to high-T_c superconductivity, the Hubbard model with repulsive interaction U played an important role. P. W. Anderson *et al.*[1] worked out the RVB theory, which predicts superconductivity in the single-band two-dimensional (2-D) case. Numerical simulations by J. Hirsch[2] and M. Imada[3] provided arguments against this point of view. Alternatively, Imada[3] suggested superconductivity in the two-band case. In this paper, I will subsequently argue that these simulations were misleading.

A further, more phenomenological approach was provided by Aharony *et al.*[4] They considered the 2-D spin-½ Heisenberg model. Localization of holes on oxygen sites between two copper spins leads to frustration effects and spin-glass behavior. Owing to the frustration effect, two holes attract each other. The model provides a good description of the insulating phase.[5] Pairing of holes may lead to superconductivity. In this paper, I will concentrate first of all on the single-band Hubbard model, for the sake of numerical feasibility in the framework of the approach described later on. In light of the simulations the two-band and Aharony models may also be expected to show similar results, but are left to future research.

2. NUMERICAL ALGORITHM

I consider the single-band Hubbard model with repulsive interaction U and hopping term t:

Interacting Electrons in Reduced Dimensions
Edited by D. Baeriswyl and D.K. Campbell
Plenum Press, New York

$$H = -t \sum_{<i,j>} \left(c_i^+ c_j + h.c.\right) + U \sum_i n_{i\uparrow} n_{i\downarrow}. \tag{1}$$

The 2-D case is studied. The method has basically been described in an earlier paper.[6] It is based on the Trotter-Suzuki transformation,[7] which maps a d-dimensional quantum system as a $d+1$-dimensional classical system described by "world lines". But compared to Ref. 6, the algorithm was changed in a crucial way leading to the results presented. Instead of having only a few elementary moves in the Monte-Carlo simulations as before, the present algorithm deals with 32 different moves to change a configuration. As the Monte-Carlo dynamics are not related to the actual physical dynamics of the system, the choice of these moves is arbitrary. The essential point is to obtain equilibrium configurations in a minimum of CPU time. To be precise: the value of a Monte-Carlo algorithm is determined by the CPU time needed to obtain, let's say, 10,000 *statistically independent* configurations from which the averaged value of the observable is calculated. One can easily have a large number of local moves per unit time, p.e. spin flips for an Ising system, without obtaining the necessary number of statistically independent configurations any faster than with an algorithm carrying out much fewer moves per unit time. Keeping this essential point in mind, one has to find a compromise for the local moves. On the one hand, new more complicated moves may cost more time which, on the other hand, may be repaid by reaching equilibrium earlier. Figure 1 shows an elementary move in the algorithm of Ref. 6. This move is modified by doubling its length in the Trotter direction and by further twists of the world lines. Extending the moves further in the Trotter direction, especially concerning hole spacial dimensions, did not lead to the desired effects. Doubling, eventually also tripling, was found to be most effective but with additional twists. The resulting speed-up of the simulations just made the subsequently described results obtainable.

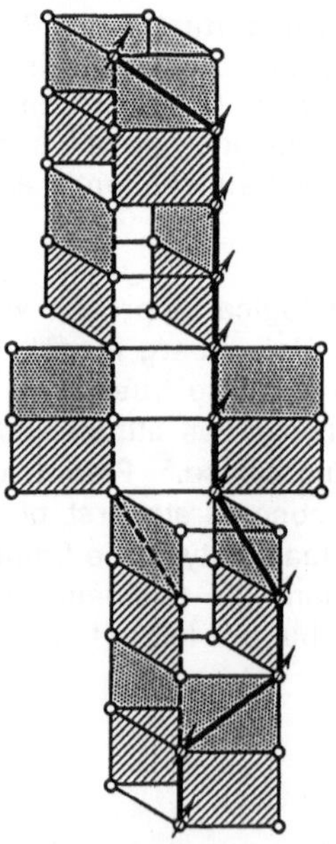

Figure 1. Elementary moves for "world lines" in the simulation. 32 different modifications are used. From Ref. 6, © Springer-Verlag 1988.

The simulations were carried out on the IBM 3090 vector processor at the Research Laboratory in Rüschlikon. The 3090 is especially favorable to vector statements of the type

$$A = A + B * C. \tag{2}$$

The more complicated moves of the type in Fig. 1 could be coded in the form of (2), thus compensating the loss of speed-up compared to more easily vectorizable simple moves as better used on p.e. a Cray supercomputer. To summarize, the algorithm described was developed especially with the IBM 3090 and its more flexible vector facilities in mind.

3. NUMERICAL SIMULATIONS

The simulations were carried out for 6x6 systems with 100 time slices and for a 12x12 system with 200 time slices. Periodic boundary conditions were imposed. U was taken $U = 4$ in units of t, temperature $T = 0.05 \times J$ with $J = 4(t^2/U)$. Taking $J \approx 1000$ K, we deal with 50 K, which is below the transition temperature. The 6x6 systems contained four holes, while the 12x12 was carried out with 14 holes.

3.1. Minus-Sign Problem

The current algorithm deals with negative transition probabilities, as probably do all known algorithms for fermionic systems of dimensions $d > 1$. To obtain an averaged value of an observable, p.e. the magnetization M, we have

$$<M> = \frac{M^+ - M^-}{Z^+ - Z^-} \tag{3}$$

where M^+ denotes the magnetization of the "positive" configuration and M^- that of the "negative". This quantity has to be normalized by the averaged sign $<\text{sign}> = Z^+ - Z^-$ denoting the positive and negative part of the partition function. In the case of *"purely"* fermionic systems, it has been shown by Hirsch[8] that the nominator and also the denominator tend to zero as $e^{-\beta N}$ (with $\beta = 1/T$, T the temperature and N the system size) leading to the well-known "zero over zero" terms. The stochastic process of a Monte-Carlo simulation leads to much larger error bars than the actual values of both quantities. Therefore, one obtains rather "obscure" expectation values, p.e. negative values for the hole-hole correlation functions, which are larger than zero by definition. Figure 2 illustrates the "minus-sign effect" in the present simulation. It shows the correlation function for holes on nearest-neighbor (n.n) and next-nearest-neighbor (n.n.n) sites, i.e. the probability of finding a pair of holes on these sites. In addition, the staggered magnetization is shown. As can readily be seen the values are quite scattered. The reason can be found by considering the averaged sign, which wiggles around zero. But on the right-hand side, i.e. after about 50,000 to 60,000 lattice sweeps (Monte-Carlo steps), a stabilization effect is seen. Figure 3 shows the "minus-sign" problem in more detail. The averaged sign is multiplied by 10 and the n.n.n. hole correlation is shown. As long as the sign fluctuates around zero, the values for the correlation function are simply nonsense. But after the 50,000-60,000 lattice runs mentioned, the sign takes a clearly positive value, leading to a stabilization of the correlation function.

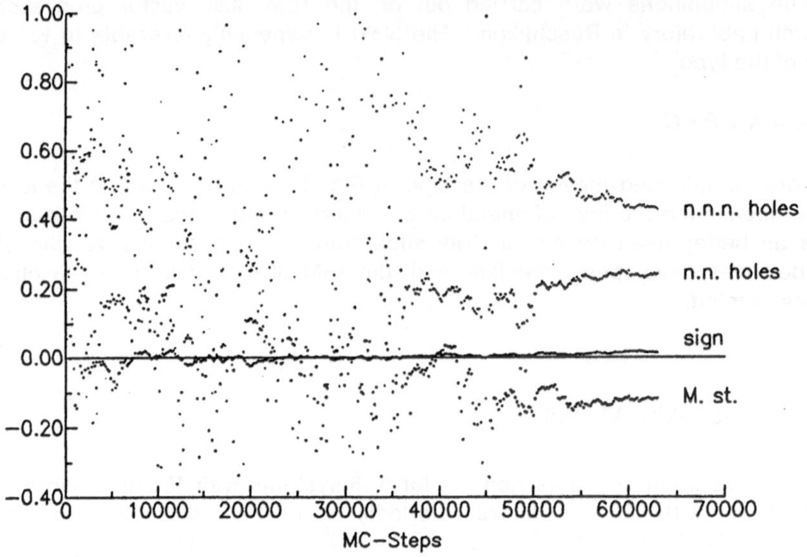

Figure 2. Minus-sign effect. Before the system has reached equilibrium, it is impossible to calculate expectation values. (6×6, $U = 4$). From Ref. 12, © Springer-Verlag 1988.

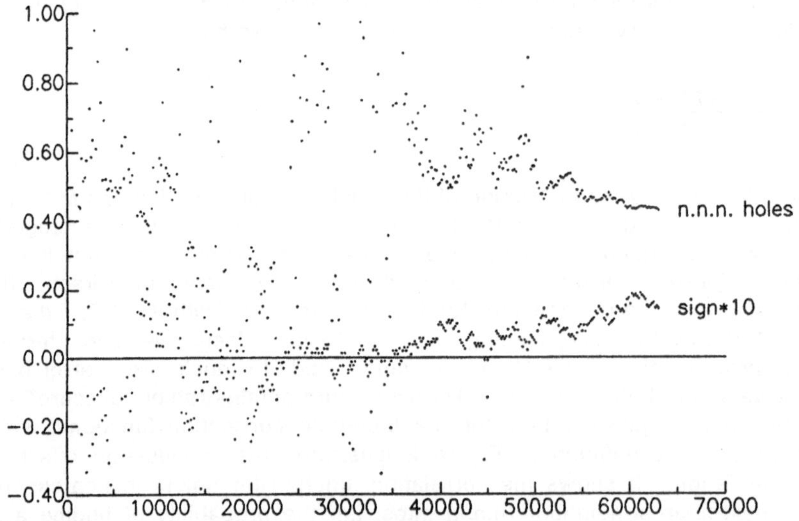

Figure 3. Minus-sign effect. Averaged sign (× 10) fluctuates around zero and destroys the expectation value for the n.n.n. hole correlation function. Sign will eventually increase above zero (equilibrium!) (6×6, $U = 4$).

Figure 4 is the continuation of Fig. 2 for much longer times up to one million lattice sweeps. We notice a clear stabilization of all expectation values, accompanied by an averaged sign different from zero. The staggered magnetization stabilizes to a value of about zero, as expected.

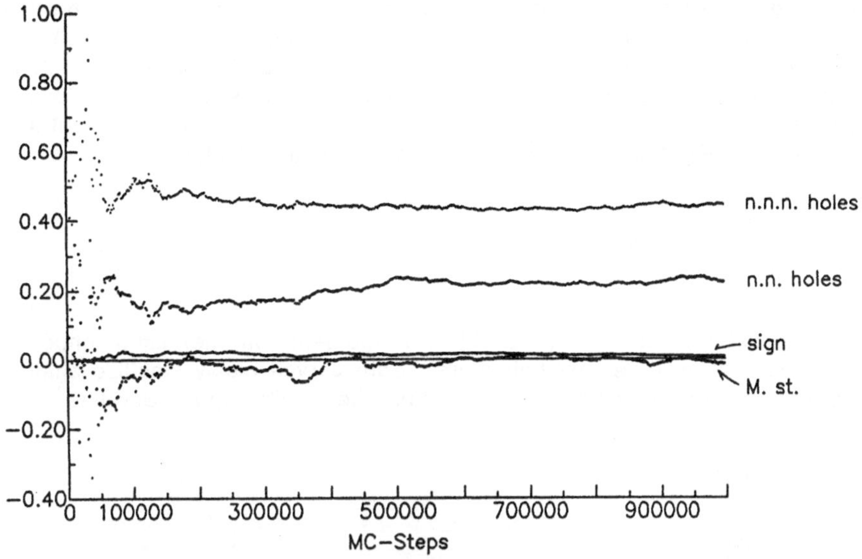

Figure 4. Stabilization effect. After equilibrium has been reached, the sign stays different from zero. Expectation values stabilize. Next-nearest-neighbor hole pairs preferred over nearest neighbors. (6×6, $U = 4$). From Ref. 12, © Springer-Verlag 1988.

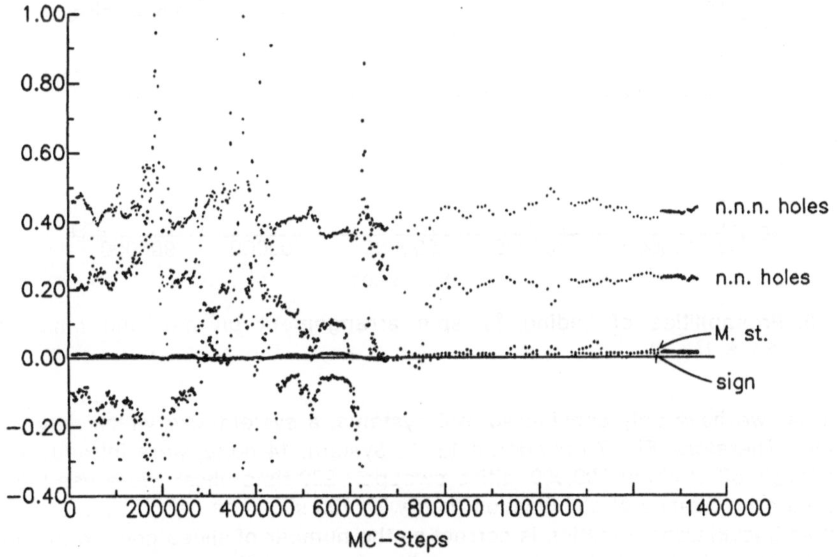

Figure 5. Stabilization effect. Random starting configuration considerably increases time needed to reach equilibrium. (6×6, $U = 4$).

One may argue at this point that the vanishing of the minus sign may depend heavily on the starting configuration. While in Figs. 2-4, two pairs of holes were introduced on n.n.n. sites, Fig. 5 shows the equivalent simulation for a random starting configuration. It shows the worst case I obtained. Only after almost 700,000 lattice sweeps did the system stabilize. The expectation values remained unchanged over the same period of time, again leading to the same result as before. Next-nearest-neighbor pairs are preferred over nearest neighbors. The correlation functions and the staggered magnetization show almost the values obtained before. Here in the second case of a random starting configuration, the holes had to find each other first before reaching equilibrium, which took more than 10 times as long as for the starting configuration where pairs had already been introduced on n.n.n. sites. It should be mentioned that usual BCS-type correlation functions for superconductivity are not easily feasible in the current approach to the simulations. They had to be left to future work. Therefore, Fig. 6 only shows correlation functions of the type

$$<\text{n.n. spins}> = \sum_{<i,j>}^{\text{n.n.}} n_{i\uparrow} n_{j\downarrow} + n_{i\downarrow} n_{j\uparrow}, \tag{4}$$

i.e. the probability of finding a ↑↓ arrangement on nearest-neighbor sites, <n.n.n spins> on next-nearest neighbors. Both curves nicely stabilize again. The n.n.n. case is suppressed compared to the n.n. The results support Sorellas[9] findings.

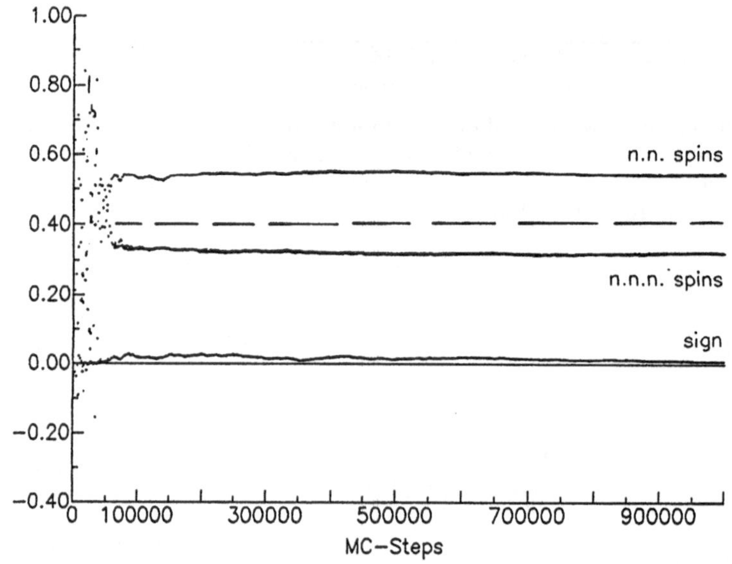

Figure 6. Probabilities of finding ↑↓ spin arrangement on n.n. and n.n.n. sites. (6×6, $U = 4$).

So far we have only considered 6x6 systems, a system size which may well be too small. Therefore, Fig. 7 provides a 12x12 system; 14 holes were introduced. The system levels off at about 200,000 lattice sweeps. 200 time slices were used, leading to an earlier occurrence of stabilization than with 100 slices as used in the 6x6. Since the Trotter-Suzuki transformation is correct as the number of slices goes to infinity, the bosonic behavior is detected earlier as we closer simulate the real system.

Again we notice that n.n.n. pairs are preferred over n.n. pairs. But the n.n. pairs are not suppressed as much as in the 6x6 system, indicating a finite size effect. The staggered magnetization is again zero. While the stabilization effect is not followed over an equally long time period as for the 6x6, the 12x12 results support the findings in the smaller system. At this point, I would like to comment on why the present effects were not obtained in earlier numerical work. First of all, people did not take the present "world-line" approaches seriously because of the the "minus-sign problem". The stabilization effect has not been seen, owing to the too short observation times and the too simple dynamics for the t used. For too large U, the effect can hardly be detected. It might well be that only intermediate values of U lead to superconductivity. Secondly, a comparable approach is the determinant method used

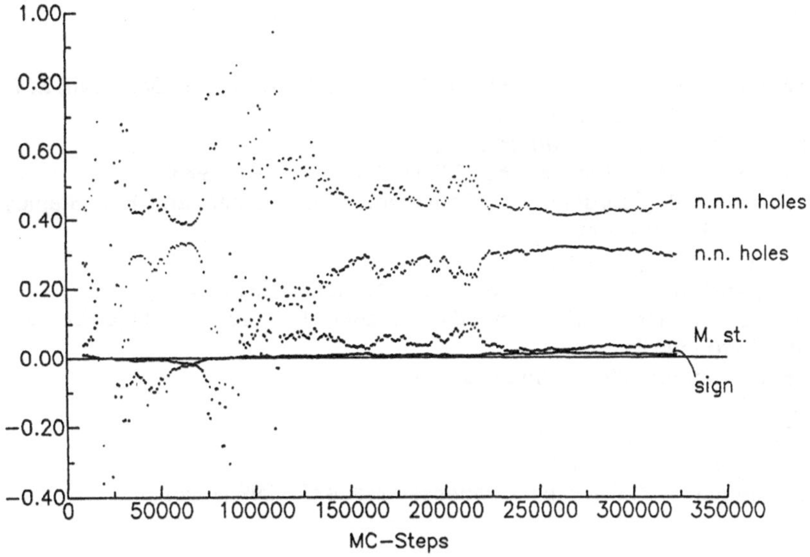

Figure 7. Same as Fig. 4, but for 12x12 system. From Ref. 12, © Springer-Verlag 1988.

by Hirsch[2] and Imada[3]. Here the fermion degrees of freedom are integrated out, leading to a determinant which has to be evaluated numerically to obtain the transition probabilities. Here is the point where the difficulties arise. For a fermionic system, the minus-sign problem translates into an ill-conditioning of the determinant at lower temperatures and larger system sizes. In principle, one again has to subtract large numbers in the computer. Therefore, the numerical problem is only shifted to a different level. For an 8x8 and $T=0.25$ system, H. De Raedt[10] calculated that the value of a determinant is of the order of 10^{53}. The ratio of values of this order has to provide a transition probability between 0 and 1. Therefore it is clear that rounding errors affect the transition probabilities, leading to a bias in the evaluation of the system through phase space. Consequently the system does not reach the part of phase space or equilibrium where the numerical difficulties due to the fermionic character disappear. While the world lines evolve correctly towards equilibrium because of correct transition probabilities, this is hard to realize in the case of the determinant method. Therefore the stabilization effect was missed. In Imada's stochastic evaluation of the determinant, the case is even worse. The basic point: the transition probabilities have to be correct, measurement of observables may be approximate.

4. CONCLUSION

Numerical simulations for 6x6 and 12x12 systems for the $U=4$, 2-D single-band Hubbard model show enhancement of next-nearest-neighbor pairs of holes and a suppression of nearest neighbors. The system shows the minus-sign effect only until it has reached equilibrium. Holes prefer next-nearest-neighbor sites, while nearest neighbors are suppressed. The present simulations were carried out using the method described in Ref. 6. The basic trick is to avoid the system's falling to lower-lying "bosonic" states by introducing a corresponding barrier, in my case an energy barrier. Reaching equilibrium leads to the stabilization effect or the vanishing of the minus-sign problem as such, as described in Ref. 6. In Sorella's approach the lower "bosonic" states are avoided by restricting the system to antisymmetric wave functions.[9] Therefore he noticed the absence of the minus-sign problem. The mathematical background of the minus-sign problem and the basic "barrier trick" as introduced in Ref. 6 are described in more detail in a separate paper.[11]

REFERENCES

1. P.W. Anderson, *Science* 235:1187 (1987); P.W. Anderson, G. Baskaran, Z. Zou, and T. Hsu, *Phys. Rev. Lett.* 58:2790 (1987).
2. J.E. Hirsch, *Phys. Rev. Lett.* 59:228 (1987).
3. M. Imada, *J. Phys. Soc. Jpn.* 56:3793 (1987), and 57:41 (1988).
4. A. Aharony, R.J. Birgenau, A. Conignlio, M.A. Kastner, and H.E. Stanley, *Phys. Rev. Lett.* 60:1330 (1988).
5. I. Morgenstern, (preprint).
6. I. Morgenstern, *Z. Phys. B – Condensed Matter* 70:291 (1988).
7. "Quantum Monte Carlo Methods," Suzuki, M. (ed.). Springer-Verlag, Berlin Heidelberg (1986).
8. G. Baskaran, (private communication).
9. S. Sorella, (these proceedings).
10. H. De Raedt, (private communication).
11. I. Morgenstern, (preprint).
12. I. Morgenstern, *Z. Phys. B – Condensed Matter* 73:299 (1988).

MONTE CARLO SIMULATION OF A THREE-BAND MODEL FOR HIGH-T$_c$ SUPERCONDUCTORS

G. Dopf, J. Rahm, and A. Muramatsu

Physikalisches Institut, Universität Würzburg
D-8700 Würzburg, F.R.G.

We consider a three-band model for the CuO_2 layers in the high-T$_c$ superconductors. The model is characterized by a Hubbard repulsion on Cu- (U_d) as well as on O-atoms (U_p), a $p-d$ hybridization t_{Cu-O}, a charge transfer energy $\Delta = \epsilon_p - \epsilon_d$ and $O-O$ hopping (t_{O-O}). In the parameter region where $< n_{Cu} > \sim 1$ for the relevant values of doping ($\delta < 0.3$), the Monte Carlo simulation shows on the one hand, a transition in the staggered magnetization for a finite concentration of holes δ_c. On the other hand, no signal for an enhanced singlet formation between Cu- and O-holes is seen. Finally we discuss the behavior of the pairing susceptibilities.

I. Introduction

It is commonly agreed that in most of the oxide superconductors the CuO_2 layers are the physically relevant elements responsible for the high critical temperature as well as other unusual properties [1]. The most significant among them is the long-range antiferromagnetic order observed in the undoped phase (e.g. La_2CuO_4 or $YBa_2Cu_3O_6$) [2]. This clear indication of strong electronic correlation led to several microscopic models based on Hubbard-type interactions to describe the electrons in the CuO_2 layers. The most simple of them is the one-band Hubbard model as proposed by P.W. Anderson at the outset [3]. Later experimental evidence [4] showed that the explicit inclusion of $2p$ orbitals on the O-sites is necessary to describe properly the charge carriers introduced by doping. However it was proposed by Zhang and Rice [5] that in the limit of strong repulsion on the Cu-sites, the low energy behavior of the system can be described by an effective one-band model.

We present here results of Monte Carlo simulations of a three-band model for the CuO_2 layers given by the following Hamiltonian:

$$H = \sum_{<i,j>\,\sigma} t_{ij}(d^\dagger_{i,\sigma}c_{j,\sigma} + h.c.) + \sum_{<j,j'>\,\sigma} \bar{t}_{jj'}(c^\dagger_{j,\sigma}c_{j',\sigma} + h.c.) +$$
$$+ \quad (\epsilon_d - \mu)\sum_{i,\sigma} n^d_{i,\sigma} + (\epsilon_p - \mu)\sum_{j,\sigma} n^p_{j,\sigma} + U_d \sum_i n^d_{i\uparrow}n^d_{i\downarrow} + U_p \sum_j n^p_{j\uparrow}n^p_{j\downarrow} \qquad (1)$$

The $p-d$ hybridization between $d_{x^2-y^2}$ orbitals on the Cu-sites and p_x or p_y orbitals on the O-sites is given by [5]

$$t_{ij} = -t_{Cu-O}(-1)^{\alpha_{ij}}\,, \qquad (2)$$

where $\alpha_{ij} = 1$ if $j = i + \frac{1}{2}\hat{x}$ or $i + \frac{1}{2}\hat{y}$ and $\alpha_{ij} = 2$ if $j = i - \frac{1}{2}\hat{x}$ or $i - \frac{1}{2}\hat{y}$. The phase factors in eq. (2) are due to the d- and p-symmetry of the Cu- and O-orbitals respectively. The O-O hopping matrix element is defined in a similar way:

$$\bar{t}_{jj'} = -t_{O-O}(-1)^{\beta_{jj'}}\,, \qquad (3)$$

Interacting Electrons in Reduced Dimensions
Edited by D. Baeriswyl and D.K. Campbell
Plenum Press, New York

where $\beta_{jj'} = 1$ if $j' = j - \frac{1}{2}\hat{x} + \frac{1}{2}\hat{y}$ or $j' = j + \frac{1}{2}\hat{x} - \frac{1}{2}\hat{y}$ and $\beta_{jj'} = 2$ if $j' = j - \frac{1}{2}\hat{x} - \frac{1}{2}\hat{y}$ or $j' = j + \frac{1}{2}\hat{x} + \frac{1}{2}\hat{y}$. The index i denotes the Cu-sites and j the O-ones. The local orbital levels are given by ϵ_p and ϵ_d and the charge transfer energy is $\Delta = \epsilon_p - \epsilon_d$. U_d and U_p are the Hubbard-couplings on the Cu- and O-sites respectively. In the course of this work we use the particle picture, i.e. the undoped situation corresponds to five particles per unit cell.

The model of eq.(1) and variations of it were studied recently by several authors both analitically [5-12] and numerically [13-15]. Among the analytical treatments one can distinguish those in the localized limit, where a strong-coupling situation ($t_{Cu-O}/U_d \to 0$) is assumed [6-10]. The Hilbert space is cut down to that of a localized spin on the Cu-sites that interacts with the charge carriers on the O-sites through an exchange interaction

$$J_{Cu-O} = t^2_{Cu-O}\left(\frac{1}{\Delta} + \frac{1}{U_d - \Delta}\right) . \tag{4}$$

Furthermore a fourth-order interaction is also taken into account [7-10]

$$J_{Cu-Cu} \sim t^4_{Cu-O}\left(\frac{1}{U_d(U_d-\Delta)^2} + \frac{1}{(U_d-\Delta)^3}\right) \tag{5}$$

that gives the antiferromagnetic exchange interaction for the Cu-spins, necessary for the description of the antiferromagnetic phase observed in the absence of doping. The strong-coupling expansion breaks down in the limit where $\Delta \sim 0$ or $\Delta \sim U_d$. Whereas $\Delta \sim 0$ still corresponds to a localized limit (recall that we are in the particle picture), $\Delta \sim U_d$ leads to a situation where large charge fluctuations take place on the Cu-sites (mixed-valence regime). To our knowledge this case could until now only be treated in the frame of a one-impurity Anderson model[11] or in a Hartree-Fock approximation [12] (but see [16]). These limitations are of course not present in the Monte Carlo simulation. We discuss below results in the localized as well as in the mixed-valence region. As a general trend it is observed that the magnetic properties of the superconducting oxides are well described in the localized regime. However a higher amplitude for the pairing susceptibilities is obtained in the mixed-valence region.

We concentrate first on the magnetic properties of the model eq.(1). Then we study the formation of singlets between Cu- and O-sites as proposed originally by Zhang and Rice [5]. For the magnetic properties it is observed that the magnetic moment on the Cu-sites as well as the staggered magnetization are higher in the localized regime than in the mixed-valence one as expected. The degree of localization is made manifest by the local magnetic moment that is independent on the concentration of dopant holes up to values of $\delta \sim 30\%$. For low enough temperatures ($\beta \sim 1/3$ in units of t_{Cu-O}) a rather sharp drop of the staggered magnetization is obtained for a finite value $\delta_c \sim 5\%$ of the concentration of the dopant holes. On the contrary, no significant changes in the magnetic properties are observed as a function of doping in the mixed-valence region. As for the formation of singlets between a Cu-site and a symmetric combination of the neighboring O-sites (see below) no signal of enhancement due to the interaction could be found.

II. The simulation

The algorithm used is the one developed by J. Hirsch [18] for a d-dimensional Hubbard model. A discrete Hubbard-Stratonovich transformation is used to decouple the interaction terms both on Cu- and O-sites. This leads to a system of free fermions interacting linearly with a fluctuating Ising field on each site. As in the case of the one-band Hubbard model, the Monte Carlo simulation is limited by temperatures $kT \sim 1/4$ from below. At lower temperatures the algorithm becomes unstable due mainly to the presence of ill-conditioned matrices that determine the statistical weight of the sampled configurations. However progress in this direction was recently reported by several groups [19-20]. Due to the limitation to rather high temperatures we do not attempt to make any statement about superconductivity but limit ourselves to magnetic properties and the formation of local singlets since they are thought to be present at temperatures of the order of J_{Cu-O}.

III. Results

The results presented here correspond if not otherwise stated to $U_p = 0$ and to $t_{O-O} = 0$ in eq.(1). All quantities are given in units of t_{Cu-O} which is set equal to one.

A qualitative difference is observed in the behavior of the system depending on whether $\Delta < U_d/2$ or $\Delta > U_d/2$. The first case corresponds to a localized regime, i.e. to a situation where $< n_{Cu} > \simeq 1$ rather independently of the degree of doping up to a value of $\delta \sim 0.3$. Since we are working in the particle picture, we define the doping as $\delta = 5 - (< n_{Cu} > +2 < n_O >)$. In the second case ($\Delta > U_d/2$), the occupation number on the Cu-sites varies in the range $< n_{Cu} > \sim 1.2 - 1.6$, the larger value corresponding to $\Delta = U_d$, the full mixed valence situation. Although these results are expected on general grounds, it is interesting to notice that this behavior is obtained for values of the interaction as low as $U_d = 4$.

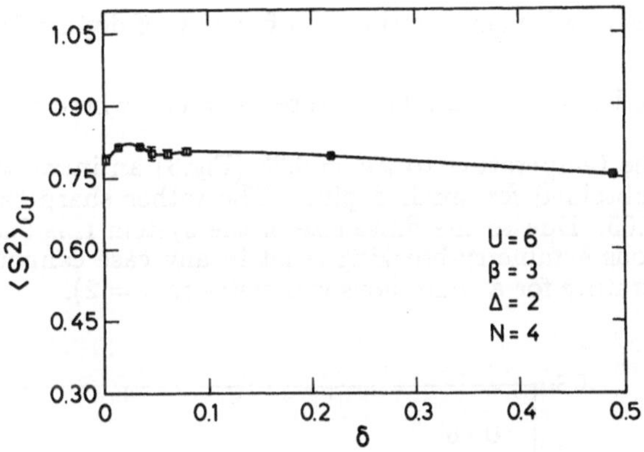

Figure 1. Local magnetic moment on Cu-sites as a function of doping.

We discuss first the magnetic behavior of the system in the localized regime since in this region we obtain results that are in agreement with experiments [2]. Figure 1 shows the local magnetic moment on the Cu-sites for a system with 16 unit cells. Up to a doping of 50% it remains rather constant at a value of $< s^2 >= \frac{3}{4} \times 0.8$. In the mixed-valence regime on the contrary, the magnetic moment decays up to a value of $< s^2 >= \frac{3}{4} \times 0.4$ for $\Delta = U_d$.

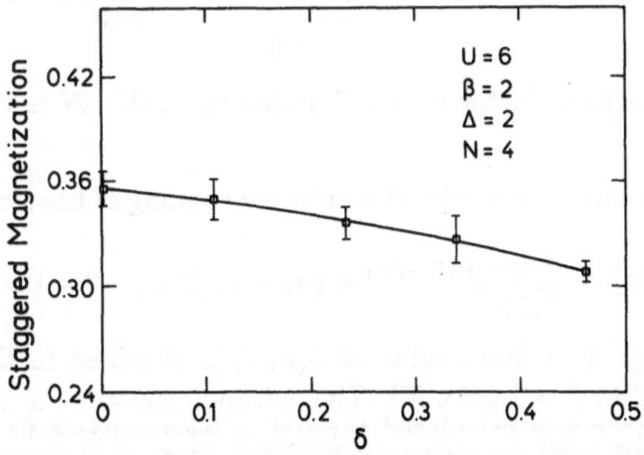

Figure 2. Staggered magnetization as a function of doping for $kT = 0.5$.

Next we consider the dependence on doping of the staggered magnetization

$$m_z = \frac{1}{N} \sum_i (-1)^i < (n_{i\uparrow} - n_{i\downarrow}) > \quad . \qquad (6)$$

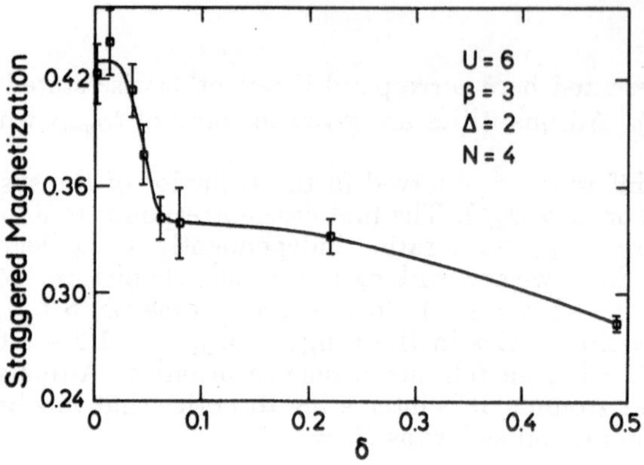

Figure 3. Staggered magnetization as a function of doping for $kT = 0.33$.

Figure 2 shows the behavior at a temperature $kT = 0.5$, where no relevant feature is observed.

As we lower the temperature to $kT = 0.33$ (Fig.3) an increase in the staggered magnetization is obtained for small doping. The rather sharp raise of the data is centered at $\delta \simeq 0.05$. Due to the finite size of the system this feature is still not a proof of spontaneous symmetry-breaking (that in any case cannot occur at a finite size and/or temperature for a continuous symmetry in $d = 2$).

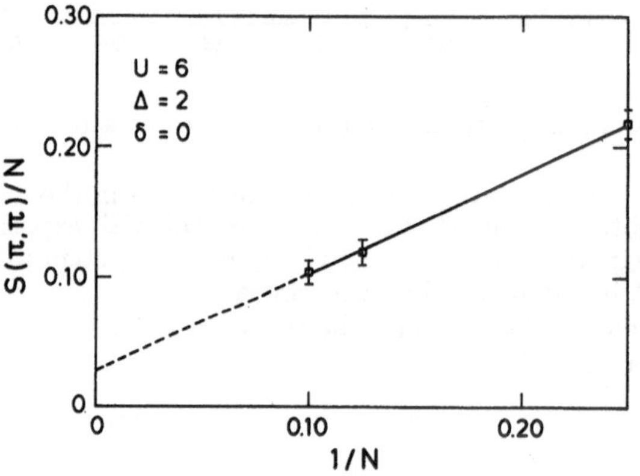

Figure 4. Structure form factor at $\delta = 0$ scaled vs. $1/N$. N is the number of unit cells.

Figures 4 and 5 show the results of a finite-size scaling of the structure form factor given by

$$S(\vec{q}) = \frac{1}{N} \sum_{i,i'} e^{i\vec{q}(\vec{R}_i - \vec{R}_{i'})} < (n_{i\uparrow} - n_{i\downarrow})(n_{i'\uparrow} - n_{i'\downarrow}) > \ . \quad (7)$$

For $\delta = 0$ (Fig. 4) a finite value of $S(\pi, \pi)$ is obtained in the limit $N \to \infty$, $\beta \to \infty$ (β is scaled as $\beta = 1.061\sqrt{N}$). The extrapolated value is $S_\infty(\pi, \pi) = 0.027 \pm 0.001$ where the error was calculated through a least-squares fit. This value is in good agreement with the ones obtained for a spin-$\frac{1}{2}$ Heisenberg antiferromagnet by Reger and Young (RY) [21] with a "world line" Monte Carlo simulation ($S_\infty^{RY}(\pi, \pi) = 0.0295$). However a word of caution is in place here since we scaled $S(\vec{q})$ as $1/N$ the data falling on a straight line. On the other hand RY used $S(\vec{q}) \sim 1/\sqrt{N}$ following the results by spin-wave theory [22] (see also [23]). Calculations for lower temperatures and various values of U_d are presently being carried out in order to ensure that the proper extrapolation was used.

For $\delta = 0.2$ (Fig.5) on the contrary, no staggered magnetization is found. These

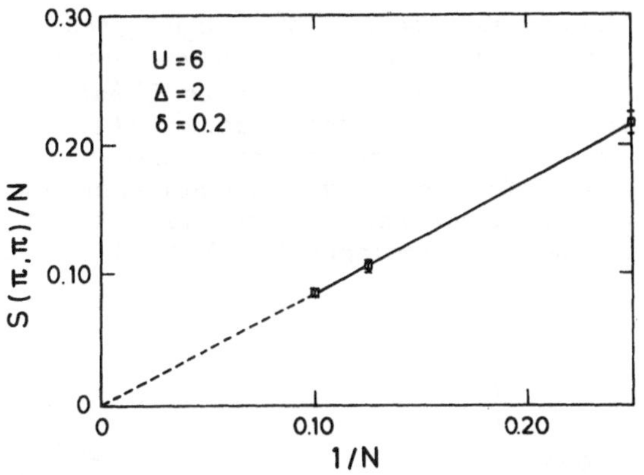

Figure 5. Same as in Fig. 4 for $\delta = 0.2$.

results suggest that a spontaneous symmetry-breaking takes place in the thermodynamic limit at $T = 0$ for the undoped system, whereas long-range order is destroyed by doping. On the basis of the results in Fig.3 this appears to happen at $\delta_c \simeq 0.05$.

Besides the magnetic properties, that are clearly visible at high temperatures in the Monte Carlo simulations, we address the question of local singlets in the system. Zhang and Rice [5] suggested that in the localized regime the dopant holes on the O-sites will strongly couple in a singlet state to the nearest neighbor hole on the Cu-site. Since the coupling for this process is given by J_{Cu-O} and assuming that the strong-coupling limit is the relevant fixed-point of the model as suggested by the reults presented above, these processes should be observable at rather high temperatures.

We calculate the propagator for those siglets defined by

$$G_s(x,\tau) = \frac{1}{N}\sum_i < \psi^\dagger(x_i + x, \tau)\psi(x_i) > , \tag{8}$$

where $\psi(x)$ corresponds to a singlet build-up by the Cu-hole and a symmetric combination of a hole on the four nearest neighbors to the Cu-site:

$$\psi(x_i) = \frac{1}{\sqrt{2}}(d_{i\uparrow}P_{i\downarrow} - d_{i\downarrow}P_{i\uparrow}) \tag{9}$$

with

$$P_{i\sigma} = \frac{1}{2}\sum_{jnn.i}(-1)^{\alpha_{ij}} c_{j\sigma} . \tag{10}$$

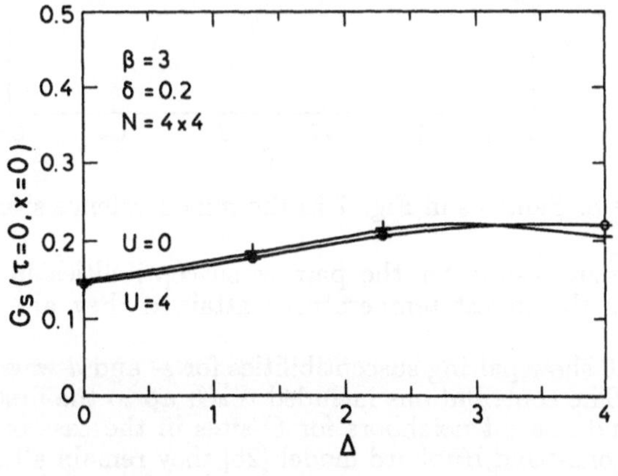

Figure 6. Squared amplitude for a local singlet as a function of the charge transfer energy $0 < \Delta < U_d$ for a doping of 20%. Circles correspond to the interacting case, crosses to the non-interacting one.

Figure 6 shows the amplitude squared of a singlet given by eq.(9) for a doping $\delta = 0.2$ both for the interacting case ($U = 4$, circles) and the non-interacting one. The non-interacting case corresponds to $U_d = 0$ in eq.(1) *but with Δ corrected in such a way that the occupation numbers both for Cu and for O are equal to the ones when* $U_d \neq 0$. In this way we take into account Hartree-type contributions. The remaining difference gives a measure of the tendency for the formation of a local singlet due to the interaction. The results show that the interaction does not give any significant enhancement in the whole parameter range ($0 < \Delta < U_d$).

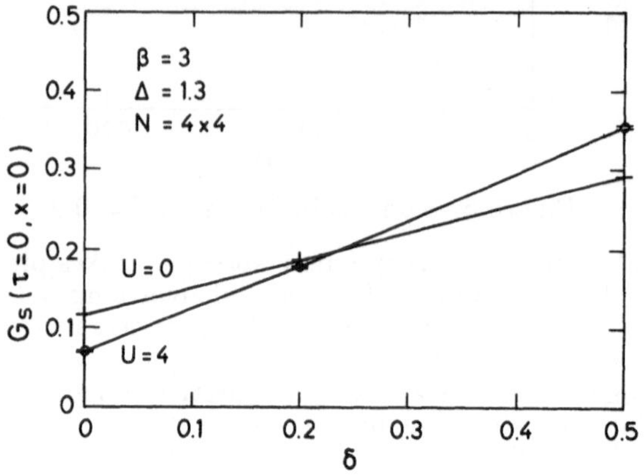

Figure 7. Squared amplitude for a local singlet in the localized regime as a function of doping.

Figure 7 shows the amplitude for the local singlet in the localized region as a function of doping. It is seen that only for $\delta > 0.3$ an appreciable enhancement is obtained. Figure 8 shows the behavior in the mixed-valence region as a function of doping. Although an enhancement is observed for $\delta > 0.1$, it is rather small up to high values of δ.

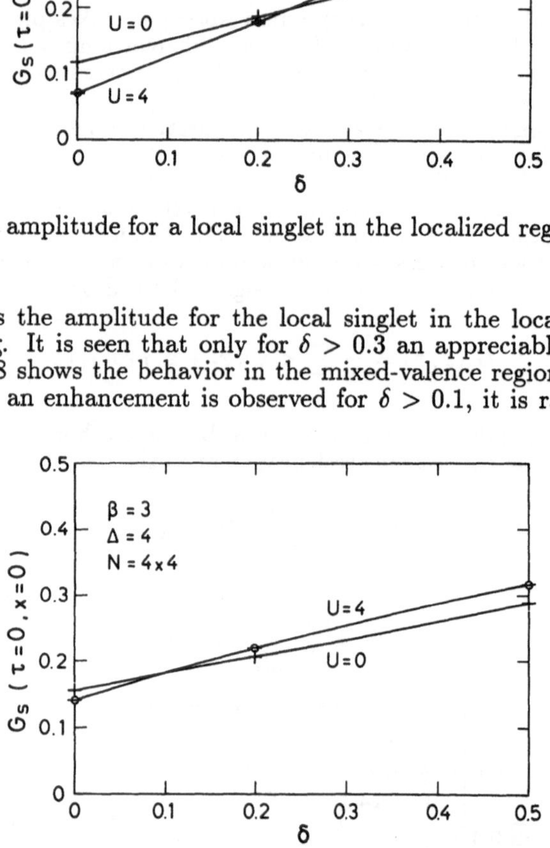

Figure 8. Same as in Fig. 7 in the mixed valence situation.

Finally we present results for the pairing susceptibilities in the mixed-valence region, since up to the lowest temperatures attained they are larger than in the localized region.

Figures 9 and 10 show pairing susceptibilities for s- and d-wave symmetry respectively for $\delta = 0.2$. The contributions included reach up to the first nearest neighbors for Cu-sites and 2nd nearest neighbors for O-sites in the case of s-wave symmetry. Similarly as in the one-band Hubbard model [25] they remain allways below the susceptibilities for $U_d = 0$.

The last figure (Fig. 11) shows the result for p-wave symmetry but now with $t_{O-O} = 1$. Surprisingly in this case and only for this symmetry an enhancement due to interaction is observed.

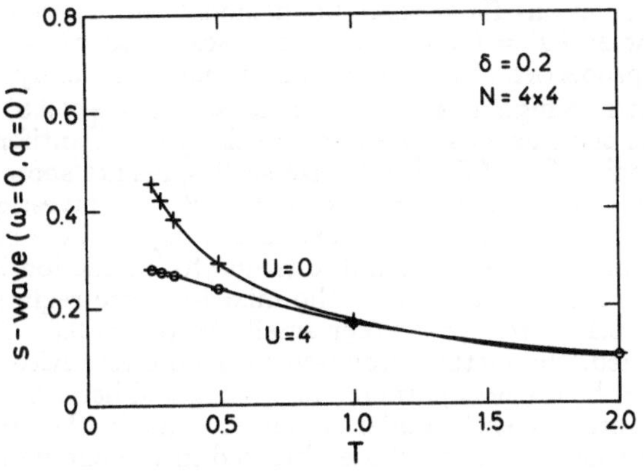

Figure 9. Pairing susceptibility for s-wave as a function of temperature in the mixed valence situation ($\Delta = U_d$).

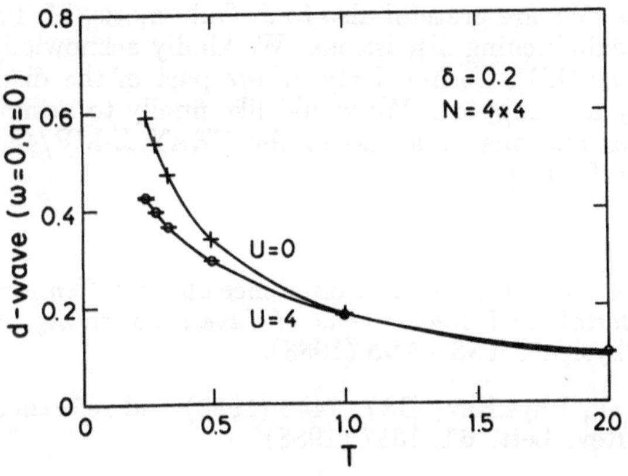

Figure 10. Same as in Fig. 9 for the d-wave symmetry.

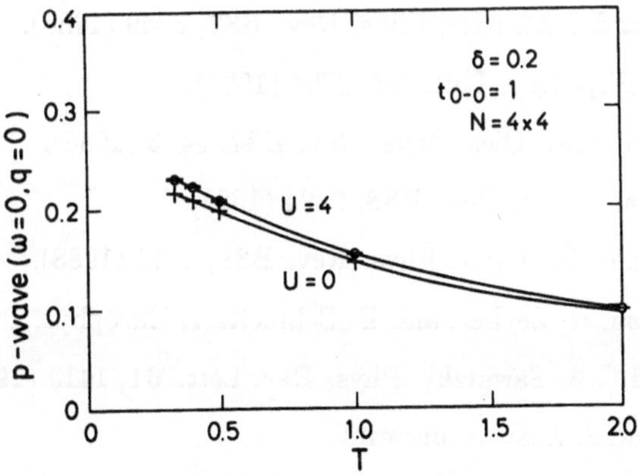

Figure 11. Pairing susceptibility for p-wave with $\Delta = U_d$ and $t_{O-O} = 1$.

IV. Summary

Monte Carlo simulations for a three-band model (eq. 1) were performed up to temperatures as low as $kT = 0.33$ (in certain cases up to $kT = 0.25$). It is found that the magnetic properties of the high-T_c materials are closely reproduced in the localized regime, where a high magnetic moment ($< s^2 >= \frac{3}{4} \times 0.8$) is obtained . Furthermore a signal of spontaneous symmetry-breaking to an antiferromagnetic phase is observed for a doping $\delta_c \simeq 5\%$. Finite-size scaling results show a finite staggered magnetization for $\delta = 0$ whereas $m = 0$ for $\delta = 20\%$. We investigated also the formation of a local singlet between a Cu-hole and a dopant hole on the neighboring O-sites. No signal for an enhanced amplitude due to interaction is found in the localized regime for $\delta < 0.3$. Although in the mixed-valence regime an enhancement is observed for $\delta > 0.1$, it remains rather small up to $\delta = 0.5$. Thus, on the basis of the results obtained, the picture proposed by Zhang and Rice [5] does not seem applicable at least in the range of parameters investigated here. Simulations for much larger values of U_d are currently under progress. Finally the results obtained for the pairing functions are similar to those obtained in the one-band Hubbard model when $t_{O-O} = 0$. However in the case of a strong direct $O - O$ hopping ($t_{O-O} = 1$), it is observed that the triplet channel (p-wave) shows an enhancement due to interaction.

Acknowledgements

We wish to thank Prof. W. Hanke for a continuous support of this work and valuable discussions. We are grateful also to J. Gubernatis, M. Imada, P. Prelovšek, and T.M. Rice for enlightening discussions. We kindly acknowledge the Institute for Scientific Interchange (ISI), Torino, Italy, where part of the discussions took place for their hospitality and support. We would like finally to acknowledge support by the HLRZ-Jülich for allowing us access to the CRAY X-MP/48 where the present calculations were performed.

References

1. Proceedings of the International Conference on *High Temperature Superconductors and Materials and Mechanisms of Superconductivity*, eds.: J. Müller and J.L. Olsen, Physica C **153 - 155** (1988).

2. Y. Endoh et al., Phys.Rev. B**37**, 7443 (1988) and references therein; M. Sato et al., Phys. Rev. Lett. **61**, 1317 (1988).

3. P.W. Anderson, Science **235**, 1196 (1987); P.W. Anderson et al., Physica C **153 - 155**, 527 (1988).

4. J.C. Fuggle et al., Phys. Rev.B**37**, 123 (1988); R.L. Kurtz et al., Phys. Rev. B**35**, 8818 (1987).

5. F.C. Zhang and T.M. Rice, Phys. Rev. B**37**, 3759 (1988).

6. V.J. Emery, Phys.Rev. Lett. **58**, 2794 (1987).

7. J. Zaanen and A.M. Olés, Phys. Rev. B**37**, 9423 (1988).

8. E.Y. Loh et al., Phys. Rev. B**38**, 2494 (1988).

9. V.J. Emery and G. Reiter, Phys. Rev. B**38**, 4547 (1988).

10. A. Muramatsu, R. Zeyher, and D. Schmeltzer, Europhys. Lett. **7**, 473 (1988).

11. H. Eskes and G.A. Sawatzky, Phys. Rev, Lett. **61**, 1415 (1988).

12. A.M. Olés and J. Zaanen, preprint.

13. J.E. Hirsch et al., Phys. Rev. Lett. **60**, 1668 (1988). J.E. Hirsch et al., Physica C **153 - 155**, 549 (1988).

14. M. Imada, J. Phys. Soc. Jpn. **57**, 3128 (1988).

15. W.H. Stephan, W.v.d. Linden, and P. Horsch, Int. J. Mod. Physics **B1**, 1005 (1988).

16. We should remark that the spin-bag idea [17] deals also with the itinerant situation, but it will not be considered here since on the one hand it is based on the one-band Hubbard model and on the other hand a comparison with the numerical results is beyond our present accuracy.

17. J.R.Schrieffer, X.G.Wen and S.C.Zhang, Phys. Rev. Lett. **60**, 944 (1988); and preprint.

18. J.E. Hirsch, Phys. Rev. B**31**, 4403 (1985).

19. J.E. Hirsch and S. Tang, preprint.

20. J. Gubernatis, private communication.

21. J.D. Reger and A.P. Young, Phys. Rev. B**37**, 5978 (1988).

22. D.A. Huse, Phys. Rev. B**37**, 2380 (1988).

23. A different scaling law was used in [24] obtaining a value of $S_\infty \simeq 0.036$.

24. P. Horsch and W.v.d. Linden, Z. Phys. B**72**,181 (1988).

25. J.E. Hirsch and H.Q. Lin, Phys. Rev. B**37**, 5070 (1988).

EXACT VALENCE-BOND APPROACH TO QUANTUM CELL MODELS

Z.G. Soos, S. Ramasesha,* and G.W. Hayden

Department of Chemistry
Princeton University
Princeton, N.J. 08544 U.S.A.

INTRODUCTION

A line between two H atoms is a simple and natural shorthand for the covalent bond of H_2. The extension to a line per spin-paired electrons in bonds provides a convenient and surprisingly powerful representation of molecular structure. Pauling's valence bond (VB) diagrams[1] are widely used throughout chemistry and increasingly in solid-state physics, from purely qualitative to quantitative applications. We emphasize in this paper the versatility of VB diagrams and their role in obtaining exact results for finite quantum cell models. Considerable progress has been made in the manipulation of VB diagrams.[2] Quantitative applications now focus on model Hamiltonians that conserve total spin and on excited states rather than on ground-state molecular properties.

A unique VB diagram can be associated with most saturated molecules. The lines (or bonds) are in fact essentially transferable in such systems, an important experimental observation that makes bond counting almost quantitative for many properties. Conjugated molecules like benzene or linear polyenes have several possible VB diagrams whose linear combinations, or "resonance", is to be found, with only the π-electrons now represented. Since smaller and smaller admixtures of higher and higher energy VB diagrams can readily be envisioned, resonance ideas can be carried to extremes and have frequently been criticized. However, neither the restriction to valence electrons nor the occurrence of small admixtures is a problem for model Hamiltonians, since the basis is inherently finite.

Linear combinations, or resonance, among VB diagrams are thus associated with delocalized electrons. The close packed structures of simple metals lead to far more bonds than available pairs of valence electrons. Pauling[3] considered fluctuating valence pairs, or resonance among various assignments of paired electrons to nearest neighbor bonds, as a representation of the normal state. The single-electron alternative due to Bloch is now so well developed, both for metals and semiconductors, that alternatives are quaint footnotes.

---

* Permanent address: Solid State and Structural Chemistry Unit, Indian Institute of Science, Bangalore, 560012, India

Interacting Electrons in Reduced Dimensions
Edited by D. Baeriswyl and D.K. Campbell
Plenum Press, New York

The same idea, as resonating valence bond (RVB) theory,[4] has emerged for the oxide superconductors, where strong if unspecified correlation between electrons appears to be necessary. Rigorous VB results for infinite systems are still restricted to special models constructed to have a VB ground state.[5] The RVB emphasis is on the correlated VB state itself, and its possible superconducting instability, rather than on the precise model and its parameters. Such issues are postponed until the relevant electrons and interactions have been identified.

From our point of view, VB diagrams are a many-electron basis for such familiar models as Heisenberg exchange in magnetic insulators, Hubbard or extended Hubbard models, or Pariser-Parr-Pople (PPP) models for interacting π-electrons.[6] Such quantum cell models abound in both physics and chemistry. They correspond to phenomenological tight-binding schemes, as illustrated by simple s-band metals or by Hückel theory for conjugated molecules, in which a specified number (N_e) of electrons or spins are distributed among N orbitals and various interactions are introduced.

As summarized in Section II, VB diagrams provide a complete many-electron basis that is particularly convenient when the total spin S is conserved, which is the case for all the models mentioned above. Since the models are finite dimensional, it follows that their exact properties can be represented in terms of VB diagrams. In Section III, we show that such correlated states are particularly important for excited states and for nonlinear properties. Diagrammatic valence bond (DVB) methods[2] then provide new insights for familiar but formerly intractable problems.

VB DIAGRAMS

We consider N_e electrons with total spin $S \leq N_e/2$ in N orbitals represented by the corners of a regular polygon in Fig. 1. While the one-dimensionality of the perimeter simplifies applications to one-dimensional models, there is in fact no restriction on the topology of the N orbitals. We may even take several orbitals to represent one atom or site. The nature of the model Hamiltonian is also left open, except for conservation of S. To construct VB diagrams, we stipulate that all bonds for spin-paired electrons connect vertices of the polygon. Singlet ($S = 0$) diagrams only contain lines and nonbonding sites. Diagrams with $S > 0$ also contain an arrow, as shown in Fig. 1 for $S = 1$, connecting 2S

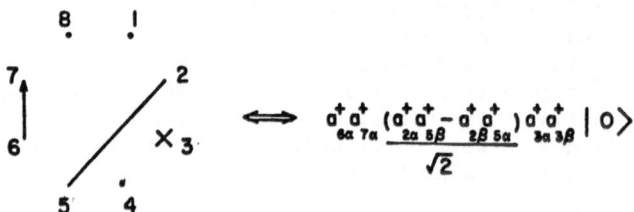

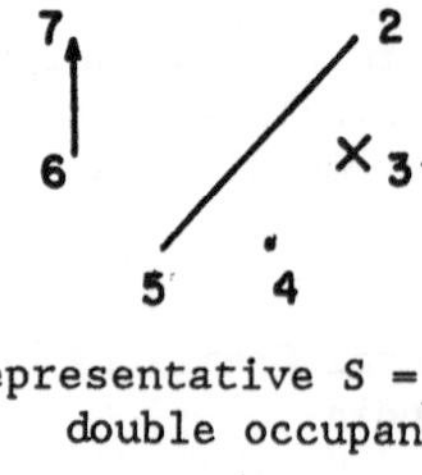

$$\Longleftrightarrow \; a^+_{6\alpha} a^+_{7\alpha} \frac{(a^+_{2\alpha} a^+_{5\beta} - a^+_{2\beta} a^+_{5\alpha})}{\sqrt{2}} a^+_{3\alpha} a^+_{3\beta} |0\rangle$$

Fig. 1. Representative $S = 1$ VB diagram for 6 electrons on 8 sites, with double occupancy in ϕ_3 and parallel spins in ϕ_6 and ϕ_7.

sites. The special case of $S = \frac{1}{2}$ is treated with the aid of a phantom site.[7] Every VB diagram manifestly conserves S by construction.

The possibilities of empty and of doubly occupied sites are represented by dots and crosses in Fig. 1, respectively. The phase factor for a pair of fermions at site 3, for example, is uniquely specified in Fig. 1 by the order $a^+_{3\alpha} a^+_{3\beta}$ of the creation operators. The Heitler-London covalent bond between sites 2 and 5 also involves a phase convention, that $a^+_{i\alpha} a^+_{j\beta}$ be positive irrespective of the order of i, j. The phase convention for the arrow is also fixed in Fig. 1 and the highest possible S_z is understood. Moreover, each fermion pair in Fig. 1 is explicitly normalized and the method of construction excludes any repeated site indices. We may consequently construct arbitrarily complicated VB diagrams $|k\rangle$ for specified values of N_e, S, and N. The overall phase is independent of the order of the product. Only the arrow for half-integral S can involve an odd number of fermions and it occurs but once.

Figure 1 explicitly demonstrates the connection between the VB diagram $|k\rangle$ and Slater determinants involving N_e-fold products of $a^+_{j\sigma}$. Each Heitler-London pair requires two Slater determinants. A diagram with r lines consequently corresponds to a fixed linear combination of 2^r Slater determinants, whose manipulation is far more difficult. By contrast, empty sites, doubly-occupied sites, or spin-parallel sites lead to a unique Slater determinant.

Rumer[8] and Pauling recognized long ago that there are too many ways of drawing n lines among $N = 2n$ vertices for $S = 0$ diagrams. <u>Legal</u> diagrams are those without intersecting lines, such as the diagrams in Figs. 1 and 2. Legal diagrams form a complete, linearly independent basis. Any diagram with intersecting lines can readily be expanded in terms of diagrams without intersecting lines. Similar rules[9] exclude intersections between the arrow in $S > 0$ diagrams and any line. Furthermore, we exclude[9] diagrams containing any singlet line p-q that encloses the arrow, such that p is smaller and q larger than any vertex in the arrow.

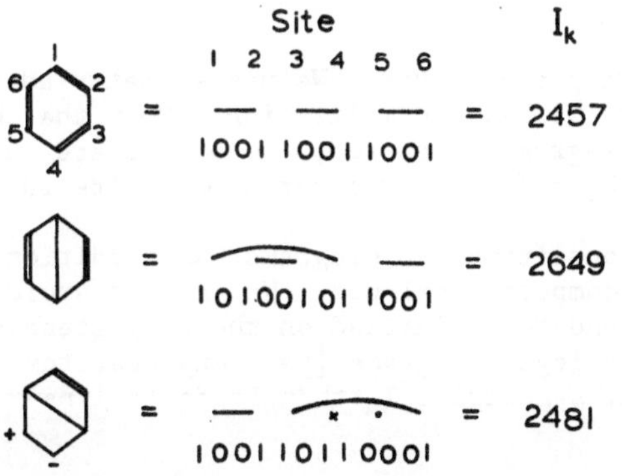

Fig. 2. Bit representation of π-electrons in a Kekulé, Dewar and ionic VB diagram of benzene.

Identities for expanding illegal diagrams in terms of legal ones can readily be derived via the Fermi commutation properties of the $a_{j\sigma}^{+}$ operators in Fig. 1 and the fact that only the operators involving the crossing lines or arrow need be considered. Another old result for singlet diagrams $|k\rangle$ and $|k'\rangle$ with r lines each is that their overlap is

$$S_{k'k} = \langle k'|k\rangle = (-2)^{i-r} \tag{1}$$

where i is the number of disconnected islands formed on superposing $|k\rangle$ and $|k'\rangle$. The generalization to $S > 0$ and to doubly or unoccupied sites is again straightforward. Legal VB diagrams consequently form complete, normal, but not orthogonal many-electron basis.

The second-quantized representation of VB diagrams in Fig. 1 facilitates the derivation of transformation rules. The bit (machine language) representation[7] in Fig. 2 affords enormous computational advantages. The N vertices of $|k\rangle$ are now converted to a 2N-digit binary number, with the first two bits describing site 1 and the last two site N. The code is 00 for dots (empty sites), 11 for crosses (doubly-occupied sites), 10 for the lower-numbered vertex of a line, and 01 for the higher-numbered vertex of the line and for all sites in the arrow. The extended rules for legal VB diagrams precisely suffice for a unique correspondence between $|k\rangle$ and a 2N-digit binary number. The generation of diagrams is thereby reduced to manipulations of 2N-digit binary numbers. The storage of diagrams as integers I_k, as indicated in Fig. 2, is very efficient. Their ordering as increasing integers allows for tree searching. Diagrams for spin-½ sites without empty or doubly occupied sites require only N digit binary numbers. Such purely covalent diagrams are the basis[10] for problems involving Heisenberg spin exchange.

Symmetry adaptation of VB diagrams is also simple. The equivalence of the Kekulé structure in Fig. 2a to another diagram with interchanged single and double bonds is the essence of resonance. The molecular point group immediately fixes the linear combinations of equivalent diagrams. Formally, we consider[11] symmetry operators R_n, n = 1, ... G, that commute with H. Then the coefficients of the diagrams

$$|k,n\rangle = R_n\,|k\rangle \qquad n = 1,2,\ ..\ G \tag{2}$$

differ by at most a phase factor. We use symmetry-adapted linear combinations in (2) without normalization. Only the smallest integer I_{min} among the G diagrams is stored; the others are regenerated as needed by operating with R_j and thereby altering the bits in a known way.

The nonorthogonality of VB diagrams has traditionally been a major obstacle to their computational use. We in fact avoid forming matrix elements $\langle k'|H|k\rangle$ and focus instead on the completeness and linear independence of the legal diagrams $|k\rangle$. Any operator that conserves N_e, S, N and additional symmetries R_n then leads to linear equations of the form

$$H\,|k\rangle = \sum_q h_{kq}\,|q\rangle \tag{3}$$

The coefficients h_{kq} depend on the model and its parameters. They form a real, sparse, nonsymmetric matrix for π-electron, spin, and solid-state models.

The extreme sparseness of h_{kq} reflects the fact that there are only about N bonds, each of which leads to 0, 1, or 2 off-diagonal elements. The dimensions of the many-electron basis for N_e, S, and N is

$$P_s = \frac{2S+1}{N+1} \binom{N+1}{\frac{N_e}{2} - S} \binom{N+1}{N-\frac{N_e}{2} - S} \qquad (4)$$

where ( ) is a binomial coefficient. The basis grows almost like 4^N. Symmetry adaptation leads in (3) to about P_s/G coupled diagrams.

Sparse matrix methods allow $P_s/G \sim 10^5$ symmetry-adapted linear combinations in (3) to be solved on a VAX 11/780 computer. Slater determinants based on S_z conservation result in matrices of $\sim 10^6$-10^7. The Rettrup algorithm provides several (~5-10) of the lowest-energy states in each symmetry subspace.[2] The left eigenvectors of h_{kq} are the desired eigenfunctions of H written as linear combinations of VB diagrams. For example, the normalized ground state $|G\rangle$ is

$$|G\rangle = \sum_k c_k |k\rangle \qquad (5)$$

and has about P_s/G nonvanishing coefficients c_k. Matrix elements[12] $\langle R|A|R'\rangle$ between correlated states are surprisingly easy to evaluate in the bit representation. The procedure is to second-quantize A, examine its effects on $|k\rangle$ as indicated in (3), and then reduce the matrix element for normalized $|R\rangle$, $|R'\rangle$ to an overlap involving (1).

We conclude this section with several generalizations of VB diagrams. For S = 1 sites, we consider[13] two vertices per site as sketched in Fig. 3. Since a line connecting the vertices of any site corresponds to a singlet, we exclude all such diagrams. The two legal diagrams for an overall singlet for four S = 1 sites are shown in Fig. 3, with the two ends connected in (b). N sites again lead to a 2N-digit

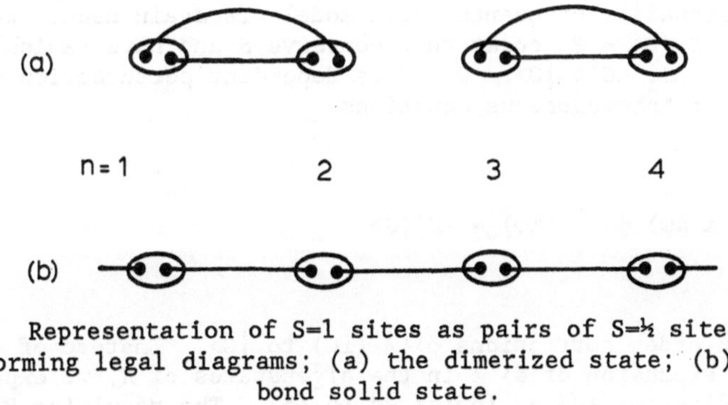

Fig. 3. Representation of S=1 sites as pairs of S=½ sites with VB singlets forming legal diagrams; (a) the dimerized state; (b) the valence bond solid state.

binary with N bits "1" and N bits "0". All the manipulations of purely covalent diagrams (no dots or crosses) still apply for the truncated basis that excludes any diagram with a line between the two vertices of a site. The reduction for S conservation is even greater for S = 1 spins.

The neutral-ionic interface[14] in organic solids containing π-electron donors (D) and acceptors (A) also leads to a truncated basis on excluding, on physical grounds, high energy D^{+2} or A^{-2} sites. Here D corresponds to a doubly-filled highest-occupied MO and A to an empty lowest unoccupied MO. Spin pairing between paramagnetic D^+ and A^- sites is represented by lines. All VB manipulations are again preserved. We expect VB diagrams to be a convenient basis for any quantum cell model that conserves S. For a given model, the complete VB basis amounts to carrying out configuration interaction (CI) to all orders in conventional one-electron schemes; that is to say, VB methods provide an efficient alternative path to exact results.

NONLINEAR RESPONSE OF CORRELATED STATES

We have thus far focused on the ground and low-lying excited states of quantum cell models that conserve S. Neither size nor dimensionality is a problem for diagram manipulations. The dimension P_s/G of the sparse matrix h_{kq} in (3) sets the computational limitation on exact solutions. VB diagrams are used today in many contexts, all different from the initial goal of accurate molecular ground states. Our recent applications[15] to excited states of conjugated polymers, such as polyacetylene or polydiacetylenes, are natural extensions of molecular π-electron theories. We now consider a VB approach to the full spectrum of excited states in analyzing static or dynamic perturbations.

Much attention has recently been given to the large and ultrafast nonlinear optical (NLO) coefficients of conjugated molecules[16] like p-nitroaniline and conjugated polymers[17] like polyacetylene or various polydiacetylenes. The response of the π-electrons to an electric field $\vec{E}$ cosωt is of primary interest. Oscillating electric or magnetic fields are special cases of perturbations $H'(\omega)$ that may be either static (ω = 0) or dynamic ($\omega > 0$). NLO coefficients are by definition the response to some order in H.

The VB basis provides a convenient new approach to such perturbations.[18] As will shortly be seen, the full spectrum of correlated eigenstates $|R\rangle$ can be included exactly even though only the ground-state $|G\rangle$ or a few excited states of H are found explicitly. The finite dimensionality of quantum cell models is again used. We consider perturbations $H'(\omega) = H' \cos\omega t$ that conserve S and have vanishing first order corrections, $\langle G|H'|G\rangle = 0$. Time dependent perturbation theory then leads to linear inhomogeneous equations

$$(H - E_G \pm \hbar\omega)\ \phi^{(1)}(\pm\omega) = -H'|G\rangle \qquad (6)$$

for the first-order corrections $\phi^{(1)}(\pm\omega)$ to $|G\rangle$. Instead of the conventional expansion of $\phi^{(1)}$ in the eigenstates of H, we expand $\phi^{(1)}$ in terms of VB diagrams $|k\rangle$ as indicated in (5). The resulting P_s linear equations for the expansion coefficients of $\phi^{(1)}$ are solved by standard methods. The extreme sparseness of the h_{kq} matrix and symmetry considerations are again advantageous. If we then consider $\phi^{(1)}$ as an eigenstate expansion in $|R\rangle$, the inner product of (6) with $|R\rangle$ leads to

$$-\langle R|H'|G\rangle = (E_R - E_G \pm \hbar\omega)\langle R|\phi^{(1)}(\pm\omega)\rangle \tag{7}$$

and $\langle R|\phi^{(1)}\rangle = C_R$ is recognized to be the expansion coefficient for $|R\rangle$. We may evaluate either side of (7) whenever the excited state $|R\rangle$ is known as a check on the VB solution for $\phi^{(1)}$ in (6).

The second-order correction to the ground-state energy is, for static H',

$$E_G^{(2)} = \langle G|H|\phi^{(1)}(0)\rangle \tag{8}$$

and reduces to matrix elements between VB diagrams. Formally, the nth-order corrections suffice for NLO coefficients to order 2n+1. The polarizability tensor $\alpha_{ij}(\omega)$ or the hyperpolarizability $\beta_{ijk}(\omega_1,\omega_2)$ thus reduce to matrix elements of $\phi^{(1)}(\pm\omega)$, $|G\rangle$ and H'. Second harmonic generation or other β_{ijk} processes are exactly given for interacting fermions in finite-dimensional models.

Higher-order corrections also obey linear inhomogeneous equations like (6), with $\phi^{(1)}$ replacing $|G\rangle$ for the $\phi^{(2)}$ correction, and so on. The usual dipole approximation for oscillating electric fields leads to $\phi_{ij}^{(2)}(\omega_1,\omega_2)$ corrections for polarizations i and j. Each component at specified frequencies can be expanded in the VB basis and found exactly from the resulting P_s linear equations. The second hyperpolarizability $\gamma_{ijk\ell}(\omega_1,\omega_2,\omega_3)$ now reduces to the sum of eight matrix elements involving $\phi^{(2)}$ and $\phi^{(1)}$, which is enormously more convenient than eight triple sums over the P_s-1 excited states.

We have recently found[18] exact NLO coefficients for the Pariser-Parr-Pople (PPP) model of linear polyenes up to $N = N_e = 12$ carbons, or up to six double bonds. Hubbard or extended Hubbard models are computationally equivalent. There are some 2.5×10^5 singlets and a symmetry reduction factor of 4 due to electron-hole and inversion (or C_2) symmetry in trans or cis polyenes. The diamagnetic susceptibilities of regular cyclic polyenes and their ions have previously been found[19] to $N = N_e = 14$, with some 2.8×10^6 singlets. NLO coefficients or other perturbations can be found exactly via the many-electron VB basis whenever the ground state $|G\rangle$ is accessible. Resonances involving p photons at low-lying states $E_R - E_G = p\hbar\omega$ also require the DVB solution for $|R\rangle$, as shown below for π-electron contributions to third harmonic generation (THG). Quite generally, then, the response to dynamic perturbations requires a few exact eigenstates in addition to frequency dependent corrections like $\phi^{(1)}(\omega)$ or $\phi^{(2)}(\omega_1,\omega_2)$.

The trans polyenes in Fig. 4 are molecular analogs of polyacetylene, with a half-filled π-band. Simple Hückel, Hubbard or PPP models have no response to fields normal to the conjugation plane, but have five

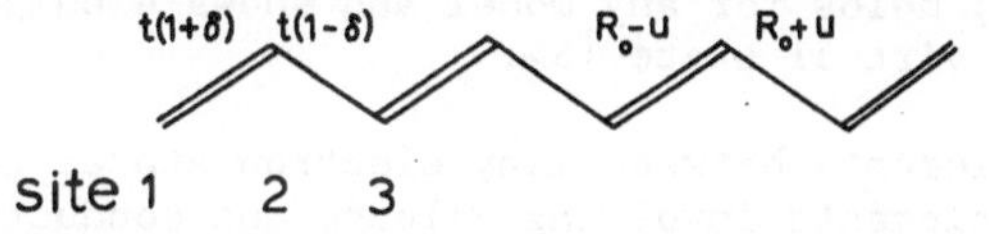

Fig. 4. Schematic representation of a 2n-site trans polyene.

independent γ-coefficients involving xxxx (parallel to chain), xxxy, xxyy, xyyy and yyyy (normal to chain). Bond alternation in |G> is represented by transfer integrals $t(1 \pm \delta)$, with $\delta = 0.07$ and $t = -2.40$ev, for partial double and single bonds. We keep the same transfer integrals for noninteracting π-electrons in Hückel models, for on-site repulsion U in Hubbard models, and for Coulomb interactions V(R) in PPP models fixed at V(0) = U by atomic data.

The THG coefficients at $\hbar\omega = 0.30$ev, which is well below the optical gap E_g, in Table 1 contrast Hückel and PPP coefficients. Except for the largest (xxxx) coefficient, molecular PPP correlations reverse the signs. The far larger magnitudes of the Hückel coefficients reflect uncorrelated motion of virtual electrons and holes induced by the field. The length, or N, dependence of the various coefficients strongly reflects both correlations and field directions.

Hubbard models provide a convenient way to vary systematically the strength of electronic correlations. We show in Fig. 5 various coefficients $\gamma(U/|t|)/\gamma(0)$ at $\hbar\omega = 0.30$ev normalized to the Hückel value. The static limit requires an extrapolation[19] in ω that would produce only small changes in Fig. 5. The xxxy, xxyy, xyyy, and yyyy coefficients all change sign with increasing U and are strongly suppressed.

The sign of the yyyy coefficient at ω=0 can readily be shown to be negative for any one-electron treatment of trans polyenes that retains electron-hole, or alternancy, symmetry between the filled valence band and the empty conduction band.[20] The Hartree-Fock limit for π-electrons is included. Since μ_y goes as $(-1)^p (n_p - 1)$ in the zero-differential-overlap approximation, $\langle r|\mu_y|r'\rangle$ vanishes unless r and r' are conjugate pairs in the valence and conduction bands. Moreover, the transition moment is independent of r. The y-polarized absorption consequently just represents the density of states. The x-polarized transitions, by contrast, are more and more strongly peaked near E_g as N increases. The Hückel entries in Table 1 consequently reflect purely energetic factors for yyyy and both energetics and matrix elements for xxxx.

The ground state |G> is a covalent A state for any of these models. Dipole-allowed transitions are to ionic B states for fields in the xy plane. The standard static (ω=0) coefficients[21] for i=x or y can be rearranged as

$$\gamma_{iiii}(0) \;\alpha\; \sum_S{}' \omega_S^{-1} \Big|\sum_R \langle G|\mu_i|R\rangle\langle R|\mu_i|S\rangle \omega_R^{-1}\Big|^2 - \sum_{TR} |\langle G|\mu_i|R\rangle|^2 |\langle G|\mu_i|T\rangle|^2 \omega_R^{-1}\omega_T^{-2} \qquad (9)$$

The first term is positive definite and involves all virtual states |S> with A symmetry except |G>. The exact excitation energies $\hbar\omega_Q = E_Q - E_G$ are measured from |G>. The second term in (9) is negative definite and involves all contributions when the virtual state |S> is |G>. The general result (9) holds for any model and shows a competition based on the nature of the virtual state |S>.

The matrix elements between many-electron states reduce in Hückel theory to matrix elements involving valence and conduction states that

Table 1. THG coefficients (10^3 atomic units) at $\hbar\omega$=0.30ev for PPP and Hückel models of N-site trans polyenes in Fig. 4.

	N	γ_{xxxx}	γ_{yxxx}	γ_{yyxx}	γ_{yyyx}	γ_{yyyy}
PPP	6	3.188	0.1271	0.1258	0.0389	0.0193
	8	10.04	0.3099	0.2650	0.0644	0.0348
	10	23.56	0.6645	0.4560	0.0906	0.0551
Hückel	6	37.29	-13.06	-4.459	-1.201	-0.1919
	8	243.7	-51.29	-13.98	-3.020	-0.3345
	10	1049	-151.2	-68.08	-6.126	-0.5110

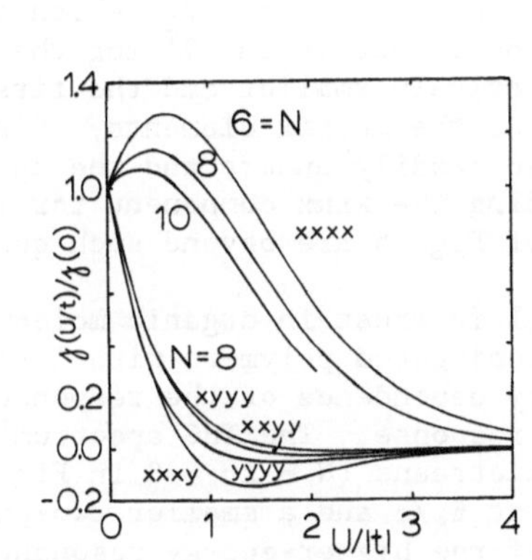

Fig. 5. Ratio of THG coefficients of Hubbard and Hückel models for trans polyenes at $\hbar\omega$=0.30ev; the xxxy, xxyy, xyyy, and yyyy coefficients are for N=8 and change sign with increasing U/|t|.

are readily found for essentially any N. All these matrix elements are identically equal for i=y in (9) due to electron-hole symmetry. The static yyyy coefficient for an N=2n site trans polyene is consequently[19]

$$\gamma_{yyyy}\ (0;\ \text{Hückel}) \quad \alpha \quad -\sum_{r=1}^{n} \omega_r^{-3} \tag{10}$$

with the sum over the n occupied valence-band states. As required physically, (10) is proportional to n for long polyenes whose Hückel gap saturates at $E_g = 2\omega_{min} = 4|t|\delta$. The negative value of $\gamma_{yyyy}(0)$ for trans polyenes is consequently a general result for one-electron models with electron-hole symmetry. The argument holds equally well for the xyyy component in Table 1, where $\langle r|\mu_x|r'\rangle$ is automatically constrained to conjugate pairs but retains an r dependence. Further work is needed to establish the negative sign for xxyy in general. Such considerations explain why De Melo and Silbey's density matrix (one-electron) approach[22] to static γ-coefficients for the PPP model of trans polyenes to N=2n=20 gave negative xxyy, xyyy, and yyyy coefficients, while the xxxy and xxxx coefficients were positive but larger than the PPP values in Table 1. Their results are intermediate between Hückel and exact PPP calculations.

The Hubbard curves for xxxy, xxyy, xyyy, and yyyy in Fig. 5 all change sign with increasing U. The xxxy coefficient vanishes at $U < |t|$, at smaller U for larger N. The xxyy and xyyy coefficients vanish for U between 1 and $2|t|$, now with U increasing with N. The yyyy coefficient vanishes around $U \approx |t|$ and has the least N dependence. A sign change also occurs in the correlated dimer, the interacting two-level system, that has two singlets with A symmetry and one with B symmetry. As has been amply documented both experimentally and theoretically, correlations[15] in Hubbard or PPP models eventually lower covalent A states below the optical gap E_g, which involves the lowest ionic B state. The ω_s denominators in (9) for the covalent A states consequently become relatively smaller and the first term wins regardless of the detailed values of the matrix elements. Since E_g goes as U for large $U/|t|$, we can also readily understand the decreasing magnitude of all γ in Fig. 5, including the xxxx component for large $U/|t|$. However; many detailed aspects of Fig. 5 are beyond such qualitative explanations.

Recent experimental interest in organic molecules with large SHG coefficients[16] and in conjugated polymers with large THG coefficients[17] emphasizes the frequency dependence of the response. The VB procedure in (6) yields the dynamic response. The THG spectrum[18] in Fig. 6 for the PPP model of trans octatetraene (N = 2n = 8 in Fig. 4) clearly shows a three-photon resonance at $E_g/3$ and a smaller two-photon peak involving 2^1A_g. The amplitudes of the higher-energy resonances are so much smaller that even modest lifetime broadening washes them out. The positions of the resonances are marked in Fig. 6. The same lifetimes merely round off the resonances involving $E_g/3$ and 2^1A_g. All these features, including the amplitudes at resonances, are accessible via the VB basis for π-electron models. All excited-state contributions to the ground-state response are included exactly to any order, without ever requiring explicit knowledge of most excited states. We anticipate many applications to interacting fermions in finite models.

SUMMARY

VB diagrams provide a simple representation of many-fermion states from a localized perspective, and as such have provided new insights over many years for many different problems. The diagrams also summarize a wealth of information about electronic structure and provide a convenient basis for quantum cell models that conserve total spin. Projections to smaller subspaces occur naturally on excluding high-energy or physically unfavorable many-electron states. In some cases, for example in excluding long-bonded diagrams relative to those with only short bonds, care must be taken with the overlap. The diagrams faithfully incorporate the physical motivation for identifying the most relevant states and for excluding others. Such choices implicitly define new models such as RVB solids or other projections. The subsequent analysis is rigorous in finite systems, as illustrated above, and there is also progress with infinite systems. For a given Hubbard, PPP or spin model, however, the complete VB analysis must coincide with single-particle states plus configuration interaction to all order. VB methods then provide a computational alternative whose convenience has been decisively increased by bit manipulations and other technical improvements.

ACKNOWLEDGEMENTS

We gratefully acknowledge partial support of this work by the National Science Foundation, DMR-8403819.

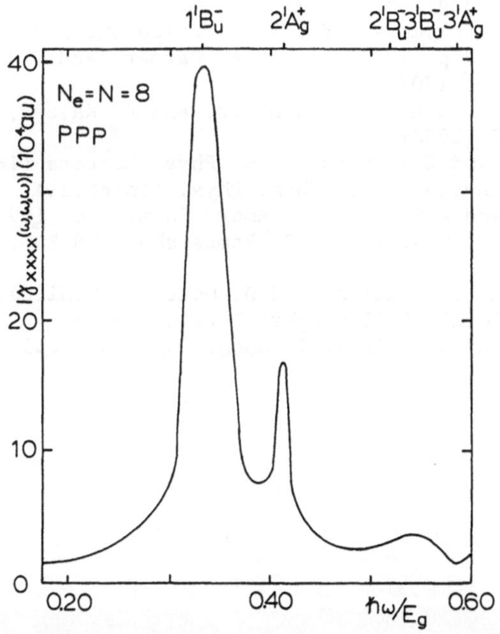

Fig. 6. Magnitude of γ_{xxxx} (ω,ω,ω) for the PPP model of trans octatetraene (N=8) for $0.20 \le \hbar\omega/E_g \le 0.60$ and E_g=4.5608ev. Two-photon resonances to A states and three-photon resonances to B states are broadened by 0.02 and 0.10ev, respectively.

REFERENCES

1. L. Pauling, J. Chem. Phys. 1, 280 (1933); L. Pauling and G. Wheland, ibid, 362 (1933).
2. Z.G. Soos and S. Ramasesha, in Valence Bond Theory and Chemical Structure (eds. D.J. Klein and S. Trinajstic, Elsevier, Amsterdam) in press.
3. L. Pauling, Phys. Rev. 54, 899 (1938).
4. P.W. Anderson, Science V235, 1196 (1987); G. Baskaran, Z. Zou, and P.W. Anderson, Solid State Commun. 63, 973 (1987).
5. I. Affleck, T. Kennedy, E. Lieb and H. Tasaki, Phys. Rev. Lett. 59, 799 (1987); Comm. Math. Phys. 115, 477 (1988); C.K. Majumdar and D.K. Ghosh, J. Math. Phys. 10, 1388 (1969); S.A. Alexander and D.J. Klein, Phys. Rev. B31, 574 (1985).
6. Z.G. Soos and D.J. Klein, in Treatise on Solid State Chemistry, Vol. 3 (ed. N.B. Hannay, Plenum, New York, 1976) p. 679.
7. S. Ramasesha and Z.G. Soos, Int. J. Quant. Chem. 25, 1004 (1984).
8. G. Rumer, Gottingen Nachr. Tech. 377 (1932).
9. S. Mazumdar and Z.G. Soos, Synth. Metals 1, 77 (1979).
10. S.R. Bondeson and Z.G. Soos, Phys. Rev. B22, 1793 (1980); Z.G. Soos, S. Kuwajima and J.E. Mihalick, ibid, B32, 3124 (1985).
11. Z.G. Soos and S. Ramasesha, Phys. Rev. B29, 5410 (1984).
12. S. Ramasesha and Z.G. Soos, J. Chem. Phys. 80, 3279 (1984).
13. K. Chang, I. Affleck, G.W. Hayden, and Z.G. Soos, J. Phys. C: Solid State Phys. (in press).
14. Z.G. Soos and S. Kuwajima, Chem. Phys. Lett. 122, 315 (1985); Z.G. Soos, S. Kuwajima, and R.H. Harding, J. Chem. Phys. 85, 601 (1986).
15. Z.G. Soos and G.W. Hayden, in Electroresponsive Molecular and Polymeric Systems (ed. T.A. Skotheim, Marcel Dekker, New York, 1988) and references therein.
16. S.J. Lalama and A.F. Garito, Phys. Rev A20, 1179 (1979); A.F. Garito, K.Y. Wong and O. Zamani-Khamiri, in Nonlinear and Electroactive Polymers (eds. P.N. Prasad and D.R. Ulrich, Plenum, New York, 1988) p. 13; D. Li, M.A. Ratner, and T.J. Marks, J. Amer. Chem. Soc. 110, 1707 (1988).
17. S. Etemad, G.L. Baker, L. Rothberg, and F. Kajzar, Proc. Mat. Res. Soc. 109, 117 (1988).
18. S. Ramasesha and Z.G. Soos, Chem. Phys. Letters (in press); Z.G. Soos and S. Ramasesha, J. Chem. Phys. (in press).
19. S. Kuwajima and Z.G. Soos, J. Amer. Chem. Soc. 109, 107 (1987).
20. G.W. Hayden, P. McWilliams, S. Ramasesha, and Z.G. Soos, in preparation.
21. D.C. Hanna, M.A. Yuratich, and D. Cotter, "Nonlinear Optics of Free Atoms and Molecules" (Springer, Berlin, 1979).
22. C.P. DeMelo and R. Silbey, J. Chem. Phys. 88, 2558, 2567 (1988).

LANCZOS DIAGONALIZATIONS OF THE 1-D PEIERLS-HUBBARD MODEL

E. Y. Loh, Jr., D. K. Campbell, and J. Tinka Gammel

Theoretical Division and Center for Nonlinear Studies
Los Alamos National Laboratory, Los Alamos, NM 87545, USA

INTRODUCTION

In contrast to the relative simplicity of independent electron theories, models describing *interacting* electrons are in general difficult to treat adequately. In their full complexity, many-electron problems involving N electronic orbitals –each of which can be empty, singly occupied with an electron of either spin, or doubly occupied – require the solution of Hamiltonian matrices of size roughly 4^N by 4^N. For a given problem, symmetries and selection rules (total spin, mirror plane or electron-hole symmetry, etc.) can be used to reduce the size of the matrix, but its growth with N will still be exponential. Often one attempts to avoid this difficulty by use of approximations involving *effective* single-particle methods – such as Hartree-Fock or Fermi liquid theory – which assume that the full problem can be treated in terms of self-consistent (or renormalized) nearly independent quasi-particle states. In Hartree-Fock, for example, one assumes that the full many-body wavefunction can be written as a single Slater determinant of one-particle wave functions. Accordingly, the problem for an N orbital system involves only an N by N matrix, albeit typically with self-consistency constraints on the parameters occurring in the Hamiltonian. Unfortunately, for strongly correlated systems such mean-field approaches can break down entirely; when the physical systems involve electronic motion in reduced dimensions, where strong quantum fluctuations can dominate the physics, this breakdown is particularly likely.

Thus in studies of "interacting electrons in reduced dimensions" one is trapped between the Scylla of exponential growth of the number of states in any exact many-body basis and the Charybdis of the failure of mean-field theories to capture adequately the effects of interactions. In the present article we focus on one technique – the Lanczos method – which, at least in the case of the 1-D Peierls-Hubbard model, appears (to continue the metaphor) to allow us to sail the narrow channel between these two hazards. In contrast to Quantum Monte Carlo methods, which circumvent the exponential growth of states by statistical techniques and importance sampling, the Lanczos approach attacks this problem head-on by diagonalizing the full Hamiltonian. Given the restrictions of present computers, this approach is thus limited to studying finite clusters of roughly 12-14 sites. Fortunately, in one dimension, such clusters are usually sufficient for extracting many of the properties of the infinite system provided that one makes full use of the ability to vary the boundary conditions. In this article we shall apply the Lanczos methodology [1,2] and novel "phase randomization" techniques [3,4] to study the 1-D Peierls-Hubbard model, with particular emphasis on the optical absorption properties, including the spectrum of absorptions as a function of photon energy. Despite the discreteness of the eigenstates in our finite clusters, we are able to obtain optical spectra that, in cases where independent tests can be made, agree well with the known exact results for the infinite system. Thus we feel that this combination of techniques represents an important and viable means of studying many interesting novel materials involving strongly correlated electrons.

THE 1-D PEIERLS-HUBBARD MODEL

Over the past several years the Peierls-Hubbard Hamiltonian [5] has emerged as an important model in which to analyze the competing (or synergetic) effects of electron-phonon (e-p) and electron-electron (e-e) interactions in a variety of quasi–one-dimensional systems, including charge transfers salts, halogen-bridged metallic chains, and conducting polymers. In the context appropriate to describe *trans*-$(CH)_x$ (the specific material on which we shall focus here), the Hamiltonian takes the form

$$H = -\sum_{\ell}(t_0 - \alpha\delta_\ell)B_{\ell,\ell+1} + \frac{K}{2}\sum_{\ell}\delta_\ell^2 + U\sum_{\ell} n_{\ell\uparrow}n_{\ell\downarrow} + V\sum_{\ell} n_\ell n_{\ell+1} \quad . \qquad (1)$$

We consider H defined on a ring of N sites and note that $c^\dagger_{\ell\sigma}(c_{\ell\sigma})$ creates (annihilates) an electron in the Wannier orbital at site ℓ, $n_{\ell\sigma} = c^\dagger_{\ell\sigma}c_{\ell\sigma}$, $n_\ell = n_{\ell\uparrow} + n_{\ell\downarrow}$, $B_{\ell,\ell+1} = \sum_\sigma(c^\dagger_{\ell\,\sigma}c_{\ell+1\,\sigma} + c^\dagger_{\ell+1\,\sigma}c_{\ell\,\sigma})$, t_0 is the hopping integral for the uniform (CH) ionic lattice, α is the electron-phonon coupling describing the modification of the hopping between adjacent sites due to the distortion of the underlying lattice, δ_ℓ is the relative displacement between the (CH) units at sites ℓ and $\ell+1$, and K represents the cost of distorting the lattice. The Coulomb repulsions among electrons are parameterized with U and V, describing the on-site and nearest-neighbor interactions, respectively. In the limit $U = 0 = V$, eqn. (1) reduces to the familiar Su-Schrieffer-Heeger (SSH) model [6] of *trans*-$(CH)_x$. Since one of our primary interests in the present study is the influence of *non-perturbative* e-e interactions, we shall typically investigate the intermediate-coupling regime by considering $U = 4t_0$, which is the full single-particle bandwidth.

For studies of the optical transitions, we shall need the current operator

$$j_{\ell,\ell+1} = i(t_0 - \alpha\delta_\ell)\sum_\sigma(c^\dagger_{\ell+1\,\sigma}c_{\ell\sigma} - c^\dagger_{\ell\sigma}c_{\ell+1\,\sigma}) \quad . \qquad (2)$$

The Fourier transform of $j_{\ell,\ell+1}$ is

$$J_q = \frac{1}{\sqrt{N}}\sum_\ell e^{-iq(\ell+\frac{1}{2})}j_{\ell,\ell+1} \quad . \qquad (3)$$

The optical-absorption coefficient $\alpha(\omega)$ is given by

$$\alpha(\omega) = \frac{1}{\omega}\sum_m |< m|J_q|0 >|^2\delta(\omega - (E_m - E_0)) \quad . \qquad (4)$$

Due to the pathologies — including violation of the f-sum rule and difficulties both with physical interpretation of the geometry [7] and with phase-randomization — that arise when we attempt to define $\alpha(\omega)$ on a closed N-site ring for $q = 0$, we will study the smallest allowed *non*zero momentum value, $q = 2\pi/N$.

THE LANCZOS METHOD

Our very large-scale diagonalization studies are made feasible by the Lanczos method [1,2], which in essence involves expressing the Hamiltonian in a cleverly chosen basis. One starts generating the basis set by normalizing some trial wavefunction, which we typically express in terms of real space occupations. Although there are advantages(including improved convergence) to starting with a good estimate of the ground state, we have usually chosen to start from a *random* wavefunction, subject only to the constraint that $S_z = 0$, where S_z is the z-component of the total spin. Since the Hamiltonian does not alter total spin, we are guaranteed to stay within the $S_z = 0$ manifold. Choosing a random starting wave function does not cost much extra computing time in most cases and, importantly, it statistically prevents one from picking a wavefunction with the wrong symmetry.

The Lanczos procedure is to generate the matrix elements of the Hamiltonian in a basis that is built up from this trial state. The basis is incremented one wavevector at a time by operating on the last basis state with the Hamiltonian and then orthonormalizing the product to all previous basis states. Notice, however, that by construction $H|i>$ will have no overlap with states numbered *higher* than $|i+1>$. Since the Hamiltonian is Hermitian, it must also be true that $H|i>$ will have overlap with no state numbered *lower* than $|i-1>$. In generating a new basis state, then, we need to orthogonalize $H|i>$ against only two states, $|i>$ and $|i-1>$, in order to form $|i+1>$, and hence the recursive procedure for generating basis states does not slow down as the basis set grows. To within roundoff errors, this approach maintains an orthogonal set of vectors.

Within this set of orthonormal vectors, that linear combination of states with the lowest expected energy forms a new estimate of the ground state. One finds this combination by diagonalizing the Hamiltonian, which is tridiagonal in this basis. The basis will probably not be complete: we generally will truncate it when we have run out of computer memory, when the estimate of the ground-state energy stops dropping, or when the residual of a new state, after orthogonalizing to other components, is negligible.

Representing the Hamiltonian in this recursively constructed basis has many advantages. First, the Hamiltonian will be tridiagonal and, so, very easy to study. More importantly, solutions for ground states and optical spectra do not require the generation of complete sets of states. For example, in working on half-filled twelve-site rings, the Hilbert space has nearly 10^6 dimensions. Formally, solutions would require complete bases, involving nearly 10^6 iterations of H acting on some $|\psi>$ — not to mention the companion orthogonalizations and normalizations of these very large wavefunctions. Even if one worked in a subspace of high symmetry, a complete subbasis would entail, at the very least, many tens of thousands of such computations. In practice, however, one can estimate the ground-state wavefunction with high precision (say, ground-state energies to five decimal places) using only a few dozen basis states — even starting from a random trial vector. Computer memory requirements are limited to storing a relatively small number of wavefunctions.

Importantly, the calculations of optical absorption spectra — and hence also gaps, conductivities, and susceptibilities — are easily carried out by generating H in a suitable basis. For the optical absorption, which is our primary present interest, we first calculate the ground state, $|\psi_0>$ using the technique described above. We then generate a new basis, using $J_q|\psi_0>$ as the first state and obtaining subsequent states by the application of H. Once the basis has been generated, the problem of calculating $\alpha(\omega)$ reduces to finding the spectral weight of this first state for the tridiagonal Hamiltonian. In essence, this amounts to determining the spectrum by measuring the moments of H^n and using the cumulant expansion. Again, convergence with the number of basis states is very good: typically, only a few dozen states are required. Even for a low-symmetry, large-space problem such as a soliton on an thirteen-site ring, fewer than 200 states are needed to achieve convergence of the spectrum to the eye. Some features of the spectra, such as the gap energy, can be determined to high precision with far smaller bases.

Given the size of the diagonalizations, the calculations are remarkably fast. To sweep a parameter range — say, a dozen different values of the dimerization — takes about twenty minutes on a CRAY supercomputer for a 12-site system. On the other hand, such a calculation requires several million words of computer memory and both the time and memory requirements grow by a factor of four for each additional site. Thus if one avoids the (slow) process of using external memory, the present limitation in system size is of order 14-15, even using the largest current computers. Though we do not report those measurements here, we have run up to 14-site lattices and could run 15-site chains on a CRAY 2 without incorporation of additional symmetries.

There are two points concerning the method which we should emphasize. First, we could clearly study somewhat (but not substantially) larger-sized systems by incorporating additional symmetries, such as restricting to a given *total* spin value (rather than just a specified value of S_z) or using mirror plane or other spatial symmetries. Indeed, a number of studies (see, e.g., [8,9]) using "valence bond diagrams" (see, e.g. [10]) have examined specific properties of systems similar in size to ours on considerably smaller computers, by making extensive use of symmetries. However, apart from the question of simplicity, the virtue of using a code that does not depend on a particular spatial symmetry is that one can study ionic geometries in which this symmetry is not present. Thus for example, it is straightforward for us to study solitons, or the effects of e-e interactions on the phonon dispersion relation, simply by re-diagonalizing H in a different (and generally non-symmetric) ionic geometry.

Second, as we noted in the introduction, for (small) finite clusters with any given boundary conditions, the lattice-size dependence in numerical calculations is great. This is very familiar in the band theory limit (weak e-e interactions), in which the density of states for a finite-size lattice is a series of delta-function spikes at the k values allowed by the boundary conditions. Since the low-energy physics of the system depends on the states near the Fermi energy, this physics is highly dependent on whether such a spike lies on or near the Fermi surface or not.

A slightly different perspective on this strong finite size dependence provides an important hint as to how to improve the situation. In an infinitely deep but finitely wide square well potential the quantum energy levels are discrete, with spacing in energy determined by the width of the well. To use a well of a given width to mimic a wider well, with more closely space energy levels, we could "play" with the walls to randomize the phase every time the electron "interacts" with the wall. In the currently relevant case of a one-dimensional system (which we will (somewhat sloppily) refer to as a "chain") the quantum phase of an electron travelling on the chain is well preserved if the chain is short. For ring boundary conditions, an electron, travelling around the ring, still "remembers" its phase after completing a full circuit. For open-ended chains, an electron, bouncing off the ends, builds up waves in the bulk. Thus, just as in the case of the energy levels in a square well, the quantum nature of these charge carriers leads to the marked sensitivity on size and boundaries.

For an infinite chain, of course, independent of the boundary conditions, an electron is not going to go away and come back with phase intact. To mimic this behavior on finite chains, one can try as in the case of the square well to "randomize" the phase artificially; techniques to do this would include changing, as examples, a local hopping or an on-site energy or a Coulomb repulsion somewhere on the chain. Results from many such calculations could be averaged together to obtain a "phase-randomized" result. In the next section, we discuss this procedure in more detail and present evidence of its efficacy.

PHASE RANDOMIZATION

Although there has been some limited earlier work on using modified boundary conditions in the context of the Hubbard model [3,4], there is as yet no provably accurate prescription for randomizing the phase effectively for arbitrary U and V. The most obvious scheme is to change the complex phase of one of the hoppings on a closed ring [3,4] . Physically, this corresponds to putting a magnetic flux through the ring and changes the locations of the momentum-space states. For $U = V = 0$, one can show, using arguments analogous to those used to prove Bloch's theorem, that this approach makes it possible to construct *exactly* larger units from smaller units by averaging over many different boundary conditions on the smaller units [11]. In Fig. 1 we demonstrate this result. To be fair, in the case of dimerization versus U the best results, which are essentially indistinguishable from the infinite-chain limit over most of the range of U, come from second-order – i.e., $1/N$ and $1/N^2$ – extrapolations from results from finite chains which are constrained to have uniform dimerization. Nevertheless, Fig. 1 does show the promise of phase randomization. Again, we note that for this bond-phase/magnetic-flux scheme, the $U = 0$ limit is recovered exactly for every lattice size.

In general,however, in the absence of any provable prescription, one must rely on the following intuitive "rules" for choosing a randomization scheme. First, whatever change is made to randomize the phase must, of course, randomize the phase effectively. We will show that this seemingly obvious point is not content-free in our later discussion. Second, the change in the system must be negligible as the lattice size is increased to infinity; for example, if only one bond or site is varied from calculation to calculation, then the effect of such a change is immaterial in the thermodynamic limit. Finally, the behavior for small lattice sizes must be illustrative of the infinite-size limit. Put another way, one must study results on various lattice sizes and still make some sort of extrapolation to the infinite chain.

Naively, it is possible to study the ground state but not to recover information about excited states using Lanczos methods. It is possible, however, to work within a manifold of different symmetry from the ground state to explore gaps of various symmetries. Further, it is also possible to study spectral distributions using the very same basis generation used to produce the ground state. One can produce and then operate on the ground state with, for example, the current operator to study the optical absorption. Then, starting with this new state, one can once again generate a basis in which the Hamiltonian is tridiagonal and study the spectral distribution of the optically-excited state in a truncated subspace. We shall discuss this approach and its results in the next section. Before proceeding, however, we should indicate both a difficulty that the optical absorption studies will pose for the bond-phase/magnetic flux "randomizing" approach introduced above and the resolution of this difficulty by use of another, previously unstudied, phase randomization technique.

For the half-filled band systems we are currently considering, in the strongly correlated limit – $U \rightarrow \infty$ in eqn. (1) – every site in the ground state is singly occupied. The current operator (describing the absorption of a photon) creates excitonic com-

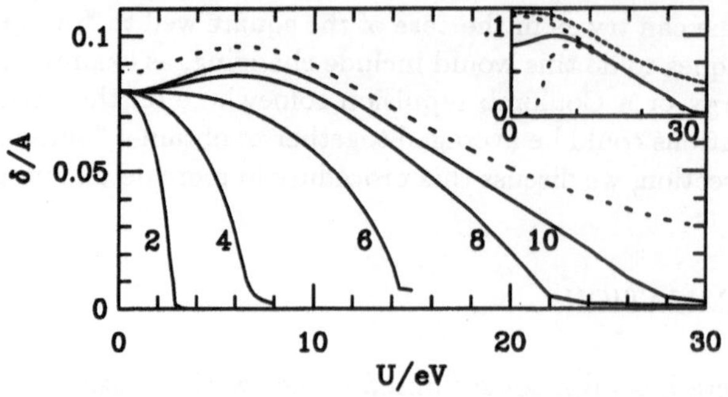

Fig. 1. Self-consistent dimerization (δ) as a function of Hubbard U for several N with 48/N different phase boundary conditions summed (solid lines) compared to the extrapolation to 48 sites from the chain constrained to have uniform dimerization (dashed line). Note the counter-intuitive increase of δ with U can be seen on systems as small as 8-sites. The phase-averaged result on 10 sites is compared in the inset with the 10-site chain (short-dashed) and 10-site ring with periodic or anti-periodic boundary conditions (dashed), where finite-size effects obscure this enhancement.

ponents – a double occupancy and a hole – each of which moves as a free particle over an effectively neutral background. Since these excitonic components are oppositely charged, they pick up opposite phases as they travel around the ring, through which we vary the magnetic flux. Thus, the effect of the flux is eliminated to lowest order and the bond-phase scheme becomes ineffective in the strong-coupling limit. To circumvent this difficulty, we shall instead multiply one of the hoppings by a real number x in the range -l $\leq x \leq 1$. This affects the double occupancy and the hole equally and effectively randomizes electronic phase in the strong-coupling limit. Note that this approach clearly incorporates not only cases of the periodic ($x = 1$) and antiperiodic ($x = -1$) rings but also the case of the open chain ($x = 0$). Thus it is in accord with our comments about "filling in" the k-values in the Brillouin zone. As we shall see in the ensuing section, this "scaled-hopping" phase randomization technique produces results in good agreement with expectations based on both strong- and weak-coupling perturbation theory arguments.

RESULTS FOR OPTICAL ABSORPTION IN 1-D CORRELATED BANDS

The optical spectra of many novel solid-state materials provide crucial insights into both the relative strength of electron-phonon (e-p) versus electron-electron (e-e) interactions and the nature of the charge carriers. The case of the dominant e-p interactions has been widely studied in applications to quasi-one-dimensional conducting polymers. The classic example is the application of the Su-Schrieffer-Heeger (SSH) model [6] to trans-$(CH)_x$. Here, the optical absorption spectrum of the ideally dimerized ground state exhibits a square-root singularity — characteristic of one-dimensional independent-electron systems — at the edge of the optical gap. Further, the nonlinear excitations — "kink" solitons, polarons, and bipolarons — produce clear signatures in the form of midgap absorptions with well-defined relative intensities (see, *e.g.*, [13] for a summary of the these features). Conversely, motivated by potential applications to certain classes of charge-transfer salts [14], the case of dominant e-e interactions has also been studied extensively [7,15,16,17,18], both within the Hartree-Fock approximation [15] and using various strong-coupling approaches [15,17,18] or diagonalization of small systems [7]. In this strong-coupling limit, the spectra — at least for weakly dimerized systems — are typically shifted to higher frequencies and do not exhibit the characteristic square-root singularity at the onset of absorption. Similarly, the characteristic midgap absorptions associated with the localized nonlinear excitations are also shifted substantially — and, in some cases, essentially removed [19] — from their positions in the e-e interaction case.

In view of these substantial qualitative differences, it is hardly surprising that in the continuing debate about the relative strength of e-p versus e-e interactions in conducting polymers, considerable recent attention has focused on the optical absorption spectra. To go beyond the purely e-p models in the the regime of weak e-e interactions, perturbation theory has been invoked [19,20], whereas for strong e-e interactions, leading-order estimates and qualitative arguments have been presented [19]. These analyses leave open the vital question of the characteristics of the optical absorption spectra for intermediate coupling, where the contributions of both the e-p and e-e interactions are expected to be significant and *a priori* neither interaction can be neglected or be treated in perturbation theory. Earlier studies in this intermediate-coupling regime, using valence-bond techniques to obtain (numerically) exact results for the full many-body problem on small finite systems [21,22,23], focused primarily on the value of the "optical gap" — more precisely, on the location of the first optically allowed ^1B_u state — and on the two-photon allowed 2^1A_g state and did not attempt to study directly $\alpha(\omega)$, the optical absorption as a function of frequency. Our present Lanczos approach, coupled with the phase-randomization/boundary-condition-averaging technique allows us for the first time to produce high-resolution spectra on small systems.

To begin our discussion we recall the analytic forms for the optical absorption spectra that are available in a wide range of limiting cases. In the absence of electronic correlations, the Peierls-Hubbard model reduces to the SSH model. Here a number of analytic results are available, particularly if one works in the continuum limit, an approximation valid if, as is the case for *trans*-$(CH)_x$, the optical gap is small compared to the bandwidth and the excitations (*e.g.*, solitons and polarons) extend over many lattice spacings. The uniform dimerization of the lattice caused by the e-p coupling opens up a gap between the valence and conduction bands, while defects in the uniform distortion appear as localized states in the gap. For the purely dimerized case [13],

$$\alpha(\omega) \quad \propto \quad 1 \ / \ \omega^2\sqrt{\omega^2 - \Delta^2} \quad , \tag{5}$$

where Δ is the gap energy. There is a characteristic square-root singularity at the gap edge due to the divergence in the density of states for transitions from the top of the valence band to the bottom of the conduction band. For the lattice model, the primary qualitative difference is that the absorption is cut off at the bandwidth $4t_0$ with another square-root singularity. Provided Δ is much less than to $4t_0$, this difference is small, however, since it occurs at high energy, where all absorptions are attenuated by the $1/\omega^2$ factor due both to vanishing matrix elements and large energy denominators. For a soliton, transitions between the midgap state and one of the two bands give rise to a contribution of the form [13]

$$\alpha_S(\omega) \propto (\omega^2 - (\Delta/2)^2)^{-1/2} \times \mathrm{sech}^2 \frac{\pi}{\Delta}(\omega^2 - (\Delta/2)^2)^{1/2} \quad , \tag{6}$$

while somewhat bleaching the interband spectrum. This midgap absorption has the same singular structure due, again, to the divergent density-of-states at the gap edge.

In the limit of large U (compared to t_0), the optical absorption is expected to be quite different. Since no two electrons, independent of spin, may occupy the same site, at half-filling, the sites are all "jammed": each has exactly one electron and none of the electrons may move. Indeed, in this limit one may prove to leading order in t_0/U the equivalence of the system to a completely filled (and therefore inert) band of *spinless* fermions [24]. If, on the other hand, one electron is removed, then the resultant hole moves as a free particle against this packed background with the same energy $\epsilon_k = -2t_0\cos(k)$ as an electron moving in an otherwise empty band. Meanwhile, although adding an extra electron to the half-filled lattice costs the large energy U, once that energy has been paid, the double occupancy also moves as a free particle, with energy $U + \epsilon_k$. Starting with the fully filled background, then, the optical excitations create double occupancies and holes in pairs at a total energy $U + 2\epsilon_k$ relative to the ground state; hence the range of the absorption should be from $U - 4t_0 < \omega < U + 4t_0$. Longer-range Coulomb repulsions produce an *attractive* interaction between the double occupancy and the hole – the standard excitonic binding mechanism– and thus skew the absorption spectrum toward lower energies. If the system is dimerized, the Brillouin zone is halved, giving rise to $\pm\sqrt{\epsilon_k^2 + (\Delta/2)^2}$ bands for the hole and $U \pm \sqrt{\epsilon_k^2 + (\Delta/2)^2}$ bands for the double occupancy. The gap which opens up in the bands opens a companion gap around the line $\omega = U$ in the optical absorption. These arguments show us where allowable transitions may occur. More detailed calculations[16,17,18], which incorporate information about the density of states and transition matrix elements, are needed to determine the shape of the spectra.

We can obtain further insight by considering a different "solvable" situation. In the limit of large dimerization – which, for purposes of illustration, we assume to be *fixed*, i.e. *not* self-consistently determined as a function of U – the hopping integral on the long bonds vanishes exponentially. For the model Hamiltonian in eqn. (1), this overlap vanishes completely for dimerization δ_0 satisfying $t_0 - \alpha\delta_0 = 0$. In this "decoupled dimer"limit, the chain may be thought of as being composed of independent, two-site systems, each having two electrons with an enhanced hopping parameter $t' = t_0 + \alpha\delta$. Further, each two-site cell has only one optical transition, at energy $\omega = U/2 + \sqrt{(U/2)^2 + (2t')^2}$, which, in strong coupling, is the absorption $\omega = U$ described above.

Armed with these analytic results for the limiting cases of strong and weak coupling and the "decoupled dimers", we investigate the optical absorption using exact diagonalizations. Specifically, we operate in a real-space basis, expressing electronic wavefunctions as products of up-electron and down-electron wavefunctions. As noted above, the only space-reducing symmetry we use is S_z, which given the form of H is in practice equivalent to conservation of the number of, individually, up and down electrons. While incorporation of other symmetries would in practice reduce the size of the Hilbert space — and, consequently, of computer time and memory requirements — such incorporation would come at a great cost in the complexity of the computer code, and, more importantly, in the generality of problems, especially with regard to lattice distortions.

We have measured the optical gap as a function of Coulomb terms U and V and lattice distortion δ. Our computed values for the gap for U and δ agree with Soos and Ramesesha [23] on both chains and rings to stated precision. (Actually, Soos and Ramesesha simply measure the gap to the lowest excited state of the correct symmetry while we measure the lowest-energy absorption. For doubly-even rings, there exists a low-lying state for which the matrix element for excitation from the ground state is zero. Hence, for these rings these two definitions of the gap are not equivalent and produce different results.)

In Fig. 2, we plot the optical gap as a function of U. The gap was calculated on 8-, 10-, and 12-site chains and then extrapolated to the infinite-chain limit using $\Delta = a+b/N+c/N^2$. Chain geometries offer the best finite-size extrapolations since all even-N chains have single-particle states which straddle the fermi energy. In contrast, the structure of the single-particle density of states varies dramatically for even and odd rings. The solid lines in the figure are the infinite-chain extrapolations for dimerized lattice distortions $\delta_\ell = (-)^\ell \delta$ of amplitudes $\delta = 0.00Å$, $0.07Å$, and $0.14Å$. The dashed lines are strong-coupling expressions for the gap to 0th, 1st, and 2nd order. Throughout, we will use the standard SSH parameter values for *trans*-$(CH)_x$: $t_0 = 2.5\text{eV}$, $\alpha = 4.1\text{eV}/Å$, and, though it does not enter here, $K = 21\text{eV}/Å^2$. Numerically differentiating the gap energy with respect to dimerization, we find that the increase of the gap due to dimerization is greatest for intermediate couplings $U \approx 4t_0 = 10\text{eV}$. All the same, even for intermediate couplings, the gap is already dominated by the contribution from e-e interactions rather than by that from the lattice distortion.

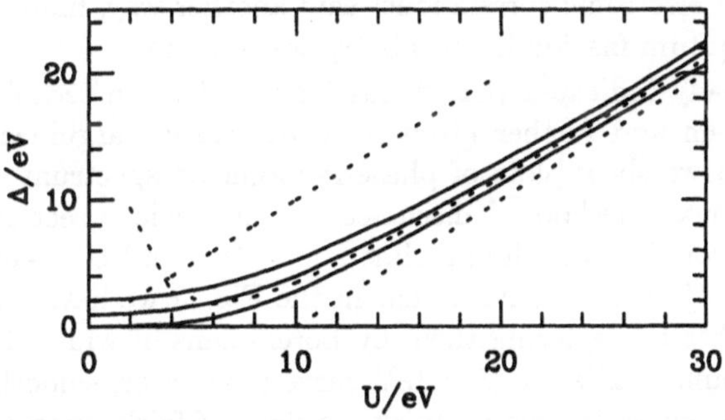

Fig. 2 Optical gap as a function of Hubbard U, extrapolated to infinite chains. The three solid lines are for $\delta = 0.00Å, 0.07Å$, and $0.14Å$, respectively. The dashed lines are the strong-coupling expansions to 0th, 1st, and 2nd order.

We have also measured the optical gap as a function of nearest-neighbor repulsion V, using 8-, 10-, and 12-site chains and extrapolating to infinite chains. While V lowers the gap for the finite chains, the extrapolated result is remarkably flat in strong coupling. To understand this result we recall our earlier remark that the effect of off-site Coulomb interactions is to provide an attraction between the double occupancy (charge e^-) and the hole (charge e^+) created in the optically excited state [17]. This attraction lowers the expectation value of the energy of that state and hence skews the absorption spectrum to lower energy. The current operator has odd parity, however, and thus a weak attraction will not be sufficient to bind the excitonic components (the double occupancy and the hole) in the optically excited state. In particular, if the lowest-energy component of the state is unbound, the double occupancy and the hole will be delocalized and the ability of the electrostatic attraction between them to decrease the gap energy will vanish with chain length. Thus we conclude that the gap will decrease with V as $N \to \infty$ only when the off-site Coulomb force is strong enough to bind the exciton.

The most difficult aspect in measuring spectra is the sparseness of the spectral peaks for finite-size systems. Clearly, there will only be a limited number of transitions for a finite-size system. One might imagine that, since the number of states in the system grows exponentially with the number, N, of spatial sites, the finite sampling of the infinite-ring spectrum presents no great difficulties. In fact, however, the number of significant transitions typically grows only linearly with N. In the limit of weak electronic correlations, for example, the electrons are essentially noninteracting and the optical transitions can be described in terms of the N different single-particle energy levels. Meanwhile, when the Hubbard U is very large, the electrons essentially become noninteracting, spinless fermions, whose transitions, again, are characterized by N different energies. Further, even in intermediate coupling (U = 10eV) and on fairly large (N = 12) rings, one gets only four absorption peaks for periodic boundary conditions for the dimerized lattice. Clearly all this depends on the symmetry of the lattice distortion as well as on the strength of the electronic interactions. In highly asymmetrical cases the *non*-randomized spectra may have reasonable numbers of peaks. However, in these asymmetric cases it is typically difficult to distill the essential physics from the spectra. In sum, we find that for any fixed set of boundary conditions the optical spectra are sampled in a (disappointingly !) sparse manner over the entire range of electronic correlations: said another way, finite-size effects appear in a decidedly quantum fashion in the absorption spectra.

We have already indicated that "phase randomization techniques" can help us resolve this problem and further given some qualitative arguments about what to expect. In Fig. 3 we show plots of phase-randomized spectrum – using the "bond phase/ magnetic flux" method [3,4] discussed in the previous section – for two values of U in the weak coupling coupling regime : U = 1eV and U = 4eV, with the bandwidth taken to be $4t_0$ = *l0*eV. Note that in the Fig. 3 we have already "smoothed" the δ-function spikes by replacing them by Lorentzians of width 0.5eV to produce a continuous spectrum. While the U = 1eV curve (the lower, smoother one) is reasonable, the $U = 4\ eV$ curve is starting to shows signs of fairly sparse structure, despite the Lorentzian smoothing. This indicates the breakdown, anticipated in our earlier discussion, of the "magnetic flux/bond phase" randomization scheme.

To transcend the limitations of this "bond phase/magnetic flux" approach, we have adopted the "scaled hopping" procedure discussed in the previous section and have randomized the phase by averaging the different spectra that are found by varying the magnitude of the "boundary" hopping –i.e., between sites 1 and N – from $-(t_0 - \alpha\delta_N)$ to $+(t_0 - \alpha\delta_N)$ in ten equal steps. Again we smooth the spectrum by replacing the δ-function spikes by Lorentzians of width 0.5eV.

In Fig. 4, we show the phase-randomized spectra produced by this technique for electrons on a 12-site ring in the strongly interacting case with U = 30eV for both a uniform lattice (δ = 0, solid line) and for a strongly distorted lattice ($\delta = 0.14\AA$, dashed line). The gross features of the spectra are predicted by strong-coupling calculations [16,18]. The numerical spectra, however, show substantially more interesting detail. The optical absorption, which to first order in the hopping is rounded, symmetric, and extends from $U - 4t_0$ to $U + 4t_0$, is shifted to higher energies in higher order. It is still centered about $\omega = U$ and so becomes skewed for finite U. In particular, the absorption has a fairly sharp onset at the gap while at high energies $\alpha(\omega)$ vanishes with a finite slope. In the presence of a strong dimerization, the gap is enhanced slightly

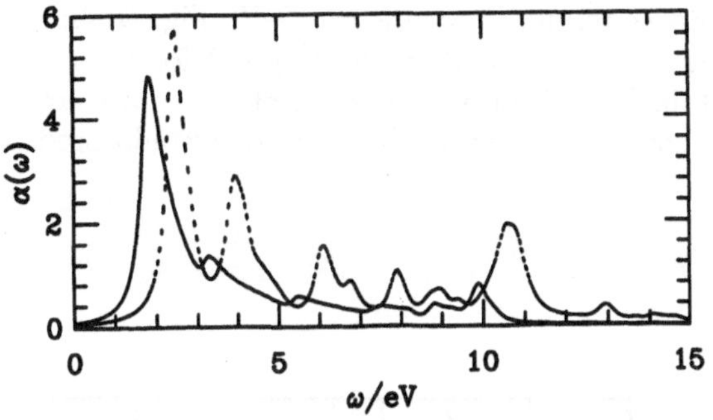

Fig. 3. The optical absorption spectra calculated using the "bond phase/ magnetic flux" phase randomization method for U = 1eV (lower smoother curve) and for U = 4eV (upper, more jagged curve) for the case in which the full band width $4t_0$ = 10eV. Note the "sparseness" of the spectrum in the latter case.

and a strong absorption appears at $U/2 + \sqrt{(U/2)^2 + (2t')^2} = 31.2$eV, as predicted by the decoupled-dimer argument. The peak due to the dimerization is decidedly asymmetric, with, again, a sharp onset on the low-energy side. The undimerized absorption is noticeably depleted on the high-energy side of the absorption peak, corresponding to the gap which opens up in the strong-coupling absorption [18] due to the gaps in the single-particle bands. There may be a companion depletion on the low-energy side, but it appears to disappear with increasing ring size.

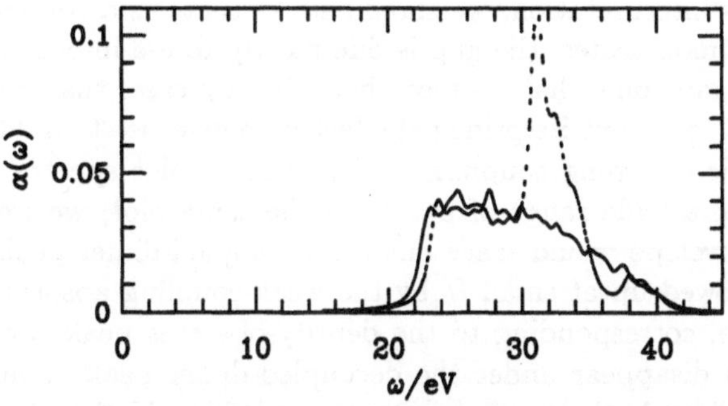

Fig. 4 The phase randomized optical spectra obtained using the "scaled-hopping" approach described in the text for 12-site rings at strong coupling (U = 30eV). The solid and dashed lines are for the uniform and dimerized ($\delta = 0.14\AA$) lattices, respectively.

For intermediate coupling $U = 4t_0 = 10\text{eV}$, Fig. 5 shows phase-randomized spectra for a uniform lattice ($\delta = 0$, solid line) and for a strongly distorted lattice ($\delta = 0.14\text{Å}$, dashed line) of 12 sites. The scale of $\alpha(\omega)$, of course, is greater than in the case of stronger coupling shown in Fig. 4, as one would predicted from the f-sum rule [7,25,26].

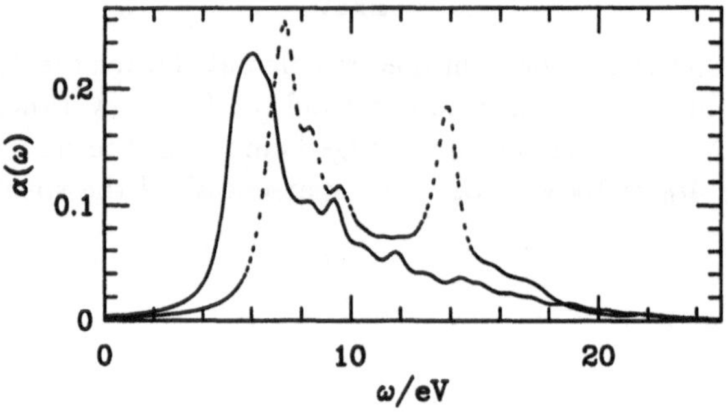

Fig. 5 The phase randomized optical spectra obtained using the "scaled-hopping" approach described in the text for 12-site rings at intermediate coupling ($U = 10\text{eV}$). The solid and dashed lines are for the uniform and dimerized ($\delta = 0.14\text{Å}$) lattices, respectively.

The results in Fig. 5 are truly hybrids of the weak- and strong-coupling absorptions. The undimerized spectrum might be thought of as strong-coupling absorption [16]— as in Fig. 4 — which has been strongly skewed to the low-energy side, as we would expect from the small value of U. A more natural picture, however, might be to associate the gap-edge peak with the square-root singularity from the diverging density-of-states, as in the noninteracting case, despite the fairly substantial value of U. Despite this interpretation, we should stress that, as a comparison of the two curves in Fig. 5 demonstrates, the gap is due mostly to e-e interactions. The dimerization raises the gap somewhat — more here, in any case, than for the $U = 30\text{eV}$ case shown in Fig. 4 — but its principal effect is to give rise to a "decoupled-dimer" peak, characteristic of strong coupling. If we were to plot spectra for strongly distorted lattices over a wide range of U, all on the same plot, we would observe two envelopes. One envelope would trace out the decoupled-dimer peak, pronounced at large U and swallowed up at small U by the weak-coupling absorption. Conversely, the other envelope, corresponding to the density-of-states peak, would dominate at small U, but then disappear under the decoupled-dimer peak at higer energies. In intermediate coupling, both structures are comparable. Notice, finally, that, apart from a slight, familiar depletion on the high-energy side of the peak, in Fig. 5 the decoupled-dimer peak appears not to bleach the undimerized spectrum but simply to increase the integrated weight of the spectrum. This is related by the f-sum rule [7,25,26] to the increase in the magnitude of the delocalization energy as the dimerization gap is opened up.

We have also investigated the effect of a nearest-neighbor V on the absorption spectrum. At half-filling, the effect in strong coupling of V on the ground state is almost exactly to reduce the effective value of U to $U - V$. For optical spectra, V only qualitatively reduces the effective value of U. As noted earlier, V is somewhat ineffectual in reducing the gap energy until it is strong enough to bind the optically excited exciton. On the other hand, the centroid of the spectrum, which we define as $\int \omega^2 \alpha(\omega) / \int \omega \alpha(\omega)$, is quite nearly equal to $U - V$. Hence, V skews the absorption toward lower energies, much as for a reduced U. Turning on V is also like a reduced U in that V suppresses the decoupled-dimer peak.

Finally, Fig. 6 shows the absorption of a neutral and of a charged *soliton* on a 11-site ring. Phase randomization has not been used to produce these spectra in part because the reduced symmetry of the problem gives richer structure but mainly because we have been unable to find a scheme which effectively randomizes electronic phases while locking the "midgap" state at midgap. As a consequence, the figure is not extremely illuminating. It does serve to show, as discussed in a number of articles, that the effect of an intermediate $U = 4t_0 = 10\text{eV}$ is to shift the charged midgap absorption (dashed line) to lower energy while shifting the neutral soliton (solid line) to higher ω, where it blends into the intergap absorption.

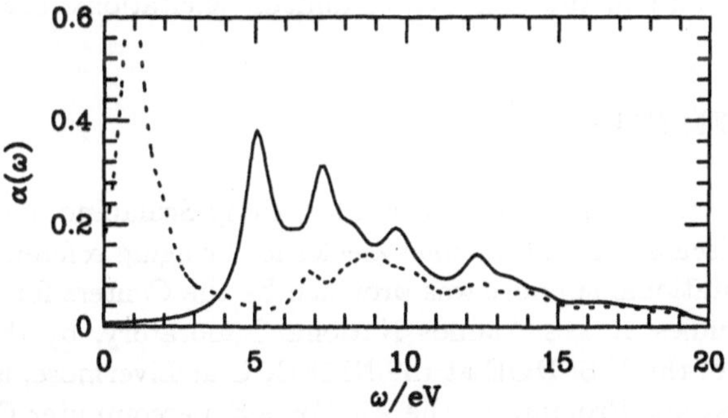

Fig. 6 The optical spectra for 11-site rings at intermediate coupling ($U = 10\text{eV}$). The solid and dashed lines are for the uniform and dimerized ($\delta = 0.14\text{Å}$) lattices, respectively.

CONCLUSIONS AND OPEN ISSUES

We believe that our results establish that the combination of Lanczos exact diagonalizations with phase randomization techniques can be used to obtain insight into many properties of strongly interacting electron systems from studies of (even small) finite-size systems.

In the case of optical absorption spectra, since electron phases interfere quantum mechanically over the entire extent of small rings or chains, small systems are strongly discrete and have very sparse spectra. Phase randomization is essential to produce high-resolution spectra from exact diagonalizations on these small (up to 12-site) rings. Simple results from limits of strong and weak electronic correlations and of large lattice distortions help us to understand results in the difficult intermediate-coupling regime. In the limit of infinite U, half-filled rings show wide, rounded absorption spectra centered about $\omega = U$. As U is decreased or V is turned on, this structure is skewed toward lower energies. In the limit of vanishing U, the skewed absorption turns into a peak at the gap edge as a square-root divergence develops in the density-of-states. For infinite U, dimerization introduces a strong peak, corresponding to the limit of decoupled two-site dimers, in the middle of the absorption. Dimerization opens up gaps in the single-particle bands and so should also open up gaps in the optical absorption around the decoupled-dimer peak. In finite U, the decoupled-dimer peak is suppressed and the gaps in the absorption are somewhat washed out, particularly on the low-energy side of the peak. In weak coupling, there are neutral- and charged-soliton absorptions at midgap. For finite U, the neutral-soliton absorption is shifted to higher energies while the charged-soliton midgap peak moves to lower ω.

Clearly much work remains. In particular, we have begun to perform these diagonalizations in momentum space to gain full control of electronic phase randomization. Our analyses of solitons and other localized defects are very preliminary, and one could indeed argue that since both our phase randomization technique and these intrinsic defects can be viewed as $1/N$ effects, it may prove very difficult to extract more reliable information about solitons than is already available. However, we hope to be able to incorporate our knowledge of important physical features — *e.g.*, midgap states — of one-dimensional electron-phonon models to a greater extent in calculating high-resolution spectra in the presence of difficult excitations, such as solitons and polarons.

ACKNOWLEDGEMENTS

We would like to thank Sumit Mazumdar, Doug Scalapino, and Zoltan Soos for several valuable discussions and Michael Ziegler for bringing references 3 and 4 to our attention. Computational support was provided by the Centers for Materials Science and Nonlinear Studies at Los Alamos National Laboratory, by the Computational Sciences Division of the U.S. DOE at the NMFECC at Livermore, and by the DOE's Supercomputing Access Program at the Air Force Supercomputer Center.

REFERENCES

1 See, e.g., the section on Lanczos diagonalization in S. Pissanetsky, Sparse Matrix Technology, London; Orlando, Academic, 1984.
2 For an early study of the application of the Lanczos method to the pure Hubbard model, see B. Fourcade and G. Spronken, Phys. Rev. B 29 (1984) 5012.
3 R. Julien and R. M. Martin, Phys. Rev. B 26 (1982) 6173.
4 A. M. Oles; G. Tréglia, D. Spanjaard, and R. Julien, Phys. Rev. B 32 2167.
5 Recent studies of the extended Peierls-Hubbard models include: S. Mazumdar and S.N. Dixit, Phys. Rev. Lett., 51 (1983) 292 and Phys. Rev. B, 29 (1984) 1824; J.E. Hirsch, Phys. Rev. Lett., 51 (1983) 296; Z.G. Zoos and S. Ramasesha, *ibid*, 51 (1983) 2374; D. K. Campbell, T. A. DeGrand, and S. Mazumdar, *ibid*, 52 (1984) 1717; J. E. Hirsch and M. Grabowski, *ibid*, 52 (1984) 1713; W.K. Wu and S. Kivelson, Phys. Rev. B, 33 (1986) 8546; S. Kivelson and W.K. Wu, *ibid*, 34 (1986) 5423; S. Kivelson and D.E. Heim, *ibid*, 26 (1982) 4278.
6 W.P. Su, J.R. Schrieffer, and A.J. Heeger, Phys. Rev. Lett., 42 (1979) 1698; Phys. Rev. B, 22 (1980) 2099
7 P. F. Maldague, Phys. Rev. B, 16 (1977) 2437.

8 S. Ramasesha and Z. G. Soos, J. Chem. Phys. 80 (1984) 3278.
9 S. Mazumdar and S. N. Dixit, Synth. Met. 28 (1989) 463.
10 S. Mazumdar and Z. G. Soos, Synth. Met. 1 (1979) 77.
11 D.K. Campbell, J.T. Gammel, and E. Y. Loh, Jr., in preparation.
12 V. Waas, J. Voit, and Hüttner, Synth. Met. 27 (1988) A21.
13 K. Fesser, A. R. Bishop, and D. K. Campbell, Phys. Rev. B, 27 (1983) 4804.
14 J. B. Torrance, B. A. Scott, and D. B. Kaufman, Sol. State Comm. 17 (1875) 1369.
15 K. Kubo, J. Phys. Soc. Japan, 31 (1970) 30. Sol. St. Comm. 17, (1975) 1369.
16 S. K. Lyo and J.-P. Gallinar, J. Phys. C., 10 (1977) 1693.
17 S. K. Lyo, Phys. Rev. B, 18 (1978) 1854.
18 J.-P. Gallinar, J. Phys. C., 12 (1979) L335.
19 D. K. Campbell, D. Baeriswyl, and S. Mazumdar, Physica 143B (1986) 533; Synth. Metals, 17 (1987) 197.
20 W.-K. Wu and S. Kivelson,Phys. Rev. B, 33 (1986) 8546.
21 S. R. Bondeson and Z. G. Soos, Chem. Phys., 44 (1979) 403.
22 S. Mazumdar and Z. G. Soos, Phys. Rev. B, 23 (1981) 2810.
23 Z. G. Soos and S. Ramasesha, Phys. Rev. B, 29 (1984) 5410.
24 J. Bernasconi, M.J.Rice, W.R. Schneider, and S. Strässler, Phys. Rev. B. 12 (1975) 1090.
25 D. Baeriswyl, J. Carmelo, and A. Luther, Phys. Rev. B, 33 (1986) 7247; 34 (1986) 8976(E).
26 S. Kivelson, T.-K. Lee, Y.-R. Lin-Liu, I. Peschel, and L. Yu, Phys. Rev. B, 25 (1982) 4173.

VARIATIONAL WAVE FUNCTIONS FOR CORRELATED LATTICE FERMIONS

Dieter Vollhardt

Institut für Theoretische Physik C
Technische Hochschule Aachen
D-5100 Aachen, Federal Republic of Germany

INTRODUCTION

Variational wave functions (VWFs) are among the very few theoretical tools that allow for straightforward, conceptually simple investigations of interacting many-body systems. They are used to describe correlations among quantum mechanical objects in an approximate, but explicit and physically intuitive,manner. They are particularly valuable in situations where standard perturbation theory fails or is not tractable. The problems in which VWFs have been employed include such diverse examples as rotons in superfluid ^{4}He [1], the plasma state of electrons in metals[2], the quantum liquids ^{3}He and ^{4}He [3-5], nuclear physics[6], superconductivity[7] and the fractional Quantum-Hall-Effect[8]. Of these the BCS-wave function is certainly the most famous. VWFs have also been used to study the possibilities for magnetic order of lattice electrons, i.e. ferromagnetism[9] and antiferromagnetism.[10-11]

At present VWFs receive a renewed interest due to the investigations of heavy fermion systems[12] and high-T_c superconductivity[13], which involve strongly correlated lattice electrons. In the absence of exact solutions of the models used for their description, e.g. the periodic Anderson model, the Hubbard model and its extentions and the Heisenberg model, VWFs provide a valuable and tractable method to study the limit of strong interactions, which is so difficult to investigate otherwise. This is clearly also reflected in the large number of groups around the world (most notably in Danmark, France, Hungary, Italy, Japan, Switzerland, USA and West-Germany) involved in such studies. In the following I will discuss the structure of variational wave functions and their applications to strongly correlated Fermi systems. Special attention is given to recent developments in the variational investigation of the Hubbard model and the antiferromagnetic Heisenberg model in reduced dimensions. Finally, a summary of the exact analytic results available today in d=1 and/or d=∞ for the simplest of these variational wave functions - the Gutzwiller wave function - and generalizations thereof will be presented.

APPLICATION TO QUANTUM LIQUIDS (^3He, ^4He)

The quantum liquids ^3He and ^4He are of particular interest in many-body physics since they represent the prototypes of Fermi (^3He)- and Bose (^4He)-liquids. In contrast to electronic systems they are neutral and extremely pure liquids, which are composed of only a single species. Microscopically ^3He is much more complicated than ^4He due to the antisymmetry of the wave function, as required by the Pauli-principle. On the other hand, Fermi statistics allows at the same time for the formulation of a very powerful phenomenological theory of Fermi liquids, namely Landau's Fermi liquid theory[14], which is not available in a Bose system. It is the existence of a Fermi surface (if it exists), which allows for well-defined elementary excitations, i.e. quasi particles, and which thereby renders possible a complete description of the low-temperature properties of a Fermi liquid (for details see ref. 15).

Microscopic investigations aiming at an understanding of the quasiparticle interaction use a combination of many-body perturbation theory, variational methods and Monte-Carlo calculations (see Krotscheck, ref. 5). These approaches have the advantage of working with systems in the thermodynamic limit and hence are independent of the size of the system. Over the last 30 years they seem to have developed into a science all by itself. The idea behind the variational method is based on the Jastrow-Feenberg approach[3-6] of constructing a VWF $|\psi>$ by applying a correlation operator $\hat{C}$ on a suitably chosen one-particle starting wave function $|\phi_0>$

$$|\psi> = \hat{C}|\phi_0> \tag{1}$$

The operator $\hat{C}$ contains the interaction between the particles in parametrized form, i.e. describes it by variational parameters. It may be expressed as

$$\hat{C} = \exp\left\{ \sum_{i<j} u_2(\mathbf{r}_{ij}) + \sum_{i<j<k} u_3(\mathbf{r}_{ij},\mathbf{r}_{ik}) + \ldots \right\} \tag{2}$$

where $\mathbf{r}_{ij}=\mathbf{r}_i-\mathbf{r}_j$, and the first term in the exponential corresponds to two-body terms, the second to three-body terms etc. These describe the purely geometrical correlations of the short-ranged hard core interaction. In addition, state dependent (e.g. spin-)interactions have to be included. Alternatively, one can write

$$\hat{C} = \left[\prod_{ij} f_2(\mathbf{r}_{ij}) \right] \left[\prod_{ijk} f_3(\mathbf{r}_{ij},\mathbf{r}_{ik}) \right] \left[S \prod_{ij} (1 + \lambda_{ij}\, \sigma_i \sigma_j) \right] \ldots \tag{3}$$

where now such a state dependent interaction has been explicitly included, with S as a symmetry operator. The wave function $|\psi>$ for the interacting system is then constructed by suitably enhancing those configurations in $|\phi>$ which are most important, i.e. which contain the favorable correlations. In this way the wave function may be adjusted. An important task is then to calculate the expectations values of on operator $\hat{0}$

$$<\hat{0}> = \frac{<\psi|\hat{0}|\psi>}{<\psi|\psi>} \tag{4}$$

Most important is $\hat{O} = \hat{H}$, with $\hat{H}$ as the Hamiltonian, such that the ground state energy can be evaluated. To this end cluster expansions and self-consistent summation techniques are employed ("Hypernetted Chain Theories").[3-6] Clearly, the VWF $|\psi\rangle$ should contain sufficient local correlations but must still be tractable.

Among the techniques employing a combination of variational methods and many-body perturbation theory the "Correlated Basis Function" (CBF)-Theory is most successful.[3-6] It is designed to improve the ground state wave function <u>and</u> describe the excited states at the same time, using a non-orthogonal basis $\{|\psi_m\rangle\}$, where

$$|\psi_m\rangle = \hat{C}|\phi_m\rangle \tag{5}$$

is constructed by letting $\hat{C}$ act on an independent particle basis $\{|\phi_m\rangle\}$. The ground state energy

$$E\{\lambda_i\} \equiv \langle \hat{H} \rangle \tag{6}$$

will depend on explicit variational parameters λ_i, with respect to which E has to be minimized

$$\left.\frac{\delta E}{\delta \lambda_i}\right|_{\overline{\lambda}_i} = 0 \tag{7}$$

This leads to an upper bound for the exact ground state energy E_0

$$E_0 \leq E\{\overline{\lambda}_i\} \tag{8}$$

A serious problem is that usually the energy <u>alone</u> is not a sensitive indicator for the quality of the VWF. The advantage of using a VWF is, however, that it is very physical and that it allows for direct improvements guided by physical requirements. In the case of liquid ^{3}He - being an enormously difficult system - one may say that although detailed quantitative agreement with experiment is still lacking, the overall progress has been remarkable and great amount of underlying physics has been understood.[5]

CORRELATED ELECTRONS

In contrast to Helium-atoms electrons are pointlike particles. It is therefore very important to distinguish between models in which electrons are confined to a lattice ("lattice models") and those in which they can move continuously in space ("continuum models"). In the latter case electrons may move in such a way as to avoid each other, which makes the bare interaction effectively much softer. Furthermore, a point-interaction (δ-function potential) has <u>no</u> effect at all in dimensions $d > 2$,[16] i.e. the particles simply do not see each other. Therefore in $d = 3$ a point-like δ-function potential is unphysical. In lattice models this is different. The lattice sites provide definite points where the interaction takes place. Therefore on a lattice even point-like particles have a finite probability for interaction, independent of the dimension of the lattice. In this way strong on-site interactions are possible.

As discussed in the introduction, VWFs designed to describe electronic systems are wellknown from standard mean field theories (e.g. Hartree-Fock) for weakly interacting (quasi-) particles. The most famous is the BCS-wave function,[7] which in its most general form (including anisotropic pairing) is given by [17]

$$|\psi_{BCS}\rangle = \prod_{k} \prod_{\sigma} [u_{k,\sigma\sigma} + \sum_{\sigma'} v_{k,\sigma\sigma'} \hat{a}^{+}_{k\sigma} \hat{a}^{+}_{-k\sigma'}] \tag{9}$$

Here $\hat{a}^{+}_{k\sigma}$ creates a particle with momentum **k** and spin σ and the 2 x 2 spin matrices $\underline{u}_k$, $\underline{v}_k$ are to be determined variationally. On the other hand, the developments in the theory of heavy fermions and high-T_C superconductivity have clearly indicated the need for an explicit inclusion of (i) the lattice, and, (ii) strong, short-range repulsive interactions. In this respect elementary mean field theories and VWFs are clearly insufficient. The generic model which takes such features explicitly into account by concentrating on the bare essentials, is the Hubbard model[18-20]

$$\hat{H}_{Hub} = \hat{H}_{kin} + \hat{H}_{int} \tag{10}$$

with

$$\hat{H}_{kin} = \sum_{ij} \sum_{\sigma} t_{ij} \hat{c}^{+}_{i\sigma} \hat{c}_{j\sigma} \tag{11a}$$

$$= \sum_{k} \sum_{\sigma} \varepsilon_k \hat{n}_{k\sigma} \tag{11b}$$

$$\hat{H}_{int} = U \sum_{i} \hat{n}_{i\downarrow} \hat{n}_{i\uparrow} \tag{12a}$$

$$= U \hat{D} \tag{12b}$$

where $c^{+}_{i\sigma}(c_{i\sigma})$ creates (destroys) a fermion at site i with spin σ and $\hat{n}_{i\sigma}=\hat{c}^{+}_{i\sigma}\hat{c}_{i\sigma}$ is the number operator. The first term, $\hat{H}_{kin}$, is the kinetic energy which describes the hopping of spins from site j to site i; it is thus of purely quantum mechanical origin. Using the momentum representation, this term may equally be expressed as shown in (11b), where ε_k is the energy dispersion and $\hat{n}^{+}_{k\sigma} = \hat{a}^{+}_{k\sigma}\hat{a}_{k\sigma}$ is the momentum distribution operator with $\hat{a}^{+}_{k\sigma}(\hat{a}_{k\sigma})$ as creation (annihilation) operators for particles with momentum **k** and spin σ. The Hubbard interaction $\hat{H}_{int}$, (12), is purely on-site and is essentially classical: it only contributes if two particles with opposite spin occupy the same lattice site. As written in (12b) this term therefore gives the total number of doubly occupied sites, $\hat{D}$, of the system, with $\hat{D} = \Sigma_i \hat{D}_i$ and $\hat{D}_i = \hat{n}_{i\downarrow}\hat{n}_{i\uparrow}$. Clearly, due to the presence of $\hat{H}_{kin}$ the ground state wave function is not an eigenfunction of $\hat{D}$. In the special case of nearest neighbor hopping we can choose $t \equiv 1$, such that

$$\hat{H}_{Hub} = \sum_{\langle ij\rangle} \sum_{\sigma} \hat{c}^{+}_{i\sigma}\hat{c}_{j\sigma} + U\hat{D} \tag{13}$$

Hence, at T = 0 the Hubbard model depends on one parameter U; for U = 0 it describes free lattice fermions and for U = ∞ localized spins.

There is another model of considerable interest - interesting in its own right - namely the Heisenberg model, describing localized Heisenberg spins $\mathbf{S}_i$ with a nearest neighbor coupling

$$\hat{H}_{Heis} = J \sum_{<ij>} \mathbf{S}_i \cdot \mathbf{S}_j \tag{14}$$

Clearly, this overall coupling constant J is arbitrary and may just as well be fixed, e.g. $J \equiv 1$. Therefore this model and its wave function depend on no parameter at all.- For half filling ($n_\uparrow = n_\downarrow = n/2$, $n = 1$) and $U \to \infty$ the Hubbard model and the Heisenberg model are related. This may be shown by second order perturbation theory[21] or, equivalently, by applying a unitary transformation to the Hubbard Hamiltonian (13) and demanding the absence of doubly occupied sites.[22] This leads to

$$\hat{H}_{eff} = -t \sum_{<ij>} \sum_{\sigma} (1-\hat{n}_{i,-\sigma}) \hat{c}^{\dagger}_{i\sigma} \hat{c}_{j\sigma} (1-\hat{n}_{j,-\sigma}) + \frac{4t^2}{U} \sum_{<ij>} (\hat{\mathbf{S}}_i \cdot \hat{\mathbf{S}}_j - \frac{1}{4} \hat{n}_i \hat{n}_j) + \hat{H}_3 \tag{15}$$

Here the first term describes the hopping of holes, i.e. of empty sites, and $\hat{H}_3$ contains processes involving three particles. In the second part one has

$$\mathbf{S}_i = \hat{c}_i^{+} \boldsymbol{\sigma} \hat{c}_i \tag{16}$$

where

$$\hat{c}_i^{+} = \begin{pmatrix} \hat{c}_{i\uparrow}^{+} \\ \hat{c}_{i\downarrow}^{+} \end{pmatrix}, \quad \hat{n}_i = \hat{n}_{i\uparrow} + \hat{n}_{i\downarrow} \tag{17}$$

and $\boldsymbol{\sigma} = \boldsymbol{\tau}/2$ with $\boldsymbol{\tau}$ as the Pauli matrices. For half-filling the first and third term vanish and, apart from a constant ($\hat{n}_i\hat{n}_j \to 1/4$), $\hat{H}_{eff}$ reduces to (14), where now the constant J is uniquely determined by the parameters of the Hubbard model as

$$J = \frac{4t^2}{U} \tag{18}$$

It may seem puzzling at first that at large, but still finite U, the Hubbard model, which always contains hopping of particles and thereby a finite number of doubly occupied sites, is related to the Heisenberg model, describing strictly localized spins. The reason is that for $U \to \infty$ doubly occupied sites in the Hubbard model become <u>virtual</u>, their density decreasing as $d \propto (t/U)^2$, such that their contribution to the ground state energy is of order t^2/U. This process is contained in J. For smaller U the two models are sufficiently different; therefore they require very different kinds of VWFs for their description.

In spite of many decades of work on these models the number of exact results for the thermodynamic limit is exceedingly small- even for ground state properties. Only in d=1 dimension is the ground state energy[23] and the spin correlation function[24] of the Heisenberg model known (and thereby that of the half-filled Hubbard-model for large U and nearest [23] and next-nearest[25] neigbors). Everything else (i.e. the explicit wave functions, other correlation functions, the momentum distribution in the case of the Hubbard model etc.)

is unknown even in d = 1 and the situation is, of course, even much worse in higher dimensions.

In this situation VWFs are one of the very few tools at hand that allow for theoretical investigations of strongly correlated many-body fermion systems. Since the Hubbard model is the generic model describing strong on-site correlations, the question is how to incorporate their effects in VWFs for this model. An important approach, which is widely used at present, goes back to Gutzwiller[9,18], who constructed a VWF (the "Gutzwiller wave function" ≙ GWF) $|\psi_G\rangle$

$$|\psi_G\rangle = g^{\hat{D}}|FS\rangle \tag{19a}$$

$$= \prod_i [1-(1-g)\hat{D}_i]|FS\rangle \tag{19b}$$

where

$$|FS\rangle = \prod_{|k|<k_F} \hat{a}^+_{k\uparrow}\hat{a}^+_{k\downarrow}|0\rangle \tag{19c}$$

is the Fermi sea with $|0\rangle$ as the vacuum, and the operator $g^{\hat{D}}$, with $0\leqslant g\leqslant 1$ as a variational parameter, reduces the amplitude of those spin configurations in $|\phi_0\rangle$ which at a given interaction U contain too many doubly occupied sites. This may also be interpreted in a different way: by noting the projection property $\hat{n}^2_{i\sigma} = \hat{n}_{i\sigma}$, the density fluctuation operator $\hat{N}^2$ is written as

$$\frac{1}{2}\hat{N}^2 = \frac{1}{2}\sum_i(\hat{n}_{i\uparrow}+\hat{n}_{i\downarrow})^2 = \frac{1}{2}\hat{N} + \hat{D} \tag{20}$$

Thus, for fixed particle number

$$g^{\hat{D}} \sim g^{\frac{1}{2}\hat{N}^2} \tag{21}$$

stating that $g^{\hat{D}}$ suppresses density fluctuations in a <u>global</u> sense.

The GWF (19) has precisely the Jastrow-Feenberg form discussed above, with $g^{\hat{D}}$ being the correlation operator which controls the main effect of the strong correlations, i.e. the suppression of double occupancies. This wave function may in principle be refined and adapted to specific physical situations by either generalizing the correlation operator itself, or - as will be discussed below - by putting the details into the starting wave function $|\phi_0\rangle$ such that

$$|\psi\rangle = g^{\hat{D}}|\phi_0\rangle \tag{22}$$

i.e. generalizing (19) to a more detailed form. Examples for VWFs containing $g^{\hat{D}}$ on top of a refined starting wave function $|\phi_0\rangle$ are

$$|\psi_{G,\,PAM}\rangle = g^{\hat{D}} \prod_{k,\sigma} [1 + a(k,\sigma)\,\hat{f}^+_{k\sigma}\hat{c}_{k\sigma}]|FS\rangle \tag{23}$$

as is used for the periodic Anderson model.[26-30] Here $|\phi_0\rangle$ introduces a hybridization between the f-electrons and the conduction electrons, while $g^{\hat{D}}$ controls the number of sites doubly occupied by f-electrons. Another example is

$$|\psi_{G,\,AFHF}\rangle = g^{\hat{D}} \prod_{k,\sigma} [u_{k\sigma}\, \hat{c}^{+}_{k\sigma} + \sigma v_{k,\sigma}\, \hat{c}^{+}_{k-Q,\sigma}]|0\rangle \tag{24}$$

where $|0\rangle$ is the vacuum. This VWF has been investigated in detail by Yokoyama and Shiba[31,32]. Here $|\phi_0\rangle$ is the usual wave function used to obtain the anti-ferromagnetic Hartree-Fock (AFHF) approximation for the Hubbard model[10,11], $u_{k\sigma}, v_{k\sigma}$ are variational functions and $Q = (\pi,\ldots\pi)$ is half the reciprocal lattice vector for the square lattice. The factor $g^{\hat{D}}$ improves the energy for small U which otherwise is non-analytic.

Probably the most wellknown of such VWFs is the "resonating valence bond" (RVB) state suggested by Anderson[33,34] in 1973 as a possible ground state of the antiferromagnetic spin-½ Heisenberg model for a two-dimensional, triangular lattice. It is made up of singlets on sites i and j

$$\hat{b}^{\dagger}_{ij} = \frac{1}{\sqrt{2}} (\hat{c}^{+}_{i\uparrow}\, \hat{c}^{+}_{j\downarrow} - \hat{c}^{+}_{i\downarrow}\, \hat{c}^{+}_{j\uparrow}) \tag{25}$$

which are then superimposed. For a distribution of bond lengths a linear combination has the form

$$\hat{b}^{+} = \sum_{k} a(k)\, \hat{c}^{+}_{k\uparrow}\, \hat{c}^{+}_{k\downarrow} \tag{26}$$

where $\Sigma_k\, a(k) = 0$ to avoid double occupancy. The RVB-state then has the form

$$|\psi_{RVB}\rangle = \lim_{g\to 0} g^{\hat{D}}\, (\hat{b}^{+})^{N/2}|0\rangle \tag{27a}$$

which may equivalently be written as

$$|\psi_{RVB}\rangle = \lim_{g\to 0} g^{\hat{D}}\, \hat{P}_{N/2}\, [u_k + v_k \hat{c}^{+}_{k\uparrow}\, \hat{c}^{+}_{-k\downarrow}]\,|0\rangle \tag{27b}$$

Clearly, $|\psi_{RVB}\rangle$, (27), is a special form of the BCS-wave function (9), namely with exactly N/2 spin-singlet Cooper pairs. Note that the GWF, (19), is <u>also</u> a special case of $|\psi_{RVB}\rangle$, (27a), namely that with $a(k) = 1$ for $k < k_F$, $a(k) = 0$ for $k > k_F$. In particular, following the proposal by Anderson[35], the RVB-wave function is used as a starting point for investigations of high-T_C superconductivity[13] in the two-dimensional Hubbard-model (plus its various extentions), when holes have been introduced to allow for hopping of the "preexisting" Cooper-pairs. So $|\psi_{RVB}\rangle$ generalizes the BCS-wave function to a lattice model where double occupancy has to be eliminated in view of the strong on-site interactions. It is interesting to note that the need for a somewhat similar generalization of $|\psi_{BCS}\rangle$ has been felt in the case of superfluid ^{3}He,[17] where Cooper pairs are made up of ^{3}He-atoms which strongly interact via a hard core interaction. As discussed by Leggett[36] a more realistic Cooper pair wave function than the model form used within BCS-theory should be employed, with the central part being effectively cut out. - The (resonating) valence bond approach is also wellknown in chemical physics where it is used for the representation of molecular structure.[37]

We see that, in principle, one may introduce more and more refinements into the GWF by custom tayloring it to the specific situation involved. However, this also creates problems: the evaluations of expectation values is becoming more and more difficult. Analytic calculations of the strongly correlated regime, if possible at all, only seem to be tractable in d = 1 anyhow (and in d = ∞, see below). On the other hand, numerical techniques - in particular variational Monte Carlo methods - are always limited to finite systems and are only able to determine a finite number of variational parameters. They cannot, in general, determine entire functions which usually appear in VWFs.

RECENT DEVELOPMENTS

In the last few years - and particularly most recently in the wake of high-T_C activities - substantial progress has been made in the refinement of VWFs described above. Here the great advantage of VWFs becomes evident, namely their explicit nature, which allows for direct improvements guided by physical insight. In most cases the evaluation of expectation values in terms of a given VWF has to be done numerically.

Important improvements of the GWF were first discussed by Stollhoff and Fulde[38] in their calculations of correlation energies of moleculs. They used a wave function containing density-density type, i.e. local, correlations between neighboring sites

$$|\psi\rangle = \prod_i \left[\prod_j (1 - \eta_i \sum_{\sigma\sigma'} \hat{n}_{i\sigma} \hat{n}_{j+i,\sigma'}) \right] |FS\rangle \tag{28}$$

where the variational parameters η_i are explicitly site dependent - a generalization suitable for finite systems. Thereby excellent correlation energies can be obtained, making this ansatz a very good approach for molecules etc. - Applying a similar idea to a VWF for the Hubbard model in d=1 Kaplan, Horsch and Fulde[39] generalized the GWF by introducing an additional variational parameter h which controls the number $\langle\hat{Q}\rangle$ of holes and doubly occupied sites sitting on neighboring sites

$$|\psi\rangle = g^{\hat{D}} h^{\hat{Q}} |FS\rangle \tag{29}$$

They had found that in the GWF, in particular for strong correlations, such pairs where not kept close enough together, thereby making it hard for a doubly occupied site to dissociate again. - In the group of Rice at the ETH-Zürich detailed numerical investigations of the effective Hamiltonian (15) have been performed. Gros, Joynt and Rice[22] obtained accurate numerical evaluations of the spin-spin and hole-hole correlation functions in d = 1 in terms of the GWF. These results yielded important insight into the appropriateness of the GWF. Furthermore, Gros[40] and Zhang, Gros, Rice and Shiba[41] tested the RVB-wave function (25) against the antiferromagnetic state $|\psi_{G,AFHF}\rangle$, (24), in d = 2, finding that in the half-filled case both states are very close in energy, lying about 5% above the estimated ground state. This is astonishing in view of the different long-range behavior of these two states. This group also investigated the pairing instabilities of generalized GWFs for less than half-filling, finding that a d-wave RVB state is lowest in energy.[40,43] - Parallel to that, but independently, Yokoyama and Shiba[43] obtained similar, very accurate investigations of the Hubbard model, i.e. of the ground state energy, correlation functions, the momentum distribution etc., in terms of the GWF (partly even in d=2,3). They also investigated[31, 32] $|\psi_{G,AFHF}\rangle$, (24), for the two-dimensional strongly interacting Hubbard model and compared it with $|\psi_{RVB}\rangle$. The conclusions are similar to those described above obtained by Rice's group (for a comprehensive discussion see ref. 45).- As explained earlier, the RVB-state introduced by Anderson[33,35] is a superposition of singlet-pairs without double occupancy of sites. The general properties of this highly interesting VWF have been discussed in detail by Anderson and co-workers[46-49] in the context of high-T_c superconductivity; within a short time an enormous literature has accummulated (see also ref. 13).

Concerning the ground state of the spin-½ two-dimensional antiferromagnetic Heisenberg model one of the most pertinent questions is whether it possesses long range order or not. This problem has been investigated in detail in terms of VWFs. Horsch and van der Linden[50] constructed a VWF which is a generalization of a state previously considered by Hulthen[23], Kasteleijn[51] and Marshall[52], i.e.

$$|\psi_{HKM}\rangle = \sum_{config.} \exp\left[-\eta \sum_{<ij>} (S_i^z S_j^z + \frac{1}{4})\right] \exp(-\eta_c N_c)(-1)^{N_A^+} |s_1, \ldots s_N\rangle \tag{30}$$

where the sum includes all configurations with $s_i = \pm ½$ and a total spin $S^z_{tot}=0$. The first exponential suppresses unfavorable spin pairs, the second exponential represents an additional suppression of clusters with equal spin. Both correlation terms act on a starting wave function whose phase is determined by N_A^+, the number of up-spins on the A-sublattice. When writing the ground state wave function in such a way as a superposition of spin-bases, Marshall[52] had proved that the expansion parameters are indeed positive definite - a fact explicitly exploited in (30). The authors found that very good ground state energies could thereby be obtained and that, in particular, this wave function implies long-range order at T = 0, in contrast to the GWF in d = 2.[50] Excited state were also considered. - A similar generalization of the Kasteleijn-Marshall wave function was introduced by Huse and Elser [53]. Their VWF has the form

$$|\psi\rangle = \sum_n e^{\tilde{H}/2} |\psi_n\rangle \tag{31}$$

where $\tilde{H}$ (which contains the variational parameters and which is related to the Hamiltonian under investigation) is an operator diagonal in the complete orthonormal set $\{|\psi_n\rangle\}: \tilde{H}|\psi_n\rangle=\lambda_n|\psi_n\rangle$. Expectation values of an operator $\hat{O}$ may then easily be evaluated by Monte-Carlo. Applying this method, which allows one to include long-range correlations and non-A-B lattices, to the Heisenberg model in d = 2, these authors also find very good ground state energies and, in particular, long range order both for the square and the triangular lattice.[53] - Most recently Tavan[54] presented a valence bond approach to the Heisenberg-model by representing the lattice in terms of triplet quasiparticles. This leads to a rather simple description of the spin liquid state and its excitations and to a hierarchy of VWFs with very good energies for the ground state in d = 1.

Purely analytic treatments of VWFs for systems in the thermodynamic limit are clearly much more difficult to perform, if possible at all. In particular, the demands on the degree of refinement of the VWF must be much more modest. On the other hand, analytic results are always fixed landmarks which allow for valuable assessments of approximate results. Of all non-mean-field type VWFs for correlated Fermi systems the Gutzwiller wave function (GWF) is the one investigated in greatest detail. Its properties are now very well understood. In spite of its simplicity an exact evaluation of expectation values in terms of the GWF was not possible until recently. Instead, the results of an approximate calculation of the ground state energy, also due to Gutzwiller[9], were used. In this approximation spatial correlations are neglected. It has been shown to be equivalent to an evaluation of matrix elements by determining the <u>classical</u> statistical weights of different spin configurations in the non-interacting wave function.[55] The Gutzwiller approximation yields very simple results in a number of physical situations (the metal-insulator transition,[56,57] normal liquid ^3He [55,58] and heavy fermions[26-30]). It allows to make contact with well established theories like Fermi liquid theory and is therefore generally acknowledged as very "physical", i.e. reasonable. However, to establish the reasonableness of the GWF itself one has to calculate expectation values without making further approximations. Numerical evaluations, first obtained in refs. 22,39,43, have been discussed above. As to analytic approaches Hashimoto[59] used a combination of analytic and numerical methods to obtain improved approximations for one- and higher-dimensional systems in the thermodynamic limit. - Writing the GWF, (19a), as

$$|\psi_G\rangle = e^{-\eta\hat{D}}|FS\rangle \tag{32}$$

where $\eta = \ln(1/g)$, perturbational approaches for $g < 1$ were used by Horsch[60], Baeriswyl and Maki[61] and Baeriswyl, Carmelo and Maki[62] to calculate the ground state energy of the Hubbard model in d = 1 for small interaction strengths U. The momentum distribution n_k was thereby obtained to second order in U.[62] - In the opposite limit, i.e. $U\to\infty$, and for n=1 Baeriswyl[63] suggested the VWF

$$|\psi\rangle = e^{-\lambda\hat{H}_{kin}/t}|\phi_{D=0}\rangle \tag{33}$$

where $|\phi_{D=0}\rangle$ is the wave function containing only those substates of the Fermi sea with no doubly occupied sites and $\hat{H}_{kin}$ is given by (11a) in d = 1. To second order in λ the expectation value of H yields[63]

$$E \approx (2\lambda t+\lambda^2 U)\,\frac{\langle\psi|(\hat{H}_{kin}/t)^2|\psi\rangle}{\langle\psi|\psi\rangle} \tag{34}$$

with an optimal value $\lambda=-t/U$. From second-order degenerate perturbation theory we know that the expectation value in (34) is precisely equivalent to that of the spin-½ antiferromagnetic Heisenberg-model. Hence one recovers (14) with $J=4t^2/U$, the exact result in this order. Unfortunately higher orders are very hard to calculate explicitly from (33).-

It would clearly be desirable to be able to perform exact analytic evaluations in terms of such VWFs for arbitrary interaction strengths and particles densities. As shown recently by Metzner and Vollhardt[64] this is indeed possible in the case of the GWF - at least in d = 1 and d = ∞ dimensions. The evaluation is made tractable by (i) a suitable choice of an expansion parameter and (ii) the introduction of special ("δ-less") contractions, which only involve anticommuting numbers. It is essential to note that D_i is a purely on-site operator, where all sites i are different. Expressing the expectation value $<\hat{O}>$, (4), of an operator $\hat{O}$ in terms of $|\psi_G>$, leads to sums over lattice sites which may be written as sums over different lattice sites only. Thus $<\hat{O}>$ takes the form

$$< \hat{O} > = \sum_{m=0}^{\infty} O_m \, (g^2-1)^m \tag{35}$$

where we see that (g^2-1) is the small parameter for expansions around g=1, not $\eta=\ln(1/g)$ as in (32). The coefficients O_m, which depend explicitly on the density n, may be obtained by standard field theoretic methods where the ensueing contractions only involve anti-commuting numbers as in a Grassmann algebra (a consequence of the lattice sites all being different).[64] In this way it is generally possible to calculate all orders of O_m in d = 1. Concerning the ground state energy E of the Hubbard model this method allowes one to obtain the expectation values of the momentum distribution $<\hat{n}_k>$ and of the Hubbard interaction $<\hat{D}>$ without approximation, leading to a E(n,U) for general n and U after minimization w.r.t. to the variational parameter g.[64] - Evaluating the ground state energy of the Hubbard model in d=1 within the hypernetted chain version of the CBF method discussed above (with $\hat{C}$ in (1) corresponding to $g^{\hat{D}}$) Fantoni, Wang, Tosatti and Yu[65] found good agreement with the exact, analytic result.[64]

The above analytic approach has been used by Gebhard and Vollhardt[66] to calculate correlation functions

$$C_j^{XY} = L^{-1} \sum_i \; <\hat{X}_i \hat{Y}_{i+j}> - <\hat{X}> <\hat{Y}> \tag{36}$$

in terms of $|\psi_G>$ for arbitrary n and U in d=1, where $\hat{X}_i, \hat{Y}_i$ are any one of the four local operators describing the spin $(\hat{S}_i)$, density $(\hat{N}_i)$, empty site $(\hat{H}_i)$ or doubly occupied site $(\hat{D}_i)$ at site i and $\hat{X}=L^{-1}\Sigma_i\hat{X}_i$ etc. Again exact analytic evaluations are possible. In particular, the spin-spin correlation function for n = 1 and U = ∞ is found as

$$C_{j>0}^{ss} = (-1)^j \; \frac{Si(\pi j)}{\pi j} \tag{37a}$$

$$= (-1)^j/2j, \; j \to \infty \tag{37b}$$

The asymptotic behavior implies a logarithmic divergence of the Fourier transform $C^{ss}(q)$ at $q = 2k_F$, i.e. at half a reciprocal lattice vector Q. This antiferromagnetic divergence of $C^{ss}(q=Q)$ is obtained with the GWF in <u>all</u> dimensions d.[66] Comparison with numerical results for C_j^{ss} and the corresponding ground state energy shows that, at least in d=1 and for $U \to \infty$, the GWF is an excellent variational wave function. This conclusion is confirmed by the hole-hole correlations obtained by $|\psi_G\rangle$ for $n < 1$ and $U = \infty$. Recently Haldane[67] and Shastry[68] recognized that the result (38a) is, in fact, the exact result for the continuum Bose gas in d=1, as obtained earlier by Sutherland.[69] Thereby they were able to prove that $|\psi_G\rangle$ is the <u>exact</u> ground state of an $S = ½$ antiferromagnetic Heisenberg model with $1/r^2$ exchange.

Most recently Metzner and the author[70] showed that exact, analytic evaluations of Hubbard-type models within the GWF are not only possible in $d = 1$ but also in $d = \infty$. In fact, in this limit VWFs of increasing refinement are found to be analytically tractable without becoming trivial. Diagrammatic evaluations of expectation values in terms of VWFs of the type shown in (22) are greatly simplified in $d = \infty$, since diagrams are made of lines, which correspond to the one-particle density matrix for the non-interacting system $P^0_{ij,\sigma} = \langle \hat{c}^\dagger_{i\sigma}\hat{c}_{j\sigma}\rangle_0$, which obeys

$$P^0_{ij,\sigma} < O\left[\frac{1}{\sqrt{d}}\right] , \quad i \neq j \tag{38}$$

This implies a collapse of those diagrams in which two vertices i and j are connected by three more separate paths, as is the case for the proper self-energy. Exact evaluations of expectation values in $d = \infty$ have sofar been possible for various starting wave functions $|\phi_0\rangle$. In particular, these results show that certain wellknown approximations used previously are exact for specific VWFs in $d = \infty$. Examples are
(i) $|\psi_G\rangle$, eq. (19): In $d = \infty$ the results of the Gutzwiller approximation for the ground state energy are obtained exactly from $|\psi_G\rangle$.[70] Furthermore, correlation functions may be evaluated without approximation.[71] The results have a RPA-type form with a renormalized coupling and show that classical counting arguments for the calculation of <u>next</u> neighbor correlations in terms of $|\psi_G\rangle$ are correct in $d = \infty$.
(ii) $|\psi_{G,AHF}\rangle$, eq. (24): In $d = \infty$ the results of the slave-boson saddle-point approximation applied by Kotliar and Ruckenstein[72] to a functional integral representation of the Hubbard model are recovered.[70] Here we have constructed the explicit <u>wave function</u>, for which this result is exact in $d=\infty$.
(iii) $|\psi_{G,PAM}\rangle$, eq. (23): In $d = \infty$ the results obtained earlier within a Gutzwiller-type approximation,[26-30] based on classical counting arguments, are recovered.[71]

From here one may go on to even more refined VWFs, which still allow for an analytic treatment. Furthermore, 1/d correction can be incorporated to open the way to finite dimensions $1<d<\infty$. Of course, the $d = \infty$ limit is not restricted to the use of VWFs.[70] The method can also be applied to the general many-body Green's function method. The first results obtained for the Hubbard model [73] and the periodic Anderson model[74] are indeed very encouraging.

CONCLUSION

Variational wave functions (VWFs) are one of the most useful and versatile tools for the investigation of strongly correlated Fermi systems. Recently, and within a very short period of time, they have been able to lead to substantial new insight into the properties of low-dimensional lattice systems, such as the Hubbard-, Heisenberg,- and periodic Anderson model. This clearly indicates that there is still a great future potential in this method. New types of VWFs are being investigated and new techniques for their evaluation are devised. To a large extend this is due to the direct, physical nature of VWFs which readily allow for refinements. In this respect it would, for example, be interesting to apply CBF-methods to the investigation of lattice systems to overcome the limitations imposed by working with small, finite systems. Analytic approaches are no longer limited to $d = 1$ dimensions but are now also tractable in $d = \infty$, which opens a new route to the investigation of finite-dimensional systems via $1/d$ expansions. Altogether it is clear that variational wave functions will continue to be an extremely helpful theoretical tool with excellent propects.

ACKNOWLEDGEMENTS

I am very grateful to my collaborators Dr. P. van Dongen, F. Gebhard and W. Metzner for discussions.

REFERENCES

1. R. P. Feynman and M. Cohen, Phys. Rev. 102 1189 (1956); R. P. Feynman, Statistical Physics (Benjamin, Reading, 1972)
2. T. Gaskell, Proc. Phys. Soc. (London) 72 685 (1958).
3. E. Feenberg, Theory of Quantum Fluids (Academic, New York, 1969).
4. C.- W. Woo, in The Physics of Liquid and Solid Helium, Part I; eds. K.H. Bennemann and J. B. Ketterson (Wiley, New York, 1976), p. 349; C.E. Campbell, in Progress in Liquid Physics; ed. C.A. Croxton (Wiley, New York, 1978), p. 213.
5. E. Krotscheck, in Quantum Fluids and Solids - 1983, AIP Conf. Proc. No. 103, eds. E.D. Adams and G.G. Ihas (AIP, New York, 1983), p. 132.
6. V. R. Pandharipande and R.B. Wiringa, Rev. Mod. Phys. 51 821 (1979).
7. J. Bardeen, L.N. Cooper and J.R. Schrieffer; Phys. Rev. 108 1175 (1957).
8. R. B. Laughlin, Phys. Rev. Lett. 50 1395 (1983).
9. M. C. Gutzwiller; Phys. Rev. A137 1726 (1965).
10. D. R. Penn, Phys. Rev. 142 350 (1966).
11. M. Cyrot, J. Physique 33 125 (1972).
12. see, for example, P.A. Lee, T.M. Rice, J.W. Serene, L.J. Sham and J.W. Wilkins, Comments Cond. Matt. Phys. XII, 99 (1986).
13. See, for example, the Proceedings of the International Conference on High Temperature Superconductors and Materials and Mechanisms of Superconductivity, Interlaken, Switzerland, 1988; eds. J. Müller and J.L. Olsen (North-Holland, Amsterdam, 1988).

14. L. D. Landau, Zh. Eksp. Teor. Fiz. 30 1058 (1956) [Sov. Phys.- JETP 3 920 (1957)]; Zh. Eksp Teor. Fiz. 32 59 (1957)[Sov. Phys.- JETP 5 101 (1957)].
15. D. Pines and P. Nozières; The Theory of Quantum Liquids, vol. I (Benjamin, New York, 1966).
16. C. Herring, in Magnetism, Vol IV, eds. G. Rado and H. Suhl (Academic, New York, 1966).
17. D. Vollhardt and P. Wölfle, The Superfluid Phases of Helium 3, (Taylor and Francis, London); to appear in 1989.
18. M. C. Gutzwiller, Phys. Rev. Lett. 10 159 (1963).
19. J. Hubbard, Proc. R. Soc. London Ser. A 276 238 (1963).
20. J. Kanamori, Prog. Theor. Phys. 30 275 (1963).
21. P. W. Anderson, in Solid State Physics, eds. F. Seitz and D. Turnbull (Academic, New York, 1963) vol. 14, p.99.
22. C. Gros, R. Joynt and T.M. Rice, Phys. Rev. B36 381(1987).
23. H. Bethe, Z. Phys. 71 205 (1931); L. Hulthén, Ark. Astron. Fyz. 26A, 11 (1938).
24. The leading algebraic behavior has been obtained by A. Luther and I. Peschel, Phys. Rev. B12 3908 (1975); the logarithmic corrections were calculated recently by T. Giamarchi and H.J. Schulz, preprint (submitted to Phys. Rev. B).
25. M. Takahashi, J. Phys. C10 1289 (1977).
26. T. M. Rice and K. Ueda, Phys. Rev. Lett 55, 995 (1985); ibid. 55 2093 (E) (1985); Phys. Rev. B34 6420 (1986).
27. C. M. Varma, W. Weber, and L.J. Randall, Phys. Rev. B33 1015 (1985).
28. B.H. Brandow, Phys. Rev. B33 215 (1986).
29. P. Fazekas, J. Magn. Magn. Mater., 63+64 545 (1987);
30. P. Fazekas and B.H. Brandow, Physica Scripta 36 809 (1987).
31. H. Yokoyama and H. Shiba, J.Phys. Soc. Jap. 56 3570(1987);
32. H. Yokoyama and H. Shiba, J. Phys. Soc. Jap. 56 3582 (1987).
33. P.W. Anderson, Mater. Res. Bull. 8 153 (1973).
34. P. Fazekas and P.W Anderson, Phil. Mag. 30 432 (1974).
35. P. W. Anderson, Science 235 1196 (1983).
36. A. J. Leggett, Nature 270 585 (1977), Phys. Rev. Lett. 39 587 (1977).
37. See the paper by Z.G. Soos, S. Ramaseska and G.W. Hayden, appearing in this volume of the NATO Advanced Research Workshop on "Interacting Electrons in Reduced Dimensions".
38. G. Stollhoff and P. Fulde, Z. Phys. 26 257 (1977).
39. T. A. Kaplan, P. Horsch and P. Fulde, Phys. Rev. Lett. 49 889 (1982).
40. C. Gros, Phys. Rev. B38 931 (1988) and Doctoral-Thesis, ETH-Zürich, 1988.
41. F. C. Zhang, C. Gros, T.M. Rice and H. Shiba, Supercond. Sci. Technol. 1 36 (1988).
42. C. Gros, R. Joynt and T.M. Rice, Z. Phys. 68 425 (1987).
43. J. Yokoyama and H. Shiba, J. Phys. Soc. Jap. 56 1490 (1987).
44. H. Yokoyama and H. Shiba, J. Phys. Soc. Jap. 57 No. 7 (1988), in press.

45. H. Shiba in Proceedings of the Workshop on Two-Dimensional, Strongly Correlated Electronic Systems, Beijing, 1988 (Gordon and Breach); in press.
46. P. W. Anderson, G. Baskaran, Z. Zou and T. Hsu, Phys. Rev. Lett. 58 2790 (1987).
47. Z. Zou and P. W. Anderson, Phys. Rev. B37 627 (1988).
48. J. M. Wheatley, T.C. Hsu and P. W. Anderson, Phys. Rev. B37, 5897 (1988).
49. S. Liang, B. Doucot and P.W. Anderson, Phys. Rev. Lett. 61 365 (1980).
50. P. Horsch and W. van der Linden, Z. Phys. 72 181 (1988).
51. P. W. Kasteleijn, Physica 18 104 (1952).
52. W. Marshall, Proc. Roy. Soc. A232 64 (1955).
53. D. A. Huse and V. Elser, Phys. Rev. Lett. 60 2531 (1988).
54. P. Tavan, Z. Phys. B 72 277 (1988).
55. D. Vollhardt, Rev. Mod. Phys. 56 99 (1984).
56. W. F. Brinkman and T. M. Rice, Phys. Rev. B2 4302 (1970).
57. T. M. Rice, Phil. Mag. B35 419 (1985).
58. D. Vollhardt, P. Wölfle and P. W. Anderson, Phys. Rev. B35 6703 (1987).
59. K. Hashimoto, Phys. Rev. B31 7368 (1985).
60. P. Horsch, Phys. Rev. B24 7351 (1981).
61. D. Baeriswyl and K. Maki, Phys. Rev. B31 6633 (1985).
62. D. Baeriswyl, J. Carmelo and K. Maki, Synthetic Metals 21 271 (1987).
63. D. Baeriswyl, in Nonlinearity in Condensed Matter, eds. R. Bishop et al., Springer Series in Solid State Sciences, vol. 69 (Springer, Berlin, 1987), p. 183.
64. W. Metzner and D. Vollhardt, Phys. Rev. Lett. 59 121 (1987), Phys. Rev. B37 7382 (1988).
65. S. Fantoni, X. Wang, E. Tosatti and L.Yu, published in ref. 13, p. 1255.
66. F. Gebhard and D. Vollhardt, Phys. Rev. Lett. 59 1472 (1987); Phys. Rev. B38 6911 (1988).
67. F. D. M. Haldane, Phys. Rev. Lett. 60 635 (1988).
68. B. S. Shastry, Phys. Rev. Lett. 60 639 (1988).
69. B. Sutherland, J. Math. Phys. 12 246, 251 (1971); Phys. Rev. A4 2019 (1971), ibid. A5 1372 (1971).
70. W. Metzner and D. Vollhardt, Phys. Rev. Lett., in press Jan. 16, 1989; (see also accompanying article in this volume).
71. F. Gebhard, P. van Dongen and D. Vollhardt, preprint RWTH/ITP-C 14/88, to be published; F. Gebhard and D. Vollhardt, preprint RWTH/ITP-C 15/88, to be published (see also the accompanying article in this volume).
72. G. Kotliar and A. E. Ruckenstein, Phys. Rev. Lett. 57 1362 (1986).
73. E. Müller-Hartmann, preprint, to be published.
74. H. Schweitzer and G. Czycholl, preprint RWTH/ITP-C 11/88, to be published.

VARIATIONAL APPROACH TO CORRELATION FUNCTIONS AND TO THE PERIODIC ANDERSON MODEL IN INFINITE DIMENSIONS

Florian Gebhard and Dieter Vollhardt

Institut für Theoretische Physik C
Technische Hochschule Aachen
D-5100 Aachen, Federal Republic of Germany

INTRODUCTION

Theoretical investigations of strongly correlated electronic systems in the thermodynamic limit, as described for example by the Hubbard model[1] or the periodic Anderson model,[2] are conventionally plagued by severe computational problems. These increase drastically with dimension and therefore one often resorts to investigations of one-dimensional systems, hoping that they are able to provide essential insight into higher-dimensional systems. Recently Metzner and Vollhardt[3] showed that, for quantum mechanical models of itinerant electrons such as the Hubbard model, the limit of infinite dimensions ($d = \infty$) provides for another useful approach. Above all, in the case of $d = \infty$ drastic calculational simplifications arise, while at the same time correlations remain non-trivial provided a proper scaling of the model with the dimension d is taken into account. –

In this paper we extend the earlier investigations in $d = \infty$ which concentrated on the ground state energy of the Hubbard model.[3] Firstly, correlation functions for Hubbard-type models are evaluated analytically in terms of the Gutzwiller wave function (GWF) without approximation.[4] This shows that non-trivial correlations may indeed exist even in $d = \infty$. Secondly, the periodic Anderson model is investigated in terms of a suitable generalization of the GWF in $d = \infty$. It is shown that the evaluations can be performed analytically, too, and that the results agree with those obtained earlier[5] by employing (semi-)classical counting arguments in the spirit of the socalled "Gutzwiller approximation".[6,7] Hence these (semi-)classical methods are found to yield the correct results in $d = \infty$ for the variational wave functions under consideration.

THE VARIATIONAL APPROACH

Here we consider variational investigations of models, whose interaction is purely on-site, i.e. only takes place between electrons of opposite spin on the same lattice site

$$\hat{H} = U \sum_{\mathbf{i}} \hat{n}_{\mathbf{i}\uparrow} \hat{n}_{\mathbf{i}\downarrow} \tag{1}$$

where $\hat{n}_{\mathbf{i}\sigma}$ is the number operators for a spin at site $\mathbf{i}$ and

$$\hat{D} = \sum_{\mathbf{i}} \hat{D}_{\mathbf{i}} = \sum_{\mathbf{i}} \hat{n}_{\mathbf{i}\uparrow} \hat{n}_{\mathbf{i}\downarrow} \tag{2}$$

Interacting Electrons in Reduced Dimensions
Edited by D. Baeriswyl and D.K. Campbell
Plenum Press, New York

is the number operator for doubly occupied sites. To evaluate expectation values $\langle \hat{O} \rangle = \langle \Psi \mid \hat{O} \mid \Psi \rangle / \langle \Psi \mid \Psi \rangle$ of operators $\hat{O}$ occurring in models containing (1), we employ generalizations of the GWF[6], i.e.

$$\begin{aligned} \mid \Psi \rangle &= g^{\hat{D}} \mid \Phi_0 \rangle & \text{(3a)} \\ &= \prod_{\mathbf{i}} \left[1 - (1-g)\hat{D}_{\mathbf{i}} \right] \mid \Phi_0 \rangle & \text{(3b)} \end{aligned}$$

Here $0 \leq g \leq 1$ is a variational parameter such that the correlation operator $g^{\hat{D}}$ suppresses unfavourable spin configurations within a suitably chosen starting wave function $\mid \Phi_0 \rangle$. For $\mid \Phi_0 \rangle = \mid FS \rangle$, the Fermi sea, (3) reduces to the original GWF[6]

$$\mid \Psi_G \rangle = g^{\hat{D}} \mid FS \rangle \tag{4}$$

A general theoretical approach to the evaluation of expectation values $\langle \hat{O} \rangle$ in terms of $\mid \Psi_G \rangle$ has recently been formulated.[8] It is based on an expansion of $\langle \hat{O} \rangle$ in powers of $(g^2 - 1)$, where the coefficients can be determined diagrammatically using standard field theoretical methods. Their evaluations is greatly simplified by use of special contractions, which only involve anticommuting numbers as in a Grassmann algebra. In this way exact, analytic evaluations are possible in $d = 1$ for arbitrary densities and interaction strengths. Diagrams in this formalism are made up of lines connecting sites $\mathbf{i}$ and $\mathbf{j}$, which correspond to

$$P^0_{\sigma,\mathbf{ij}} = \langle \hat{c}^+_{\mathbf{i}\sigma} \hat{c}_{\mathbf{j}\sigma} \rangle_0 \tag{5}$$

where $\langle \hat{O} \rangle_0 = \langle FS \mid \hat{O} \mid FS \rangle$, and of point vertices. –

In $d = \infty$ $P^0_{\sigma,\mathbf{ij}}$ has the property[3]

$$P^0_{\sigma,\mathbf{ij}} \leq \mathcal{O}\left(\frac{1}{\sqrt{d}}\right) \qquad \mathbf{i} \neq \mathbf{j} \tag{6}$$

which causes a collapse of diagrams in which vertices $\mathbf{i}$ and $\mathbf{j}$ are connected by three or more separate paths. This leads to tremendous simplifications in the diagrammatic calculation of expectation values and thereby allows for exact evaluations of the ground state energy of the Hubbard model in $d = \infty$ in terms of increasingly refined variational wave functions.[3]

CORRELATION FUNCTIONS

There are four independent correlation functions for Hubbard-type models

$$C^{XY}_{\mathbf{f}} = \frac{1}{L} \sum_{\mathbf{h}} \langle \hat{X}_{\mathbf{h}} \hat{Y}_{\mathbf{h+f}} \rangle - \langle \hat{X} \rangle \langle \hat{Y} \rangle \tag{7}$$

where $\langle \hat{X} \rangle = L^{-1} \sum_{\mathbf{i}} \hat{X}_{\mathbf{i}}$, with L as the number of lattice sites, and $\hat{X}_{\mathbf{i}}$, $\hat{Y}_{\mathbf{i}}$ correspond to the spin $\hat{S}^z_{\mathbf{i}}$ $(= \hat{n}_{\mathbf{i}\uparrow} - \hat{n}_{\mathbf{i}\downarrow})$, density $\hat{N}_{\mathbf{i}}$ $(= \hat{n}_{\mathbf{i}\uparrow} + \hat{n}_{\mathbf{i}\downarrow})$, hole $\hat{H}_{\mathbf{i}}$ $(= (1 - \hat{n}_{\mathbf{i}\uparrow})(1 - \hat{n}_{\mathbf{i}\downarrow}))$ or double occupancy $\hat{D}_{\mathbf{i}}$ $(= \hat{n}_{\mathbf{i}\uparrow}\hat{n}_{\mathbf{i}\downarrow})$. These correlation functions were recently calculated by us in terms of the GWF for $d = 1$ in closed form.[9]

Exact evaluations are also possible in $d = \infty$. To be able to exploit the simplifications in the diagrammtic calculation of (7) arising from (6), we have to separate out those parts of the graphs in which vertices are connected by three or more separate paths. This procedure will now be examplified in the case of the spin-spin correlation function $C^{XY}_{\mathbf{j}}$. Noting that the external vertices of diagrams contributing by $C_{\mathbf{j}}^{XY}$ are only connected to two lines each, we introduce a 4-point vertex function $\Gamma_{\sigma,\sigma'}(1,2,3,4)$, where $\mathbf{i} \equiv \mathbf{f}_i$ etc. , such that $C_{\mathbf{j}}^{SS}$ is given by

$$C_{\mathbf{j}}^{SS} = \frac{1}{L}\sum_{\sigma}\sum_{\mathbf{f}}\Big\{-(P_{\sigma;\mathbf{f},\mathbf{f}+\mathbf{j}})^2 + P_{\sigma;\mathbf{f},\bar{1}}P_{\sigma;\mathbf{f},\bar{2}}\Gamma_{\sigma,\sigma}(\bar{1},\bar{2},\bar{3},\bar{4})P_{\sigma;\bar{3},\mathbf{f}+\mathbf{j}}P_{\sigma;\bar{4},\mathbf{f}+\mathbf{j}}$$
$$-P_{\sigma;\mathbf{f},\bar{1}}P_{\sigma;\mathbf{f},\bar{2}}\Gamma_{\sigma,-\sigma}(\bar{1},\bar{2},\bar{3},\bar{4})P_{-\sigma;\bar{3},\mathbf{f}+\mathbf{j}}P_{-\sigma;\bar{4},\mathbf{f}+\mathbf{j}}\Big\}$$
$$+\delta_{\mathbf{j},0}\Big[n - 2\bar{d}\frac{1-g^2}{g^2}\Big] + \frac{1}{L}\sum_{\mathbf{f}}\Big(n_{\mathbf{f},\uparrow} - n_{\mathbf{f}+\mathbf{j},\downarrow}\Big)^2 - \Big[\frac{1}{L}\sum_{\mathbf{f}}\Big(n_{\mathbf{f},\uparrow} - n_{\mathbf{f},\downarrow}\Big)\Big]^2 \tag{8}$$

where $n_\sigma = n_\uparrow + n_\downarrow$ is the particle density and $\bar{d} = \langle\hat{D}\rangle/L$ is the density of doubly occupied sites.[9] Here a full line between two vertices $\mathbf{f}$, $\mathbf{g}$ is taken as a dressed fermion line, which contains all self-energy insertions, and corresponds to a factor $P_{\sigma;\mathbf{f},\mathbf{g}}$.[3] As usual indices with a bar ($\bar{\mathbf{h}}$, $\bar{1} = \bar{\mathbf{f}}_1$, etc.) are summed over. The (two-particle) irreducible 4-point vertex $\Gamma^*_{\sigma,\sigma'}(1,2,3,4)$ is implicitly obtained from $\Gamma_{\sigma,\sigma'}$ as

$$\Gamma_{\sigma,\sigma'}(1,2,3,4) = \Gamma^*_{\sigma,\sigma'}(1,2,3,4) - \Gamma^*_{\sigma,\sigma'}(1,2,\bar{3},\bar{4})P_{\sigma';\bar{3},\bar{1}}P_{\sigma';\bar{4},\bar{2}}\Gamma_{\sigma',\sigma'}(\bar{1},\bar{2},3,4)$$
$$- \Gamma^*_{\sigma,-\sigma'}(1,2,\bar{3},\bar{4})P_{-\sigma';\bar{3},\bar{1}}P_{-\sigma';\bar{4},\bar{2}}\Gamma_{-\sigma',\sigma'}(\bar{1},\bar{2},3,4) \tag{9}$$

Clearly, all vertices contained in Γ^* are now connected with each other by three or more separate paths and hence collapse in $d = \infty$, such that

$$\Gamma^*_{\sigma,\sigma'}(1,2,3,4) \equiv \delta_{1,2}\delta_{1,3}\delta_{1,4}\tilde{\Gamma}^*_{\sigma,\sigma'}(1) \tag{10}$$

which defines a local quantity $\tilde{\Gamma}^*_{\sigma,\sigma'}(\mathbf{j})$. This implies $\Gamma_{\sigma,\sigma}(1,2,3,4) \equiv \delta_{1,2}\delta_{3,4}\tilde{\Gamma}_{\sigma,\sigma'}(1,3)$ for the vertex-function itself. The quantities $\tilde{\Gamma}^*_{\sigma,\sigma'}(\mathbf{j})$ are easily determined from the value of the correlation function at $\mathbf{j} = 0$.

The above formalism is valid for arbitrary starting wave functions $|\ \Phi_0\rangle$ in (3). In particular, for the GWF with $|\ \Phi_0\rangle = |\ FS\rangle$ and $n_\uparrow = n_\downarrow = n/2$ all calculations are greatly simplified due to translational invariance and spin-symmetry. For example $\tilde{\Gamma}^*_{\sigma,\sigma'}(\mathbf{f})$ reduce to mere constants ($\tilde{\Gamma}^*_{\sigma,\sigma} = \tilde{\Gamma}^*_{-\sigma,-\sigma} \equiv \Gamma_1$, $\tilde{\Gamma}^*_{\sigma,-\sigma} = \tilde{\Gamma}^*_{-\sigma,\sigma} \equiv \Gamma_2$), such that the equations for $\Gamma_{\sigma,\sigma'}$, (9), are easy to solve. The momentum-dependent spin-spin correlation function is then found as[4]

$$C^{SS}(\mathbf{k},n,g) = \frac{C_0(\mathbf{k},n)}{1 - V_S(g,n)C_0(\mathbf{k},n)} \tag{11}$$

where $C_0(\mathbf{k},n) = C^{SS}(\mathbf{k},n,1)$ is the result for the noninteracting case ($g = 1$) and

$$V_S = \frac{1}{n-2\bar{d}} - \frac{1}{n-2\bar{d}_0} \tag{12}$$

The mean double occupancy $\bar{d}$ has been calculated earlier as[3]

$$\bar{d}(g,n_\uparrow,n_\downarrow) = \frac{1 + n(g^2-1) - \sqrt{1-(1-g^2)[n(2-n)+g^2(n^2-4\bar{d}_0)]}}{2(g^2-1)} \tag{13}$$

where $\bar{d}_0 = n_\uparrow n_\downarrow$ and $n = n_\uparrow + n_\downarrow$. It is interesting to note that in $d = \infty$ the correlation function C^{SS} does not reduce to some trivial result, as might have been expected. The RPA-like structure of (11) is due to the summation of dressed bubble-diagrams, which are the only ones that survive in $d = \infty$. (This finding is expected to be true even for the exact solution of the Hubbard model in $d = \infty$). The quantity V_S corresponds to a renormalized coupling constant. — Other correlation functions, such as C^{NN}, C^{DN}, C^{DD}, are obtained by the same procedure and are found to have the same structure as that of C^{SS}.[4] –

The Fourier transformation of (11) into position space ($C^{SS}(\mathbf{k}) \to C^{SS}_{\mathbf{j}}$) is a delicate problem in view of the integration involved and the two limits $L \to \infty, d \to \infty$, which do not commute. Nonetheless, $C_0(\mathbf{k}, n)$ can be calculated explicitly in $d = \infty$ [4] and the Fourier retransformation is achieved by noting that for almost all $\mathbf{k}$–values $C_0(\mathbf{k}, n)$ assumes the average value $\langle C_0(\mathbf{k}, n)\rangle_{\mathbf{k}} = n(1 - n/2)$, such that (11) can be expanded in the quantity $C_0(\mathbf{k}, n) - \langle C_0(\mathbf{k}, n)\rangle_{\mathbf{k}}$, which is small for almost all $\mathbf{k}$. In zeroth order one obtains $C^{SS}_{\mathbf{j}=0} = n - 2\bar{d}$, while the nearest-neighbor contribution to $\mathbf{j} = 0$ is obtained by the first order term as

$$C^{SS}_{\mathbf{j}=n.n.}(g, n) = \frac{C^{SS}_{\mathbf{j}=n.n.}(1, n)}{[1 - V_S(g, n)\langle C_0(\mathbf{k}, n)\rangle_{\mathbf{k}}]^2} \tag{14a}$$

$$= \left(\frac{n - 2\bar{d}}{n - 2\bar{d}_0}\right)^2 C^{SS}_{\mathbf{j}=n.n.}(1, n) \tag{14b}$$

This shows that, at least for nearest-neighbors, the correlation function for the interacting case is proportional to that for the non-interacting case. This result had been derived earlier[5] by (semi-)classical counting arguments in the spirit of the Gutzwiller approximation[6,7]. Here we are able to show that for the GWF this result is exact in high dimensions ($d = \infty$). In fact, numerical investigations[5] indicate that (14) is well obeyed even in $d = 2$. Hence we learn that results for high dimensions are able to give essential insight into the behavior of much lower-dimensional systems.

PERIODIC ANDERSON MODEL (PAM)

This model is generally considered appropriate for a description of the physics of heavy fermion systems.[2] It considers almost localized f-shell electrons on a periodic lattice (with lattice sites essentially singly occupied due to the strong intraatomic Coulomb repulsion), in a sea of conduction electrons (c) with a finite hybridization between c's and f's: In the absence of orbital degeneracy it has the form

$$\hat{H}_{PAM} = \sum_{\mathbf{k}\sigma} \epsilon_c(\mathbf{k}) \hat{c}^+_{\mathbf{k}\sigma} \hat{c}_{\mathbf{k}\sigma} + \sum_{\mathbf{k}\sigma} \epsilon_f(\mathbf{k}) \hat{f}^+_{\mathbf{k}\sigma} \hat{f}_{\mathbf{k}\sigma} + U \sum_{\mathbf{i}} \hat{n}^f_{\mathbf{i}\uparrow} \hat{n}^f_{\mathbf{i}\downarrow} - \sum_{\mathbf{k}\sigma} \left[V_{\mathbf{k}} \hat{f}^+_{\mathbf{k}\sigma} \hat{c}_{\mathbf{k}\sigma} + hc\right] \tag{15}$$

The first two terms describe the band behavior of c- and f-electrons, the third is the Hubbard term (1) for the f's (being the only interaction term in $\hat{H}_{PAM}$) and the last term produces the quantum mechanical mixing. For $U = 0$ (15) is easily diagonalized and yields for less than half filling ($n_\uparrow + n_\downarrow \leq 2$)[10,11]

$$|\Psi\{a^0_\sigma(\mathbf{k})\}\rangle = \prod_\sigma \prod_{|\mathbf{k}| \leq k^\sigma_F} \left[1 + a^0_\sigma(\mathbf{k}) \hat{f}^+_{\mathbf{k}\sigma} \hat{c}_{\mathbf{k}\sigma}\right] |cFS\rangle \tag{16}$$

where $|cFS\rangle$ is the Fermi sea of c-electrons and

$$a^0_\sigma(V_{\mathbf{k}}, \epsilon_f(\mathbf{k})) \equiv a^0_\sigma(\mathbf{k}) = \frac{2V_{\mathbf{k}}}{\epsilon_f(\mathbf{k}) - \epsilon_c(\mathbf{k}) + \left[(\epsilon_f(\mathbf{k}) - \epsilon_c(\mathbf{k}))^2 + 4V^2_{\mathbf{k}}\right]^{1/2}} \tag{17}$$

For a variational investigation of the interacting problem, a wave function of the form given by (3) has been suggested[10,11,12],

$$| \Psi_{PAM}\rangle = g^{\hat{D}^f} \mid \Psi\{a_\sigma(\mathbf{k})\}\rangle \quad (18)$$

where now $\hat{D} \equiv \hat{D}^f = \sum_\mathbf{i} \hat{n}^f_{\mathbf{i}\uparrow}\hat{n}^f_{\mathbf{i}\downarrow}$ and $a^0_\sigma(\mathbf{k})$ has been replaced by some function $a_\sigma(\mathbf{k})$, which has to be determined variationally.

To evaluate expectation values in terms of $| \Psi_{PAM}\rangle$ it is essential to note that the Gutzwiller correlation operator acts only on the f-electrons. Therefore it is possible to express all expectation values in terms of the f-electron self-energy S_σ, which itself depends only on the properties of the f-electrons. In this situation one may simply take over the results obtained earlier within the ordinary GWF[3] for the Hubbard model. In particular, construction of the proper self-energy S^*_σ via the Dyson equation leads to a collapse of diagrams in $d = \infty$, such that S^*_σ becomes site-diagonal. The remaining local quantity is obtained by summing the skeleton expansion, which can be performed exactly in $d = \infty$. The results for the Hubbard model, using the GWF, (4), and for the PAM in terms of $| \Psi_{PAM}\rangle$ are then very similar. For example, the relation between $\overline{d}^f$, the average double occupancy for f-electrons, and the variational parameter g is given by (13) with $\overline{d} \rightarrow \overline{d}^f$ and $n_\sigma \rightarrow n^f_\sigma$. The main differences are that the number of f-electrons, n^f_σ, is not conserved, i.e. depends on g and $a_\sigma(\mathbf{k})$, and that fermion lines $P^0_{\sigma,\mathbf{ij}}$, (5), for the f's have a more complicated structure. In spite of this the expectation value $\langle \hat{H}_{PAM}\rangle$ can be calculated exactly in $d = \infty$. Thereby we derive the results obtained earlier by (semi-)classical counting arguments à la Gutzwiller[10-13] (for the most complete treatment see ref. 14), which are therefore seen to be strictly correct in high dimensions. In $d = \infty$ the correlation operator in (18) leads to a simple renormalization of one-particle quantities, e.g.

$$\begin{aligned} V_\mathbf{k} &\rightarrow \tilde{V}_\mathbf{k} = \sqrt{q_\sigma} V_\mathbf{k} \\ \epsilon_f(\mathbf{k}) &\rightarrow \tilde{\epsilon}_f(\mathbf{k}) = q_\sigma \epsilon_f(\mathbf{k}) + shift \\ a_\sigma(\mathbf{k}) &\rightarrow \tilde{a}_\sigma(\mathbf{k}) = \sqrt{q_\sigma} a^0_\sigma(\tilde{V}_\mathbf{k}, \tilde{\epsilon}_f(\mathbf{k})) \end{aligned} \quad (19)$$

where q_σ is the correlation-induced reduction in the kinetic energy (the "discontinuity" of the momentum distribution in the case of the Hubbard model), which for $U = \infty$ is given by

$$q_\sigma = \frac{1 - (n^f_\uparrow + n^f_\downarrow)}{1 - n^f_\sigma} \quad (20)$$

This shows that one may indeed construct an effective Hamiltonian defined in terms of the renormalized quantities $\tilde{V}_\mathbf{k}, \tilde{\epsilon}_f(\mathbf{k})$, as anticipated by Rice and Ueda[10]. The results derived above for $d = \infty$ also agree with those obtained by a slave boson approach to the PAM via a path integral method.[15]

CONCLUSION

Using variational wave functions explicit analytic results have been obtained for correlation functions and the periodic Anderson model in the limit of infinite dimensions. Correlations are found to remain non-trivial. Semi-classical counting methods, frequently used for an approximate evaluation of expectation values in lower dimensional systems, were shown to become exact in $d = \infty$.

ACKNOWLEDGEMENTS

We thank Dr. P. van Dongen and W. Metzner for useful discussions.

REFERENCES

1. M. C. Gutzwiller, Phys. Rev. Lett. **10** (1963); J. Hubbard, Proc. R. Soc. London **A276** 238 (1963); J. Kanamori, Prog. Theor. Phys. **30** 275 (1973)
2. See, for example, P. A. Lee, T. M. Rice, J. W. Serene, L. J. Sham and J. W. Wilkins, Comments Cond. Matt. Phys. XII, 99 (1986)
3. W. Metzner and D. Vollhardt, Phys. Rev. Lett. , in press (January 16, 1989); see also the accompanying article by these authors in this volume of the NATO proceedings
4. For details, see F. Gebhard, P. van Dongen and D. Vollhardt, preprint RWTH/ITP-C 14/88 , to be published
5. F. C. Zhang, C. Gros, T. M. Rice and H. Shiba; Supercond. Science and Techn. **1** 36 (1988)
6. M. C. Gutzwiller, Phys. Rev. **A137** 1276 (1965)
7. D. Vollhardt, Rev. Mod. Phys. **56** 99 (1984)
8. W. Metzner and D. Vollhardt, Phys. Rev. Lett. **59** 121 (1987); Phys. Rev. **B37** 7382 (1988)
9. F. Gebhard and D. Vollhardt, Phys. Rev. Lett. **59** 1472 (1987); Phys. Rev. **B38** 6911 (1988)
10. T. M. Rice, K. Ueda, Phys. Rev. Lett. **55** 995 (1985); Phys. Rev. **B34** 6420 (1986)
11. B. H. Brandow, Phys. Rev. **B33** 215 (1986)
12. C. M. Varma, W. Weber and L. J. Randall, Phys. Rev. **B33** 1015
13. P. Fazekas and B. H. Brandow, Physica Scripta **36** 809 (1987)
14. V. Z. Vulović and E. Abrahams, Phys. Rev. **B36** 2614 (1987)
15. G. Kotliar and A. E. Ruckenstein, Phys. Rev. Lett. **57** 1362 (1986)

THE HUBBARD MODEL IN INFINITE DIMENSIONS

Walter Metzner and Dieter Vollhardt

Institut für Theoretische Physik C
Technische Hochschule Aachen
D-5100 Aachen, Federal Republic of Germany

INTRODUCTION

Investigations of classical spin systems in the limit of infinite dimensions ($d=\infty$) are well appreciated as a useful concept in statistical physics.[1] An exact solution in $d=\infty$ often provides the correct qualitative behavior of solutions in lower dimensions. In particular, it may serve as a starting point for systematic investigations of finite dimensional systems via 1/d-expansions. Exact solutions for systems with localized spins in $d=\infty$ are generally related to mean field solutions, since fluctuations become unimportant in infinite dimensions. In the case of fermionic systems with mobile degrees of freedom, such as the Hubbard model,[2] mean field solutions are known too, but explicit results for infinite dimensions have so far not been obtained. In fact, exact results for the Hubbard model are only available in $d=1$.[3] Therefore many fundamental questions are still open, for example, whether such quantum mechanical systems become "trivial" in $d=\infty$ and whether their exact solution corresponds to some type of mean field solution. - To answer at least parts of these questions we will first discuss a scaling of the Hubbard model such that in the limit $d=\infty$ a non-trivial model results. We will then investigate this model by calculating the correlation energy in weak coupling and by applying variational techniques.[4]

SCALING OF THE HUBBARD MODEL

The Hubbard model is given by[2]

$$\hat{H} = \hat{H}_{kin} + U \sum_i \hat{n}_{i\uparrow}\hat{n}_{i\downarrow} \tag{1a}$$

$$\hat{H}_{kin} = \sum_\sigma \sum_{i,j} t_{ij}\, \hat{c}^\dagger_{i\sigma}\, \hat{c}_{j\sigma} = \sum_{k,\sigma} \varepsilon_k \hat{n}_{k\sigma} \tag{1b}$$

where $\hat{H}_{kin}$ is the kinetic energy operator written in position space and momentum space, respectively. For nearest neighbor

Interacting Electrons in Reduced Dimensions
Edited by D. Baeriswyl and D.K. Campbell
Plenum Press, New York

hopping on a hypercubic lattice in d dimensions the energy dispersion has the form

$$\varepsilon_k = -2t \sum_{n=1}^{d} \cos k_n \tag{2}$$

where $\mathbf{k} = (k_1, \ldots k_d)$. In d=∞ the central limit theorem yields the corresponding density of states (DOS) as

$$N(E) = \frac{\exp[-(E/2t\sqrt{d})^2]}{2t\sqrt{\pi d}} \tag{3}$$

In d=∞ N(E) is seen to be a Gaussian without van-Hove singularities. To obtain a finite DOS[5] a scaling $t \to t^* = t/\sqrt{2d}$ is required, such that $N(E) \to N^*(E) = \exp(-E^2/2)/\sqrt{2\pi}$, where $t \equiv 1$. The average kinetic energy is then also finite, i.e. $\bar{\varepsilon}_0(n) = -2N^*(E_F)$. In this way one obtains a non-trivial model with competing potential and kinetic energy which are both of the same order of magnitude even in d=∞. The naive scaling $t \to t/d$ would imply $\bar{\varepsilon}_0 = 0$, making (1) a localized model ab initio.

CORRELATION ENERGY

For small U the correlation energy E_c of the Hubbard model can be determined by Goldstone perturbation theory.[6] The second order contribution to the ground state energy in d dimensions for an equal number of up- and down spins ($n_\uparrow = n_\downarrow = n/2$) is effectively given by a 3d-dimensional integral over the Brillouin zone

$$E_2 = \frac{LU^2}{(2\pi)^{3d}} \int \prod_{i=1}^{4} d^d k_i \, \frac{n^0_{k_1} n^0_{k_2} (1-n^0_{k_3})(1-n^0_{k_4})}{\varepsilon_{k_1} + \varepsilon_{k_2} - \varepsilon_{k_3} - \varepsilon_{k_4}} \, \delta(k_1+k_2+k_3+k_4) \tag{4}$$

where momentum conservation has been written explicitly and $n^0_k = 1$ for $k<k_F$ and zero elsewhere. L is the number of lattice sites. Writing the denominator z as $\exp(-\lambda z)$ and integrating over λ and expressing the δ-function as $\delta(\mathbf{k}) = (2\pi)^{-d} \Sigma_f \exp(i\mathbf{k}\cdot\mathbf{f})$, one obtains

$$E_2/LU^2 = -\int_0^\infty d\lambda \sum_f [F^+(\lambda,f)]^2 [F^-(\lambda,f)]^2 \tag{5}$$

where $F^\pm(\lambda,f)$ are the Fouriertransforms of $n^0_k \exp(\lambda\varepsilon_k)$ and $(1-n^0_k)\exp(-\lambda\varepsilon_k)$, respectively. Since $\Sigma_f[F^\pm(\lambda,f)]^2$ is finite for all d and since the number of nearest neighbors of site f=0 increases proportional to d, $F^\pm(\lambda;f')$ vanishes at least as $1/\sqrt{d}$ for $d\to\infty$ for f′ nearest neighbor to f=0. Analogous arguments hold for next-nearest neighbors etc. Consequently, only the term with f=0 contributes to (5) (i.e. $\delta(k_1+k_2+k_3+k_4) \to 1$ in (4)), such that the momentum-integrals in $F^\pm(\lambda,0)$ now depend on the momenta only via the corresponding <u>energies</u>. By use of the DOS $N^*(E)$ one finally finds

$$E_2/LU^2 = -\int_0^\infty d\lambda \, e^{2\lambda^2} P^2(E_F - \lambda) P^2(-E_F - \lambda) \tag{6}$$

with P(x) as the probability function. The result for E_2 in d=∞ is therefore seen to be particularly simple. The reason for this is that for an arbitrary choice of momenta meeting at a vertex the corresponding energies are randomized by Umklapp-processes which makes them mutually independent. This allows one to go from a momentum integration to an energy integration. - The result for E_2 as a function of density n is shown in Fig. 1 together with the results in d = 1,2,3. The result for d=3, which can only be obtained by Monte-Carlo integration, is obviously very well approximated by that in $d=\infty$, which is so simple to calculate. For n not too close to n=1 even the result for d=2 is well approximated by that in $d=\infty$, while the case d=1 is quite distinct from all finite dimensions. - The existence of a finite U^2-contribution to E_c for small U shows that the mean field solution obtained within unrestricted Hartree-Fock <u>cannot</u> be exact in $d=\infty$ since in that solution an asymptotic expansion yields $E_2 = 0$.

VARIATIONAL WAVE FUNCTIONS

Simplifications similar to those described above also occur in the evaluation of expectation values $<\hat{O}> = <\Psi|\hat{O}|\Psi>/<\Psi|\Psi>$ in terms of generalized Gutzwiller wave functions

$$|\Psi> = g^{\hat{D}} \, |\Phi_0> \tag{7}$$

in $d = \infty$. Here $|\Phi_0>$ is an arbitrary, not necessarily translational invariant one-particle wave function, $g \in [0,1]$ is a variational parameter and $\Sigma_i \, \hat{n}_{i\uparrow} \hat{n}_{i\downarrow} = \hat{D}$, the interaction part of (1a), counts the doubly occupied sites. The actual Gutzwiller wave function (GWF)[8] is obtained with $|\Phi_0>$ as the simple Fermi sea.

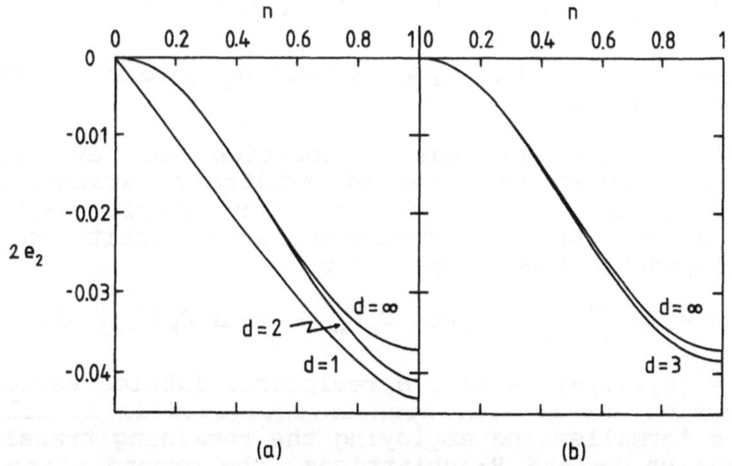

Fig. 1. The second order contribution to the correlation energy $E_2 = e_2(LU^2/|\bar{\epsilon}_0(1)|)$ as a function of density n for various dimensions d; (a) $d=1,2,\infty$, (b) $d=3,\infty$.

Expectation values $<\hat{O}>$ can be expanded in powers of (g^2-1) by using standard field-theoretical methods and their diagrammatic representation in analogy with the Green's function approach to a Φ^4-theory.[10] In particular, the ensueing diagrams may be conveniently expressed by the "self energy" S.[4] Thereby the one-particle density matrix $P_{\sigma fh} \equiv <\hat{c}^\dagger_{f\sigma}\hat{c}_{h\sigma}>$, which determines the momentum distribution and thus the kinetic energy in (1a) and the interaction term in (1a) are determined on the same footing:

$$P_{\sigma fh} = P^0_{\sigma fh} + 1_{fh}\,[(1-g^2)(P^0_\sigma S_\sigma)_{ff} - (S_\sigma)_{ff}]\,/(1+g)^2$$

$$+\ [(P^0_\sigma-(1+g)^{-1})S_\sigma(P^0_\sigma-(1+g)^{-1})]_{fh} \tag{8a}$$

$$<\hat{D}> = \frac{g^2}{1-g^2}\,\frac{1}{L}\ \mathrm{Tr}(P^0_\sigma S_\sigma) \tag{8b}$$

Here $P^0_{\sigma fh} = <\phi_0|\hat{c}^\dagger_{f\sigma}\hat{c}_{h\sigma}|\phi_0>$ and P^0_σ, P_σ, S_σ are taken as matrices with elements $S_{\sigma fh}$ etc. and $\mathrm{Tr}(\ ) = \Sigma_f(\ )_{ff}$.

For large d the evaluation of diagrams is again greatly simplified, since - as in the case of $F^\pm$ in (5) - $P^0_{\sigma fh}$ vanishes as least as $1/\sqrt{d}$ for $f\neq h$. We now consider two vertices **f**,**h** in a diagram which are connected by three or more separate paths. The evaluation of the diagram involves the lattice sum over **f**,**h** and all the other vertices. Since the contributions from $f\neq h$ are suppressed at least as $1/\sqrt{d}$, only the on-site (f=h) terms remain in d=∞, i.e. the two vertices collapse into a <u>single</u> vertex. In particular, the proper self-energy $S^*_{\sigma fh}$, defined as the sum over all one-particle irreducible diagrams, now becomes diagonal, i.e. $S^*_{\sigma fh} \equiv \delta_{fh} S^*_{\sigma f}$. S* can be written as a sum over all skeleton diagrams with dressed lines $\bar{P} = P^0 + P^0 S P^0$. In d=∞ all the vertices of a skeleton diagram are then seen to collapse into one vertex. The structure of the collapsed skeleton diagrams is so simple that an exact summation is possible, yielding

$$S^*_{\sigma f} = -\frac{1}{2\,\bar{P}_{\sigma ff}}\,[1 - \sqrt{1 + 4(1-g^2)\bar{P}_{\sigma ff}\bar{P}_{-\sigma ff}}\ ] \tag{9}$$

This, together with the Dyson equation, determines S* and S for given P^0 and g.

One is now in the position to evaluate the Hubbard-Hamiltonian in terms of arbitrary variational wave functions of the form (7) in d=∞. For example, $|\Phi_0>$ in (7) may be choosen as a Hartree-Fock spin density wave with antiferromagnetic long range order

$$|\Psi_{AF}> = g^{\hat{D}} \prod_\sigma \prod_{k<k_F} \left[\cos\theta_k \hat{a}^+_{k\sigma} + \sigma \sin\theta_k \hat{a}^+_{k+Q}\right]|0> \tag{10}$$

where $Q = (\pi,\ldots,\pi)$ is half a reciprocal lattice vector. This wave function has already been investigated in detail. Using the above formalism and employing the remaining translational invariance on A- and B-sublattices, the ground state energy $E\{g;\theta_k\} = <\hat{H}>$ is obtained as a functional of g and θ_k. The minimization of E can be performed in closed form, i.e. without assuming any particular form for the function θ_k. The optimal θ_k is in general different from the Hartree-Fock form[7] which has been used in numerical calculations.[11] E may equivalently be expressed as a function of the sublattice magnetization $m \equiv |\langle \hat{n}_{i\uparrow} - \hat{n}_{i\downarrow} \rangle|$ and the density of doubly occupied sites $\bar{d} \equiv L^{-1} \langle \hat{D} \rangle$ as

$$E(m,\bar{d})/L = q\bar{\varepsilon}_{HF}(m) + U\bar{d} \tag{11}$$

where $\bar{\varepsilon}_{HF}(m)$ is the result for the kinetic energy within unrestricted Hartree-Fock,[7] and

$$q = \frac{4(n-2\bar{d})\sqrt{\bar{d}\,(\bar{d}+1-n)} + 2(2\bar{d}+1-n)\sqrt{(n-2\bar{d})^2 - m^2}}{\sqrt{(n^2-m^2)\,[\,(2-n)^2 - m^2]}} \tag{12}$$

is a renormalization factor. This result is identical to the result of a slave boson approximation by Kotliar and Ruckenstein.[12] Here we have found an explicit wave function, $|\Psi_{AF}\rangle$, for which this result is exact in $d=\infty$.[4] Assuming $m = 0$, $|\Psi_{AF}\rangle$ in (10) reduces to the GWF [8] and (12) becomes identical to the formula of the much-employed, semi-classical Gutzwiller approximation.[8,13] Therefore the Gutzwiller approximation provides an exact evaluation of the GWF in $d = \infty$.[4]

For $n = 1$ the minimization of the energy $E(m,\bar{d})$ obtained with the more general wave function (10) yields $m > 0$ (antiferromagnetic insulator) and $q < 1$ for all $U > 0$, in contrast to earlier attempts to generalize the Gutzwiller approximation to antiferromagnetism.[14] For $U \to \infty$, $|\Psi_{AF}\rangle$ approaches the Neél state. For $0.85 < n < 1$ there exists a bounded regime in U with $m > 0$, while for smaller densities m is always zero. The fact that the minimum of E is attained for $g < 1$, shows that the additional correlations, introduced by the factor $g^{\hat{D}}$ into the antiferromagnetic starting wave function in (10), are important for all n and $U > 0$.

SUMMARY

The Hubbard model has been investigated in the limit of infinite spatial dimensions. The most important finding is that the model possesses a non-trivial limit, where diagrammatic calculations are substantially simpler than in finite dimensions. The weak coupling correlation energy is found to be a very good approximation for $d = 3$. As to variational wave functions, the results of the wellknown Gutzwiller approximation[8,13] and of a recent slave boson approach[12] have been recovered by evaluating a class of variational wave functions exactly in $d = \infty$. - The concept derived here is equally applicable to the evaluation of correlation functions[15] and to the general Green's function approach for the Hubbard model and related models.[16]

References

1. See, for example, H. E. Stanley in Phase Transitions and Critical Phenomena, vol. 3 , eds. C. Domb and M.S. Green (Academic Press, London, 1974), p. 485.
2. M. C. Gutzwiller, Phys. Rev. Lett. 10 159 (1963); J. Hubbard, Proc. R. Soc. London, A276 238 (1963); J. Kanamori, Prog. Theor. Phys. 30 275 (1963).
3. E. H. Lieb and F.Y. Wu, Phys. Rev. Lett. 20 1445 (1968).
4. W. Metzner and D. Vollhardt, Phys. Rev. Lett., in press (January 16, 1989).
5. U. Wolff, Nucl. Phys B225 391 (1983).
6. W. Metzner and D. Vollhardt, Phys. Rev. B (in press).
7. D. R. Penn, Phys. Rev. 142 350 (1966); M. Cyrot; J. Physique 33 125 (1972).
8. M. C. Gutzwiller, Phys. Rev. A137 1726 (1965).
9. For a review, see the accompanying article by D. Vollhardt appearing in this volume of the proceedings.
10. W. Metzner and D. Vollhardt, Phys. Rev. Lett. 59 121 (1987); Phys. Rev. B37 7382 (1988).
11. H. Yokoyama and H. Shiba, J. Phys. Soc. Jpn. 56, 3582 (1987).
12. G. Kotliar and A.E. Ruckenstein, Phys. Rev. Lett. 57 1362 (1986) .
13. For a detailed discussion, see D. Vollhardt, Rev. Mod. Phys. 56, 99 (1984) ; D. Vollhardt, P. Wölfle and P. W. Anderson, Phys. Rev. B35 6703 (1987).
14. T. Ogawa, K. Kanda and T. Matsubara, Prog. Theor. Phys. 53 614 (1975); F. Takano and M. Uchinami, Prog. Theor. Phys. 53 1267 (1975); J. Florencio and K.A. Chao,Phys. Rev. B14 3121 (1976).
15. See the accompanying article by F. Gebhard and D. Vollhardt appearing in this volume of the proceedings.
16. E. Müller-Hartmann, preprint; G. Czycholl, preprint.

CORRELATION FUNCTIONS FOR THE TWO-DIMENSIONAL HUBBARD MODEL

José Carmelo

Instituto Nacional de Investigaçao Cientifica
Centro de Fisica de Matéria Condensada
1699 Lisboa Codex, Portugal
and Departamento de Fisica
Universidade de Evora
7001 Evora Codex, Portugal

Dionys Baeriswyl*

CENG, Département de Recherche Fondamentale
Service de Physique/Groupe DSPE
85X, 38041 Grenoble Cedex, France

Xenophon Zotos

Institut für Theorie der kondensierten Materie
Physikhochhaus, Postfach 6380
7500 Karlsruhe 1, FRG

INTRODUCTION

The role of electronic correlations in the interplay of magnetism and superconductivity represents one of the major problems in the theory of high temperature superconductors. The two-dimensional single-band Hubbard model provides the simplest framework for discussing this fundamental question[1]. It has been argued that this model catches the essential physics of a CuO_2 plane[2], but there is presently no agreement on this point[3]. There is no doubt, however, that the model is well suited for displaying a rich variety of phenomena such as the metal-insulator transition, the competition between ferromagnetism and antiferromagnetism or superconductivity.

In this short note we study (equal-time) correlation functions of the two-dimensional Hubbard model in the small U limit, using the Gutzwiller ansatz for the many-electron ground state. Results for the one-dimensional case have already appeared[4,5]. Full details for the square lattice will be published elsewhere[6]. Our variational procedure has the advantage of treating an infinite system, at zero temperature, and the disadvantage of

*Permanent address : Institut für Theoretische Physik, ETH-Hönggerberg, 8093 Zürich, Switzerland

Interacting Electrons in Reduced Dimensions
Edited by D. Baeriswyl and D.K. Campbell
Plenum Press, New York

attributing weights to different configurations in a rather crude way, thereby neglecting subtle correlation effects, especially for large U[7]. The variational approach is somehow complementary to the Quantum Monte Carlo method, which is unbiased in the choice of configurations, but restricted to small systems and relatively high temperatures[8].

APPROACH

The Hubbard Hamiltonian

$$H = -t \sum_{\langle i,j \rangle \sigma} \left(c^{+}_{i\sigma} c_{j\sigma} + h.c. \right) + U \sum_{i} n_{i\uparrow} n_{i\downarrow}, \tag{1}$$

where $c^{+}_{i\sigma}$ creates an electron at site i with spin σ and $n_{i\sigma} = c^{+}_{i\sigma} c_{i\sigma}$, depends on a single parameter U/t. Similarly, the Gutzwiller ansatz for the ground state,

$$\Psi = e^{S} \Psi_0, \tag{2}$$

where Ψ_0 is the exact solution for U = 0 and the operator

$$S = -\frac{1}{2} \eta \sum_{i} n_{i\uparrow} n_{i\downarrow} \tag{3}$$

reduces double occupancy, depends on the single variational parameter η, which is fixed by minimizing the energy. Expanding the expectation value of the Hamiltonian (1) up to second order in η and minimizing with respect to η one finds that η is simply proportional to U/t. A comparison with both a systematic perturbation expansion[9] and numerical simulations[8,10] shows that this expansion makes sense for $U \lesssim 4t$. Furthermore the Gutzwiller wavefunction accounts for about 80% of the correlation energy in this parameter range[6].

According to the Linked Cluster Theorem the expectation value of any operator A is given by

$$\langle A \rangle \equiv \frac{\langle \Psi | A | \Psi \rangle}{\langle \Psi | \Psi \rangle} = \left\langle \Psi_0 | e^{S} A e^{S} | \Psi_0 \right\rangle_c , \tag{4}$$

where only connected diagrams occur in the expansion in powers of S. Specifically, we consider correlation functions

$$S(m,n) = \left\langle O^{+}_{m} O_{n} \right\rangle - \left\langle O^{+}_{m} \right\rangle \left\langle O_{n} \right\rangle, \tag{5}$$

where the operators O_n represent, for example, the spin density,

$$O_n = \sigma_z(n) = \sum_{\sigma} \sigma c^{+}_{n\sigma} c_{n\sigma}, \tag{6}$$

an extended singlet pair

$$O_n = C_{es}(n) = \frac{1}{2} \sum_{n'\sigma}{}' \sigma\, c_{n\sigma} c_{n',-\sigma}, \tag{7}$$

(summation over neighboring sites) or other important quantities like the charge density, the hole density, the bond order or an extended triplet pair[6]. Expanding to first order in η, we find

$$S_\sigma(m,n) = 2P(m,n)Q(m,n) + 2\eta \sum_i P(i,m)Q(i,m)P(i,n)Q(i,n) \tag{8}$$

for the spin-spin correlations and

$$S_c(m,n) = \frac{1}{2} \sum_{m'n'}{}' \{P(m,n)P(m',n') + P(m',n)P(m,n')$$

$$-\eta \sum_i [P(i,n)P(i,n')Q(i,m)Q(i,m') + (Q \leftrightarrow P)]\} \tag{9}$$

for the pair-pair correlations where the functions

$$P(m,n) = \langle \Psi_0 | c^+_{m\sigma} c_{n\sigma} | \Psi_0 \rangle, \tag{10a}$$

$$Q(m,n) = \delta_{m,n} - P(m,n) \tag{10b}$$

correspond to equal-time Green's functions.

RESULTS AND DISCUSSION

In one dimension the correlation functions P(m,n) are very simple analytical expressions, allowing the summations in Eqs.(8) and (9) to be performed analytically[5]. In two dimensions the correlation functions have to be evaluated numerically, except for the special case of a half-filled band. It is instructive to calculate the Fourier transforms

$$S(\vec{q}) = \sum_m e^{i\vec{q}(\vec{x}_m - \vec{x}_n)} S(m,n) \tag{11}$$

where $\vec{x}_m$, $\vec{x}_n$ are vectors of the square lattice (lattice constant 1) and $\vec{q}$ belongs to the first Brillouin zone. The spin-spin correlation function $S_\sigma(\vec{q})$ assumes its maximum at (π,π), which is shown in Fig.1 as a function of the electron density $n = N_e/N_a$ for different values of U.

As expected, the antiferromagnetic fluctuations increase with increasing U, and they are strongest for a half-filled band. The correlation function for extended singlet pairing, $S_c(\vec{q})$, has its maximum at $\vec{q} = 0$, which is shown in Fig. 2. In contrast to on-site (singlet) pairing which is suppressed by correlation effects[6], the fluctuations associated with extended singlet pairing are seen to be enhanced as U increases. Furthermore, they remain large within a substantial region close to the half-filled band (n = 1). A comparison of Figs. 1 and 2 indicates that the delicate balance between antiferromagnetism and superconductivity may indeed sensitively depend on band-filling.

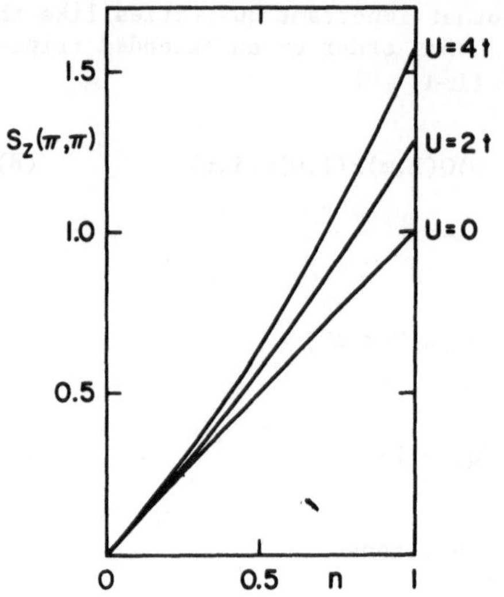

Fig.1. Antiferromagnetic correlations as functions of density and interaction strength.

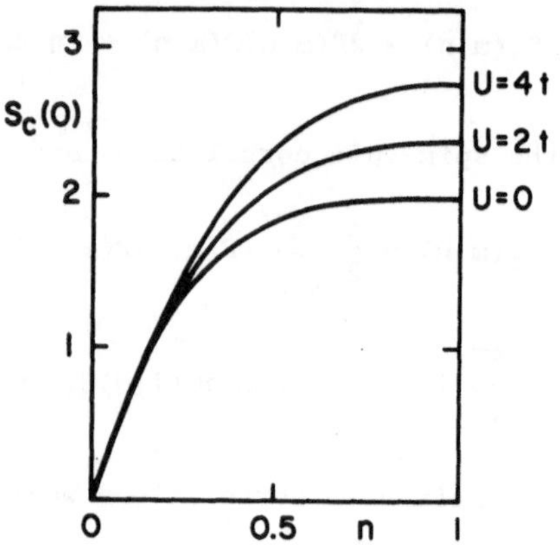

Fig.2. Superconducting correlations as functions of density and interaction strength.

ACKNOWLEDGMENT

We have greatly profitted from both the warm hospitality and the computer facilities of the Institute for Scientific Interchange in Torino, where most of the present work was performed.

REFERENCES

1. P.W. Anderson, Science 235, 1196 (1987)
2. F.-C. Zhang and T.M. Rice, Phys. Rev. B37, 3759 (1988)
3. V.J. Emery and G. Reiter, Phys. Rev. B38, 4547 (1988)
4. F. Gebhard and D. Vollhardt, Phys. Rev. Lett. 59, 1472 (1987) ; Phys. Rev. B38, 6911 (1988)
5. J. Carmelo and D. Baeriswyl, Physica C 153-155, 1281 (1988)
6. J. Carmelo, D. Baeriswyl and X. Zotos, manuscript in preparation
7. G. Stollhoff, Z. Phys. B69, 61 (1987)
8. J.E. Hirsch and H.Q. Lin, Phys. Rev. B37, 5070 (1988)
9. W. Metzner and D. Vollhardt, Phys. Rev. B, in press
10. S. Sorella, E. Tosatti, S. Baroni, R. Car and M. Parrinello, Int. J. Mod. Phys. B 1, 993 (1988)

DENSITY FUNCTIONAL CALCULATIONS FOR STRONGLY CORRELATED SYSTEMS

O. Gunnarsson and O.K. Andersen

Max-Planck Institut für Festkörperforschung

D-7000 Stuttgart 80, W.Germany

A. Svane

Institute of Physics, University of Aarhus

DK-8000 Aarhus C, Denmark

1. INTRODUCTION

For atoms and small molecules there are rather accurate *ab initio* many-body methods, like the configuration interaction method. For solids such methods are often too complicated, and there are relatively few *ab initio* many-body calculations. These have usually been performed for rather simple systems, such as diamond and Si and free electron like material. These calculations have usually been based on the "local approach",[1] the GW approximation,[2] or Monte Carlo methods.[3] For more complicated systems many-body calculations have almost exclusively been based on model Hamiltonian approaches. The density functional formalism provides an alternative method for including many-body effects in *ab initio* calculations. This formalism requires the solution of a one-particle equation and the calculations are therefore relatively simple. It is therefore possible to treat complicated systems with, say, 50 atoms per unit cell without shape approximations for the potential.[4] Calculations have been made for interfaces,[4] impurities[5] and amorphous solids.[6]

The many-body effects are usually treated in the so-called local spin density (LSD) approximation. Although there are several proposals for more accurate approximations, there is no systematic procedure for improving the LSD approximation, and in many cases where the accuracy of the LSD approximation is insufficient also other available approximations give unsatisfactory results. The accuracy of the LSD approximation is therefore a crucial issue. In this paper we therefore give some examples of how the LSD approximation works, ranging from cases where the agreement with experiment is quite good to cases where the LSD approximation fails qualitatively. We also show a few applications based on a different approximation, the so-called self-interaction corrected (SIC)-LSD approximation.

Interacting Electrons in Reduced Dimensions
Edited by D. Baeriswyl and D.K. Campbell
Plenum Press, New York

2. THE DENSITY FUNCTIONAL FORMALISM

The density functional formalism is based on a theorem by Hohenberg and Kohn,[7] which states that all ground-state properties are functionals of the ground-state density $n(\mathbf{r})$. In particular the energy of a system with an external potential $v_{ext}(\mathbf{r})$ can be written as

$$E_v[n] = \int v_{ext}(\mathbf{r})n(\mathbf{r})d^3r + F[n], \tag{1}$$

where $F[n]$ is a universal functional of the density. Hohenberg and Kohn[7] further showed that $E_v[n]$ is minimized by the exact ground-state density.

To simplify the task of finding good approximations to $F[n]$, Kohn and Sham[8] split off the kinetic energy $T_0[n]$ of *noninteracting* electrons with the density n

$$F[n] = T_0[n] + \frac{e^2}{2}\int \frac{n(\mathbf{r})n(\mathbf{r}')}{|\mathbf{r}-\mathbf{r}'|}d^3rd^3r' + E_{xc}[n]. \tag{2}$$

The second term is the Hartree electrostatic energy and the last term, the so-called exchange-correlation energy, by definition contains all remaining contributions to the energy. Kohn and Sham[8] showed that in this approach we only need to find approximations for $E_{xc}[n]$, and that the minimization of the energy functional requires the solution of a *one-particle* problem

$$\{\frac{-\hbar^2}{2m}\nabla^2 + v_{ext}(\mathbf{r}) + e^2\int \frac{n(\mathbf{r}')}{|\mathbf{r}-\mathbf{r}'|}d^3r' + v_{xc}(\mathbf{r})\}\psi_i(\mathbf{r}) = \varepsilon_i\psi_i(\mathbf{r}), \tag{3}$$

where

$$v_{xc}(\mathbf{r}) = \frac{\delta E_{xc}}{\delta n(\mathbf{r})}. \tag{4}$$

The density is then given by

$$n(\mathbf{r}) = \sum_i^{occ} |\psi_i(\mathbf{r})|^2, \tag{5}$$

and the total energy can be expressed in terms of ε_i and $n(\mathbf{r})$.

The eigenvalues ε_i enter the theory as Lagrange parameters, and they have not been shown to have a physical meaning in general. It is, nevertheless, common to associate ε_i with excitation energies, although this is not in general correct. This was demonstrated by, for instance, Almbladh and Pederoza,[9] who calculated the exact eigenvalues for a Be atom and found that the absolute value of the 1*s* eigenvalue (115.1 eV) deviates from the experimental 1*s* binding energy (123.6 eV). Another example was provided by a simple model of a semiconductor, where the deviation between the eigenvalues and the low-lying excitation energies was found to be very small but nonzero.[10] For the one-dimensional Hubbard model, on the other hand, there is a qualitative difference between the eigenvalues and the excitation energies.[10] In the following we nevertheless follow the common approach of associating eigenvalues with excitation energies, since there is often no other approximation to the excitation energies available.

In the Kohn-Sham approach (2) all contributions to the energy are treated exactly except the exchange-correlation energy $E_{xc}[n]$. The most commonly used approximation for $E_{xc}[n]$ is the local density (LD) approximation[8,11]

$$E_{xc}[n] = \int \epsilon_{xc}(n(\mathbf{r}))n(\mathbf{r}), \tag{6}$$

where $\epsilon_{xc}(n(\mathbf{r}))$ is the exchange-correlation energy per electron of a homogeneous system with the density $n(\mathbf{r})$. The LD approximation is often generalized to a local spin density (LSD) approximation, where the spin densities $n_\sigma(\mathbf{r})$ are used as input parameters.

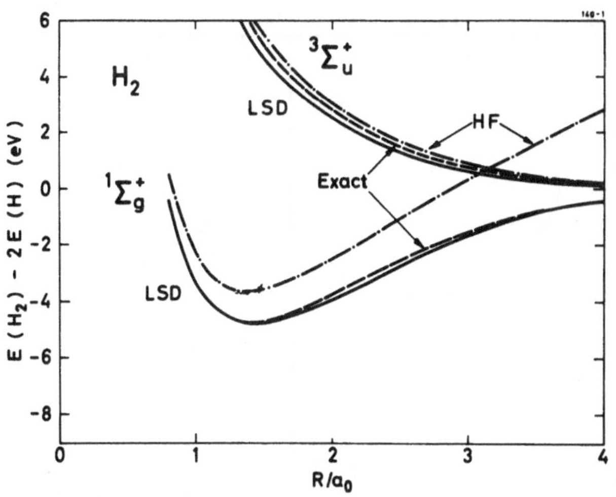

Fig. 1. The energy of the ${}^1\Sigma_g^+$ and ${}^3\Sigma_u^+$ states of the H_2 molecule as a function of the molecular separation R. The LSD and HF approximations are compared with the exact results. (From Gunnarsson and Johansson[13]).

The L(S)D approximation becomes exact if the density variations are weak and spatially small. This condition is almost never fulfilled for systems of interest. Nevertheless, the L(S)D approximation has turned out to give surprisingly accurate results in many situations. This can be partly understood by observing that the LSD approximation is actually much better than suggested by the arguments above. In particular, it can be shown that the L(S)D approximation fulfills a sum rule exactly.[12]

3. GROUND-STATE PROPERTIES

It was originally expected that the L(S)D approximation would not give a satisfactory description of chemical bonding.[8] One of the first tests, with a sufficient numerical accuracy, of the description of chemical bonding was a calculation for the H_2 molecule.[13] In Fig. 1 we compare results in the LSD and Hartree-Fock (HF) approximations with exact results. For the ${}^1\Sigma_g^+$ state the LSD approximation is close to the experimental result over the whole range of distances and it is substantially better than the HF approximation. For the ${}^3\Sigma_u^+$ state both approximations are fairly close to the exact result. The binding energy is predicted to be 4.8 eV compared with the exact result 4.75 eV. The HF and Heitler-London approximations give 3.64 eV and 3.75 eV, respectively. It should be noted, however, that for large distances the LSD approximation incorrectly predicts an broken symmetry state, with a spin up polarization on one atom and a spin down polarization on the other atom. These results are typical for *s* bonded systems. For *p* and *d* bonded systems the error in the binding energy is normally larger.[14]

As an example of the bonding properties of solids, we show in Fig. 2 results for the cohesive energy, the Wigner-Seitz radius and the bulk modulus of 26 metals.[15] These calculations were performed without including spin-polarization effects. Including spin-polarization improves the description of the Wigner-Seitz radius and the bulk modulus for the magnetic systems.[16] The cohesive energy of the 3*d* metals is overestimated by typically 0.1 Ryd, while the other systems are rather accurately described. The errors in the Wigner-Seitz radius are generally small. The bulk modulus is rather well described for the *sp* and 4*d* systems, while there are substantial errors for the 3*d* metals, which are partly due to the neglected magnetic effects.

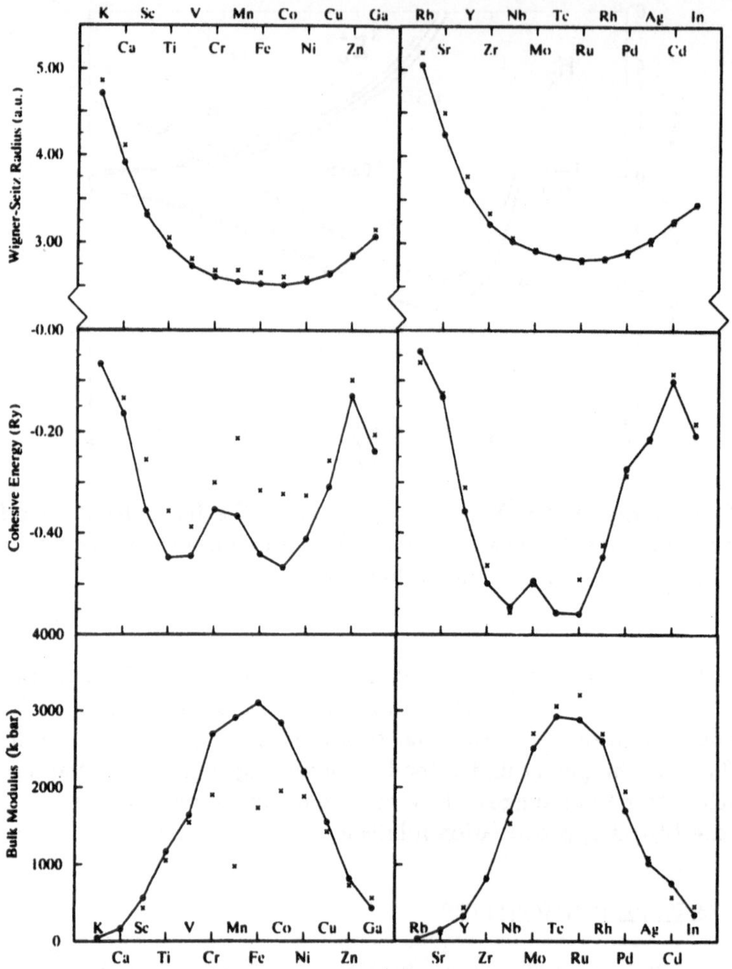

Fig. 2. Cohesive properties. Top row: Cohesive energy (Ry/atom). Middle row: Wigner-Setz radius (a.u.). Bottom row: Bulk modulus (Kbar). The circles show calculated results and the crosses measured results. (From Moruzzi et al[15].)

4. COMPARISON OF EIGENVALUES AND EXCITATION ENERGIES

In this section we assume that the energy eigenvalues can be directly compared with experimental excitation energies. Although there is no rigorous support within the theory for this approach, this is the normal procedure within band structure theory, and it is therefore important to assess its accuracy. Almost all of these calculations use the LSD approximation. Since the corresponding exchange-correlation potential is based on data for the homogeneous electron gas, we may expect large errors for systems where the correlation effects are very different from those in the homogeneous electron gas. For instance, in semiconductors or insulators, the presence of a band gap strongly influences the correlation effects and it is not very surprising that the LSD approximation predicts band gaps which may be a factor of 2 too small.

For metallic systems we expect the importance of the correlation effects to depend on U/W, where U is an effective Coulomb interaction and W is the band width. It is therefore interesting to study the series of $3d$ metals. The $3d$ band width, defined as the square root of the second moment, is 4.6 eV, 4.1 eV, 3.6 eV and 2.7 eV for Fe, Co, Ni and Cu, respectively, according to band structure calculations[19]. From Auger

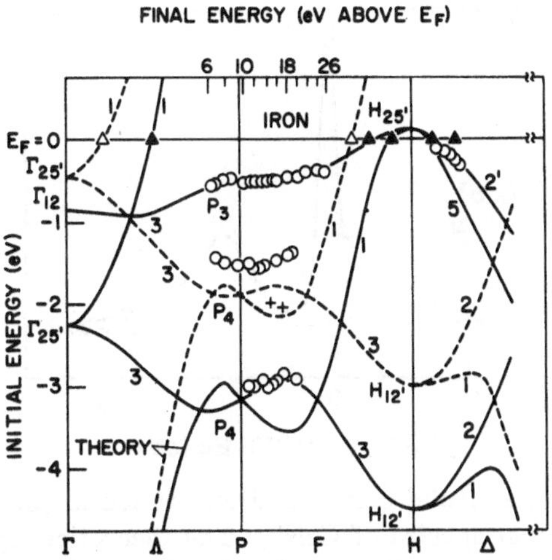

Fig. 3. Experimental and theoretical band structure for Fe. Solid and dashed lines show the majority and minority bands calculated by Callaway and Wang[17]. The circles and crosses show states observed in photoemission and the triangles data deduced from de Haas-van Alphen data. (From Eastman et al[18]).

experiments U was deduced to be 8 eV for Cu, 4 eV for Ni and small for Fe and Co[20]. Thus U/W is small for Fe but ≈ 1 for Ni. Band structure calculations have been compared with angular resolved photoemission data by Eastman et al[18]. They found that the ratio of the theoretical to experimental band width is 1.1, 1.2 and 1.45 for Fe, Co and Ni, respectively. Fe, Co and Ni are ferromagnets and there is therefore an energy splitting between spin up and spin down electrons. The ratio between the theoretical and experimental splitting was found to be 1.0, 1.2 and 2.2 for Fe, Co and Ni, respectively[18]. Thus Fe appears to be quite well described by band theory while for Ni the accuracy is less satisfactory. To further illustrate the description of Fe, we compare in Fig. 3 the theoretical and experimental band structure of Fe. For Cu there is a good agreement between band structure calculation and photoemission experiments,[21] in spite of the large value of U/W. The reason is that the 3d band of Cu is essentially filled. In photoemission at most one 3d hole is created and the strong Coulomb interaction is not effective. In Auger emission, on the other hand, two 3d holes may be created, and there is therefore qualitative deviation between Auger experiments and the predictions of band theory,[22] due to the strong Coulomb interaction between the holes.

An even more extreme example is provided by the Ce compounds, where $U \sim 6$ eV and the 4f band width $W \leq 1$ eV.[24] As an example we show an experimental 4f spectrum for $CeNi_2$ in Fig. 4. The spectrum is spread out over about 8 eV in qualitative disagreement with band theory, which predicts a single peak at the Fermi energy with the width 1 eV or less for Ce compounds. The figure also illustrates that the spectrum can be fairly well described by the Anderson impurity model. We therefore later discuss to what extent the LSD approximation can be used to calculate the parameters of the Anderson model.

A remarkable application of the LSD approximation is the calculation of the Fermi surface of UPt_3 by Wang et al.[25] UPt_3 is a heavy fermion system and the correlation effects should be very important. This is, for instance, illustrated by the fact that the calculated effective masses are about a factor 20 too small. Nevertheless, the calculated Fermi surface is in rather good agreement with experiment, with deviations of the same order as is found in "favourable" cases like the noble metals.[26] The reasons for this success have been discussed by Zwicknagl,[27] who concluded that the LSD approximation

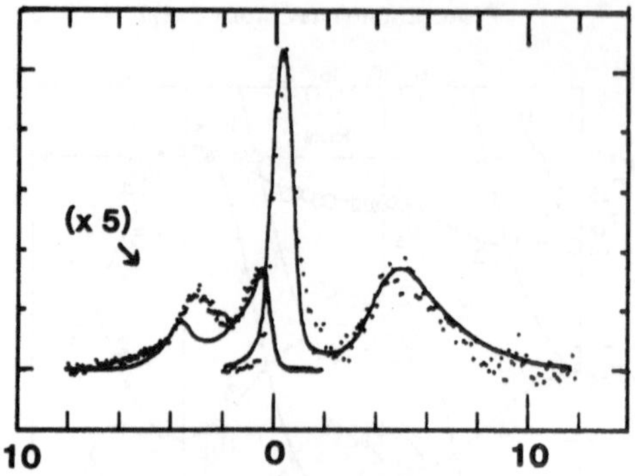

Fig. 4. The theoretical (solid curve) and experimental (dots) photoemission and Bremsstrahlung Isochromat spectra of $CeNi_2$. The theoretical spectrum is based on a calculation in the Anderson impurity model. The photoemission part of the spectrum has been multiplied by a factor 5 for clarity. (From Allen et al[23])

would not give good results for systems where the Kondo temperature is larger than the crystal field splitting. It is interesting that it is essential that the $5f$ electrons are treated as itinerant for UPt_3. If they were treated as core electrons the good agreement for the Fermi surface would be lost. On the other hand, it was found that for UPd_3 good Fermi surfaces were obtained only if the $5f$ electrons were treated as core electrons.[28] Finally, we notice that even if the exact exchange-correlation potential were known, the Fermi surface, derived from the corresponding eigenvalues, would in general differ from the experimental Fermi surface[29].

5. LOCALIZATION DUE TO THE COULOMB INTERACTION

The LSD approximation can for some systems describe aspects of the localization which takes place in a Mott insulator. Consider, for instance, the $3d$ oxides. In Fig. 5 we show the lattice parameter for these systems[31]. MnO, FeO, CoO and NiO are Mott insulators (or generalizations of this concept[34]). The corresponding loss of the $3d$ contribution to the cohesion shows up as an increase in the lattice parameter, relatively to the parabolic shape one would normally expect for a $3d$ system. The calculations properly predict an antiferromagnetic solution for MnO and the systems to the left of MnO, resulting in a small $3d$ contribution to the cohesion. The Wigner-Seitz radius therefore has a very similar trend both according to theory and experiment. Apart from MnO, which has a half-filled $3d$ shell, the band gap is, however, very small or zero.[35] In a similar way, the LSD approximation can also describe the loss of $5f$ contribution to the cohesion in Am (see Fig. 5).[32] Some problems with the LSD approximation are illustrated by the $\alpha - \gamma$ transition in Ce. This transition is normally ascribed to the Kondo effect ("Kondo volume collapse" model).[36] In a study of the properties of Ce as a function of the lattice parameter, it was found that in the LSD approximation Ce becomes ferromagnetic at about the lattice parameter of γ-Ce.[37] This leads to a transfer of electrons from bonding states at the bottom of the band to less bonding states at higher energies, resulting in a loss of $4f$ cohesion. Since, however, the $4f$ band of Ce has the filling $\approx \frac{1}{14}$, only a small fraction of the $4f$ cohesion is lost, in contradiction to current understanding of the γ phase.[36] This suggests that the situation with a half-filled band is particularly favourable for the LSD approximation.

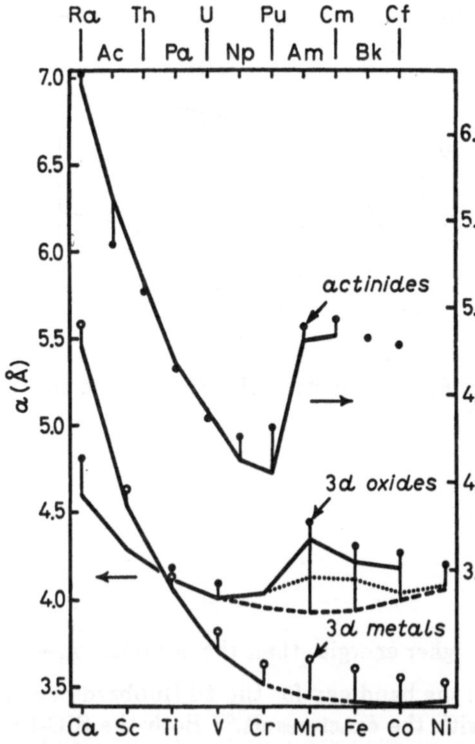

Fig. 5. Experimental and theoretical lattice parameters for the 3*d* metals,[30] the 3*d* monoxides[31] and the actinide metals.[32] For the elemental metals the lattice parameter was determined for the fcc structure with the proper atomic volume. The circles show experimental results, the full line theoretical results for the nonferromagnetic or antiferromagnetic phase, the dotted line calculations for the ferromagnetic phase and the dashed line the calculations for the nonmagnetic phase. (From Andersen et al[33])

An alternative to the LSD approximation is provided by the self-interaction corrected (SIC) LSD approximation, where the nonphysical interaction of an electron with itself is subtracted. For a solution i with the spin σ of the SIC-LSD equations, the potential is then

$$v_{xc,\sigma,i}^{SIC}(\mathbf{r}) = v_{xc,\sigma}^{LSD}(n_\uparrow(\mathbf{r}), n_\downarrow(\mathbf{r})) - \int d^3r' \frac{n_i(\mathbf{r}')}{|\mathbf{r}-\mathbf{r}'|} - v_{xc,\uparrow}^{LSD}(n_i(\mathbf{r}), 0) \tag{7}$$

where $v_{xc,\sigma}^{LSD}(n_\uparrow, n_\downarrow)$ is the LSD XC potential and $n_i(\mathbf{r})$ is the density corresponding to this solution. The SIC potential is orbital dependent, which may lead to broken symmetry solutions, where, for instance, one orbital localizes on each site. Let us assume that there is one electron and one orbital with the degeneracy M per atom. We now form a trial solution, where the ith one-particle solution of the Kohn-Sham equation is assumed to be the m_ith orbital on the l_ith site. The Hartree plus exchange-correlation potential for the ith solution is then zero on the l_ith site and repulsive on all other sites. In the first iteration this solution expands somewhat and it obtains weight on neighboring sites. This raises the potential on the central site and lowers it on the neighbouring sites. If the Coulomb interaction is large, the solutions, however, stay essentially localized when the problem is iterated to convergence. This solution reduces the Coulomb interaction at the cost of an increase in kinetic energy, as one expects for a Mott insulator.[38] This approach works also for a band filling which differs from $\frac{1}{2}$ (here $1/M$), since the $M-1$ unoccupied orbitals feel a repulsive potential on all sites and

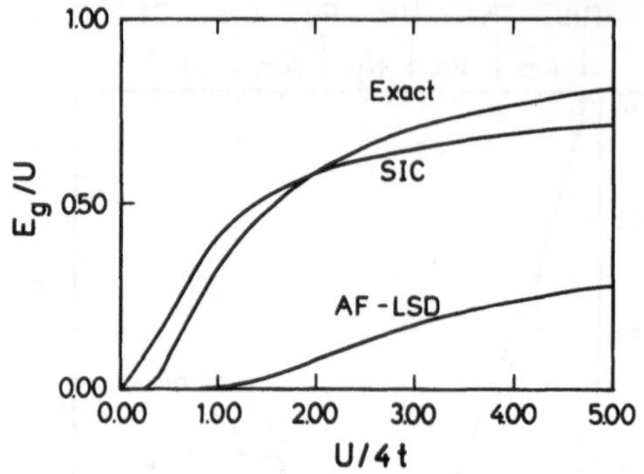

Fig. 6. The band gap E_g as a function of $U/(4t)$ in the 1d Hubbard model. (From Svane and Gunnarsson[38]).

form extended states at higher energies than the occupied ones.[38]

In Fig. 6 we compare the band gap for the 1d Hubbard model in the SIC-LSD and LSD approximations[38] with the exact result.[39] Both the SIC-LSD and LSD approximations give an antiferromagnetic (AF) solution. While the AF-LSD solution properly gives a gap for all values of U, the gap is much too small. The SIC-LSD approximation, on the other hand, gives a gap of essentially the right magnitude, although the gap goes to zero too slowly for small values of U.

The calculations for the 1-, 2- and 3-dimensional Hubbard models show that the SIC-LSD approximation gives a rather accurate description of the magnetic moment of these systems, while the LSD approximation greatly underestimates the moment.[40] The situation seems to be similar for the high T_c superconductor La_2CuO_4. For this system the LSD approximation fails to give antiferromagnetism, and the calculated Stoner parameter is at least a factor 2 too small.[41] We have performed SIC-LSD calculations for the 2-band, 2-dimensional Hubbard model of La_2CuO_4, deducing the parameters from *ab initio* calculations.[40] These calculations predict a much stronger tendency to antiferromagnetism in the SIC-LSD than in the LSD approximation, and the magnetic moment ($\sim 0.8\mu_B$) in the SIC-LSD approximation is actually too large compared with experiment ($\sim 0.4\mu_B$[42]). The SIC-LSD approximation gives a gap of about 2 - 2.5 eV,[40] compared with the experimental result of about 2 eV.[43]

6. CALCULATION OF PARAMETERS FOR THE ANDERSON MODEL

As indicated in Sec. 3 (Fig. 4) the Anderson model[44] can describe the spectra of some strongly correlated systems where the LSD approximation breaks down. It is then interesting to ask if the LSD approximation can, nevertheless, be used for these systems for calculating the parameters in the Anderson model. Work along these lines have been performed for $4f$ systems,[45,46] dilute magnetic semiconductors[47,48] and high T_c compounds.[41,49] Here we focus on the calculation of hopping matrix elements.

The Anderson model is defined by the Hamiltonian

$$H = \sum_{k,\sigma} \varepsilon_k n_{k\sigma} + \sum_{m,\sigma} \varepsilon_m n_{m\sigma} + \sum_{k,m,\sigma} (V_{km} \psi^\dagger_{m\sigma} \psi_{k\sigma} + h.c.)$$

$$+ \frac{1}{2} U \sum_{m,m',\sigma,\sigma'}{}' n_{m\sigma} n_{m'\sigma'}. \tag{8}$$

Here the conduction states of the system are described by ε_k, where k is a combined wave vector and band index. The energy of the localized $4f$-level is ε_m, where m is an orbital index. The hopping is described by V_{km}, and the Coulomb interaction between the localized electrons by U. The spin label is σ.

To solve the *impurity* Anderson model we do not need to know the individual hopping matrix elements but just particular combinations[24]

$$\sum_k V^*_{km} V_{km'} \delta(\varepsilon - \varepsilon_k) \equiv |V_m(\varepsilon)|^2 \delta_{mm'}, \tag{9}$$

where the $\delta_{mm'}$ follows if m and m' belong to different irreducible representations or different partners of the same irreducible representation of the point group of the impurity site. The quantity $|V_m(\varepsilon)|^2$ can be expressed in terms of the density of states, $\rho_m(\varepsilon)$, projected onto the localized orbital $|m>$[48]

$$\pi |V_m(\varepsilon)|^2 = -Im\{[\int d\varepsilon' \frac{\rho_m(\varepsilon')}{\varepsilon - \varepsilon' - i0}]^{-1}\}. \tag{10}$$

The use of (10) only requires a sensible definition of the localized orbital $|m>$ and an accurate calculation of $\rho_m(\varepsilon)$. In particular, it is not necessary to use a tight-binding fit of a band structure calculation. The ε dependence of $|V_m(\varepsilon)|^2$ is essential. Thermodynamic properties are essentially determined by the value of $|V(\varepsilon)|^2$ close to the Fermi energy ε_F, while electron spectra are influenced by $|V(\varepsilon)|^2$ over a larger energy range and can reflect structures in $|V(\varepsilon)|^2$.[50]

Here we focus on the valence photoemission spectrum and the static $T = 0$ susceptibility, $\chi(0)$. For these quantities the hopping $f^0 \to f^1$ is particular important. It has been shown that the $4f$ wave function, needed to obtain $\rho_m(\varepsilon)$, should then be calculated for a potential corresponding to an $4f^1(5d6s)^3$ configuration.[51]

In Fig. 7 we present results for α- and γ-Ce obtained from the $4f$ projected density of states $\rho_{4f}(\varepsilon)$ calculated for a periodic Ce metal in the LSD approximation.[46] For simplicity we have neglected the m dependence. From (9) we can interpret $|V(\varepsilon)|^2$ as a density of states $\rho(\varepsilon)$ times an averaged hopping matrix element squared $v(\varepsilon)^2$,

$$|V(\varepsilon)|^2 = |v(\varepsilon)|^2 \rho(\varepsilon). \tag{11}$$

Since we expect the $4f$-orbital on one site to primarily couple to the $6s$, $6p$ and $5d$ orbitals on the neighbouring sites, we compare the results for $|V(\varepsilon)|^2$ with the projected density of states $\rho_{spd}(\varepsilon)$ onto the $6s$, $6p$ and $5d$ orbitals. We can see that the structures in $\rho_{spd}(\varepsilon)$ are essentially reflected in $|V(\varepsilon)|^2$, although the energy dependence of $v(\varepsilon)$ cannot be neglected.

To test the calculated magnitude of $|V(\varepsilon)|^2$, we have compared with the experimental valence and core level photoemission spectrum of α- and γ-Ce, the static, $T = 0$ susceptibility of α-Ce and, since $\chi(0)$ is not accessible for the high temperature phase, the quasi-elastic line width Γ in neutron scattering for γ-Ce. From a comparison with these experimental data, we have concluded that the calculated $|V(\varepsilon)|^2$ is a factor $\frac{4}{3}$ too large.[46] It is an interesting question if this overestimate is due to a the LSD potential

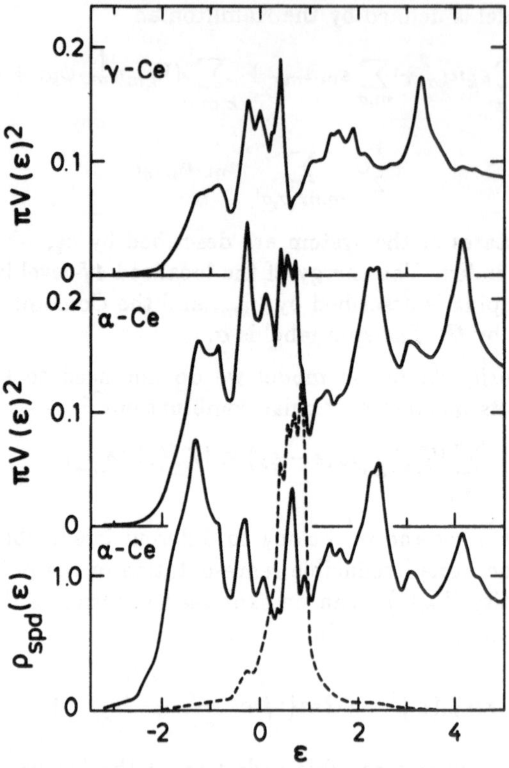

Fig. 7. The hybridization $|V_{4f}(\varepsilon)|^2$ (Eq.(9)) for γ- and α-Ce as a function of the energy ε measured relative to the Fermi energy $\varepsilon_F = 0$. The lower part of the figure shows the projected density of states $\rho_{spd}(\varepsilon)$ onto the 6*s*, 6*p* and 5*d* orbitals. The dashed curve shows the 4*f* density of states multiplied by a factor $\frac{1}{7}$. The results have been broadened by a Lorentzian with the full width 0.07 eV. All energies are in eV. (From Gunnarsson et al[46]).

not being sufficiently attractive or if it is due to renormalization effects beyond a density functional calculation of this type. From a comparison with experiment, we further concluded that the calculation gives an accurate description of the change of $|V(\varepsilon)|^2$ in going from α- to γ-Ce.

REFERENCES

1. See, e.g., S. Horsch, P. Horsch and P. Fulde, Phys. Rev. **B 28**, 5977 (1983).
2. M.S. Hybertsen and S.G. Louie, Phys. Rev. Lett. **55**, 1418 (1985); R.W. Godby, M. Schlüter and L.J. Sham, Phys. Rev. Lett. **56**, 2415 (1986); W. von der Linden and P. Horsch, Phys. Rev. **B 37**, 8351 (1988).
3. See, e.g., S Fahy, X.W. Wang and S.G. Louie, Phys. Rev. Lett. **61**, 1631 (1988).
4. P. Blöchl, G.P. Das, N.E. Christensen and O.K. Andersen (to be publ.)

5. See, e.g., F. Beeler, M. Scheffler, O. Jepsen and O. Gunnarsson, Phys. Rev. Lett. **54**, 2525 (1985); F. Beeler, O.K. Andersen and M. Scheffler, Phys. Rev. Lett. **55**, 1498 (1985); A. Svane, Phys. Rev. Lett. **60**, 2693 (1988).
6. S.K. Bose, S.S. Jaswal, O.K. Andersen and J. Hafner, Phys. Rev. **B 37**, 9955 (1988).
7. P. Hohenberg and W. Kohn, Phys. Rev. **136**, B864 (1964).
8. W. Kohn and L.J. Sham, Phys. Rev. **140**, A1133 (1965).
9. C.-O. Almbladh and A.C. Pederoza, Phys. Rev. **A 29**, 2322 (1984).
10. O. Gunnarsson and K. Schönhammer, Phys. Rev. Lett. **56**, 1968 (1986).
11. See, e.g., U. v Barth and A.R. Williams: In *Theory of the Inhomogeneous Electron Gas*, ed. by S. Lundqvist and N.H. March (Plenum, New York 1983).
12. O. Gunnarsson and B.I. Lundqvist, Phys. Rev. **B 13**, 4274 (1976).
13. O. Gunnarsson and P. Johansson, Int. J. Quant. Chem. **10**, 307 (1976); O. Gunnarsson, P. Johansson, S. Lundqvist and B.I. Lundqvist, Int. J. Quant. Chem. Symp. **9**, 83 (1975).
14. R.O. Jones and O. Gunnarsson, Rev. Mod. Phys. (to be published).
15. V.L. Moruzzi, J.F. Janak and A.R. Williams, *Calculated Electronic Properties of Metals*, Pergamon, New York, 1978; see also Ref. 30 and 33.
16. J.F. Janak and A.R. Williams, Phys. Rev. **B 14**, 4199 (1976); U.K. Poulsen, J. Kollar and O.K. Andersen, J. Phys. **F 6**, L241 (1976); O.K. Andersen, J. Madsen, U.K. Poulsen, O. Jepsen and J. Kollar, Physica B **86-88**, 249 (1977).
17. J. Callaway and C.S. Wang, Phys. Rev. **B 16**, 2095 (1977).
18. D.E. Eastman, F.J. Himpsel and J.A. Knapp, Phys. Rev. Lett. **44**, 95 (1980).
19. O.K. Andersen and O. Jepsen, Physica B&C **91**, 317 (1977).
20. E. Antonides, E.C. Janse and G.A. Sawatzky, Phys. Rev. **B 15**, 1669 (1977).
21. P. Thiry, D. Chandesris, J. Lecante, C. Guillot, R. Pinchaux and Y. Petroff, Phys. Rev. Lett. **43**, 82 (1979).
22. See, e.g., Antonides et al[20] and references therein.
23. J.W. Allen, S.J. Oh, O. Gunnarsson, K. Schönhammer, M.P. Maple, M.S. Torikachvili and I. Lindau, Adv. Phys. **35** (1986) 275.
24. O. Gunnarsson and K. Schönhammer: Phys. Rev. Lett. **50**, 604 (1983); Phys. Rev. **B 28**, 4315 (1983), and in: *Handbook on the Physics and Chemistry of Rare Earths*, eds. K. Gschneider, L. Eyring and S. Hüfner, Vol. 10 (North Holland, Amsterdam 1987), p. 103.
25. C.S. Wang, M.R. Norman, R.C. Albers, A.M. Boring and W.E. Pickett, H. Krakauer and N.E. Christensen, Phys. Rev. **B 35**, 7260 (1987).
26. N.E. Christensen, O.K. Andersen, O. Gunnarsson and O. Jepsen, J. Magn. Magn. Mater. (in press)
27. G. Zwicknagl, J. Magn. Magn. Mater. (in press)
28. M.R. Norman, T. Oguchi and A.J. Freeman, J. Magn. Magn. Mat. **69**, 27 (1987).
29. K. Schönhammer and O. Gunnarsson, Phys. Rev. **B 37**, 3128 (1988).

30. A.R. Mackintosh and O.K. Andersen, in *Electrons at the Fermi Surface*, edited by M. Springford (Camebridge, 1980), p 149.

31. O.K. Andersen, H.L. Skriver, H. Nohl and B. Johansson, Pure Appl. Chem. **52**, 93 (1979).

32. H.L. Skriver, O.K. Andersen and B. Johansson, Phys. Rev. Lett. **41**, 42 (1978); Phys. Rev. Lett. **44**, 1230 (1980).

33. O.K. Andersen, O. Jepsen and D. Glötzel, in *Highlights of Condensed-Matter Theory*, 1985, LXXXIX Corso, Soc. Italiana di Fisica,Bologna, p. 59.

34. J. Zaanen, G.A. Sawatzky and J.W. Allen, Phys. Rev. Lett. **55**, 418 (1985).

35. K. Terakura, A.R. Williams, T. Oguchi and J. Kübler, Phys. Rev. Lett. **52**, 1830 (1984).

36. J.W. Allen and R.M. Martin, Phys. Rev. Lett. **49**, 1106 (1982).

37. D. Glötzel, J. Phys. F **8**, L163 (1978).

38. A. Svane and O. Gunnarsson, Phys. Rev. **B 37**, 9919 (1988).

39. E.H. Lieb and F.Y. Wu, Phys. Rev. Lett. **20**, 1445 (1968).

40. A. Svane and O. Gunnarsson, Europhys. Lett. **7**, 171 (1988).

41. J. Zaanen, O. Jepsen, O. Gunnarsson, A.T. Paxton, O.K. Andersen and A. Svane, Physica C **153-155** 1636 (1988).

42. D. Vaknin, S.K. Sinha, D.E. Moneton, D.C. Johnson, J. Newsam, C.R. Safinya and H.E. King, Jr., Phys. Rev. Lett. **58**, 2802 (1987).

43. J.M. Ginder, R.M. Roe, Y. Song, R.P. McCall, J.R. Gaines, E. Ehrenfreund, A.J. Epstein, Phys. Rev. **B 37**, 7506 (1988).

44. P.W. Anderson, Phys. Rev. **124**, 41 (1961). P.W. Anderson: Phys. Rev. **124**, 41 (1961).

45. O. Sakai, H. Takahashi, H. Takeshige and T. Kasuya, Solid State Commun. **52** (1984) 997; R. Monnier, L. Degiorgi and D.D. Koelling, Phys. Rev. Lett. **86**, 2744 (1986); L. Degiorgi, T. Greber, F. Hullinger, R. Monnier, L. Schlapbach and B.T. Thole, Europhys. Lett. **4** (1987) 755; J.M. Wills and B.R. Cooper, Phys. Rev. **B 36**, 3809 (1987); A.K. McMahan and R.M. Martin, to be published

46. O. Gunnarsson, N.E. Christensen and O.K. Andersen, J. Magn. Magn. Mat.

47. H. Ehrenreich, Science **235** (1987) 1029, and references therein.

48. O. Gunnarsson, O.K. Andersen, O. Jepsen and J. Zaanen, *Ab initio* calculation of the parameters in the Anderson model, *in:* Core Level Spectroscopy in Condensed Systems, ed., J. Kanamori and A. Kotani, Springer, Berlin (1988) p. 82; submitted to Phys. Rev.

49. M. Schlüter, M.S. Hybertsen and N.E. Christensen, Physica B&C; C.F. Chen, X.W. Wang, T.C. Leung and B.N. Harmon, (to be publ.), A.K. McMahan, R.M. Martin and S. Sapathy (to be publ.)

50. O. Gunnarsson and K. Schönhammer, Phys. Rev. Lett. **41**, 1608 (1978); O. Gunnarsson, K. Schönhammer, D.D. Sarma, F.U. Hillebrecht and M. Campagna, Phys. Rev. **B 32**, 5499 (1985); J.C. Fuggle, O. Gunnarsson, G.A. Sawatzky and K. Schönhammer, Phys. Rev. **B 37**, 1103 (1988).

51. O. Gunnarsson and O. Jepsen, Phys. Rev. **B 38**, 3568 (1988).

DIMENSIONALITY EFFECTS IN *trans*-$(CH)_x$

P. Vogl * and D. K. Campbell

Theoretical Division and Center for Nonlinear Studies
Los Alamos National Laboratory, Los Alamos, NM 87545, USA

INTRODUCTION

In a conference proceedings devoted to "interacting electrons in reduced dimensions" it might at first seem inappropriate to discuss a calculation of the full three-dimensional (3-D) structure of *trans*-polyacetylene [*trans*-$(CH)_x$]. This impression is only strengthened by the very extensive literature on one-dimensional (1-D) models of *trans*-$(CH)_x$, much of it focusing on the seemingly more relevant—at least in the present context—issue of the relative importance of electron-electron (e-e) versus electron-phonon (e-p) interactions [for reviews, see[1–6]]. Nonetheless, as we hope will become apparent, the (often fairly small) interchain couplings that are present in the real solid state materials can alter profoundly expectations based on strictly 1-D models. For *trans*-$(CH)_x$ this is particularly the case. Thus, if nothing else, our contribution serves as a reminder that the issue of dimensionality can in fact be absolutely crucial in determining the properties of even highly anisotropic, "low"-dimensional materials, independent of which interactions appear to be dominant. To begin our discussion, we start from an analysis of the ground state of crystalline *trans*-$(CH)_x$.

THE GROUND STATE OF CRYSTALLINE, 3-D *trans*-$(CH)_x$

To calculate the ground state properties of crystalline *trans*-$(CH)_x$, we have employed local-density-functional theory in both spin-polarized (LSDA) and spin-unpolarized (LDA) forms,[7] using the *ab-initio* pseudopotential plane-wave method.[8,9] LDA has been shown to reproduce accurately structural properties of a broad range of atoms,[10,11] molecules,[12] and solid-state systems,[13,14] including semiconductors, insulators, and van der Waals crystals[15] from first principles. In most cases, quantities involving the total energy, such as lattice constants, bulk moduli, phonon frequencies, charge densities and structural stabilities, have been found to agree within a few percent of experimental values.

Our method consists of applying the formalism of Ihm *et al.*[9] in the plane-wave representation, evaluating the exchange-correlation energy using the functional given by Painter[16], and solving the Kohn-Sham equations [17,18] self-consistently. We use an approach based on the Hellman-Feynman theorem [see Ihm *et al.*[9]] to study the dependence on ionic positions. The full technical details are available in the literature[19,20,21].

---

* Permanent Address: Institut für Theoretische Physik, Universität Graz, Graz A-8010, Austria

Interacting Electrons in Reduced Dimensions
Edited by D. Baeriswyl and D.K. Campbell
Plenum Press, New York

Most recent experimental structural data indicate that *trans*-$(CH)_x$ has a monoclinic unit cell with $P2_1/c$ symmetry.[22] The unit cell contains two transplanar C_2H_2 chain units, with one shorter and one longer carbon-carbon bond in each chain corresponding to a dimerized structure. The precise relationship between the two polyacetylene chains in the unit cell has remained controversial. Two possible configurations have been proposed: (1) $P2_1/a$, with the dimerization in phase on adjacent chains,[23]; and (2) $P2_1/n$, with the dimerization out of phase on adjacent chains[24]. In order to characterize completely either of these two structures, several constants must be known. The include (with the experimental values in parentheses): the lattice constants a (4.18 Å), b (7.34 Å), c (2.455 Å), the monoclinic angle β (90.5 deg) between a and c, two setting angles ϕ_C (57 deg) and ϕ_H (57 deg) between the plane containing the carbon and hydrogen chain, respectively, and the b-axis, the projection p (0.68 Å) of the C-C distance within a chain onto the $a - b$ plane, the distance ℓ_{C-H} (1.09 Å) between the carbon and hydrogen atoms within a chain, the dimerization amplitude u_C (0.052 Å) of the C-C bonds in the c-direction, and the displacement u_H (0.025 Å) of the hydrogen atoms relative to their adjacent carbon atoms.[23,25] Many of these quantities are apparent in Fig. 1, in which we depict the unit cell of *trans*-$(CH)_x$ in the $P2_1/a$ structure.

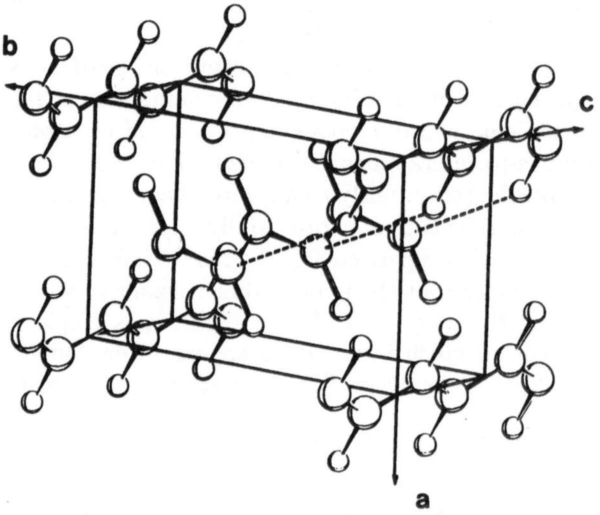

Fig. 1. Conventional unit cell of *trans*-$(CH)_x$ in the $P2_1/a$ structure. The dashed lines are drawn to emphasize a characteristic feature of this structure: the hydrogen atoms of one chain point approximately towards the carbon atoms of the neighboring chains.

For both the (purely theoretical) self-consistent values of these parameters and for the (actual) experimental values, our calculations show that the $P2_1/a$ structure is preferred. For the experimental parameters, it is lower in total energy than the $P2_1/n$ structure by 0.01 ± 0.003 eV per $(C_2H_2)_2$ unit. This difference is consistent with the similarity of the two structures and is of the order of optical phonon energies in solids. Parenthetically, we note that the self-consistent values of the parameters are in good agreement with the experimental values, except in the (important !) case of the dimerization, for which the self-consistent value is substantially smaller than the measured value. Possible origins and consequences of this discrepancy are discussed in detail elsewhere [19,20,21]; here we observe simply that this discrepancy does not affect any of the later conclusions of this article.

It is instructive to interpret our results in the context of several earlier qualitative arguments that were advanced in favor of one or the other structure for crystalline *trans*-CH_x. On the basis of the tight-binding SSH-model,[26] including (constant) interchain hopping matrix elements at each site, it was found[27] that an *alternating* dimerization on neighboring chains (as in the $P2_1/n$ structure (see, e.g., [21])) is energetically favorable, provided there is particle-hole symmetry.[27] Alternatively, if one allows the inter-chain hopping integrals to alternate in sign, then one can argue[27] that the $P2_1/a$ structure is favored. Based on the electronic charge distributions predicted by the LDA [19,20,21] one can show that there is non-negligble interchain (chemical) bonding interaction between carbon and hydrogen atoms on adjacent chains. This interaction violates particle-hole symmetry. Each H-atom on one chain points approximately towards a carbon p-orbital on one of the neighboring chains. These adjacent atoms are shown in Fig. 1 by dashed lines. The distance between the H-atom on one chain and its nearest carbon neighbor on another chain is 2.9 Å in the $P2_1/a$ structure, and 3.1 Å in the $P2_1/n$ structure. This difference provides a crucial additional bonding energy that favors the $P2_1/a$ structure.

Within the full LDA approach, we can calculate a number of aspects of the electronic structure, including spin polarization and charge densities, interchain electron-phonon interactions, and density of states. Here we focus on the basic electronic band structure, since this will be essential to our later discussion. In Fig. 2a we show the LDA results for the electronic band structure for the $P2_1/a$ phase of crystalline *trans*-$(CH)_x$. The figure caption provides details. This figure shows that the interchain interactions alter crucially the symmetry and dimensionality of the band states close to the minimum gap.

To gain a more intuitive feel for these interchain effects, it is useful to think in terms of the simple tight-binding models, in which the transverse couplings and band structures are immediately related. Following the standard tight-binding results, we can equate the width of the splitting in the band edge states near the $\vec{k}$-point B in Fig. 2a to the transverse band width, $4t_\perp$. With this identification, we obtain an effective interchain hopping matrix element $t_\perp = 0.12$ eV, compared to $t_\parallel = 2.5$ eV for the intra-chain coupling matrix element. These values are roughly consistent with experimental anisotropy factors of the order of 100 in the conductivity, [28] since the ratio of transverse to longitudinal conductivity is proportional to $(t_\perp/t_\parallel)^2$. A tight-binding analysis of optical experiments in *trans*-$(CH)_x$ under pressure[29] has led to estimates for $4t_\perp = 0.3$ eV, which is also consistent with our first-principles results. The interchain effects are very sensitive to crystalline order; by increasing the distance between the two chains by 5%, the splittings in the band edge states decrease by 31%. This result suggests that in materials with substantial amorphous regions or large numbers of extrinsic defects, the strong constraints of the crystalline phase may be somewhat relaxed. In particular, since the computed energy difference between the $P2_1/a$ and $P2_1/n$ phases is very small, it is possible that both phases may appear in actual samples, depending on crystal growth and preparation conditions, as well as on levels of extrinsic defects and impurities.

Significantly, in view of the underestimate of the dimerization in the self-consistent parameters, it can be shown that [19,20,21] the crucial band-edge splittings are surprisingly insensitive to the value of the dimerization. Since these band-edge states play an essential role both for low-lying excitations and in the formation of intrinsic defects this observation is most crucial. In the next section, we turn to the study of these intrinsic defects.

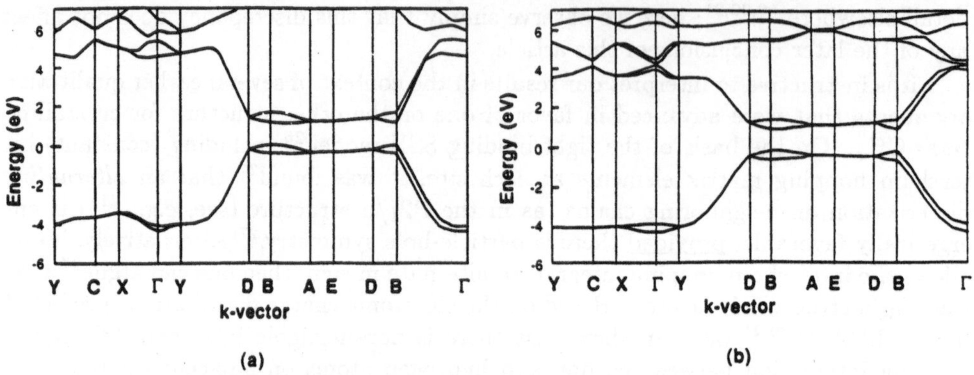

Fig. 2. Comparison of calculated electronic band structures within (a, left) the density-functional approach and (b, right) the multi-orbital tight-binding model used to compute the electronic structure of the intrinsic defects. In both cases the zero of energy has been chosen to lie within the energy gap. The dimerization has been set to the experimental value of $u_C = 0.05$ Å ; in (a) the other structural parameters have been optimized self-consistently. The conduction bands include a k-independent self-energy correction of $\Sigma_{el} = 0.74$ eV (see discussion in [19,20,21]). In (b), the various two-body matrix elements have been optimized to fit the structure predicted by the full LDA results. The similarity of the bands close to the minimal energy gap lends credence to the tight-binding calculations.

INTRINSIC DEFECTS IN *trans*-$(CH)_x$

Within the adiabatic approximation, charge carriers in strictly one-dimensional deformable systems will always self-trap due to the electron-phonon interaction. Typically, the self-trapped charge and the accompanying atomic displacement pattern form a polaron, although in the case of 1-D *trans*-$(CH)_x$ the degeneracy of the ground state implies that "kink" solitons can also be formed. In three dimensions, on the other hand, the electron-phonon interaction is able to localize charge carriers only in the limit of strong coupling, which is not thought to be relevant for polyacetylene. Thus in general, carriers in intrinsically 3-D materials remain delocalized. Emin[30] has given an excellent discussion of this general behavior, as well as of exceptions, in a wide range of materials. Intuitively, the situation is analogous to a particle in an attractive well. In 1-D, there is a bound state for any well-depth, whereas in 3-D, a localized state occurs only when the potential strength exceeds a certain threshold value.

Given our conclusion that the states near the band edge in crystalline *trans*-$(CH)_x$ exhibit splittings due to interchain couplings, it is natural to ask whether this interchain interaction is sufficiently strong to render the material intrinsically 3-D and hence unlikely to support polarons (see, e.g.,[31]) or whether charge carriers self-trap, as predicted by most of the (primarily 1-D) theoretical models studied thus far.

Unfortunately, a fully self-consistent density functional calculation of an extended defect in a crystal as complicated as that of *trans*-$(CH)_x$ is at present not feasible, even with modern supercomputers. Therefore, we have mapped the first-principles calculations for the perfect *trans*-$(CH)_x$ crystal onto an accurate, multi-orbital, three-dimensional tight-binding model. Although empirical, this approach is, we believe, quite accurate; indeed, it has recently been demonstrated that the electronic structures of a wide variety of solids can be successfully predicted with semi-empirical tight-binding models containing only universal Hamiltonian matrix elements.[32,33] Following these ideas, we have determined the matrix elements by initializing the fitting procedure with the universal tight-binding parameters of Ref. 33. Subsequently, these parameters have been fine-tuned to obtain quantitative agreement with our first-principles local density calculations. It turns out that a nearest-neighbor tight-binding model with a single *s*-state and three *p*-states per site, plus the interactions between the H and the adjacent C atoms on neighboring chains (as indicated by the dashed lines in Fig. 1), can account *quantitatively* for all of the following electronic structure properties, as calculated by the LDA: (i) the electronic band structure in the gap region; (ii) the atomic character of the bands as obtained from integrated charge densities in each valence band; and (iii) the deformation potentials, i.e. the changes of the band edge states upon dimerization, throughout the Brillouin zone. In Fig. 2b we show the calculated tight-binding band structure for the $P2_1/a$ structure of *trans*-$(CH)_x$. The lower energy bands differ somewhat from the LDA-bands in Fig. 2a, but all major features of the bands near the Fermi energy which are relevant for the defect calculations discussed below are accurately reproduced.

In connection with this tight-binding fit, it is important to note that the splittings of the band edge states are predominantly caused by the $t_{C-pH-p\sigma}$ interchain matrix element; the contribution of the $t_{C-pH-s\sigma}$ matrix element to the band edge states is zero by symmetry. This geometrical fact that the interchain coupling is mediated by a particular carbon-hydrogen overlap has especially important consequences for comparisons with previous theoretical attempts to take interchain couplings into account within a tight-binding approximation. In particular, simple "two-chain" models[6,27,34–36] can *not* fit our full 3-D tight-binding band structure, since the geometrical linkages dictated by the manner in which the hydrogen and carbon orbitals overlap causes adjacent carbons on a given chain to interact effectively with *different* chains in the unit cell. Thus although one can concoct a straightforward tight-binding scheme with a single effective interchain coupling, it does not reduce to a two-chain model. We shall return to this point in the next section.

To study the influence of the interchain interaction on intrinsic defects in 3-D crystalline *trans*-$(CH)_x$, we consider frozen-in displacement patterns of the ions corresponding to "polaron-like" and "bipolaron-like" configurations. We then investigate the electronic structure corresponding to these configurations to see whether there are any *intra*-gap electronic "bound" states. If there are no such states and all electronic states are *d*elocalized, then we can conclude that there are no localized, self-trapped intrinsic defects and that the 3-D couplings have destablized the nonlinear excitations characteristic of 1-D systems.

To avoid possible confusion about the interpretation of our results, we should define these various configurations and the corresponding lattice distortions more precisely. In the 1-D continuum version of the SSH model, the lattice displacement order-parameter $\Delta(x)$ associated with a polaron is shown schematically in Fig. 3a for both the "single site" and "multi-site" polaron-like lattice distortions. In Fig. 3b we sketch schematically the corresponding distortions themselves.

In Fig. 4a, we show the displacement order parameter $\Delta(x)$ corresponding to the bipolaron-like displacement pattern shown in Fig. 4b. This displacement corresponds to a very weakly bound kink/antikink pair, the "separation" of which is called d_{BP} in the Fig. 4a.

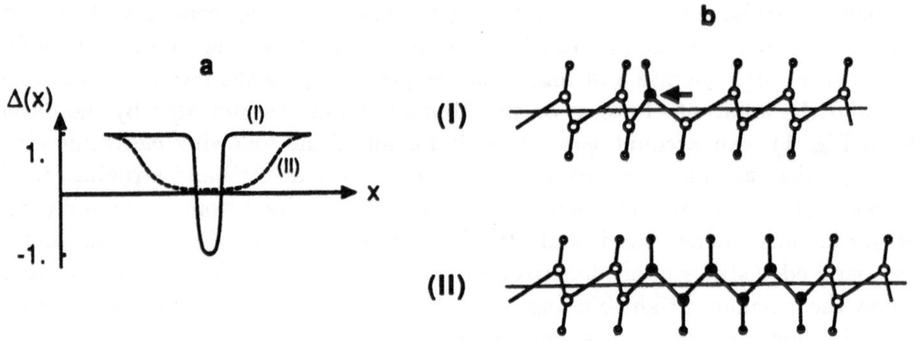

Fig. 3. Schematic atomic displacement patterns corresponding to the "polaron-like" defect discussed in the text: (a) sketches of the "continuum limit" order parameter for the case of the (strongly distorted) single-site polaron (solid curve, labeled I) and extended polaron (dashed curve, labeled II); (b) the atomic positions (with the dimerization amplitude greatly exaggerated for clarity) corresponding to (I) a localized, "single-site" polaron and (II) an extended polaron.

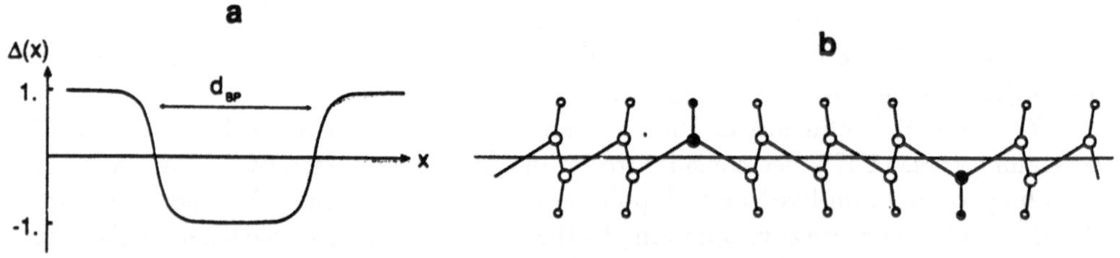

Fig. 4. Schematic atomic displacement pattern corresponding to the "bipolaron-like" defect discussed in the text: (a) a sketch of the "continuum limit" order parameter for the bipolaron, showing the effective separation, d_{BP}, of the kink and antikink; (b) the atomic positions (with the dimerization amplitude greatly exaggerated for clarity) corresponding to a bipolaron, i.e. a pair of confined kink anti-kink solitons.

In the limit of infinite displacement $\delta u_C = \infty$, the approximate particle-hole symmetry guarantees that even the single-site "polaron-like" defect (Fig. 3a(I) and 3b(I)) must eventually develop two bound states close to mid-gap, since such a displacement simply corresponds to breaking the bonds. Similarly, a bipolaronic displacement with complete bond-order reversal—such as depicted in Fig. 4b—will also lead to near-mid-gap levels in the limit of infinite separation of the kink and anti-kink, since this situation represents two separated solitons. These limiting cases are not relevant to real defects, of course.

To examine a spatially localized defect in an otherwise perfect crystal with translational invariance it is most convenient to use the scattering theoretic Green function technique developed by Koster and Slater.[37] The details of our specific application are available in the literature.[19,20,21]

POLARON-LIKE DEFECTS

In our calculation of "polaron-like" configurations, we shall use the distortions shown schematically in Figs. 3(b). Note that in all cases, we shall assume that the distortions occur on a single chain and are embedded in an otherwise perfect three-dimensional crystal. In single *trans*-$(CH)_x$ chains, theoretical studies based on the SSH model predict that a polaron is the lowest energy state available to a single added electron (or hole).[38,39,40]

Let us start with the single-site polaron-like distortion (Fig. 3b(I)). Recent calculations show[41] that the bound states in the energy gap produced by such a "single-site" defect are only slightly shallower than those produced by an extended defect. Thus by increasing the distortion in the former defect, we can simulate a spatially more extended lattice distortion. In Fig. 5, we have plotted the bound states in the energy gap as a function of the displacement of the single-site polaron. The dashed lines are single-chain calculations, where all interchain matrix elements are set equal to zero, while the full three-dimensional calculations are indicated by full lines. The position of the polaron bound state energies in a simple 1-D continuum model are also shown. As one can deduce from this figure, a 10% distortion δu_C – corresponding to a "reversal" of the dimerization, since this is only about 5% in the ground state – leads to bound state energies close to those of the continuum model, *provided* we turn off all interchain couplings. Once the interchain coupling is included, however, *no bound states appear in the energy gap* up to distortions exceeding 35% of the bond length, which are physically unattainable.

BIPOLARON-LIKE DEFECTS AND KINK SOLITONS

SSH-type calculations in single chain models of *trans*-$(CH)_x$ predict that photoexcited electron-hole pairs form bipolarons which rapidly decay into separated kink and antikink solitons.[42,43] In Fig. 4 we depict the displacement pattern corresponding to separated kink and anti-kink solitons. The distance between the two undimerized C-H pairs in this figure is 2.5 lattice constants along the c-axis. We have performed three-dimensional Green function calculations with frozen-in "bipolaron-like" distortions having various separations d_{BP} on a single chain in otherwise perfect crystalline *trans*-$(CH)_x$ in the $P2_1/a$ structure. The results for the binding energies as a function of d_{BP} are summarized in Fig. 6. As in Fig. 5, the dashed lines show the results with all interchain matrix elements set equal to zero, while the full lines represent the defect levels in the three-dimensional crystal.

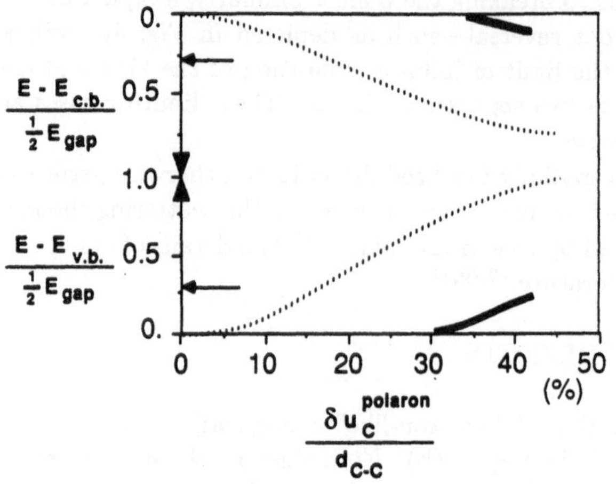

Fig. 5. The intragap electronic levels as a function of distortion for a "single-site" polaron, as depicted in Fig. 3b(I). The solid curve is with the actual interchain coupling predicted by our calculations; the dashed curve is the idealized purely 1-D result. The arrows indicate the position of the bound polaron states in a continuum SSH-type 1-D model.

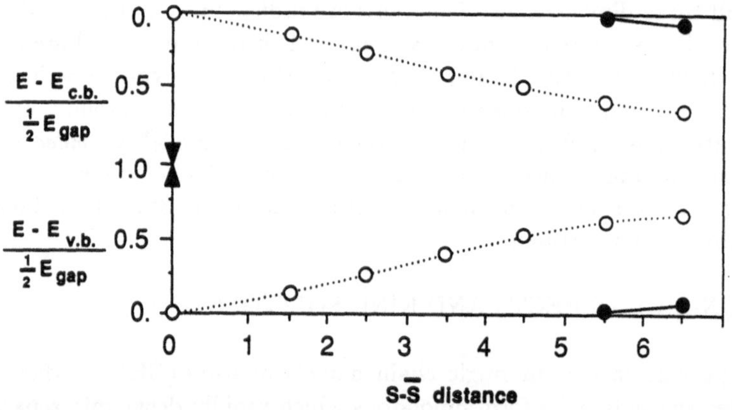

Fig. 6. The intragap electronic levels as a function of separation, here called "$s - \bar{s}$ separation" but called d_{BP} in the text for a bipolaron-like defect described in Fig. 4. The solid circles are with interchain coupling; the open circles are the purely 1-D result. The connecting lines are drawn only to guide the eye. In our calculations, the distorted chain is embedded in the 3-D crystal.

As in the case of the polaron, interchain coupling strongly impedes the formation of bound states in the energy gap of three-dimensional *trans*-$(CH)_x$. Only for separations exceeding 6 lattice constants (12 carbon sites in chain direction) between the kink- and antikink-soliton on one chain does a pair of shallow bound states appear in the energy gap. For larger separations, two bound states remain in the gap and, in the limit of infinite kink-antikink separation, lie close to the gap center. Due to the mixing of carbon and hydrogen wave functions and the consequent breaking of particle-hole symmetry, they do not lie exactly in the center of the gap even in the limit of infinite separation. For more realistic separations, the bound states are quite shallow; by extrapolating our results we estimate the binding energies to barely exceed room temperature thermal energies for distances smaller than 8-10 lattice constants, for example.

Although bipolaron lattice configurations do produce intragap electronic bound states, the stability of genuine bipolarons requires more. First, one must analyze the overall energetics of bipolaron formation. Apart from the energy to distort the particular chain on which the bipolaron resides, the formation of a separated kink-antikink pair costs interchain bonding energy, since the dimerization between the kinks is out of phase with the dimerizations on the adjacent chains. We estimate from our total energy calculations that the maximum kink separation attainable in a perfectly ordered crystal is of the order of 50 lattice constants in chain direction. Since the electronic wave function in a kink-bound state is expected to extend over 5-10 lattice constants,[26] this interchain confinement to 50 lattice constants strongly limits the formation of a bipolaron. In the $P2_1/a$ structure, it is probably less than that due to the stronger interchain coupling. Second, if two electrons are localized in the bipolaronic intra-gap levels, Coulomb repulsion effects will further reduce the total effective binding energy and hence tend to destabilize the bipolaron .

Although these results do not entirely preclude the possibility of formation of bipolarons in perfect crystalline *trans*-$(CH)_x$, they do call into question previous discussions of the energetics of these nonlinear excitations. In particular, our results appear to make it more likely that thermal fluctuations or direct Coulomb repulsion between the two electrons in the bipolaron can destabilize this localized intrinsic defect even if it is formed.

A 3-D, ONE-BAND, TIGHT-BINDING MODEL

For purposes of studies of more complex properties of *trans*-$(CH)_x$, it is most valuable to have a model which, while faithfully reflecting the 3-D character of the band edge found in the quantitative calculations, is simpler and physically more transparent than either the full-fledged LDA calculation of the multi-orbital tight-binding fit. In this section we develop a simple 3-D generalization of the 1-D SSH model which takes into account the interchain interaction in an approximate, but qualitatively correct way. This model should be suitable for studies of dimensionality effects on optical absorption and dynamical properties (such as the formation of breathers) which are likely to remain inaccessible to first-principles calculations.

By analogy to SSH, we shall include explicitly only one C-p orbital per atom; in the 1-D case, this would lead to the familiar π-molecular orbital. The C-p orbital interacts both with its (two) neighboring C-atoms on the same chain and with (certain of) the C-atoms on adjacent chains. As noted above and suggested by Fig. 1, since this interchain coupling in the actual material is mediated by the hydrogen atoms, the resulting effective carbon-carbon interaction across the chains couples only *particular* C-atoms. This is illustrated specifically in Fig. 7, in which the interchain couplings appropriate for *trans*-$(CH)_x$ in the $P2_1/a$ structure are shown on a sketch of the unit

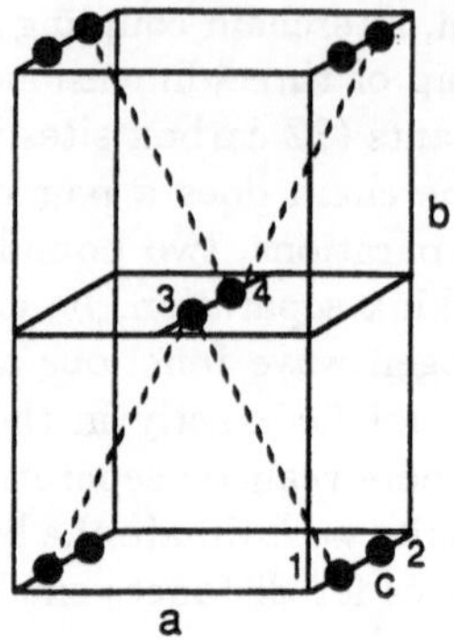

Fig. 7. Schematic depiction of the (assumed orthorhombic) unit cell of *trans*-$(CH)_x$ in the $P2_1/a$ structure. Only the positions of the C-atoms are shown: further, since the effects of their detailed "zig-zag" positions along the chain can be incorporated into the tight-binding parameters, only the projections onto the c-axis are shown. The dominant interchain Hamiltonian matrix elements are indicated by dashed lines. They originate in the short-range electron-ion interaction that is mediated by the hydrogen atoms.

cell. Since the monoclinic angle between a and c differs very little from 90 degrees (see above), we shall henceforth consider for simplicity an orthorhombic cell.

We can label the atomic positions in the crystal by $\mathbf{x_i} + \mathbf{R}$, where $\mathbf{R}$ denotes the lattice vectors. The atoms labeled $1 = 1, 2, 3, 4$ are characterized by $\mathbf{R} = 0$. Taking into account the alternating short and long bonds and consequent alternation of the Hamiltonian matrix elements, we can write

$$\begin{aligned} &\langle \mathbf{x}_i|H|\mathbf{x}_i\rangle = 0 \ \forall i \ , \\ &\langle \mathbf{x}_1|H|\mathbf{x}_2\rangle = t_{long} \ , \ \langle \mathbf{x}_i|H|\mathbf{x}_2 - (0,0,c)\rangle = t_{short} \ , \\ &\langle \mathbf{x}_3|H|\mathbf{x}_4\rangle = t_{long} \ , \ \langle \mathbf{x}_3|H|\mathbf{x}_4 - (0,0,c)\rangle = t_{short} \ . \end{aligned} \tag{1}$$

The nonzero *inter*chain Hamiltonian Hmatrix elements are given by

$$\begin{aligned} &\langle \mathbf{x}_1|H|\mathbf{x}_3\rangle = \langle \mathbf{x}_1|H|\mathbf{x}_3 - (a,0,0)\rangle = t_{\perp a} \ , \\ &\langle \mathbf{x}_2|H|\mathbf{x}_4 - (0,b,0)\rangle = \langle \mathbf{x}_2|H|\mathbf{x}_4 - (a,b,0)\rangle \simeq t_{\perp a} \ . \end{aligned} \tag{2}$$

The interchain matrix elements on the first and second lines of eq. (2) are not required to be equal by symmetry, but since the difference is very small we choose to neglect it here.

In **k**-space, one finds the non-zero Hamiltonian matrix elements to be

$$\begin{aligned} &\langle 1|H_k|2\rangle = \langle 3|H_k|4\rangle = t_{long} + e^{ik\cdot(0,0,-c)}t_{short} \ , \\ &\langle 1|H_k|3\rangle = t_{\perp a} + e^{ik\cdot(-a,0,0)}t_{\perp a} \ , \\ &\langle 2|H_k|4\rangle = e^{ik\cdot(0,-b,0)}t_{\perp a} + e^{ik\cdot(-a,-b,0)}t_{\perp a} \ . \end{aligned} \tag{3}$$

where $\mathbf{k} = (2\pi\kappa_x/a, 2\pi\kappa_y/b, 2\pi\kappa_z/c)$ with the dimensionless numbers $\kappa_x, \kappa_y, \kappa_z$ in the range (-0.5,0.5). The four eigenvalues of this Hamiltonian are

$$E_k = \pm\{T_{||}^2 + T_{\perp}^2 \pm 2T_{||}T_{\perp}cos(\pi\kappa_y)\}^{1/2} , \tag{4}$$

where

$$T_{||}^2 = t_{short}^2 + t_{long}^2 + 2t_{short}t_{long}cos(2\pi\kappa_z) , \tag{5a}$$

$$T_{\perp}^2 = 4t_{\perp a}^2 cos^2(\pi\kappa_x) . \tag{5b}$$

The minimum energy gap occurs at the k-point $\kappa = (0, 0, 0.5)$, where

$$E_{\kappa=(0,0,0.5)} = \pm|t_{long} - t_{short}| \pm 2t_{\perp a} . \tag{6}$$

The total band width is given by $2(T_{\perp}+T_{||})$. In Fig. 8 we show the band structure that results from this simplified one-orbital tight-binding model. In Fig. 8 we have assumed that $t_{long} = t_0 - \alpha u_c$ and $t_{short} = t_0 + \alpha u_c$ and used the values $t_0 = 2.5eV$, $\alpha = 7.5eV/\AA$, $u_c = 0.05\AA$, and $t_{\perp a} = 0.12eV$. The agreement between the band structure shown in Fig. 8 and those obtained from both the full first-principles calculation (Fig. 2a) and the multi-orbital tight-binding fit (Fig. 2b) is quite satisfactory, particularly given the simplicity of the one-orbital model.

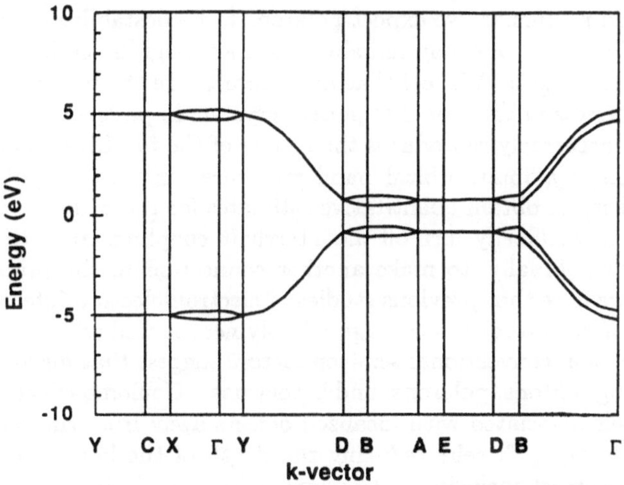

Fig. 8. The band structure calculated from the 3-D SSH-type model using the parameters discussed in the text. This should be compared with the results of the full LDA calculation (Fig. 2a) and of the multi-orbital, tight-binding model (Fig. 2b).

As an illustration of the utility of this simple model, we calculate the band structure energy E_{BS} – that is, the energy summed all occupied band states– per unit cell. In order to obtain an analytic expression, we evaluate this sum at only one special k-point – $\kappa = (0.25, 0.25, 0.25)$ in the irreducible wedge [44]. For the parameters appropriate to *trans*-$(CH)_x$, this introduces an error of less than 10 %. Since the interchain interaction is much smaller than the intrachain interaction, we can expand E_{BS} with respect to $t_{\perp a}/t_{\parallel}$, where $t_{\parallel}^2 = t_{short}^2 + t_{long}^2$. To lowest order in $t_{\perp a}/t_{\parallel}$, this gives

$$E_{BS} = \frac{1}{N} \sum_{nk}^{\propto c} E_k \approx -4t_{\parallel} - \frac{2t_{\perp a}^2}{t_{\parallel}} . \tag{7}$$

where N is the number of unit cells in the crystal. Carrying out a similar calculation for the $P2_1/n$ structure leads to an expression for E_{BS} of the same form as eq. (7), but with the (smaller) transverse hopping appropriate to the $P2_1/n$ case replacing $t_{\perp a}$. Thus this simple 3-D SSH-type model does capture the correct energy ordering of the two structures and, for the expected values of the parameters, gives a fairly accurate quantitative fit as well. We are therefore hopeful that it will prove a useful first approximation in more detailed studies of the effects of dimensionality on the properties of *trans*-$(CH)_x$.

DISCUSSION AND CONCLUSIONS

As we remarked at the outset, our 3-D LDA calculation is in a sense complementary to the many recent detailed 1-D studies of the effects of correlations in conducting polymers. Dimensionality, like electron-electron interactions, can have a profound effect on the properties of many of the recently discovered novel materials. We believe that our LDA results, although by no means definitive, strongly suggest that the effect of three-dimensionality is essential to understanding the behavior of crystalline *trans*-$(CH)_x$. Among our present results, perhaps the most striking is the conclusion that the localized intrinsic defects predicted by the 1-D models are substantially altered by 3-D effects. In particular, we expect polarons to be destabilized by the interchain interactions while bipolarons may remain stable, although much less robust that the 1-D models would suggest. We feel that these results are the best presently available calculations of defects in the true 3-D perfect crystal. In particular, our multi-orbital tight-binding fit accurately reproduces the results of the full LDA calculation, including realistic geometry, multi-orbital band structure, and electron-phonon coupling constants. Further, we obtain quantitative estimates for the behavior of intragap levels which, when we artifically turn off the interchain coupling, correctly reproduce the earlier 1-D results. Finally, to make another connection to the primary subject of the conference, we note that previous studies of electron-electron interaction effects in 1-D models of $(CH)_x$ and related conjugated polymers as well as substantial literature on Coulomb effects in conventional semiconductors suggest that for localized intrinsic defects – including solitons, polarons, and bipolarons – Coulomb effects typically drive the intragap levels associated with localized defects away from the center of the gap toward the band edges, thereby reducing the depth of the bound states. Thus, we believe, that the correct inclusion of Coulomb effects would further destabilize these localized excitations.

In conclusion we come to the essential issue of the applicability of our results to presently existing conducting polymers. Here we should re-emphasize that our calculations have been carried out for perfectly-ordered crystalline, three-dimensional *trans*-$(CH)_x$. At present, the chain lengths even in highly oriented, crystalline Graz/Durham *trans*-$(CH)_x$ are smaller than 40 C-H units[23,45] and there is a high density of broken bonds. In the morphologically complex Shirakawa material, many chains are in a highly disordered environment near the fibril surface;[46,47] it is thus conceivable that observables such as optical excitations in this material exhibit properties which are more characteristic of single-chains of *trans*-$(CH)_x$ than of a three-dimensionally ordered crystal. The ultimate explanation of the various physical phenomena associated with the intra-gap levels observed in real samples of *trans*-$(CH)_x$ will require considerable further study and is obviously beyond the scope of any single article. We hope, however, that our present LDA calculations and the simplified 3-D SSH-type model represent essential foundations for these further investigations of these exciting novel materials.

ACKNOWLEDGEMENTS

We are grateful to Dionys Baeriswyl, Richard Friend, Günter Leising, and Otto Sankey for many enlightening discussions and to the staff of the Institute for Scientific Interchange for providing such a pleasant venue for the meeting. We thank the Centers for Materials Science and Nonlinear Studies for computational support at Los Alamos National Laboratory and the Computational Sciences Division of the U.S. DOE for computational support at the National Magnetic Fusion Energy Computing Center at Lawrence Livermore National Laboratory. PV is grateful to the Max Kade Foundation for support of his visit to Los Alamos, where this work was begun. DC thanks the Institut für Theoretische Physik of the Karl-Franzens Universität in Graz for its hospitality during the completion of the work.

REFERENCES

1. For the most recent in the series of conference proceedings reviewing all aspects of this and related issues, see Syn. Met. 27 (1988), 28 (1989), and 29 (1989).
2. A.J. Heeger, S. Kivelson, J. R. Schrieffer, and W-P. Su, Rev. Mod. Phys. 60, 781 (1988).
3. D. Baeriswyl, D.K.Campbell, and S. Mazumdar, in *Theory of Pi-Conjgated Polymers*, edited by H. Kiess (Springer, Heidelberg, 1989), in press.
4. Lu Yu, *Solitons and Polarons in Conducting Polymers*, (World Scientific, Singapore, 1988).
5. A. A. Ovchinnikov and I. I. Ukrainskii, Sov. Sci. Rev. B, Chem. 9, 125 (1987).
6. S. A. Brazovskii and N. N. Kirova , Sov. Sci. Rev. A, Phys. 5, 100 (1984).
7. *Theory of the Inhomogeneous Electron Gas*, ed. by S. Lundquist and N.H. March (Plenum, New York, 1983), and references therein.
8. D.R. Hamann, M. Schlüter, and C. Chiang, Phys. Rev. Lett. 43, 1494 (1979).
9. J. Ihm, A. Zunger, M.L. Cohen, J. Phys. C 12, 4409 (1979).
10. O. Gunnarson and R.O. Jones, Physica Scripta 21, 394 (1980).
11. The LDA or LSDA does not become exact in the limiting case of an hydrogen atom; the error in the wave function is very small however, as shown by O. Gunnarson, B.I. Lundquist, and J.W.Wilkins, Phys. Rev. B 10,1319 (1974).
12. B. Delley, P.Becker, B.Gillon, J. Chem. Phys.80, 4286 (1984); B. Delley, A.J. Freeman, and D.E. Ellis, Phys. Rev. Lett. 50, 488 (1983).
13. J. Callaway and N. H. March, Solid State Physics 38, 136 (1984).
14. A discussion of recent work on structural properties with this method is given by, e.g., S. Fahy, S.G. Louie , Phys. Rev. B 36, 3373 (1987) and references therein.
15. S. B. Trickey, F. R. Green, Jr., and F. W. Averill, Phys. Rev. B 8 4822 (1973).
16. G.S. Painter, Phys. Rev. B 24, 4264 (1981).
17. W. Kohn and L.J. Sham, Phys. Rev. 140, A1133 (1965).
18. O. Gunnarson and B.I. Lundquist, Phys. Rev. B 13, 4274 (1976).
19. P. Vogl, D.K. Campbell, and O. Sankey, Syn. Met. 28, D513 (1989).
20. P. Vogl and D. K. Campbell, Phys. Rev. Lett. 62, 2012 (1989).

21. P. Vogl and D. K. Campbell, "First-Principles Calculations of the Three-Dimensional Structure and Intrinsic Defects in trans- Polyacetylene", Phys. Rev. B, to appear.
22. J. P. Pouget, in *Electronic Properties of Polymers and Related Compounds*, edited by H. Kuzmany, M. Mehring, and S. Roth, (Springer, Heidelberg, 1986), Springer Series in Solid State Sciences, Vol. 63, p. 26.
23. H. Kahlert, O. Leitner, and G. Leising, Syn. Met. 17, 467 (1987).
24. C. R. Fincher, C.-E. Chen, A.J. Heeger, A. G. MacDiarmid, and J.B. Hastings, Phys. Rev. Lett. 48, 100 (1982).
25. R.H. Baughman and G. Moss, J. Chem. Phys. 77, 6321 (1982).
26. W.-P. Su, J.R. Schrieffer, and A.J.Heeger, Phys. Rev. Lett. 42, 1698 (1979); Phys. Rev. B 22, 2099 (1980); Phys. Rev. B 29, 2309 (1984).
27. B. Horovitz, Phys. Rev B 12, 3174 (1975); D. Baeriswyl and K.Maki, Phys. Rev. B 28, 2068 (1983) and Syn. Met. 28 D507 (1989).
28. G. Leising, R. Uitz, B. Ankele, W. Ottinger, and F. Stelzer, Mol. Cryst. Liqu. Cryst. 117, 327 (85); W. Ottinger, G. Leising, H. Kahlert, p. 63 in *Electronic Properties of Polymers and Related Compounds*, edited by H. Kuzmany, M. Mehring, and S. Roth, (Springer, Heidelberg, 1986), Springer Series in Solid State Sciences, Vol. 63.
29. D. Moses, A. Feldblum, E. Ehrenfreund, A. J. Heeger, T.C. Chung, and A. G. MacDiarmid, Phys. Rev. B 26, 3361 (1982).
30. D. Emin, in *Handbook of Conducting Polymers*, edited by T.A. Skotheim (Dekker, New York, 1986), Vol. 2, p. 915 and references cited therein.
31. D. Emin, Phys. Rev. B 33, 3973 (1986).
32. W.A. Harrison, *Electronic structure and the Properties of Solids*, (Freeman, San Francisco, 1980).
33. J.A. Majewski and P. Vogl, Phys. Rev. B 35, 9666 (1987).
34. Yu. N. Gartstein and A. A. Zakhidov, Sol. State Comm. 62, 213 (1987); (erratum) *ibid.* 65, ii (1987).
35. K. Maki, Syn. Met. 9, 185 (1984).
36. J. C. Hicks and Y. R. Lin-Liu, Phys. Rev. B 30, 6184 (1984).
37. G. F. Koster and J. C. Slater, Phys. Rev 95, 1167 (1954); 96, 1208 (1954).
38. D.K. Campbell and A.R. Bishop, Phys. Rev. B 24, 4859 (1981).
39. R. Chance, D.S. Boudreaux, J.L. Bré das, and R. Silbey, in *Handbook of Conducting Polymers*, edited by T.A. Skotheim (Dekker, New York, 1986), Vol. 2, p.825.
40. W.P. Su, in *Handbook of Conducting Polymers*, edited by T.A. Skotheim (Dekker, New York, 1986), Vol. 2, p. 757.
41. S. Phillpot, D. Baeriswyl, A.R. Bishop, P. Lomdahl, Phys. Rev. B 35, 7533 (1987).
42. A.R. Bishop, D.K. Campbell, P.S. Lomdahl, B. Horovitz, and S.R. Phillpot, Phys. Rev. Lett. 52, 671 (1984).
43. R. Ball, W. P. Su, and J. R. Schrieffer, Jour. de Phys. Coll. 44, C3-429 (1982).
44. H.J.Monkhorst and J.D. Pack, Phys. Rev. B 13, 5188 (1976).
45. B. Ankele, G. Leising, and H. Kahlert, Solid State Comm. 62, 245 (1987).
46. T. Ito, H. Shirakawa, S. Ikeda, J. Polymer Sci. Polymer Ed. 12, 11 (1974).
47. G. Leising, H. Kahlert, and O. Leitner, in *Handbook of Conducting Polymers*, edited by T.A. Skotheim, (Dekker, New York, 1986), Vol. 1, p. 56.

THE HUBBARD MODEL FOR ONE-DIMENSIONAL SOLIDS

Anna Painelli* and Alberto Girlando**

*Dept. of Chemical Physics, Padova University, Padova, Italy
**Inst. of Chemical Physics, Parma University, Parma, Italy

In recent years several unambiguous experimental evidences have been found stressing the relevance of electron-electron (e-e) interaction in one-dimensional (1D) solids, either charge-transfer salts[1] or polymers.[2] However there is not agreement between physicists on how to model e-e interaction in polymers, so that the effects of such interaction on the dimerization amplitude and on the band gap are still controversial.[3-13] The Hubbard model[14] has been generally adopted to investigate the role of e-e interaction in 1D solids. It is the solid state counterpart of the Pariser-Parr-Pople (PPP) model, widely tested and successfully applied in molecular physics.[15] However, the electrons in isolated molecules experience an unscreened Coulomb repulsion, whereas in 1D solids the interaction between the electrons along a chain may be screened due to the presence of mobile electrons on neighboring chains. The applicability of Hubbard model to systems with highly screened potential still has to be settled.

The one-electron part of the hamiltonian for 1D solids is generally written in the tight-binding approximation,[16] accounting for just one orbital (ψ_i) on each site. As we have recently stressed,[10] this picture requires a very realistic choice of ψ_i as the frontier atomic (or molecular) site orbital, and also requires that S, the overlap integral between orbitals on adjacent sites, is a small but non-negligible quantity. Being S non-negligible, one is not allowed to disregard the non-orthogonality of the ψ_i basis functions and has to orthogonalize them to give a φ_i basis set. This makes the problem rather difficult. However the smallness of S makes possible a simplification: In fact S can be conveniently chosen as an expansion parameter, in such a way that only terms of the order or greater than S^n are retained. On this basis we shall briefly discuss the relative magnitudes of the various e-e repulsion integrals in the two opposite limits of very long and very short range repulsion potential.

If the range of the potential is very large - much larger than the range of φ_i - one can adopt the old Mulliken approximation[17] which expresses the off-diagonal elements of the interelectronic potential [$V(i,j,m,n) = \langle \varphi_i(r)\varphi_j(r') | V(r-r') | \varphi_m(r')\varphi_n(r) \rangle$ with $i \neq n$ and/or $j \neq m$] in terms of the diagonal ones ($i=n$ and $j=m$). It is not difficult to show that such an approximation yields the vanishing, to any S order, of all the off-diagonal $V(i,j,m,n)$. In other terms, the zero-differential overlap approximation of PPP or Hubbard model is completely satisfactory.

Interacting Electrons in Reduced Dimensions
Edited by D. Baeriswyl and D.K. Campbell
Plenum Press, New York

In the case of a short range potential, the problem becomes more difficult and, to make it easier, we consider just the extreme limit of a δ-function potential. In a previous paper[10] we have written down the e-e hamiltonian correct to the order S, having chosen as ψ_i the carbon $2p\pi$ Slater atomic orbital (that relevant, e.g., to polyacetylene). In this paper we extend the calculation to the next order in S.[18]

The first question to be settled is the order of magnitude of the overlap between orbitals on next nearest neighbor sites, $S' = \langle\psi_i(r)|\psi_{i+2}(r)\rangle$. Fig. 1 shows the S'/S^2 curve as a function of S.[19] In the same figure we have also drawn the $S(S)$ straight line. The crossing of the two curves ($S \sim 0.25$) separates the region where S' is larger than S^3 from that where it is of the same order or smaller than S^3. Therefore, in the small S region terms of the order of S' have to be retained being non-negligible in respect to S^2. This result is not surprising,[10] since at large distance (small S) the overlap integral between atomic orbitals decays nearly exponentially with the site distance.

We now turn attention to the matrix elements of the e-e potential in the non-orthogonal basis: $\mathscr{V}(i,j,m,n) = \langle\psi_i(r)\,\psi_j(r')|\,V(r-r')\,|\psi_m(r')\,\psi_n(r)\rangle$. To the S^2 order, terms relevant to sites more distant than next nearest neighbors can be neglected. Moreover, for a δ-function potential, $\mathscr{V}(i,j,m,n)$ does not depend on the order of i, j, m, n indices, so that the only terms we have to consider are the following ones: $\mathscr{U} = \mathscr{V}(i,i,i,i)$, $\mathscr{X} = \mathscr{V}(i,i,i,i+1)$, $\mathscr{W} = \mathscr{V}(i,i,i+1,i+1)$, $\mathscr{Y} = \mathscr{V}(i,i+1,i+1,i+2)$, $\mathscr{X}' = \mathscr{V}(i,i,i,i+2)$, $\mathscr{Z} = \mathscr{V}(i,i+1,i+2,i+2)$, $\mathscr{W}' = \mathscr{V}(i,i,i+2,i+2)$. The $\mathscr{U}$, $\mathscr{X}$, $\mathscr{W}$ terms have already been calculated[10] and $\mathscr{X}'$, $\mathscr{W}'$ can be obtained from the $\mathscr{X}$ and $\mathscr{W}$ values at double lattice spacing. $\mathscr{Y}$ and $\mathscr{Z}$ are three center integrals and can be evaluated in the elliptic coordinate space, making resort to some algebraic trick proposed by Hirschfelder *et al.*[20] In Fig. 2 we report in $\mathscr{U}S$ units the $\mathscr{X}$, $\mathscr{W}$, $\mathscr{Y}$, and $\mathscr{X}'$ curves as functions of S (the $\mathscr{Z}$ and $\mathscr{W}'$ curves are not reported being always lower than $\mathscr{Y}$ and $\mathscr{X}'$ ones). It turns out[10] that, if $\mathscr{U}$ is taken as the energy unit, $\mathscr{X}$ is the only term of order S (it actually becomes negligible at $S \gtrsim 0.7$, but this region is certainly beyond the range of applicability of the proposed expansion). In Fig. 3 the $\mathscr{W}$, $\mathscr{Y}$, and $\mathscr{X}'$ curves are shown as functions of S, taking $\mathscr{U}S^2$ as energy unit. To the S^2 order, besides $\mathscr{X}$, also $\mathscr{W}$ term has to be accounted for, whereas all the others are negligible.

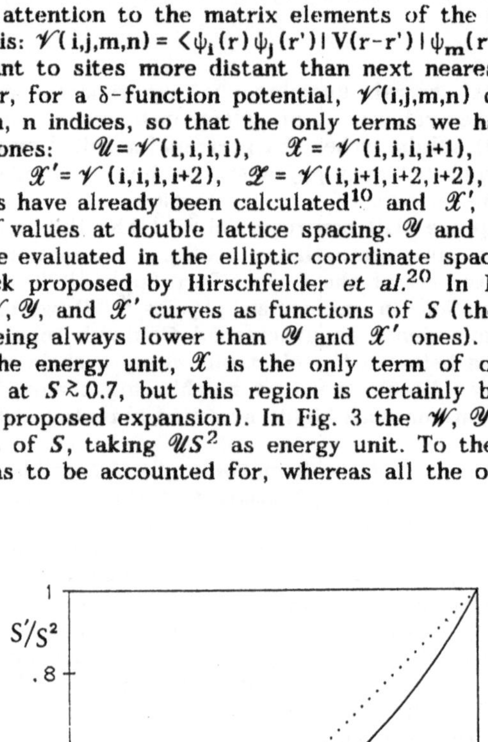

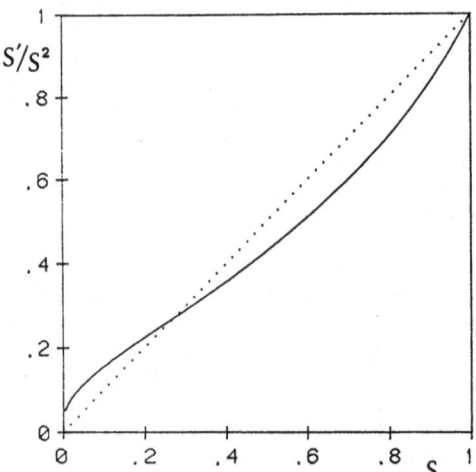

Fig. 1 Value of the next nearest neighbor overlap integral (S') in respect to S^2, as function of S. The dotted line shows the S function itself.

The orthonormalization of the basis set up to the S^2 order can be carried out as follows:

$$\varphi_i = \left(1+\frac{3}{4}S^2\right)\psi_i - \frac{1}{2}S\left(\psi_{i+1}+\psi_{i-1}\right) - \left(\frac{1}{2}S' - \frac{3}{8}S^2\right)\left(\psi_{i+2}+\psi_{i-2}\right) \quad (1)$$

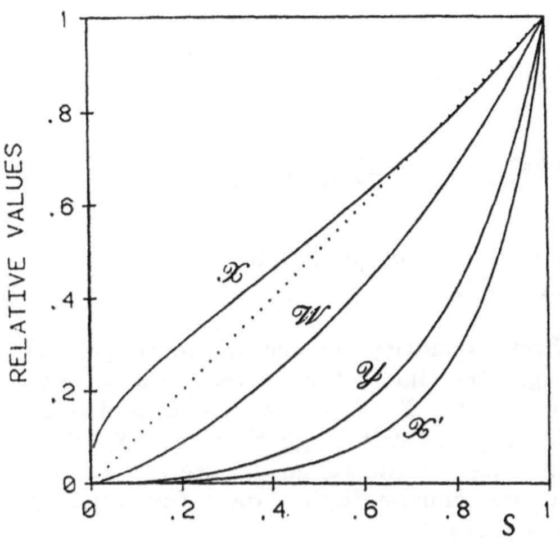

Fig. 2. Values, in $\mathcal{U}S$ units, of various integrals in the non-orthogonal basis, as functions of S. The dotted line shows the S function.

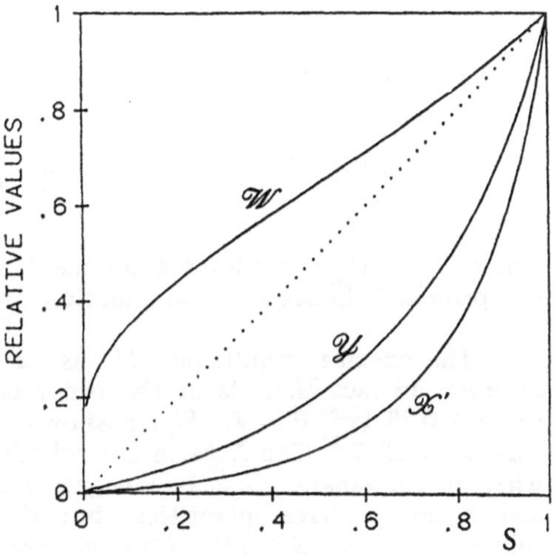

Fig. 3. Values, in $\mathcal{U}S^2$ units, of various integrals in the non-orthogonal basis, as functions of S. The $\mathcal{X}$ term is not shown, being of the order or greater than 1.

We remind that S' term has to be introduced only for small S ($S \lesssim 0.25$). Eq. 1 allows one to express the matrix elements of V(r-r') in the orthogonal basis [V(i, j, m, n)] in terms of the $\mathcal{V}$(i, j, m, n). To the first order in S,[10] there are two non-negligible V(i, j, m, n), namely $U = V(i,i,i,i) = \mathcal{U}$ and $X = V(i,i,i,i+1) = \mathcal{X} - S\mathcal{U}/2$. Fig. 4 shows the X value (SU units) as a function of S. For $S \gtrsim 0.2$, X is negligible in respect to U, so that one ends up with the simple Hubbard (U, t) model. On the other hand for smaller S, X is non-negligible and is actually negative, indicating an *effective attraction* between site and bond charges. This rather surprising result is a consequence of the orthogonalization of the basis functions, being $\mathcal{X}$ actually positive.

Going to the second order in S, the non-negligible terms are the following ones: $U = \mathcal{U}(1+3S^2) - 4S\mathcal{X}$; $X = \mathcal{X} - S\mathcal{U}/2$; $W = V(i,i,i+1,i+1) = \mathcal{W} - 2S\mathcal{X} + S^2\mathcal{U}/2$; $Y = V(i,i+1,i+1,i+2) = -S\mathcal{X} + S^2\mathcal{U}/4$; $X' = V(i,i,i,i+2) = -S\mathcal{X}/2 - (S'/2 - 3S^2/8)\mathcal{U}$. The e-e hamiltonian can therefore be written as follows:

$$
\begin{aligned}
H \; = \; & 2U\sum_{i} a^{+}_{i,\alpha} a^{+}_{i,\beta} a_{i,\beta} a_{i,\alpha} \\
& + 2X\sum\nolimits'_{i,\sigma} \left(a^{+}_{i,\sigma} a_{i+1,\sigma} + \text{H.c.}\right)\left(a^{+}_{i,\sigma'} a_{i,\sigma'} + a^{+}_{i+1,\sigma'} a_{i+1,\sigma'}\right) \\
& + 4W\sum_{i,\sigma,\sigma'} \left(a^{+}_{i,\sigma} a_{i,\sigma} a^{+}_{i+1,\sigma'} a_{i+1,\sigma'}\right) \\
& + 2W\sum_{i,\sigma,\sigma'} \left(a^{+}_{i,\sigma} a_{i+1,\sigma} + \text{H.c.}\right)\left(a^{+}_{i,\sigma'} a_{i+1,\sigma'} + \text{H.c.}\right) \qquad (2) \\
& + 2Y\sum_{i,\sigma,\sigma'} a^{+}_{i,\sigma} a_{i,\sigma} \left(a^{+}_{i+1,\sigma'} a_{i-1,\sigma'} + \text{H.c.}\right) \\
& + 2Y\sum_{i,\sigma,\sigma'} \left(a^{+}_{i,\sigma} a_{i+1,\sigma} + \text{H.c.}\right)\left(a^{+}_{i,\sigma'} a_{i-1,\sigma'} + \text{H.c.}\right) \\
& + 2X'\sum\nolimits'_{i,\sigma} \left(a^{+}_{i,\sigma} a_{i+2,\sigma} + \text{H.c.}\right)\left(a^{+}_{i,\sigma'} a_{i,\sigma'} + a^{+}_{i+2,\sigma'} a_{i+2,\sigma'}\right)
\end{aligned}
$$

where $a_{i,\sigma}$ ($a^{+}_{i,\sigma}$) annihilates (creates) a spin σ electron in the φ_i orbital, and the primed Σ indicates a summation over $\sigma \neq \sigma'$.

The on-site repulsion U is a finite quantity in the whole range of interest. In fact $U/\mathcal{U}$ is of the order or greater than 1 for $\mathcal{X}/\mathcal{U}\,S > 3/4$, i.e., for $S \lesssim 0.75$ (cf. Fig. 2). Fig. 5 shows the X, W, Y, X' curves (S^2U units) as functions of S.[21] The X term is negligible only at fairly large S values ($S \gtrsim 0.3$), whereas for smaller S it is a negative quantity. In the range of interest W, X', and Y are positive quantities, but they are non-negligible only for small S values ($S \lesssim 0.25$, $S \lesssim 0.15$, $S \lesssim 0.05$, respectively).

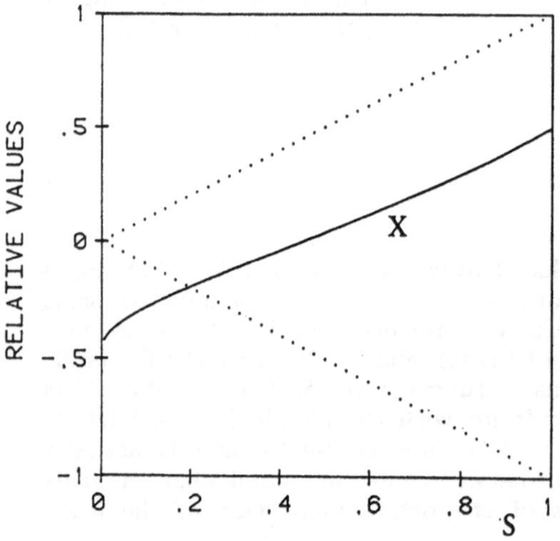

Fig. 4. Value, in US units, of the X integral in the orthogonal basis For comparison, the $\pm S$ functions are also reported (dotted lines).

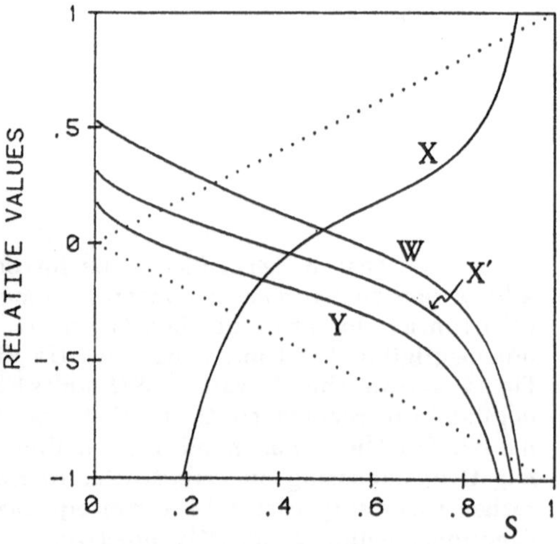

Fig. 5. Values, in US^2 units, of various integrals in the orthogonal basis.

In conclusion, for long range potentials the zero-differential overlap approximation of Hubbard model is satisfactory. This result is not surprising as zero-differential overlap approximation has a very long history in molecular physics.[15] For short range potentials the problem is more difficult and we have written down the two-electrons hamiltonian to the order S and S^2, assuming a δ-function potential. At the first order in S, the simple Hubbard (U, t) model is not adequate for $S \lesssim 0.2$. In this range in fact a new term, X, representing the effective attraction between bond and site charges, has to be introduced. The simple Hubbard model is regained at larger S, but in this regime the first order expansion may become inadequate. To the S^2 order, the simple Hubbard model is regained for $S \gtrsim 0.3$. At smaller S other terms, X, W, X', and Y have to be taken into account. Moreover, in writing the one-electron part of the hamiltonian, one has to be consistent with the two-electrons part. As it is suggested by Fig. 1, in going to the second order of S, for $S \lesssim 0.25$, besides t (the nearest neighbor charge transfer integral), one should introduce also t' (the next nearest neighbor charge transfer integral). Up to now only Hubbard models with at most the inclusion of X and/or W terms have been investigated and X has always been assumed positive. Our work clearly indicates that in the infinite screening limit X is actually negative, and also suggests that, in going to the second order in S, other terms besides W may become important.

ACKNOWLEDGMENTS

We thank D.K.Campbell, Z.G.Soos, W.-P.Su, X.Sun, and J.Voit for useful discussions and correspondence. This work has been supported by the Ministero della Pubblica Istruzione and by the National Research Council (CNR) of Italy through its Centro Studi Stati Molecolari Radicalici ed Eccitati.

REFERENCES

1. S.Mazumdar and S.N.Dixit, Phys.Rev.B 34:3683 (1986) and references therein.
2. Z.G.Soos and G.W.Hayden, in: "Electroresponsive Polymeric Systems", T.Skotheim ed., M.Dekker, New York (1988), p.197 and references therein.
3. S.Kivelson, W.-P.Su, J.R.Schrieffer, and A.Heeger, Phys.Rev.Lett. 58:1899 (1987).
4. C.Wu, X.Sun, and K.Nasu, Phys.Rev.Lett. 59:72 (1988).
5. D.Baeriswyl, P.Horsch, and K.Maki, Phys.Rev.Lett. 60:70 (1988).
6. J.T.Gammel and D.K.Campbell, Phys.Rev.Lett. 60:72 (1988).
7. S.Kivelson, W.-P.Su, J.R.Schrieffer, A.Heeger, Phys.Rev.Lett. 60:73 (1988).
8. A.Painelli and A.Girlando, Solid State Commun. 66:273 (1988).
9. Z.G.Soos and G.W.Hayden, Mol.Cryst.Liq.Cryst. 160:421 (1988).
10. A.Painelli and A.Girlando, Synth.Metals 27:15 (1988) and Phys. Rev. B, in press.
11. D.K.Campbell, J.T.Gammel, and E.Y.Loh, Phys.Rev.B, in press.
12. J.Voit, Diagonal and Off-Diagonal Electronic Interactions and Phonon Dynamics in an Extended Model of Polyacetylene, this book; J. Voit, Synth.Metals, in press.
13. V.Waas, J.Voit, and H.Büttner, Synth.Metals, in press.
14. P.W.Anderson, Phys.Rev. 115:2 (1959); J.Hubbard, Proc.R.Soc.London A276:238 (1963) and Phys.Rev.B 17:494 (1978).
15. L.Salem, "The Molecular Orbital Theory of Conjugated Systems", Benjamin, New York (1966).
16. N.W.Ashcroft and N.D. Mermin, "Solid State Physics", Holt-Saunders, London (1976), chap. 10.
17. R.S.Mulliken, J.Chim.Phys. (Paris) 46:675 (1949).

18. Since for large intersite distances multi-center integrals are strongly dominated by the exponential decay of the site orbitals, we believe that the results obtained with the carbon $2p\pi$ Slater orbitals are applicable, at least for small S, to any choice of atomic or molecular site orbitals.
19. R.S.Mulliken *et al.*, J.Chem.Phys. 17:1248 (1949).
20. J.Hirschfelder, H.Eyring, and N.Rosen, J.Chem.Phys. 4:121 (1936).
21. The large absolute values of X, W, Y and X' terms in the $S \gtrsim 0.75$ regime are due to the vanishing of U in this regime, where the S expansion is clearly unphysical. Moreover in evaluating X' we have always retained the S' terms, even if they should be neglected for $S \gtrsim 0.25$. The results are not qualitatively modified.

OFF-DIAGONAL COULOMB INTERACTIONS IN THE EXTENDED PEIERLS-HUBBARD MODEL: EXACT DIAGONALIZATION RESULTS

D.K. Campbell, J. Tinka Gammel, and E.Y. Loh, Jr.

Theoretical Division and Center for Nonlinear Studies
Los Alamos National Laboratory, Los Alamos, NM 87545, USA

INTRODUCTION

As evidenced by the many contributions to these proceedings, the role of electron-electron interactions in solid-state systems continues to be the subject of intense investigation and debate. Much of the recent discussion has been stimulated by experimental work on exciting novel materials, including high-temperature superconducting copper oxides, "heavy-fermion" systems, organic synthetic metals, and halogen-bridged transition-metal chains. Unlike conventional metals, for which standard single-electron (band) theories describe quantitatively the electronic structures and excitations, these new materials are currently thought to have properties dominated by many-body effects arising from strong electron-electron interactions. It is essential to have tractable models that capture the essence of both electron-phonon (e-p) and electron-electron (e-e) interactions and that represent faithfully their synergetic, or competing, effects.

For conjugated polymers, the extended Peierls-Hubbard model has been widely accepted as correctly incorporating the effects of both e-p and e-e interactions. In the absence of e-e interactions, the ground state of $(CH)_x$ is the $2k_F$ bond-order wave (BOW) – dimerization/bond alternation – predicted by Peierls theorem. It is now well established [1] that for the weak e-p coupling appropriate to $(CH)_x$, the on-site Coulomb repulsion actually enhances dimerization up to fairly large values ($U > 6t_0$). These results are still widely regarded as counter to the conventional wisdom that Coulomb interactions should suppress the build-up of charge anywhere, on the sites or on the bonds. Since the extended Peierls-Hubbard model incorporates only (lattice) site-diagonal parts of the electron-electron interactions (U, V), and omits the off-diagonal bond-charge repulsions (W, X), it is natural to ask whether this model adequately describes the full consequences of e-e interactions. In the specific context of $(CH)_x$, the potential importance of this omission is readily recognized [2]. Intuitively, the bond-charge repulsion should suppress dimerization, since it opposes the build-up of charge on the bonds. Thus, the absence of W and X terms in the standard extended Peierls-Hubbard models suggests *a priori* that these models may artificially favor the continuation of dimerization in the half-filled band into the region of intermediate to strong Coulomb interaction.

More generally, the omission of terms such as W and X raises significant questions about the appropriateness of Hubbard models for describing e-e interactions in the

Interacting Electrons in Reduced Dimensions
Edited by D. Baeriswyl and D.K. Campbell
Plenum Press, New York

whole class of novel solid state materials. Thus this issue is extremely important and must be investigated in a thorough and definitive manner.

MODEL

To analyze the effects of off-diagonal Coulomb terms in quasi-one-dimensional systems, we consider the modified Peierls-Hubbard Hamiltonian containing all nearest neighbor interactions [2,3,4,5]

$$H = -\sum_{\ell}(t_0 - \alpha\delta_\ell)B_{\ell,\ell+1} + \frac{K}{2}\sum_{\ell}\delta_\ell^2 + U\sum_{\ell} n_{\ell\uparrow}n_{\ell\downarrow} + V\sum_{\ell} n_\ell n_{\ell+1}$$
$$+ X\sum_{\ell} B_{\ell,\ell+1}(n_\ell + n_{\ell+1}) + W\sum_{\ell}(B_{\ell,\ell+1})^2 \quad , \qquad (1)$$

where $n_{\ell\sigma} = c_{\ell\sigma}^\dagger c_{\ell\sigma}$ and $B_{\ell,\ell+1} = \sum_\sigma(c_{\ell\ \sigma}^\dagger c_{\ell+1\ \sigma} + c_{\ell+1\ \sigma}^\dagger c_{\ell\ \sigma})$. In H, t_0 is the hopping integral for the uniform lattice, α is the electron-phonon coupling, δ_ℓ is the relative displacement between the ions at sites ℓ and $\ell+1$, K represents the cost of distorting the lattice, and U, V, X, and W model the electron-electron interactions. U represents the on-site Coulomb repulsion, and V the nearest-neighbor repulsion. X is a "mixed" term involving both on-site and bond-charge effects, and W is the pure bond-charge repulsion. The presence of such terms follows directly from the original many-body Hamiltonian, including Coulomb interactions; the explicit derivations are available in the literature [2,5,6,7]. U, V, and all the longer-ranged diagonal terms will always contain contributions dependent only on the range of the electron-electron potential, whereas W and X are suppressed by the atomic orbital overlap. This is the origin of the familiar result that for narrow bands the diagonal Coulomb terms are dominant. However, even if the band is not narrow, if the potential is not strongly screened, one still expects the diagonal terms to be more important numerically than the off-diagonal terms. When the off-diagonal terms are not *a priori* negligible, the central issue is the extent to which they produce results qualitatively different from those predicted in their absence. To answer this question correctly, it is clear that – whatever the relative values of V, X, and W – one must anticipate that $U > V, X, W$ and hence must adopt a method that gives correct results in this parameter regime. Since in many materials one expects $U \simeq 4t_0$, to be certain of the results one should use (numerically) exact many-body methods known to be reliable in the intermediate-coupling regime, though examining various limiting cases gives useful insight into the exact results. Indeed, to gain insight into our later exact diagonalization results, we begin with an analysis of the strong-coupling limit.

STRONG COUPLING PERTURBATION THEORY

In the limit that $U \to \infty$, double occupancy of any site is energetically not allowed. The resultant spin-Peierls Hamiltonian [1b,8] is

$$H_{eff} = \frac{K}{2}\sum_{\ell}\delta_\ell^2 + N\cdot V + \sum_{\ell}\Big(\frac{t_{\ell,X}^2}{U-V} - W\Big)(4\vec{S}_\ell\cdot\vec{S}_{\ell+1} - 1) \qquad (2)$$

where $t_{\ell,X} = ((t_0 - 2X - \alpha\delta_\ell)$. The first effect of W, then, is to suppress antiferromagnetism. Further, to lowest order in δ_ℓ, the spin coupling is proportional to

$$t_{\ell,X}^2 - W(U-V) \simeq t_0'^2 - 2\alpha' t_0'\delta_\ell + 0(\delta_\ell^2) \ ,$$

where $t_0' \equiv \sqrt{(t_0-2X)^2 - W(U-V)}$ and $\alpha' = \alpha(t_0-2X)/t_0'$. For given t_0 and X, since the effective t_0' decreases with W, the effective electron-phonon coupling $\pi\lambda' \equiv 2\alpha'^2/Kt_0'$ increases with W due both to the increase in α' and the decrease in t_0'. This leads to the initially surprising conclusion that, for large U, the dimerization should initially increase with W, the bond-charge repulsion.

When W reaches a critical value $W_c \simeq (t_0-2X)^2/(U-V)$, the spin coupling becomes ferromagnetic. The spins will then tend to align, forming a ferromagnetic SDW with $4\vec{S}_\ell \cdot \vec{S}_{\ell+1} \simeq 1$. The minimum energy configuration then has $\delta_\ell = 0$, and the ground state is an undimerized ferromagnetic SDW. The resulting phase diagram is shown in Fig. 1a.

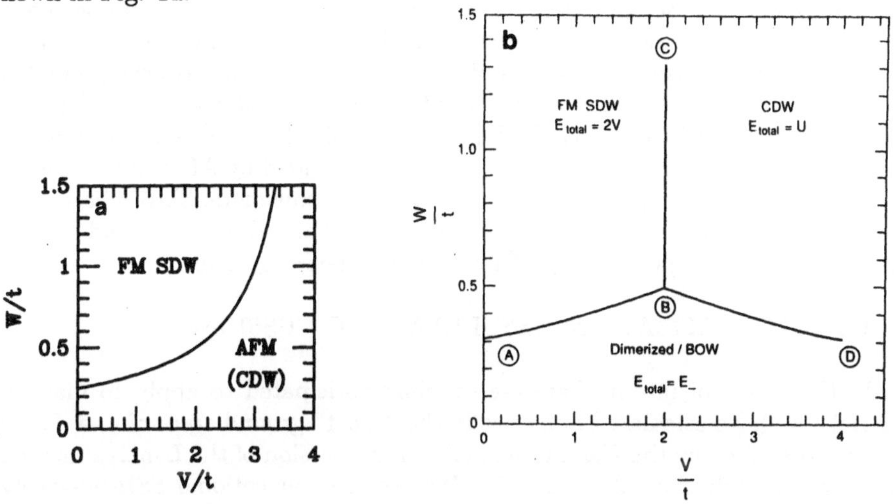

Fig. 1. The phase diagram as a function of V/t and W/t for intermediate coupling $U = 4t = 4t_0 - 8X$ (a) from strong-coupling perturbation theory and (b) for the half-filled dimer.

ANALYTIC RESULTS FOR A DIMER

The "dimer" – two electrons on two sites – provides surprisingly accurate insight into many aspects the behavior of larger systems. For comparison with the numerical results, we use periodic boundary conditions, so that $c_3^+ = c_1^+$ and similarly for all other operators. The Hamiltonian in (1) then assumes the explicit form

$$\begin{aligned} H_d = -2t_0 B_{1,2} + U(n_{1\uparrow}n_{1\downarrow} + n_{2\uparrow}n_{2\downarrow}) + 2Vn_1n_2 \\ + 2XB_{1,2}(n_1+n_2) + 2W(B_{1,2})^2 \quad . \end{aligned} \tag{3}$$

Since there is no possibility of bond alternation with only one bond, no explicit electron-phonon coupling appears. Further, one has $n_1 + n_2 \equiv 2$; thus just as in the case of the $U \to \infty$ limit, the X term simply renormalizes the hopping to $t \equiv t_0 - 2X$.

Simple counting shows that there are six possible states for two electrons on two sites. As basis states we choose

$$|1,1> \equiv c_{2\uparrow}^+ c_{1\uparrow}^+|0>, \; |1,-1> \equiv c_{2\downarrow}^+ c_{1\downarrow}^+|0>, \; |1,0> \equiv \frac{1}{\sqrt{2}}(c_{2\downarrow}^+ c_{1\uparrow}^+ + c_{2\uparrow}^+ c_{1\downarrow}^+)|0>,$$

$$|0,0> \equiv \frac{1}{\sqrt{2}}(c_{2\downarrow}^+ c_{1\uparrow}^+ - c_{2\uparrow}^+ c_{1\downarrow}^+)|0>,$$

$$|S> \equiv \frac{1}{\sqrt{2}}(c_{1\uparrow}^+ c_{1\downarrow}^+ + c_{2\uparrow}^+ c_{2\downarrow}^+)|0>, \; |A> \equiv \frac{1}{\sqrt{2}}(c_{1\uparrow}^+ c_{1\downarrow}^+ - c_{2\uparrow}^+ c_{2\downarrow}^+)|0> .$$

As the notation suggests, the first three states are, respectively, the $S_z = +1, -1$, and 0 components of the spin triplet state and, by the spin symmetry of H, must all have the same energy. The state $|0,0>$ is a spin-singlet made entirely from singly-occupied states, while $|S>$ and $|A>$ are, respectively, spin-singlets made from symmetric and antisymmetric combinations of the two states involving double occupancy. Trivial algebra leads to the following results: (1) the triplet states are degenerate eigenstates of H_d, with eigenvalue $E_T = 2V$; (2) the state $|A>$ is also an eigenstate, with eigenvalue U; and (3) the states $|0>$ and $|S>$ are coupled, with the two eigenvalues $E_\pm = 8W + U/2 + V \pm \Delta/2$ where $\Delta = ((U-2V)^2 + (8t_0 - 16X)^2)^{1/2}$. The ground state "phase diagram" for the dimer follows immediately by comparing these eigenenergies. For $U \equiv U = 4t$ (intermediate coupling), this diagram as a function of W/t and V/t is shown in Fig. 1b. For small values of W, the state E_- is always the ground state. This state, which in the absence of e-e interactions is just the $k = 0$ band state, corresponds to the "dimerized/BOW" phase and indeed, in the larger systems shows non-zero values of the dimerization. In the region bounded by AB and BC, the triplet state is the ground state; this corresponds to the ferromagnetic SDW phase found in larger systems. Finally, in the region bounded by BC and BD, the ground state is the state $|A>$; this corresponds to the CDW phase in larger systems.

EXACT DIAGONALIZATION: RESULTS AND DISCUSSION

In the region of intermediate e-e coupling anticipated to apply to many novel materials, we have calculated numerically the "exact" ground state of 4, 6, 8, and 10 site rings described by the Hamiltonian (1) using a version of the Lanczos method [9]; details will be published elsewhere [5]. We use the conventional SSH-polyacetylene parameters $\alpha = 4.1\text{eV}/\text{Å}, K = 21\text{eV}/\text{Å}^2$, and $t_0 = 2.5\text{eV}$. We focus on the phase diagram as a function of W and V at fixed $U = U_0$. Surprisingly, even for intermediate coupling the primary effect of X is merely to renormalize t_0. The details of results including the X term are reported elsewhere [5].

Fig. 2 shows the phase diagram for an 8-site ring; comparison with 4-, 6-, and 10-site results, as well as the analytic strong coupling and dimer predictions, suggests this diagram reflects the infinite-ring behavior. The phase boundaries in Fig. 2 in general reflect a "first order transition" in the dimerization order parameter, δ_0: that is, there is a sudden qualitative change in the nature of the ground state, and δ_0 drops almost immediately from a finite value to zero. However, for short segment of the BOW/CDW boundary near $W = 0$ — the range is roughly $0 < W < 0.1$ — the transition becomes second order. Except for this short segment, the dimerized phase has non-zero dimerization on its boundary.

In Figs. 3 and 4 we show the optimal dimerization δ_0 versus W when $U_0 = 4t_0 (= 10\text{eV})$, $V = 0$ and $V = 3\text{eV}$, respectively. Note the distinction between "Jahn-Teller" ($4N$) and "non-Jahn-Teller" ($4N+2$) systems persists even away from the band theory limit. It also suggests that systems with $N \geq 8$ are near the converged large N behavior. Incidentally, the dotted regions of the 4-site ring results reflect the dimerization observed in the BOW phase. However, the actual ground state of the 4-site system at values of W in these dotted regions is a different, small-ring phase, which does not appear in the larger rings. Thus the solid line for the 4-site system, which shows the dimerization going to zero at relatively small values of W, although strictly correct, is essentially an artifact of the small system size. The dotted line, which explicitly ignores this small-ring phase and plots the dimerization assuming the BOW state remains the ground state until the transition to a ferromagnetic SDW, shows more clearly the true finite size effects on W_c and δ_0 vs. W.

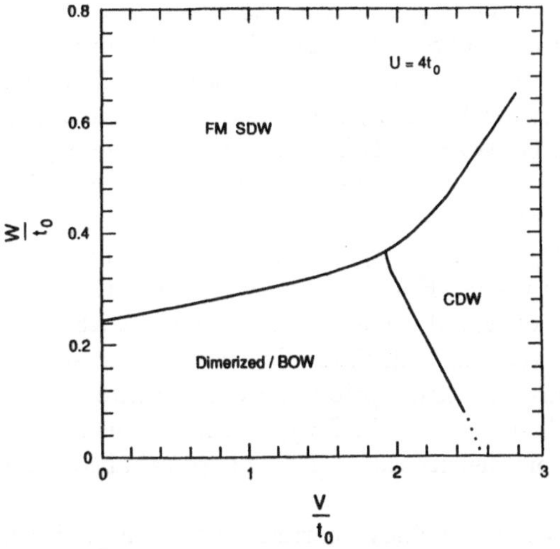

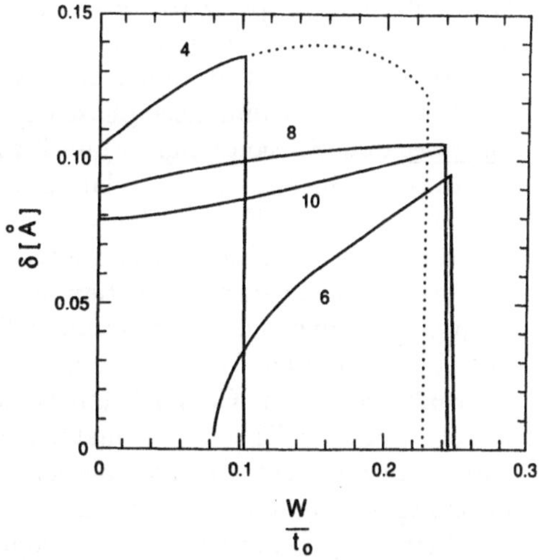

Fig. 2 (left). The phase diagram as a function of V/t_0 and W/t_0 for intermediate coupling $U = 4t_0 = 10\text{eV}$, and $X = 0$. Phase boundaries are plotted for a 8-site ring. The ground-state changes discontinuously across solid lines, smoothly across the dotted line.

Fig. 3 (right). Dimerization as a function of W/t_0 for $U = 4t_0 = 10\text{eV}$ and $V, X = 0$ for 4-, 6-, 8-, and 10-site rings. For the 4-site ring, the dotted line gives results for the lowest energy dimerized state even though for this small ring the ground state is not dimerized for intermediate values ($0.10 < W/t_0 < 0.225$) of W/t_0.

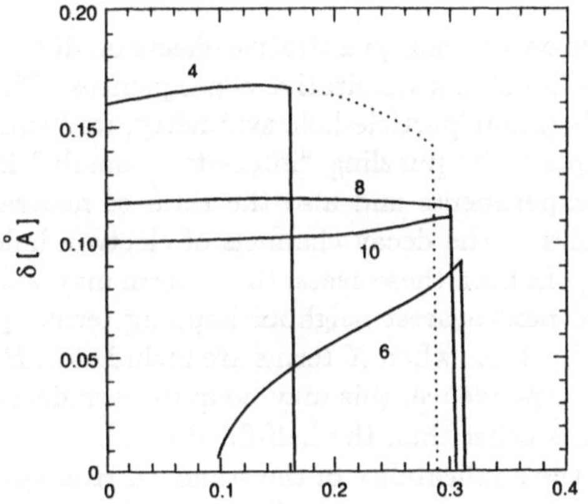

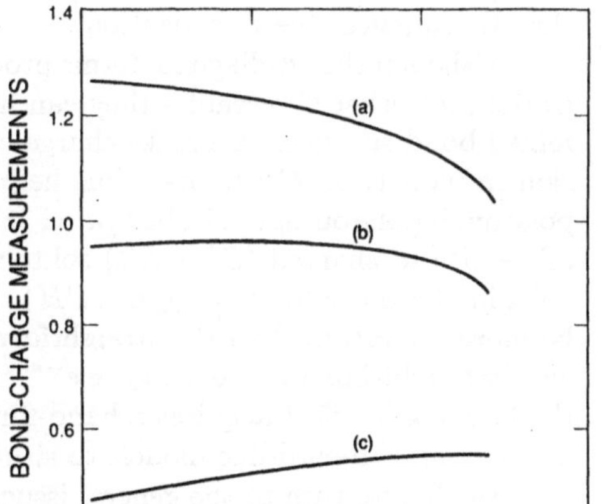

Fig. 4 (left). Dimerization as a function of W/t_0 for $U = 4t_0 = 10\text{eV}$, $V = 3\text{eV}$, and $X = 0$ for 4-, 6-, 8-, and 10-site rings. The 4-site ring is not dimerized for intermediate values ($0.16 < W/t_0 < 0.29$) of W/t_0.

Fig. 5 (right). Bond-charge measurements on the 8-site ring as a function of W/t_0 for $U = 10\text{eV}$ and $V = 0$: a) bond-charge correlation (multiplied by 1/2), b) average bond charge, and c) alternating bond charge, which is proportional to the dimerization.

The dimerized/BOW phase persists for a substantial range of Coulomb repulsion, both diagonal and off-diagonal. Indeed, dimerization increases monotonically with W before dropping rapidly to zero in a "first order phase transition" at $W = W_c$ in agreement with the strong coupling arguments above. In particular, as shown in Fig. 3, even for $V = 0$, W does not destroy dimerization until $W_c = 0.6\text{eV} \simeq 0.25t_0$. For $0 < V < U_0/2$, the dimerized phase persists until still larger values of W, and again W increases dimerization (slightly) until the BOW/FM SDW boundary is reached (Fig. 4). Note the increase in W_c relative to $V = 0$ as expected from the strong coupling arguments above. The existence of real materials with $W > W_c$ remains an open question; it is conceivable that the recently observed organic ferromagnetic materials may be modeled using parameters in this range. However, for $(CH)_x$ and the other conjugated polymers, the experimentally observed dimerization requires, within the model Hamiltonian, $W < W_c$. Importantly, one still finds dimerization for strong, internally consistent Coulomb interactions; the assumption of weak e-e interactions is not required. This is fortunate, for in the case of $(CH)_x$, such an assumption appears inconsistent with both observed spin density ratios [10] and optical absorption involving neutral and charged solitons [1f].

Increasing W, of course, must suppress bond-charge correlations. In Fig. 5, the bond-charge correlation, average bond charge, and alternating bond charge are plotted as a function of W for intermediate coupling, $U_0 = 4t_0$ and $V = 0$, on the 8-site ring. As expected, the bond-charge correlation, $\langle B^2 \rangle = \frac{1}{N}\sum B^2_{l,l+1}$, which couples directly to W in the Hamiltonian, is suppressed monotonically as W is turned on. Counter to simple intuition, however, the average bond charge, $\langle B \rangle = \frac{1}{N}\sum B_{l,l+1}$, does not fall off as dramatically as the correlation and indeed, for small W, the bond charge stays remarkably flat. Meanwhile, the alternating bond charge, which is related to the dimerization by $\langle B' \rangle = \frac{1}{N}\sum(-)^l B_{l,l+1} = K\delta/\alpha$, increases with W, as we have seen earlier. In sum, although W does suppress bond-charge correlations, the effects on average and alternating bond charge are quite different, and, in particular, we observe that W enhances the dimerization.

Although the off-diagonal terms produce only minor quantitative effects on dimerization, for other observables they can have important qualitative consequences. The mixed bond-site term X breaks charge conjugation/particle-hole symmetry; its inclusion in models of $(CH)_x$ may thus help explain the puzzling "intensity anomaly" in polaron/bipolaron optical absorption [11] experiments and also the ratio of neutral $(S^\circ - S^\circ)$ to charged $(S^+ - S^-)$ soliton pairs in the decay channels of electron-hole pairs in photo-excitation of trans-$(CH)_x$ [1f]. In both these cases, the X term may well be more important than the straightforward next-nearest neighbor hopping term, t_2; within a tight-binding model, $t_2 \propto e^{-\kappa_0 a} t_0$. Further, when X terms are included in H, the hopping in effect acquires a band-filling dependence; this may be quite significant in applying Hubbard-like models to situations other than the half-filled band.

Finally, we turn to the general issue of the applicability of the standard Hubbard model. Here it is useful to place our work in the context of several recent articles that, in one way or another, have revisited the original analyses [6,7,12] that justified the "zero differential overlap" approximation that neglects off-diagonal terms. The simple, intuitive arguments for e-e interactions suppressing dimerization, based on first-order weak-coupling perturbation theory [2], are not born out by more detailed calculation. Although our study focused explicitly on a (numerically) exact solution for) short-ranged Coulomb effects, involving only on-site and nearest neighbor interactions, our results are consistent with two recent variational studies [13] involving (approximate) solutions for the full (screened) Coulomb interaction. For the expected region of parameters, the variational results on the full Coulomb problem are consistent with early (Gutzwiller) variational calculations in the pure Hubbard model [14]. In summary, the familiar Hubbard and extended Hubbard models remain valid and useful theoretical starting points for understanding the role of electron-electron interactions in a variety of novel real materials.

ACKNOWLEDGEMENTS

We would like to thank Alberto Girlando, Sumit Mazumdar, Eugene Mele, and Anna Painelli for useful discussions. We are happy to acknowledge the Centers for Materials Science and Nonlinear Studies for computational support at Los Alamos National Laboratory and the Computational Sciences Division of the US DOE for computational support at the NMFECC at Livermore.

REFERENCES

1. Recent studies of the extended Peierls-Hubbard models include: (a) S. Mazumdar and S.N. Dixit, Phys. Rev. Lett., 51 (1983) 292 and Phys. Rev. B, 29 (1984) 1824; (b) J.E. Hirsch, Phys. Rev. Lett., 51 (1983) 296; (c) Z.G. Zoos and S. Ramasesha, *ibid*, 51 (1983) 2374; (d) D. K. Campbell, T. A. DeGrand, and S. Mazumdar, *ibid*, 52 (1984) 1717; (e) J. E. Hirsch and M. Grabowski, *ibid*, 52 (1984) 1713; (f) W.K. Wu and S. Kivelson, Phys. Rev. B, 33 (1986) 8546; S. Kivelson and W.K. Wu, *ibid*, 34 (1986) 5423; (g) S. Kivelson and D.E. Heim, *ibid*, 26 (1982) 4278.
2. S. Kivelson, W.P. Su, J.R. Schrieffer, and A.J. Heeger, Phys. Rev. Lett., 58 (1987) 1899.
3. D. Baeriswyl, P. Horsch, and K. Maki, Phys. Rev. Lett., 60 (1988) 70.
4. J. T. Gammel and D. K. Campbell, Phys. Rev. Lett., 60 (1988) 71.
5. D.K. Campbell, J.T. Gammel, and E.Y. Loh, Jr., Phys. Rev. B, 38 (1988) 12043 (RC); and to be published. Synthetic Metals, 27, (1988) A9.
6. J. Hubbard, Proc. Roy. Soc. London A, 276 (1963) 238.
7. R.G. Parr, J. Chem. Phys., 20 (1952) 1499; R. Pariser and R.G. Parr, *ibid*, 21 (1953) 767; J.A. Pople, Proc. Roy. Soc. London A, 68 (1955) 81.
8. J.W. Bray, L.V. Interrante, I.S. Jacobs, and J.C. Bonner, in "Extended Linear Chain Compounds", edited by J.S. Miller, Plenum, New York, 1983, Vol. 3, p. 355.
9. See, e.g., the section on Lanczos diagonalization in Pissantsky, "Sparse Matrix Technology", London; Orlando, Academic, 1984.
10. A.J Heeger and J.R. Schrieffer, Sol. St. Comm., 48 (1983) 207; H. Thomann, L.R. Dalton, M. Grabowski, and T.C. Clarke, Phys. Rev. B, 31 (1985) 3141.
11. K. Fesser, A.R. Bishop and D.K. Campbell, Phys. Rev. B, 27 (1983) 4804.
12. L. Salem, "Molecular Orbital Theory of Conjugated Systems", W. A. Benjamin, New York, 1966; R.G. Parr, D.P. Craig, and I.G. Ross, J. Chem. Phys., 18 (1950) 1561.
13. C. Wu, X. Sun, and K. Nasu, Phys. Rev. Lett., 59 (1987) 831; A. Painelli and A. Girlando, submitted to Solid State Commun.; Synthetic Metals, 27, (1988) A15.
14. D. Baeriswyl and K. Maki, Mol. Cryst. Liq. Cryst., 118 (1985) 1.

ELECTRON CORRELATION AND PEIERLS INSTABILITY*

Xin Sun[a,b,c)], Jun Li[c)] and Chang-qin Wu[a,c)]

a)Center of Theoretical Physics, CCAST(World Lab.)
b)International Center for Theoretical Physics Trieste, Italy
c)Department of Physics, Fudan University, Shanghai, China+

INTRODUCTION

The Peierls instability[1] is the characteristic of one-dimensional electron-lattice systems. Due to the electron-lattice interaction, $2k_F$ phonons of one-dimensional lattice are condensed to form a lattice distortion with period $\lambda=1/2k_F$. This distortion opens a gap on the Fermi surface, which lowers the electron energy in the occupied part of the band. For small lattice distortion, the electron energy gain is greater than the elastic energy of lattice distortion. Therefore, one-dimensional lattice is unstable against any arbitrarily small electron-lattice interaction. In the case of half-filled band, $2k_F=1/(2a)$, where a is the lattice constant, and the period of distortion is $\lambda=2a$, it is the dimerization. The prototype of half-filled one-dimensional system is polyacetylene, and it seems that the Peierls instability is the most direct way to understand the origin of the dimerization in the polymers. However, in this interpretation, only electron-lattice interaction is considered, the electron-electron interaction is completely neglected. Then a fundamental question arises: What is the effect of electron interaction on the dimerization? It is a knotty problem and there is a sharp dispute to answer this question.

So far the Hubbard model or the extended Hubbard model[2]

$$H_H = U\sum_{i,s} n_{i,s}n_{i,-s} + \sum_{\substack{i,j\\s,s'}} V_{ij}n_{i,s}n_{j,s'} \qquad (1.1)$$

has been commonly used to describe the electron interaction, where $n_{i,s}$ is the electron density operator on site i with spin s, U is the strength of on-site repulsion, V_{ij} is the inter-site repulsion. Since this model is very successful in many different

*This work was supported in part by the Natural Science Foundation of China.
+Mailing Address.

systems, it is also taken into the study of the effect of electron interaction on the dimerization of polymers. In recent years, based on this model, many different techniques have been developed to investigate this problem: unrestricted Hartree-Fock approximation[3], Gutzwiller variational calculation[4], Monte Carlo simulation[5], renormalization group theory[6], valence bond method[7], perturbation theory[8] and others. All these different theories got the same behavior in the dependence of the amplitude u of the dimerization on the strength U of electron interaction. The curve u(U) first goes up and reaches its maximum in $U \sim 4t_o$, which is the band width, then it falls down quickly when U increases further. Such behavior shows that dimerization is initially enhanced by electron interaction. Furthermore, the electron interaction in polyacetylene is about $U \sim 2t_o$, where the amount of dimerization is much greater than that without electron interaction. It means that the main part of the dimerization in polyacetylene is induced by electron-electron interaction rather than electron-lattice interaction. This result completely changes the previous explanation about the origin of the dimerization. In order to understand the physical mechanism of such enhancement, S. Mazumdar and S. Dixit have given an argument based on the resonance of valence bond.[7]

However, this point of view is challenged recently and a sharp dispute appears. Kivelson, Su, Schrieffer and Heeger argue that the interaction between electrons is repulsive, which is unfavorable to the distortion away from the uniform structure, then the electron interaction will provide additional stiffness against lattice distortion and tend to oppose the dimerization induced by electron-lattice interaction[9]. So they declare an opposite opinon that the electron interaction will reduce the dimerization. They point out that the extended Hubbard model only keeps diagonal terms of Coulomb interaction, which are site-charge repulsions; but, the electron interaction contains other repulsions which are missed in the extended Hubbard model, one of them is the bond-charge repulsion

$$W\,(C^+_{i,s}C_{i+1,s} + h.c.)(C^+_{i,s'}C_{i+1,s'} + h.c.)\;, \qquad (1.2)$$

which is off-diagonal term of Coulomb interaction. Actually it is the exchange term associated with the nearest site-charge repulsion $Vn_{i,s}n_{i+1,s'}$. In the case of δ-function interaction, W=V. Adding this bond-charge repulsion W to the extended Hubbard model, they find that the dimerization is suppressed by electron interaction.

Immediately, many other groups presented their opposing comments on KSSH's result.[10] There were two main points in these comments. First, KSSH used an extremely short-range interaction -- the $\delta(r)$ interaction, and, then, W=V; it is unreasonable. For real screening, W is smaller than V, it can not qualitatively change the behavior of Hubbard model. Second, KSSH's theory only made first order perturbation, in which on-site repulsion U was excluded, U would enter in second order. Since U is always much larger than W, and the numerical studies have shown that U alone initially enhances dimerization. So KSSH's result is questionable. These groups insist that the electron interaction should enhance the dimerization.

From these debates it can be seen that their divergences come from different descriptions of electron interaction. One side uses the extended Hubbard model to describe electron interaction, then they draw the conclusion that the dimerization is enhanced by electron interaction. Other side says the extended Hubbard model is not enough to reflect the repulsion between electrons, besides the site-charge repulsion, there exists the bond-charge repulsion also. They take both the site-charge repulsion and the bond-charge repulsion to describe the electron interaction, then get the opposite conclusion. It is obvious that the site-charge repulsion and the bond-charge repulsion are only some terms of Coulomb interaction, the whole interaction between electrons contains many other terms also. Therefore, in order to clarify this dispute and get a reliable answer, we should first give a general description for the electron interaction, which will be done in the next section. Then, the CBF method (Correlated Basis Functions) will be used to solve the many-body problem with the electron interaction. The results are discussed in the last section, then we can see how the above-mentioned dispute to be clarified.

ELECTRON-ELECTRON AND ELECTRON-LATTICE INTERACTIONS

It is well known that the Hubbard model is a good approximation for the systems with narrow bandwidth, where the overlap between different sites are small, then the off-diagonal terms of electron interaction can be neglected comparing to the diagonal terms. But, it is not the case for polyacetylene, which bandwidth is about 10 eV and even larger than U, then the off-diagonal terms can compete with the two-center diagonal terms V_{ij}. It is true that the on-site repulsion U is always much larger than other terms even in polymers. However, it should be noticed what we are studying now is the dimerization, which is related to the bond length, a quantity existing between the nearest neighbor sites rather than on the site. Since U is on-site repulsion, it is only in the second order to be involved in the bond length. So it can not directly affect the dimerization. The interaction terms connecting nearest neighbor sites will directly affect the dimerization and they play more important role. For this reason, although U is much bigger than other terms, the effect of other terms on the dimerization could surpass U. Therefore, the crucial fact to determine the dimerization is the competition between off-diagonal terms and two-center diagonal terms. The wider is the bandwidth, the more effective are the off-diagonal terms. Besides the bond-charge repulsion W, there are many other off-diagonal terms, what is their role in dimerization should also be considered. In order to get a reliable conclusion on the effect of electron interaction on the Peierls instability, it is needed to include all the interaction terms. The results based on taking only some terms of interaction would be doubtable.

Thus we should take a general formula to describe the electron interaction, it is the screened Coulomb interaction[11]

$$V(x-x') = \left\{ U_0/\sqrt{1+(x-x')^2/a^2} \right\} \exp(-\beta|x-x'|/a), \qquad (2.1)$$

where U_0 is the interaction strength and β is the screening factor, which determins the interaction range $\Lambda=a/\beta$. This general interaction includes all diagonal and off-diagonal term. Acturally, the interaction (2.1) can be written in second quantized representation with the Wannier function $\phi_n(x)$ as the basis

$$H' = \sum_{\substack{ijlm \\ ss'}} V(i,j,l,m) C^+_{i,s} C^+_{j,s'} C_{l,s'} C_{m,s} \qquad (2.2)$$

where $V(i,j,l,m)$ is the matrix elements of the interaction $V(x-x')$

$$V(i,j,l,m) = \int dx \int dx' \phi^*_i(x) \phi^*_j(x') V(x-x') \phi_l(x') \phi_m(x) \qquad (2.3)$$

The one-center term $V(i,i,i,i)$ is the on-site repulsion U, the two-center diagonal term $V(i,j,j,i)$ is the site-charge repulsion V_{ij} , the exchange term $V(i,j,i,j)$ is the bond-charge repulsion W, and remained off-diagonal term of two-center $V(i,i,i,j)$ is the site-bond interaction X.[9] Obviously, the extended Hubbard model and the KSSH's model are different approximations of the interaction (2.1) by neglecting all or some off-diagonal terms. So, the general interaction (2.1) will be able to give a reliable answer to the above dispute and find the limitations of extended Hubbard model and KSSH's model.

For the electron-lattice interaction, it should also be considered what is a proper description for a system with wide bandwidth. Here , the tight binding approximation is no longer a good approximation, then the SSH (Su-Schrieffer-Heeger) model[12] should be improved. Generally, the Hamiltonian of electron-lattice interaction can be written as

$$H_o = \sum_i \left[-(\hbar^2/2m)\nabla_i^2 + \sum_l v(x_i - X_l) \right] , \qquad (2.4)$$

where $v(x_i - X_l)$ is the potential produced by the atom on X_l exerting on the electron at x_i .

In the light of the above analysis, the whole Hamiltonian of the one-dimensional electron-lattice system reads

$$H = H_o + \frac{1}{2}\sum_{i,j} V(x_i - x_j) + \frac{1}{2}K \sum_i (X_{l+1} - X_l - a)^2 \qquad (2.5)$$

the last term in (2.5) is the elastic energy of the lattice distortion. As usually, in this stage the adiabatic approximation has been taken and the quantum movement of atoms is neglected.

Due to the Peierls instability, atoms' positions X_l can not be fixed. It is the main task of our work to determine the lattice distortion and, then, to get the dependence of dimerization $u(U_o, \beta)$ on the interaction strength U_o and the interaction range Λ.

The Hamiltonian (2.5) is a many-body problem, the key quantity of many-body system is the electron correlation function g(1,2), from which the energy and other properties of the system, including the dimerization u, can be obtained. In the next section we will use CBF (Correlated Basis Functions)[13] method to calculate the correlation function.

CORRELATED BASIS FUNCTIONS METHOD

The eigen-equation of non-interacting Hamiltonian H_o is

$$H_o \phi_k(x) = \varepsilon_k \phi_k(x) , \qquad (3.1)$$

where $\mathcal{E}_k$ is energy spectrum and $\phi_k(x)$ is the Bloch wave function. Then, the Wannier wave function is

$$\phi_n(x) = \sum_k e^{ikna}\phi_k(x) \ . \tag{3.2}$$

The ground state of the interacting Hamiltonian H can be formally written as

$$\Psi(1,2,\ldots,N) = D[\phi_k] \cdot \exp[u(1,2,\ldots,N)] \tag{3.3}$$

where $D[\phi_k]$ is the wave function of the ground state of H_0, $u(1,2,\ldots,N)$ is an unknown function, which represents the electron correlation induced by electron interaction, and can be determined by variation. The correlation of the whole system consists of two-body, three-body and multi-body correlations, and $u(1,2,\ldots,N)$ can be decomposed as

$$u(1,2,\ldots,N) = \frac{1}{2!}\sum_{i,j} u_{i,j} + \frac{1}{3!}\sum_{i,j,k} u_{i,j,k} + \cdots \tag{3.4}$$

where $u_{i,j}$, $u_{i,j,k}$ ... are two-body, three-body ... correlation factors. In the case of half-filled band, each atom has only one electron, and the interaction range Λ is the order of lattice constants a, so the electron density is not high. Meanwhile, since the electron interaction is repulsive, there is no condensation. Therefore, it is rare for three or more electrons to be gathered closely, then the two-body correlation $u_{i,j}$ is the dominant part, the three-body and mulit-body correlations can be neglected. In this approximation, the ground state (3.3) is simplified to

$$\Psi(1,2,\ldots,N) = D[\phi_k] \cdot \exp\Big[\frac{1}{2}\sum_{i,j} u_{i,j}\Big] \ . \tag{3.5}$$

It is the Jastrow wave function.[14] The energy of ground state is

$$\begin{aligned}
E &= \langle\Psi|H|\Psi\rangle / \langle\Psi|\Psi\rangle \\
&= \sum_{occ.} \mathcal{E}_\mu + \frac{1}{2}\int d1\int d2\,[P(1)-n_0]V(1,2)[P(2)-n_0] \\
&+ \frac{1}{2}\int d1\int d2 P(1)P(2)V(1,2)\,[g(1,2)-1] \\
&+ \frac{\hbar^2}{8m}\int d1\int d2 P(1,2)(\nabla_1 u_{12})^2 \\
&+ \frac{\hbar^2}{8m}\int d1\int d2\int d3 P(1,2,3)(\nabla_1 u_{12})(\nabla_1 u_{13}) \\
&+ \frac{1}{2}K\sum_l (X_{l+1}-X_l-a)^2 \ ,
\end{aligned} \tag{3.6}$$

where n_0 is the average density of electrons, and $P(1,2,\ldots,n)$ is the n-particle distribution function

$$P(1,2,\ldots,n) = \frac{N!}{(N-n)!}\int|\Psi(1,2,\ldots,N)|^2\, d(n+1)\ldots dN \Big/ \int|\Psi(1,2,\ldots,N)|^2\, d1d2\ldots dN \ . \tag{3.7}$$

The one-particle distribution P(1) is the electron density, the two-particle distribution can be expressed as

$$P(1,2) = P(1)P(2)g(1,2) , \tag{3.8}$$

where g(1,2) is the electron correlation function.

In the energy (3.6), the first term is kinetic energy of non-interacting electrons, second term the electrostatic energy, third term the exchang energy, fourth and fifth terms the correlation energy, last term the elastic energy. Since the three- and multi-body correlations are much less important than the two-body correlation, the n-particle distribution function (n>2) can be expanded in terms of the two-body correlation function g(1,2) and its convolution integrals (the convolution approximation[15]).

The correlation function g(1,2) can be obtained by using the CBF method. First introduce a parameter ξ before the correlation factor u_{ij} in the wave function (3.5), then, distribution $P(1,2,...,n/\xi)$ will depend on the parameter ξ.

Differentiate $P(1,2,...,n/\xi)$ with respect to ξ , then integrate it, the integral equations of $P(1/\xi)$ and $P(1,2/\xi)$ can be established as

$$P(1/\xi) = P(1/0)\exp\Big\{ \int_0^\xi d\xi' \int d2\ u_{12}\ P(1,2/\xi')/P(1/\xi') + \frac{1}{2}\int_0^\xi d\xi' \int d2 \int d3\ u_{23} \left[P(1,2,3/\xi')/P(1/\xi') - P(2,3/\xi')\right]\Big\}, \tag{3.9}$$

$$P(1,2/\xi) = P(1,2/0)\exp\Big\{\xi u_{12} + \int_0^\xi d\xi' \int d3(u_{13}+u_{23})P(1,2,3/\xi') /P(1,2/\xi') + \frac{1}{2}\int_0^\xi d\xi' \int d3 \int d4\ u_{34}\left[P(1,2,3,4/\xi') - P(1,2/\xi')P(3,4/\xi')\right]/P(1,2/\xi')\Big\} , \tag{3.10}$$

where $P(1/0)$ and $P(1,2/0) = P(1/0)P(2/0)g(1,2/0)$ are the density and two-partical distribution function of non-interacting electrons,

$$P(1/0) = 2\sum_k |\phi_k(x_1)|^2 , \tag{3.11}$$

$$g(1,2/0) = 1 - \frac{1}{2}\left|2\sum_k \phi_k^*(x_1)\phi_k(x_2)\right|^2 \Big/ P(1/0)P(2/0). \tag{3.12}$$

Thus P(1) and P(1,2) can be calculated by solving the combined integral equations (3.9) and (3.10).

RESULTS AND DISCUSSIONS

All the formulations have been estabilished in the last section, then the variational procedure can preceed in the following way: For the given interaction strength U_0 and screening factor β , first solve non-interacting eigen-equation (3.1) to get ε_k , ϕ_k, P(1/0) and P(1,2/0), all these quantities are dependent on the dimerization u. Substitute P(1/0), P(1,2/0) and trial correlation factor u_{ij} into the combined integral equations (3.9) and (3.10), the electron density P(1) and correlation function g(1,2) can be calculated by numerical iterations. Then, substitute P(1) and g(1,2) into Eq.(3.6) to get the energy, which depends on u and u_{ij}. Finally, minimize the energy with respect to u and u_{ij}, the dimerization u can be determined. Choosing different U_o and β, we can get the dependence of the dimerization $u(U_o,\beta)$ on the interaction strength U_o and screening factor β.

In order to make the calculation more transparent, a square well is taken as the lattice potential $v(x_i-X_l)$, which center sits at X_l with height v_o and width b. In the case of half-filled band,

$$X_n = a\left[n+(-1)^n u\right] . \tag{4.1}$$

One of the results $u(U_o,\beta)$ is shown in the Fig.1 , which parameters are v_o=40eV, b=0.6Å, a=1.22Å, and K=48.9eV/Å^2, then the bandwidth $4t_o$=14.9eV, electron-lattice coupling constant λ=0.33 . There are several curves in Fig.1 ,they correspond to different β. For small β, here $\beta<2$, the curve $u(U_o)$ goes up first, after it reaches its maximum, the curve falls down quickly. The behavior of these curves are similar as that of Hubbard model. However, when β is big enough, here $\beta\geqslant 2$, the curve declines monotonically. These results tell us that, if the screening β is small, the dimerization is initially enhanced by electron interaction; but if β is big, the dimerization will be suppressed by electron interaction. The conclusion is:

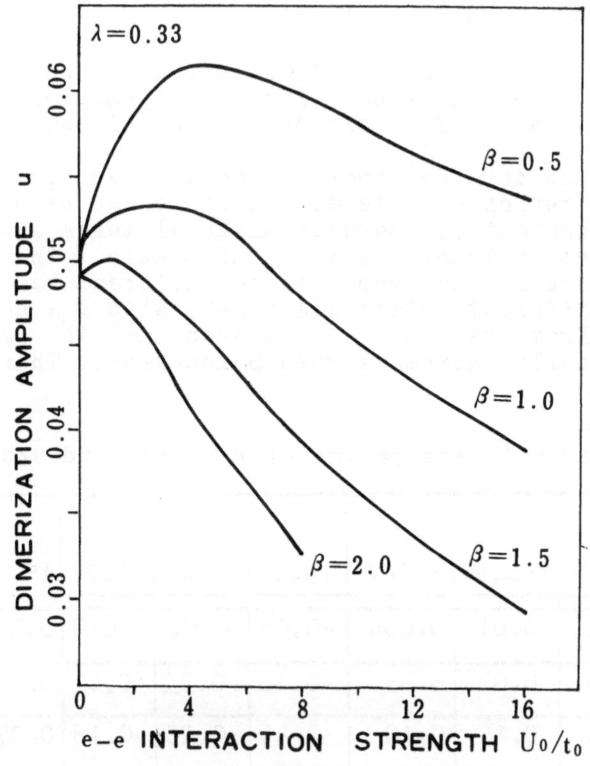

Fig. 1. The dependence of dimerization u on electron interaction.

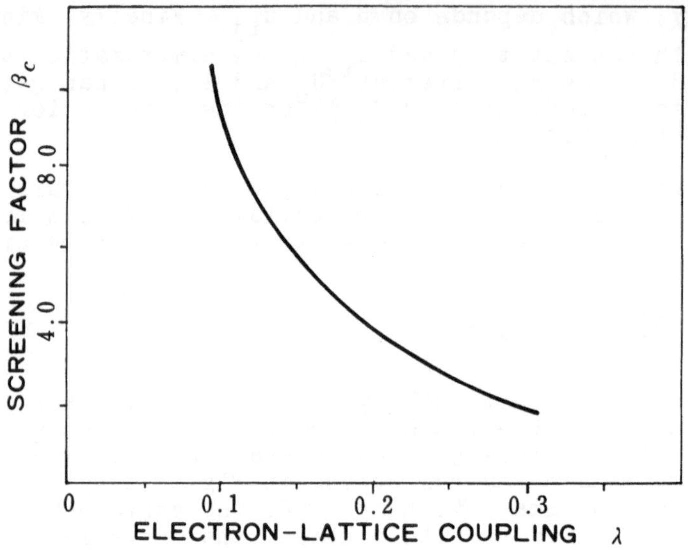

Fig. 2. The dependence of β_c on λ .

1. There exists a critical length Λ_c for the interaction range $\Lambda=a/\beta$. If the range Λ is longer than Λ_c, the dimerization is enhanced by electron interaction; Otherwise, if the range is shorter than Λ_c, the dimerization is reduced by electron interaction. It is the criterion to judge the effect of electron interaction on the dimerization.

2. The critical length Λ_c is dependent on the bandwidth and the electron-lattice coupling λ. The smaller is λ or bandwidth, the bigger is β_c. This dependence is shown in Fig. 2.

The reason for the dimerization to have opposite behavior in different region of interaction range can be understood by looking the competition between diagonal terms and off-diagonal terms of electron interaction. In the table 1 the ratios W/V, X/V and V/U are listed, where three different bandwidthes come from three different potentials v(x-X) with v_0= 40, 60, 80 eV and b=0.6Å. From the table, it is seen that the ratios W/V and X/V monotonically increases when β increases. Therefore, when

Table 1. The ratios of W/V, X/V and V/U.

β	W/V			X/V			V/U		
	15.0eV	11.4eV	8.7eV	15.0	11.4	8.7	15.0	11.4	8.7
1.0	0.02	0.01	0.004	-0.06	0.002	0.013	0.39	0.37	0.35
3.0	0.10	0.05	0.03	-0.18	-0.05	-0.01	0.14	0.11	0.10
5.0	0.26	0.16	0.10	-0.45	-0.23	-0.16	0.07	0.05	0.04
7.0	0.43	0.31	0.22	-0.77	-0.54	-0.49	0.05	0.03	0.02

the interaction range Λ is long, the off-diagonal terms are very small and can be neglected comparing to the diagonal terms, then it becomes the extended Hubbard model, so the dimerization is enhanced by electron interaction. In the opposite case, the interaction range is short, the off-diagonal terms become important, the extended Hubbard model is no longer a good approximation, the bond-charge repulsion and other off-diagonal terms make the dimerization reduced.

The table 1 also shows that the wider is the bandwidth, the bigger is the ratios W/V. It tells us, if a system has a wide bandwidth, its off-diagonal terms will become more effective, and we should go beyound the Hubbard model.

Based on the above analysis, it would not be difficult to clarify the abovementioned dispute about the dimerization. The side starting from the extended Hubbard model has neglected all the off-diagonal terms, implicitely, they were talking the electron interaction with long-range interaction, so their result was enhancement. The other side took the δ-function interaction, it is extremely short-range interaction, thus they got the opposite result that the electron interaction reduce dimerization. All these results are consistent with the criterion given in the conclusion 1.

Finally, we would give the particular result for the polyacetylene. Experiments indicate that a=1.22Å, its bandwidth is

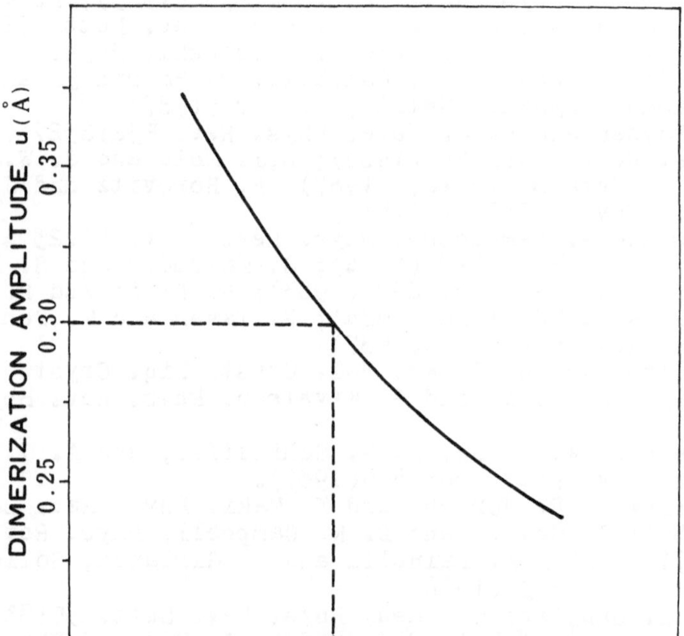

Fig. 3. The dependence of dimerization u on the screening factor β in polyacetylene.

10eV, the strength of electron interaction $U \sim 5$eV, the electron-lattice coupling $\lambda=0.2$. With these parameters the dimerization without electron interaction is $u_o=0.025\text{Å}$, and the dependence of dimerization $u(\beta)$ on the screening factor β is shown in Fig. 3. The experiment has shown the dimerization of polyacetylene is $u=0.03\text{Å}$, it indicates that the screening factor of polyacetylene is $\beta_{PA}=1.7$ and the dimerization is enhanced 20% by the electron interaction.

ACKNOWLEDGMENTS

The authors are grateful to Dr. D.Campbell and Dr. E.Loh for their helpful discussions and suggestions. One of us(X.S.) would like to thank Prof. M.Tosi and the International Center for Theoretical Physics for the hospitality during his visit there.

REFERENCES

1. R. E. Peierls, "Quantum theory of solids," Clarendon, Oxfords(1955).
2. J. Hubbard, Proc. Roy. Soc. London, 276A:238 (1963).
3. K. R. Subbaswamy and M. Grabowski, Phys. Rev. B24:2168(1981); S. Kivelson and D. Heim, Phys. Rev. B26:4278 (1982); H. Fukutome and M. Sasai, Prog. Theor. Phys. 61:1 (1983).
4. P. Horsch, Phys. Rev. B24:7351 (1981); D. Baeriswyl and K. Maki, Phys. Rev. B31:6633 (1985); Synthetic Metals, 17:13 (1987).
5. T. E. Hirsch, Phys. Rev. Lett. 51:296 (1983); D. K. Campbell, T. DeGrand and S. Mazumdar, Phys. Rev. Lett. 52:1717 (1984); J. E. Hirsch and M. Grabowski, Phys. Rev. Lett. 52:1713 (1984); D. K. Campbell, D. Baeriswyl and S. Mazumdar, Synthe. Metals, 17:197 (1987).
6. G. W. Hayden and E. J. Mele, Phys. Rev. B32:6527 (1985); Phys. Rev. B34:5484 (1986); E.J. Mele and G. W. Hayden, Synth. Metals, 17:107 (1987); B. Horovitz and J. Solyom, Phys. Rev. B32:2681 (1985).
7. Z. Soos and S. Ramasesha, Phys. Rev. Lett. 51:2374 (1983); Phys. Rev. B29:5410 (1984); S. Mazumdar and S. N. Dixit, Phys. Rev. Lett. 51:292 (1983); S. Dixit and S. Mazumdar, Phys. Rev. B29:1824 (1984); P. Tavan and K. Schulten, Phys. Rev. B36:4337 (1987).
8. S. Kivelson and W. K. Wu, Mol. Cryst. Liq. Cryst. 118:9 (1985); W. K. Wu and S. Kivelson, Phys. Rev. B33:8546 (1986).
9. S. Kivelson, W. P. Su, J. R. Schrieffer, and A. J. Heeger, Phys. Rev. Lett. 58:1899(1987).
10. D. Baeriswyl, P. Horsch, and K. Maki, Phys. Rev. Lett. 60:70 (1988); J. Gammel and D. K. Campbell, Phys. Rev. Lett. 60:71 (1988); A. Painelli and A. Girlando, Solid State Commun. 66:273 (1988).
11. C. Wu, X. Sun, and K. Nasu, Phys. Rev. Lett. 59:831 (1987).
12. W. P. Su, J. R. Schrieffer, and A. J. Heeger, Phys. Rev. Lett. 42:1698 (1979).
13. E. Feenberg, "Theory of Quantum Fluids," Pure and Applied Phys. Series Vol. 31, Academic, New York (1969); S. Chakravarty and C. W. Woo, Phys. Rev. B13:4815 (1976).
14. R. Jastrow, Phys. Rev. 98:1479 (1955).
15. F. Y. Wu and M. Chien, J. Math. Phys. 11:1912 (1970).

INSTABILITIES IN HALF-FILLED ONE-DIMENSIONAL SYSTEMS: VALENCE BOND ANALYSIS

Anna Painelli* and Alberto Girlando**

*Dept. of Chemical Physics, Padova University, Padova, Italy
**Inst. of Chemical Physics, Parma University, Parma, Italy

1. INTRODUCTION

Theoretical investigations on one-dimensional (1D) half-filled systems started many years ago, yet they still attract considerable attention, due to the wide variety of properties and phenomena discovered in the continuously growing family of synthetic 1D compounds, such as charge transfer (CT) salts, 1D polymers or mixed valence inorganic chains.[1] In this paper we focus attention on the ground state instabilities of chains with alternating or uniform on-site energy, by analyzing the interplay between the electronic kinetic energy and several types of interactions, namely, the electron-electron (e-e) one (both intra- and intersite), the electron-molecular vibration (e-mv) and the electron-lattice phonon (e-lph) couplings. To our knowledge none of the so far proposed theoretical models has simultaneously taken into account all the above interactions.

The electronic part of the problem is described in terms of the extended Hubbard hamiltonian and the model is solved exactly for small clusters (open chains up to 10 and rings up to 14 sites) through the diagrammatic valence bond (DVB) technique.[2] The infinite chain results are obtained by extrapolation. It has been already pointed out that the DVB diagrams, besides offering direct insight into the physics of the problem, are the natural basis for treating site-diagonal e-e interactions.[2] In our model, the on-site e-e interaction is accounted for by the usual Hubbard U, whereas the intersite one is introduced in terms of unscreened Coulomb repulsion between point charges on the molecular sites. Even if this choice is not necessarily adequate for every real system, it is certainly consistent with the adopted extended Hubbard model.[3] In any case the results are not strongly dependent on the form of the chosen electrostatic potential, as they generally compare well with those obtained by adopting a mean field approach.

In introducing the electron-phonon coupling, we distinguish between phonons modulating on-site energies (e-mv coupling) and those modulating hopping integrals (e-lph coupling). The coupling is set forth in the Herzberg-Teller scheme, which is an adiabatic, perturbative approach. The adiabatic approximation is thus the only one required for the solution of the proposed 1D Peierls-Hubbard model, apart from the unavoidable inaccuracies implied in the extrapolation of small cluster results to the infinite chain.

2. CHAINS WITH ALTERNATING ON-SITE ENERGY

We first consider systems with alternating on-site energy. Mixed stack CT crystals, made up of electron-donor (D) and acceptor (A) organic π-molecules alternately arranged along the chain, are taken as reference experimental systems. A wide amount of experimental data on ionicity (average charge on the sites), stack structure and phase transitions of these systems has been collected,[4] in particular after the discovery of the so-called neutral-ionic (N-I) phase transition in tetrathiafulvalene-chloranil (TTF-CA).[5] Of course, our model is applicable also to different real systems such as $-(A-B)_x^-$ conjugated polymers, as long as they can be described in terms of the Hubbard model.

Electronic hamiltonian

The extended Hubbard hamiltonian for a regular (uniform distance among sites) chain with alternating on-site energy (mixed stack, ms) is written as:[6,7]

$$\begin{aligned} H_{ms} &= -[(\varepsilon_A - \varepsilon_D)/2] \sum_i (-1)^i n_i + U \sum_i a^+_{i,\alpha} a^+_{i,\beta} a_{i,\beta} a_{i,\alpha} \\ &\quad + t \sum_{i,\sigma} (a^+_{i,\sigma} a_{i+1,\sigma} + \text{H.c.}) + \sum_{i>j} (-1)^{i+j} V_{ij} \rho_i \rho_j \\ &= -\delta \sum_i (-1)^i n_i + \frac{U}{2} \sum_i n_i (n_i - 1 + (-1)^i) \\ &\quad + t \sum_{i,\sigma} (a^+_{i,\sigma} a_{i+1,\sigma} + \text{H.c.}) + \sum_{i>j} (-1)^{i+j} V_{ij} \rho_i \rho_j \end{aligned} \tag{1}$$

where $\varepsilon_{A,D}$ is the A, D site energy, U the on-site repulsion energy assumed equal for A and D sites, t the CT integral and V_{ij} the electrostatic energy between i and j sites with unit charge. Moreover, $a_{i,\sigma}$ ($a^+_{i,\sigma}$) is the annihilation (creation) operator for electrons with spin σ, $n_i = a^+_{i,\sigma} a_{i,\sigma}$ is the number operator and $\rho_i = 2 - n_i$ for D (i odd) and $\rho_i = n_i$ for A (i even) sites. In the second equality we define δ, the difference between electronic affinity of the A site and the ionization potential of the D site, namely $\delta = (A_A - I_D)/2 = (\varepsilon_A - \varepsilon_D + U)/2$. In the following we assume $U \to \infty$ and also $\varepsilon_D \to \infty$, so that $I_D = \varepsilon_D - U$ is finite.[8] The second form of the hamiltonian shows the equivalence between this assumption and the neglect of states with doubly ionized sites, a fairly good approximation for ms CT crystals.

It is convenient to express the results of calculations in terms of the following electronic parameters (here and henceforth $\sqrt{2}\,|t|$ is taken as energy unit):

$$\gamma = \delta + V\alpha/2 = -(I_D \quad A_A)/2 + V\alpha/2 \tag{2}$$

$$\varepsilon_c = V(\alpha \quad 1) \tag{3}$$

where $V = V_{i,i+1}$ and α is a generalized Madelung constant, $\alpha = \Sigma'_{i,j}(-1)^{i+j} V_{ij} / N V$ (N is the number of sites). In Eqs. 2, 3 γ represents the balance between the cost of ionizing a DA pair and the gain in Madelung energy, whereas ε_c represents the stabilization of a D^+A^- pair in the ionic lattice due to intersite e-e interaction. We take ε_c as a measure of the strength of the latter interaction.

By exploiting the symmetry of the problem,[6] one classifies the zero-wavevector eigenstates as symmetric (A_1) or antisymmetric (A_2) in respect to the inversion center residing on the sites. Fig.1, upper part, shows the energy gap between lowest A_1 and A_2 singlets (Δ_{ss}) as a function of γ in the absence of intersite e-e interaction ($\varepsilon_c = 0$). In the lower part of the Figure we show the corresponding degree of ionicity $\rho = \langle \rho_i \rangle$ for the ground state of the system. Clearly, there is a critical γ value ($\gamma^* \sim 0.42$) where the system switches from a non-degenerate to a degenerate ground state. At the same critical γ the $\rho(\gamma)$ curve has maximum slope, separating the N and I regimes. Thus, N and I regimes are characterized by the non-degeneracy and degeneracy, respectively, of the singlet ground state.

The physical origin of the above described behavior can be easily understood in terms of the DVB real space representation.[6,7] In fact, in the infinite stack limit the fully N and fully I diagrams are completely decoupled,[9] and cannot contribute at the same time to the ground state. Moreover, whereas there is just one fully N diagram (A_1 symmetry), in the infinite chain limit all other diagrams combine to give functions of both A_1 and A_2 symmetry. Therefore the N regime ($\gamma < \gamma^*$), characterized by the contribution of the fully N diagram to the ground state, is non-degenerate and has A_1 symmetry. On the contrary in the I regime ($\gamma > \gamma^*$) there is no contribution of the fully N diagram, and the A_1 and A_2 subspaces become identical: the ground state is degenerate. It is also easy to show that the asymmetry of the N-I interface for $U > 0$ ($\gamma^* \neq 0$, $\rho \sim 0.63$) is due to the spin degrees of freedom.[6]

We now proceed to examine the effect of intersite e-e interaction on the N-I interface. Fig. 2 shows the $\rho(\gamma)$ curves for different values of ε_c; the triangles represent the results of the direct solution of the hamiltonian of Eq. 1, whereas the full line refers to calculations in which the

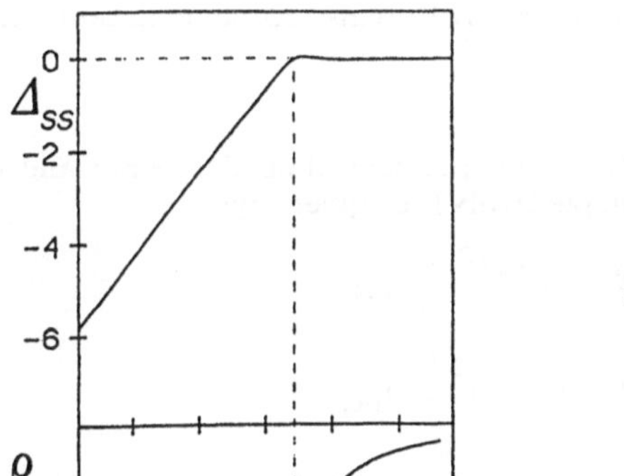

Fig. 1. Upper panel: energy gap between lowest A_1 and A_2 singlets as a function of γ for $\varepsilon_c = 0$. Lower panel: the corresponding degree of ionicity (ρ).

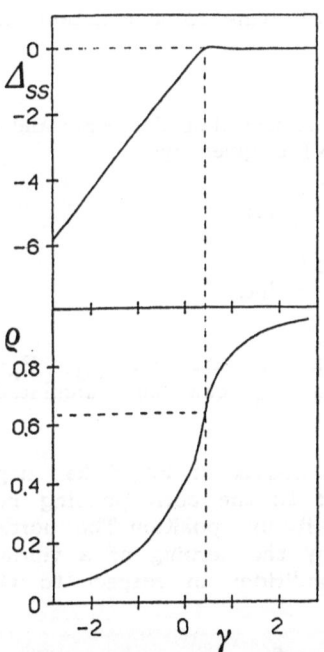

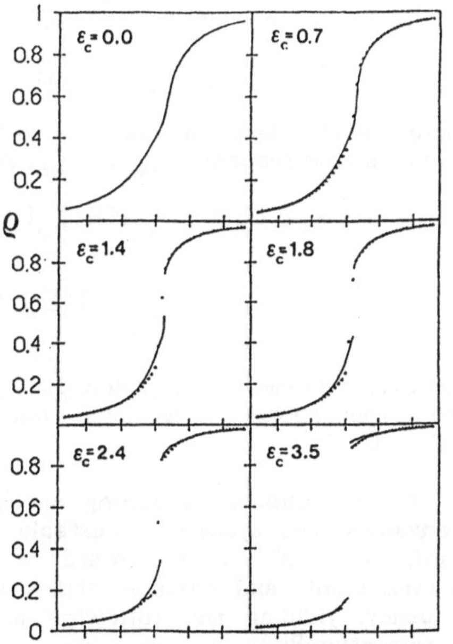

Fig. 2. Effect of intersite e-e interaction (ε_c) on $\rho(\gamma)$ function (triangles).The full line shows the results of a mean field treatment of e-e interaction.

intersite e-e interaction is treated in mean field.[6] By increasing ε_c there is an increase of the slope of the $\rho(\gamma)$ curve near the N-I borderline, and beyond a critical ε_c value ($\varepsilon_c^* \sim 2.0$) there is a discontinuous jump between N and I ground states.[9] In other words, intersite e-e interaction favors an uneven distribution of the charges on the sites, and beyond ε_c^* we have forbidden ionicity regions. Thus, the N-I boundary appears to be a first order boundary with a critical point.

Electron-phonon couplings

The electron-phonons couplings are introduced by expanding the electronic hamiltonian of Eq.1 on the vibrational coordinates.[10] Since phonons are treated in the harmonic approximation, the expansion is consistently carried out to the second order. The effects of second order coupling can be accounted for by simply renormalizing some parameters,[10] so that the e-mv and e-lph coupling hamiltonians can be written as ($\hbar = 1$):

$$H_{e\text{-}mv} = N^{-1/2} \sum_\mu Q_\mu \sqrt{\omega_\mu}\, g_\mu \sum_i (-1)^i n_i \tag{4a}$$

$$H_{e\text{-}lph} = 2N^{-1/2} \sum_\nu U_\nu \sqrt{\omega_\nu}\, g_\nu \sum_i (-1)^i b_i \tag{4b}$$

where Q_μ and U_ν are the zone-center crystal normal coordinates modulating the on-site energies and the bond energies (CT integrals), respectively, ω_μ and ω_ν are the corresponding vibrational frequencies, whereas g_μ and g_ν are the e-mv and e-lph linear coupling constants, as usually defined.[7] In Eq. 4b $b_i = (1/2)\sum_\sigma (a^+_{i,\sigma} a_{i+1,\sigma} + \text{H.c.})$ defines the bond order operator.

The Herzberg-Teller treatment gives the ground state energy corrected for the contribution of Eq. 4 perturbing terms, and the second derivatives of this energy in respect to the normal coordinates yield the following force constant matrix:

$$F_{\alpha\alpha'} = \omega_\alpha^2 \delta_{\alpha\alpha'} - \chi \sqrt{\omega_\alpha \omega_{\alpha'}}\, g_\alpha g_{a'} \tag{5}$$

where χ is the electronic response to the e-ph perturbation. For e-mv and e-lph couplings this response (χ_v and χ_b, respectively) is given by:[11]

$$\chi_v = (2/N) \sum_F |\langle G|\Sigma_i(-1)^i n_i|F\rangle|^2 / \omega_{FG} \tag{6a}$$

$$\chi_b = (8/N) \sum_F |\langle G|\Sigma_i(-1)^i b_i|F\rangle|^2 / \omega_{FG} \tag{6b}$$

where $|G\rangle$ represents the ground state, $|F\rangle$ an excited state, and $\omega_{FG} = E_F - E_G$. Given the above expressions, both χ_v and χ_b can be calculated by DVB method.

If the coupling is strong enough, the F matrix of Eq. 5 has negative eigenvalues: the system is unstable in respect to the corresponding normal coordinates, and relaxes towards a new equilibrium position. The borderline between stable and unstable states is given by the zeroing of a vibrational frequency, yielding the following stability conditions in respect to Q_μ or U_ν relaxation:[6,7]

$$\varepsilon_{sp} < \chi_v^{-1}, \quad \text{with} \quad \varepsilon_{sp} = \Sigma_\mu g_\mu^2/\omega_\mu \tag{7a}$$

$$\varepsilon_d < \chi_b^{-1}, \quad \text{with} \quad \varepsilon_d = \Sigma_\nu g_\nu^2/\omega_\nu \tag{7b}$$

where ε_{sp} and ε_d are the small polaron binding energy and the lattice distortion energy, respectively. The relaxation of Q_μ modes yields a redistribution of the charges on the sites, and is therefore related to N-I instability. On the contrary, the relaxation of U_ν modes corresponds to stack dimerization. Therefore, from Eqs. 7 it immediately turns out that the $\chi_v^{-1}(\rho)$ and $\chi_b^{-1}(\rho)$ curves, as they result from DVB calculations, can be interpreted as 0 K phase diagrams for N-I and regular-dimerized (r-d) stack instabilities, respectively. In fact, by putting on the ordinate axis the ε_{sp} or ε_d values, the stable states are identified as those lying below the χ_v^{-1} or χ_b^{-1} curves, respectively. We shall examine the N-I and r-d stack instabilities in the order.

In dealing with the N-I instability, rather than discussing the just mentioned phase diagrams, we shall approach the problem from a different side, which offers a better insight into the role played by e-mv coupling.[7] From the total hamiltonian, sum of the electronic (Eq.1), of the e-mv (Eq.4a), and of the purely vibrational hamiltonians, it is easy to derive the equations of motion for the Q's and evaluate their equilibrium position in the ground state: $\langle Q_\mu \rangle = N^{1/2}(1-\rho)\omega_\mu^{-3/2} g_\mu$. In the adiabatic limit this expectation value can be substituted in the e-mv hamiltonian: $H_{e-mv} = (1-\rho)\varepsilon_{sp}\sum_i (-1)^i n_i$, and this hamiltonian can be absorbed in the purely electronic one (Eq. 1) by renormalizing δ: $\Delta = \delta - \varepsilon_{sp}(1-\rho)$. Thus the problem has a self-consistent solution: from the known $\rho(\gamma, \varepsilon_c, \varepsilon_{sp}=0)$ curves (Fig.2), one evaluates $\delta = \Delta + \varepsilon_{sp}(1-\rho)$ for each ε_{sp} and constructs the $\rho(\gamma, \varepsilon_c, \varepsilon_{sp})$ curves, as exemplifed in Fig. 3, left side. By increasing ε_{sp} the original $\rho(\gamma, \varepsilon_c)$ curve becomes more and more steep, until its slope becomes negative. The negative slope corresponds to an instability region which in fact coincides with that defined by the above derived $\varepsilon_{sp} \geq \chi_v^{-1}$ condition.[7] Thus, provided ε_{sp} is large enough, one has forbidden ionicity regions (dashed part of the curve in Fig.3, right side) even if $\varepsilon_c < \varepsilon_c^*$. In other words, the e-mv coupling cooperates with intersite e-e repulsion in favoring an uneven distribution of the charges among the sites. Another aspect put in evidence by the right side of Fig.3 is the presence of a bistability region, i.e., a region where two stable states (N and I) coexist for the same value of γ. Although one has to be careful about the subtleties involved in going to the thermodinamic limit,[12] Fig.3 immediately suggests the possibility of hysteresis in the first order N-I phase transition.[13] However, the situation is more complex, due to the presence of the r-d stack instability we are going to examine.

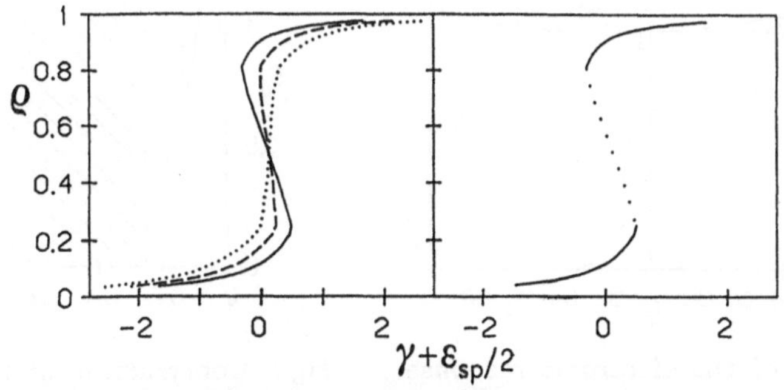

Fig. 3. Effect of e-mv coupling on the $\rho(\gamma, \varepsilon_c=1.8)$ curve. The left side shows the curves for ε_{sp}= 0.0, 1.0 and 2.0 (dotted, dashed and full line, respectively). The right side evidences the instability region (dotted part) of the ε_{sp}= 2.0 curve.

Focusing attention on r-d stack instability, we report in Fig. 4 the $\chi_b^{-1}(\rho)$ curves obtained by the DVB method for different ε_c values (continuous lines). As already discussed, these curves can be interpreted as 0 K phase diagrams for the r-d stack interface: putting ε_d on the ordinate axis, the stable states lie below the curves. Thus on the N side the system may be stable or unstable in respect to dimerization, depending on the value of ε_d, but on the I side the system is intrinsically unstable. The latter result follows directly from the fact that the I ground state of a regular stack is degenerate, and the e-lph coupling mixes the two degenerate subspaces. A second point worth noting in Fig. 4 is that the intersite e-e repulsion affects the r-d stack interface: as ε_c increases, it becomes easier for the system to dimerize. This somewhat unexpected result can be understood by adopting a mean field picture,[7] which shows that the electronic energy gain upon dimerization increases if the charges on the molecular sites are allowed to reorganize following dimerization. This energy gain diverges at the N-I interface, where the system gives an infinite response to any perturbation (in this case, the dimerization) able to induce ionicity variations. The results of this mean field approach are shown as dashed lines in Fig. 4.

Since, as we have seen, e-mv and e-e interactions cooperate, it might appear that also e-mv coupling favors the dimerization instability. However, this

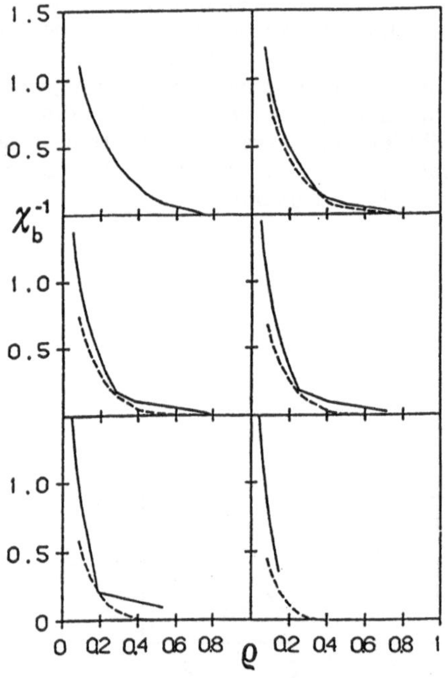

Fig. 4. Inverse of the electronic response to e-lph perturbation (χ_b) as function of ρ for the same ε_c values as Fig. 2. The dashed line gives the results of a mean field approach (see text).

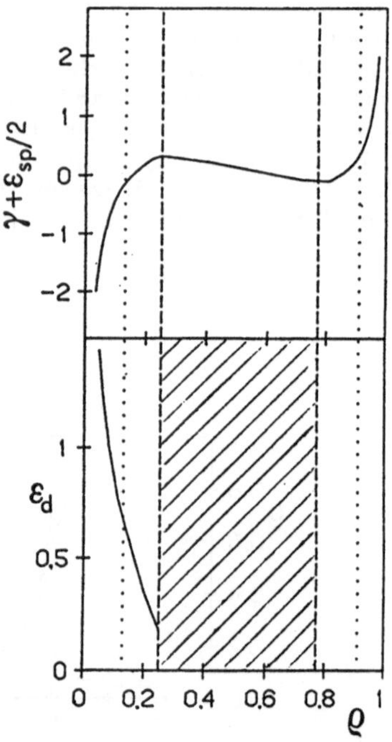

Fig. 5. Construction of the complete phase diagram for a ms chain with ε_c = 1.8 and ε_{sp} = 1.3 (see text). In the lower panel the dashed area indicates the forbidden ionicity region.

does not occur. Our previously published result[7] is incorrect, due to the omission of the molecular phonon energy in the derivation.

Zero-temperature phase diagram

Summarizing the results of the preceding section, we can say that for regular ms, the N-I boundary is a first order boundary with a critical point. On the other hand, the dimerization instability is characterized by a symmetry breaking driven by lattice phonons, and is expected to give rise to a continuous, second order phase transition. Moreover, the I regime is intrinsically unstable towards dimerization. In order to understand what happens in real systems where both N-I and r-d instabilities coexist, we construct the complete phase diagram as exemplified in Fig. 5 for a system with $\varepsilon_c = 1.8$ and $\varepsilon_{sp} = 1.3$. In the upper part we report the relevant $\rho(\gamma)$ curve defining the N-I interface, namely the forbidden ionicity region and the bistability regions. These regions are then projected in the lower panel, which concerns the r-d interface, thus obtaining the full 0 K phase diagram. The shaded area indicates the forbidden ionicity region, whereas the dotted lines delimit the bistability regions.

Now suppose ρ is increased due to a lattice contraction induced by the action of temperature or pressure. If ε_d crosses the χ_b^{-1} curve in the N stable regime, one expects a second order dimerization transition. On the other hand, if ε_d is low enough, the forbidden ionicity region is reached first, and one observes a first order transition to an I, dimerized stack. Finally, if ε_d crosses the χ_b^{-1} curve in the bistability region, we have a complex situation, with at least three minima in the potential surface. It is difficult to precisely state the evolution of the phase transition, but the above described picture strongly points towards a first order phase change, very sensitive to external conditions and characterized by the possible coexistence of different phases. This situation corresponds to what has been experimentally observed in the famous temperature or pressure induced Nr-Id phase transition of TTF-CA,[5,14] whose microscopic parameters, as estimated from optical data,[15,16] approximately match those of Fig. 2 (with $\varepsilon_d \sim 0.1 - 0.2$).

The different behavior displayed under pressure by two other quasi N ms CT crystals, DBTTF-TCNQ and TTF-2,5Cl_2BQ, can be understood in terms of our DVB phase diagram. For DBTTF-TCNQ, the forbidden ionicity region shrinks in respect to that of Fig. 5 due to the lower values of ε_c and ε_{sp}:[7,16] thus by increasing ρ, ε_d crosses the χ_b^{-1} curve before reaching the bistability region. The system is expected to undergo a Nr-Nd transition, as observed.[17] For TTF-2,5Cl_2BQ there is no forbidden ionicity region,[16] and indeed under pressure the ionicity increases continuously from N to I, accompanied by a dimerization of the stack.[17,18]

In addition to these particular cases, the above phase diagram allows one to rationalize the various types of phases and phase transitions observed in ms CT crystals. For the variability ranges estimated for ε_c, ε_{sp}, and ε_d,[7] one indeed predicts (Fig. 4) that low ionicity crystals exhibit a regular stack structure at all temperatures, whereas I systems are always dimerized at sufficiently low temperatures. Intermediate ionicity systems are very unlikely to be found due to the presence of forbidden ionicity regions (Figs. 2, 3). Finally, the different behavior of the few observed intermediate ionicity crystals, which are dimerized at ambient conditions, and of highly ionic systems, which dimerize only at low temperatures,[4] has been rationalized by explicit calculation of the energy gain upon dimerization.[7] It turns out that such an energy gain is maximum at intermediate ρ, whereas it goes to zero as $\rho \to 1$. A useful comparison can be done with the Peierls instability of segregated stack systems: at intermediate ρ the charge degrees of freedom are strongly involved, and one has an electronic Peierls transition. On the contrary, as $\rho \to 1$ only spin degrees of freedom are involved, and the transition can be assimilated to a spin Peierls one.[6,7]

3. CHAINS WITH UNIFORM ON-SITE ENERGY

Half-filled chains with uniform on-site energy can be obviously considered as a particular case of the above considered systems with alternating on-site energy. Several classes of compounds can be described by the model we are going to discuss, namely 1:1 segregated stack CT crystals (e.g., KTCNQ, TTFBr), polymers (e.g., polyacetylene), or halogen bridged transition metal complexes (e.g., $[Pt(en)_2][Pt(en)_2X_2]ClO_4$, with X = Cl, Br, I and en = ethylendiamine; Pt-Cl, Pt-Br and Pt-I for short). The latter class of compounds can be assimilated to a half-filled chain with uniform on-site energy if one adopts a crystal field-like approach,[19,20] where the charge on the halogen atoms is assumed fixed, so that the halogens have the only effect of producing an electrostatic potential at the metal sites. All these different kinds of compounds have been widely investigated,[1,21-23] and it is known that in general they have a ground state with either a bond dimerization (a charge density wave on the bonds, bCDW) or a non-uniform distribution of the charges on the sites (a site charge density wave, sCDW).[22] Despite the many theoretical models developed to deal with these systems,[21-23] in our opinion a unitary picture able to account for the observed different ground states is still lacking.

Electronic hamiltonian: comparison with mixed stacks

We write the extended Hubbard hamiltonian for a regular chain with uniform on-site energy (e.g., segregated stack, ss) as follows:[20]

$$H_{ss} = U \sum_i a^+_{i,\alpha} a^+_{i,\beta}\, a_{i,\beta} a_{i,\alpha} + t \sum_{i,\sigma} (a^+_{i,\sigma} a_{i+1,\sigma} + \text{H.c.}) + \sum_{i>j} (-1)^{i+j} V_{ij}\, q_i q_j \tag{8a}$$

where q is the operator for the charge on the sites, defined as $q_i = z_i - n_i$. In the last expression z_i is the charge of the vacuum state, and is obviously different for the different systems, for instance: KTCNQ, $z_i = 0$; TTFBr, $z_i = 2$; $(CH)_x$, $z_i = 1$; Pt-X chains, $z_i = 4$. If one introduces for the ss systems a "degree of ionicity" operator, $\rho_i = 2 - n_i$ for odd sites and $\rho_i = n_i$ for even sites, analogous to that defined for ms, the above hamiltonian becomes:

$$H_{ss} = V\alpha \sum_i (-1)^i n_i + U \sum_i a^+_{i,\alpha} a^+_{i,\beta}\, a_{i,\beta} a_{i,\alpha} + t \sum_{i,\sigma} (a^+_{i,\sigma} a_{i+1,\sigma} + \text{H.c.}) + \sum_{i>j} (-1)^{i+j} V_{ij}\, \rho_i \rho_j \tag{8b}$$

A comparison with Eq. 1 shows that this hamiltonian is equivalent to that of ms, but the role of on-site energy alternation is played by $V\alpha$, which is a measure of intersite e-e interactions. Thus we can adopt for ss the same set of parameters as for ms, namely γ, ε_c, ε_{sp}, ε_d, and U, but with a different meaning for γ. For ms γ is the balance between the cost of ionizing a DA pair and the Madelung energy; for ss it is interpreted as the balance between on-site and intersite electron-electron interaction ($\gamma = \delta + V\alpha/2 = U/2 - V\alpha/2$). On the other hand ρ, the degree of ionicity of ms, is now a measure of sCDW amplitude. Finally, it is worth reminding that the $U \to \infty$ limit adopted to describe ms CT crystals is not adequate to describe ss systems; the results we will present henceforth are for finite U.[24]

The formal equivalence between ss and ms hamiltonians allows us to make extensive and sound comparisons between different systems and to extract the fundamental and unifying features of the underlying physics.[20] In particular we can show, in terms of the real space DVB representation, that the sCDW-bCDW interface of ss shares the same physics as the N-I interface of ms. As in the case of ms systems, in fact, in ss we have fully I diagrams, degenerate and, in the infinite stack limit, completely decoupled from the fully N diagrams; the γ variation drives the system from a N regime (no contribution of fully I diagrams to the ground state) to an I regime (no contribution of fully N diagrams). As before, in the I regime the lowest A_1 and A_2 singlets are degenerate and the system is unstable towards bCDW. However, in ss systems or more generally in systems with finite U, there are two fully N diagrams, left and right ($\rho=0$, $\rho=2$), with doubly occupied and empty sites alternating in opposite order. These two diagrams are completely decoupled in the infinite chain limit and cannot contribute contemporaneously to the ground state. In the N regime therefore the system has to choose between two equivalent but different ground states, with the contribution of fully N left or right diagram: the system undergoes a spontaneous symmetry breaking giving rise to a sCDW condensation.

It is important to recognize that the just described equivalence between ms and ss systems is not complete. In fact Eq. 8b hamiltonian is a broken symmetry hamiltonian, being actually relevant to states with larger electronic density on odd sites. It therefore offers an adequate description of ss systems in the sCDW regime or at the N-I interface, where states with $\rho<1$ and $\rho>1$ are actually decoupled. On the contrary it gives an improper picture of ss systems in the bCDW regime, as it arbitrarily chooses one of the two coupled configurations, yielding the fictitious result of a finite sCDW amplitude ($\rho<1$), whereas the systems is not expected to spontaneously break the site-equivalence symmetry. To confirm this finding, we have performed explicit DVB calculations on regular ss systems, analyzing the three relevant subspaces, namely A_1, totally symmetric, A_2, antisymmetric in respect to inversion on the sites, and B_1, antisymmetric in respect to inversion on the bonds.[20] Fig. 6 shows the energy gap between lowest A_1 and A_2 (left) and lowest A_1 and B_1 (right) subspaces, as a function of γ, for two different U values, 4 and 12 (we recall that $\sqrt{2}\,|t|$ is taken as energy unit). It turns out immediately that there is a critical γ value (γ^*) above which the A_1 and A_2 subspaces are degenerate (the system is intrinsically unstable towards bCDW) and below which the A_1 and B_1 subspaces are degenerate (system unstable towards sCDW). The fundamental role played by γ, the balance between on-site and intersite e-e interactions, in setting up the sCDW-bCDW interface has already been recognized.[25] However, the parallel drawn here with the N-I interface suggests that the interface is not symmetric around $\gamma = 0$. Fig. 6 indeed shows that γ^* is slightly shifted towards positive values, although further calculations are needed to confirm this finding.

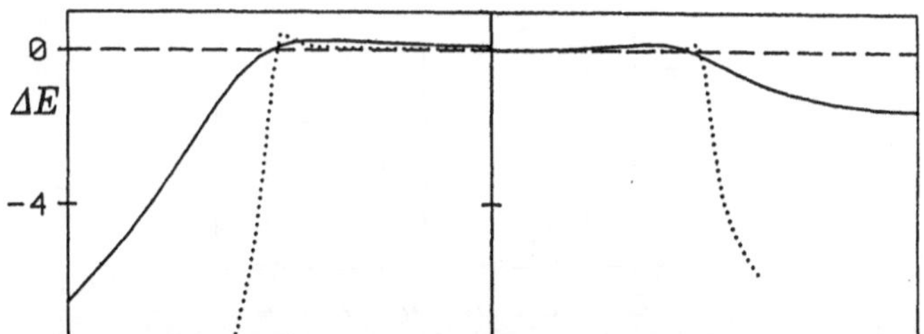

Fig. 6. Energy gap between lowest A_1 and A_2 (left panel) and A_1 and B_1 (right panel) states for ss systems with U=4 (full line) and U=12 (dotted line).

Electron-phonon couplings and 0 K phase diagram

We can now proceed to investigate the effects of finite strength e-mv and e-lph couplings on the sCDW-bCDW interface, following the methods already described. Having recognized the formal equivalence between ss and ms hamiltonians for $\gamma < \gamma^*$, in this region we can evaluate the amplitude of the sCDW as a function of γ by simply making resort to DVB calculations on ms with finite U (U=4).[24] The effect of intersite e-e interaction is introduced in mean field approximation. Considering that $\gamma = U/2 - V\alpha/2$ and $\varepsilon_c = V(\alpha - 1)$, at fixed U (and α) only one γ value is allowed for ss with a given ε_c. From the family of $\rho(\gamma)$ curves at different ε_c one therefore extracts the $\gamma(\rho)$ curve for ss systems, as shown in Fig. 7, upper panels. By introducing the e-mv and e-lph couplings in the usual manner, we construct the zero-temperature phase diagrams for sCDW and bCDW formation (middle and lower panels of Fig. 7, respectively). The left part of Fig. 7 shows that systems far from sCDW-bCDW interface (e.g., $\gamma = -1$) are rather stable in respect to both e-ph perturbations (the stable states lie below the curves). Near the interface (e.g., $\gamma = 0$, right side of Fig.7) the e-ph couplings become more effective, and the system may undergo an increase in the sCDW amplitude, driven by e-mv interaction, or/and a stack dimerization driven by e-lph coupling. We remark again that the previously found cooperation between e-mv and e-lph coupling in setting up bCDW is a spurious result, and the lower part of Fig.7 is accordingly modified in respect to that already published.[20]

In the $\gamma > \gamma^*$ (bCDW) regime, the construction of the phase diagram requires DVB calculations on dimerized ss systems. Some results have already been published,[20] but they are still rather preliminary. On the other hand, quantitative comparisons with real systems are difficult, as precise values of some of the relevant parameters are not available. Therefore in the following we shall limit

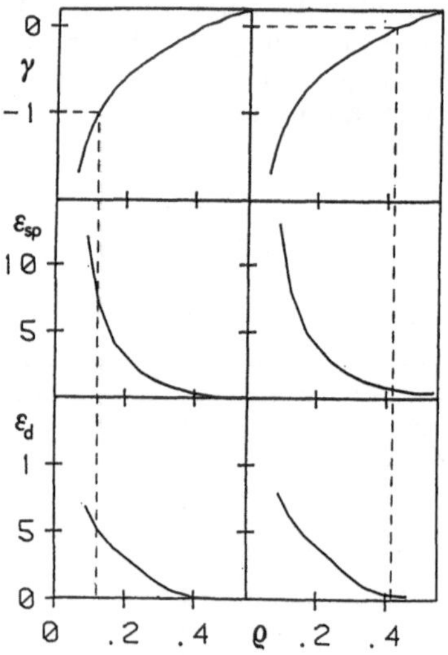

Fig. 7. Upper panels: γ vs ρ for a sCDW stack with U = 4. The middle and lower panels present the Q_μ and U_ν relaxation phase diagrams, respectively, for a system with $\gamma = -1$ (left side) and $\gamma = 0$ (right side).

ourselves to a qualitative discussion of the importance of the various parameters in determining the ground state geometry of the three different classes of compounds mentioned at the beginning of this section. According to our model, the ground state geometry is determined by the interplay of γ (= $U/2 - V\alpha/2$), ε_{sp} and ε_d; as we have seen, U and ε_d favor bCDW, whereas $V\alpha$ and ε_{sp} favor sCDW formation.

For $(CH)_x$ the large values of U and ε_d in comparison to ε_{sp} clearly point towards a bCDW ground state, as indeed experimentally found. On the other hand, the different ground state observed for ss CT salts and Pt-X chains (bCDW and sCDW, respectively) cannot be explained in terms of differences in U, ε_d and ε_{sp}. In fact, U and ε_d are of the same order of magnitude in the two systems; recent estimates[20] of e-mv coupling strength in Pt-X chains gives $\varepsilon_{sp} = 0.8 - 1.1$, of the same order of magnitude or slightly larger than for ss CT salts.[10] Therefore the different ground state of the two classes of compounds has to be ascribed to a difference in $V\alpha$. Indeed, the large polarizability of the organic π-molecules of CT salts in respect to the Pt-X complexes yields the prediction of a lower $V\alpha$ for the former systems. A second indication of the relevance of the $V\alpha$ term in determining the ground state geometry comes from the comparison of Pt-Cl, Pt-Br and Pt-I chains, where the sCDW amplitude decreases with increasing size of the halogen ion:[26] U, ε_{sp} and ε_d remain practically the same, but the polarizability increases with the halogen size so that $V\alpha$ decreases.

4. CONCLUSIONS

In this paper we have illustrated the use of DVB technique to investigate the instabilities of half-filled 1D chains with alternating or uniform on-site energy, showing how the real space representation of the method helps to develop a unified view of the different instabilities. We have seen that the resulting four dimensional 0 K phase diagram is able to account for the stability ranges of the various phases and for the observed phase transitions of ms CT crystals.[6,7] Such comprehensive view has still to be developed for 1D systems with uniform on-site energy, since for these systems both DVB calculations and the experimental estimate of the microscopic parameters need further work. Yet, some important qualitative trends start to emerge, putting in evidence the importance of electron-electron interactions. We have already stressed in the Introduction that our choice of unscreened Coulomb potential between point charges on the sites is not necessarily adequate to describe e-e interaction in every real system. On the other hand, consistency with the adopted model is mandatory, so that if one wants to investigate systems with highly screened electron-electron potentials, a non-Hubbard model, with site-bond repulsion terms included, would be the appropriate choice.[3] Work in this direction is in progress.

ACKNOWLEDGMENTS

We thank Z.G. Soos for giving us the original version of the DVB computer program and for stimulating discussions and correspondence. We also acknowledge helpful discussions with D.Baeriswyl, D.K.Campbell, S.Mazumdar, N.Nagaosa, M.J.Rice and J.Voit. This work has been supported by the Ministero della Pubblica Istruzione and by the National Research Council (CNR) of Italy through its Centro Studi Stati Molecolari Radicalici ed Eccitati.

REFERENCES

1. See for instance: "Extended Linear Chain Compunds", J.S. Miller ed., Plenum, New York (1983), Vol.1-3; Proceedings of the Int. Conf. on Science and Technology of Syntetic Metals, H. Shirakawa and T.Yamabe eds., Synt.Metals 17-19 (1987).
2. S. Ramasesha and Z.G. Soos, J.Chem.Phys. 80: 3278 (1984) and Int.J.Quantum Chemistry 25: 1003 (1984). See also: Z.G. Soos, Exact Valence-Bond Approach to Quantum Cell Models, this book.
3. A. Painelli and A. Girlando, Synth.Metals 27: 15 (1988); A. Painelli and A. Girlando, The Hubbard Model for One-Dimensional Solids, this book.
4. A.Girlando, A.Painelli and C.Pecile, Mol.Cryst.Liq.Cryst. 120: 17 (1985); A.Girlando, A.Painelli and C.Pecile, J.Chem.Phys. 89: 494 (1988) and references therein.
5. J.B.Torrance, A.Girlando, J.J.Mayerle, J.C.Crowley, V.Y.Lee, P.Batail and S.J. La Placa, Phys.Rev.Lett. 47: 1747 (1981).
6. A. Girlando and A. Painelli, Phys.Rev.B 34: 2131 (1986).
7. A. Painelli and A. Girlando, Phys.Rev.B 37: 5748 (1988).
8. Z.G.Soos, R.H.Harding and S.Ramasesha, Mol.Cryst.Liq.Cryst. 125: 59 (1985).
9. Z.G.Soos and S.Kuwajima, Chem.Phys.Lett. 122: 315 (1985).
10. A.Painelli and A.Girlando, J.Chem.Phys. 84: 5665 (1986).
11. A.Girlando and A.Painelli, Physica B 143: 559 (1986).
12. R.B. Griffiths, in: "Phase Transitions and Critical Phenomena", C.Domb and M.S.Green eds., Academic Press, London (1972), chap. 2.
13. D.L.Landau and E.M.Lifshitz, "Statistical Physics", Addison-Wesley, Reading (1969).
14. H.Hanfland, A.Brillante, A.Girlando and K.Syassen, Phys.Rev.B 38: 1456 (1988) and references therein.
15. A.Painelli and A.Girlando, J.Chem.Phys. 87: 1705 (1987).
16. A.Painelli and A.Girlando, Synth.Metals 19: 509 (1987).
17. A.Girlando, C.Pecile, A.Brillante and K.Syassen, Synth.Metals 19: 503 (1987).
18. M.Hanfland, K.Reimann, K.Syassen, A.Brillante and A.Girlando, Synth.Metals 27: 549 (1988).
19. S.Ichinose, Solid State Comm. 50: 137 (1984).
20. A.Painelli and A.Girlando, Synth.Metals 29: 181 (1989).
21. S.Kivelson, Phys.Rev.B 28: 2653 (1983) and references therein.
22. S.N.Dixit and S.Mazumdar, Phys.Rev.B 29: 1824 (1984) and references therein.
23. D.Baeriswyl and A.R.Bishop, Physica Scripta T19: 239 (1987) and J.Phys.C 21: 339 (1988).
24. A.Painelli and A.Girlando, Synth.Metals 27: 121 (1988).
25. S.Mazumdar and D.K. Campbell, Phys.Rev.Lett. 55: 2067 (1985) and Synth.Metals 13: 163 (1986).
26. R.J.H.Clark, in "Advances in Infrared and Raman Spectroscopy", Wiley-Hayden, London (1984), Vol.11.

QUANTUM MONTE CARLO STUDY OF NEUTRAL-IONIC TRANSITION

Naoto Nagaosa

Department of Applied Physics
University of Tokyo, Bunkyo-ku
Tokyo 113, Japan

INTRODUCTION

Quasi one-dimensional (1D) electronic systems are characterized by
(i) strong electron-electron and electron-phonon interactions,
(ii) large quantum as well as thermal fluctuations, and
(iii) nonlinear topological soliton excitations.

In this paper, we discuss the distinguished phenomenon, i.e., neutral-ionic transition[1], which shows all the above features. It is a phase transition in some mixed stack charge transfer compounds, e.g., p-TTF-Chloranil. These materials are composed of the weakly interacting chains which consist of alternating donor(D) and acceptor(A) molecules. The transition is schematically explained as follows. In the neutral phase, the HOMO (Highest Occupied Molecular Orbital) of the donor is doubly occupied (D^0) while the LUMO (Lowest Unoccupied Molecular Orbital) of the acceptor is empty (A^0). In the ionic phase, on the other hand, both the HOMO and LUMO are singly occupied to result in the configuration D^+A^-. The relative stability of these two phases is determined by the competition between the energy difference of LUMO and HOMO and the Madelung energy. The former prefers the neutral phase while the latter the ionic phase. By applying the pressure and/or decreasing the temperature, the lattice constant decreases and the Madelung energy increases. This results in the transition from the neutral to ionic phase (NI transition). On this transition, in particular in p-TTF-Chloranil, extensive experimental investigations have been performed. (1)Optical spectra[2], (2)X-ray diffraction[3], (3)electric conductivity, (4)specific heat, (5)ESR[4], (6)dielectric constant have been measured for various temperatures and pressures. These physcal quantities show anomalous behaviour near the transition as follows:
(i) The electric conductivity is enhanced up to about $1\Omega^{-1}\mathrm{cm}^{-1}$ under pressure (10kbar), which is 10^9 10^{16} times larger than the ordinary insulating organic crystals. The activation

Interacting Electrons in Reduced Dimensions
Edited by D. Baeriswyl and D.K. Campbell
Plenum Press, New York

energy for the conductivity (0.1eV) is much smaller than the excitonic absorption energy (0.7eV).

(ii) The coexistence of the neutral and ionic molecules is detected by the optical spectra near the transition temperature under pressure, which is in contrast to the first order like transition under the ambient pressure[2].

(iii) The dimerization of the lattice is observed in the X-ray diffraction experiment below the transition temperature (ionic phase)[3].

(iv) The magnetic susceptibility shows the Curie law below the transition temperature (ionic phase), while it is very small in the high temperature phase (neutral phase). The narrowing of the resonance peak toward the transition temperature from below is observed, which is interpreted in terms of the spin soliton[4].

The oversimplified classical model mensioned above fails to explain these variety of phenomena. The itineracy of the electron is essential, and the phenomenon should be described in terms of the interacting (quasi) one-dimensional (1D) electron system, which we will discuss below[5].

MODEL AND ITS PROPERTIES

We start with the following Hamiltonian.

$$H = -t \sum_{i\,s}(C_{is}C_{i+1s} + h.c.)(t+u_i-u_{i+1}) + \frac{\Delta}{2}\sum_i(-1)^i n_i$$

$$+ \frac{U}{2}\sum_{i\,s} n_{is}n_{i-s} + V\sum_i n_i n_{i+1} + \frac{2}{S}\sum_i (u_i-u_{i+1})^2 \qquad (1)$$

where $C_{is}(C_{is})$ is the annihilation (creation) operator of the electron on i th site with spin s . We place donor on odd i site and acceptor on even i site. The Coulomb repulsion is taken into account in terms of the on-site (U) and nearest neighbor site (V) interactions. The effective site energy difference Δ is given by

$$\Delta = U - I + A - 4V \qquad (2)$$

where I is the ionization energy of the donor molecule and A is the affinity for electron of the acceptor molecule. The displacement u_i of the i th molecule modulates the hopping integral, which gives rise to the electron-lattice interaction. We have investigated this model by the quantum Monte Carlo simulation and obtained the following results[5].

In the Absence of the Electron-Lattice Interaction. (S=0)

(i) There are two phases which are separated by continuous or discontinuous transition. In the absence of V, the transition is always continuous. When V is larger than some critical value,

the transition becomes discontinuous with the jump in the degree of charge transfer.
(ii) The two phases are characterized by the presence or absence of the gap E_{MAG} in the magnetic excitations. In the large U/V and/or large U/Δ phase, the low lying states are essentially those of the Heisenberg chain, and the generalized susceptibility for the spin density wave (SDW) and bond order wave (BOW) diverges with the same critical exponent as in the Heisenberg chain. This is the ionic phase.
(iii) As we increase Δ, the magnetic gap E_{MAG} begins to rise from zero at some critical point $\Delta = \Delta_C$. In this neutral phase, the generalized susceptibilities remain finite because of the gap.

			EXCITATION ENERGY
(a)	$D^+A^-D^+A^-D^+A^-D^+A^-D^+A^-D^+A^-D^+A^-D^+A^-D^+$	IONIC	
(b)	$D^0A^0D^0A^0D^0A^0D^0A^0D^0A^0D^0A^0D^0A^0D^0A^0D^0$	NEUTRAL	
(c)	$D^+A^-D^+A^-D^+A^-D^+A^-D^+A^0D^0A^0D^0A^0D^0A^0D^0$	1-DW	V/2
(d)	$D^+A^-D^+A^-D^+A^-D^+A^0D^0A^-D^+A^-D^+A^-D^+A^-D^+$	1-CT Ex.=2-DW	V
(e)	$D^+A^-D^+A^-D^+A^0D^0A^0D^0A^-D^+A^-D^+A^-D^+A^-D^+$	2-CT Ex.=2-DW	V
(f)	$D^+A^-D^+A^0D^+A^-D^+A^-D^+A^-D^+A^-D^0A^-D^+A^-D^+$	e-h pair =4-DW	2V

Fig. 1 Ground and excited states at the transition point. The ionic (a) and neutral (b) ground states are degenerate. The excited states with one (d) and two (e) charge transfer (CT) excitons are degenerate. In this case, the domain wall (DW) between the neutral and ionic phase (c) is the more fundamental elementary excitation. The electron-hole pair (f) is also regarded as 4-DW state.

(iv) The charge transfer gap E_{CT} is defined by the excitation energy of the infinitely separated electron-hole pair. E_{CT} decreases towards the transition point $\Delta = \Delta_C$ from the both sides. When the transition is continuous, E_{CT} touches zero at $\Delta = \Delta_C$. This means the appearance of the metallic state. When the transition is discontinuous, on the other hand, E_{CT} remains nonzero although it has a sharp minimum at $\Delta = \Delta_C$.
(v) The number of the charge transfer exciton is not a good quantum number at the transition, because the energy of the neutral and ionic states are degenerate (Fig.1). In this case, the domain wall between the neutral and ionic phase is the more fundamental elementary excitation.

Effects of the Electron-Lattice Interaction

(i) In the presence of S, the dimerization occurs in the ionic phase with the broken inversion symmetry. As a result, the gap is introduced to the excitation spectrum. As we increase S, the discontinuity of the transition decreases and at last the continuous transition results.
(ii) Comparing with the experimental results and our simulation results, we determine the values of the parameters appropriate to p-TTF-Chloranil as follows:
t =0.2eV, V=0.7eV, U=1.5eV, S=0.2eV.
(iii) Two types of soliton exist. The spin soliton has spin 1/2 and fractional charge, while the charge soliton has no spin and also the fractional charge. The excitation energy E_{SS} (E_{CS}) of the spin (charge) soliton is calculated to be 0.1eV (0.06eV) at the transition point for the parameters listed above. Exactly at the discontinuous transition point, the charge soliton is dissociated into two domain walls which separate the neutral and ionic phases (NIDW). If the transition is continuous, on the other hand, the charge soliton state merges smoothly into the ground state while the spin soliton into the polaron state in the neutral phase. The schematic view of the energies of the spin and charge soliton is shown in Fig.2.

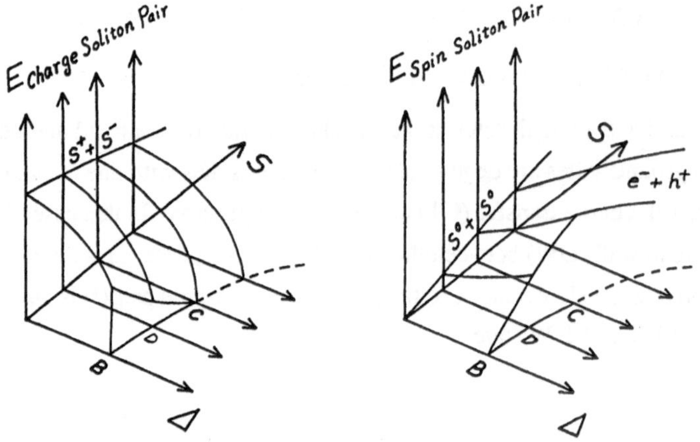

Fig. 2. Schematic view of the energies of the charge and spin solitons. As we increase S beyond the point C in the basal plase, the transition becomes continuous (dashed line).

DISCUSSION AND CONCLUSIONS

In the previous section, we have investigated the purely one-dimensional model. In real experimental situations, the three-dimensional (3D) interchain interactions are present, and the temperature is finite. If the former prevails, the transition is discontinuous while the latter reduces the discontinuity and at last makes the transition continuous[6]. Under ambient pressure, the transition temperature Tc is 84K and the thermal excitation of NIDW's ,whose energy is $E_{CS}/2 = 0.03eV$, is not enough to make the transition continuous. Under pressure (10kbar), on the other hand, the transition temperature is around the room temperature and the thermal excitation of NIDW's results in the continuous transition accompanied by the coexistence of ionic and neutral molecules. The confinement of the NIDW pair is gradually released as the energy difference between the neutral and ionic phases decreases, which leads to the increase in the thermal population of the NIDW's around the transition. The peak in the electric conductivity at the transition may be related to this increase in NIDW's. The NIDW alone, however, can not contribute to the DC conductivity as is explained below. The current due to the motion of NIDW' s or the charge soliton is accompanied by the inversion of the polarization, which vanishes after the completion of the inversion. To obtain the DC conductivity, another type of the soliton, i.e., spin soliton, should also contribute. The activation energy for the conductivity is, therefore, the formation energy of the spin soliton 0.1eV, which is in good agrement with the experiment.

In conclusion, we have presented a model to describe the neutral-ionic transition. The kink structures, i.e., solitons and domain wall, play essential roles in the phase transition, and the various anomalous experimental results can be consistently interpreted in this picture.

REFERENCES

1. J.B.Torrance et al., Phys. Rev. Lett. 47, 1747 (1981).
2. Y.Tokura et al., Solid State Commun. 43, 757 (1982).
3. Y.Kanai et al., Synthetic Metals 10,157 (1984/85).
4. T.Mitani et al., Phys. Rev. Lett. 20, 842 (1984).
5. N.Nagaosa, Solid State Commun. 51, 179 (1986).
   N.Nagaosa and J.Takimoto, J. Phys. Soc. Jpn. 55, 2737 and 2747 (1986).
   N.Nagaosa, J. Phys. Soc. Jpn. 55, 2756 (1986).
6. N.Nagaosa, J. Phys. Soc. Jpn. 55, 3488 (1986).

DIAGONAL AND OFF-DIAGONAL ELECTRONIC INTERACTIONS AND PHONON DYNAMICS IN AN EXTENDED MODEL OF POLYACETYLENE

Johannes Voit
Theoretische Physik 1 and BIMF, Universität Bayreuth
D-8580 Bayreuth, Fed. Rep. Germany

INTRODUCTION

The basic physics of polyacetylene [1] is nowadays believed to be contained in the model proposed by Su, Schrieffer, and Heeger (SSH) [2] describing a one-dimensional (1D) tight-binding model of uncorrelated electrons whose hopping amplitude is modulated by acoustic phonons. One important extension of this model, motivated by a variety of experiments [3] which are in qualitative disagreement with the simple one-electron picture of SSH, consists in the inclusion of electron-electron interactions often parametrized by an (extended) Hubbard model with an electronic on-site repulsion U (and a nearest neighbor interaction V). An established result of earlier theoretical studies was that dimerization was *increased* by weak interactions U and $V < U/2$. [4,5,6]. It therefore came as a surprise that recent work [7,8,9] claimed that important terms in the electronic interactions were missing and that upon their inclusion, dimerization would be *decreased* by electronic correlations. Most of the discussion [8,9] was centered on the relative magnitude of interaction parameters appropriate to real conducting polymers. Another point of debate is the influence of quantum fluctuations of the phonons on the dimerization. Several authors [10] concluded that the low energy properties of the SSH model with finite ionic mass were determined by the Gross-Neveu ($\omega_{2k_F} \to \infty$) limit. However, it is clear that for low enough phonon frequencies, a crossover to effective semiclassical (Peierls) behaviour *must* occur. Accurate conditions for this crossover as well as its possible dependence on electronic interactions are not known today. Moreover, many of the previous studies of the influence of electronic interactions were performed in the adiabatic limit for the phonons. Despite the low phonon frequencies in real systems, justifying in principle a mean field theory, it is not clear to what extend the above results hold when the phonon dynamics is retained. In fact, a previous study of incommensurate one-dimensional (1D) metals [11] demonstrated that the interplay of repulsive electron-electron and attractive electron-phonon interactions depends crucially on the phonon frequency. A general understanding of these issues is of prime importance; once the cooperation and competition of various interactions in the establishment of the dimerized state is understood, one can search for different e.g. superconducting ground states in these models. Such a programme should, on the one hand, help to clarify the discrepancy between a recent theory suggesting that $(CH)_x$ should be a high-temperature superconductor [12] and experiment [1]; on the other hand, it might provide some guideline to the chemists' attempts to synthesize a superconducting polymer.

This situation calls for a theory treating Coulomb and electron-phonon interaction with phonons of arbitrary frequency in a unified manner. It is the purpose of the present paper to present a novel approach to the SSH-model allowing for an explicit discussion of the cooperation between different interactions and of the role of the phonon frequency. We work in

Interacting Electrons in Reduced Dimensions
Edited by D. Baeriswyl and D.K. Campbell
Plenum Press, New York

the continuum limit and are therefore limited to asymptotic low energy properties in weak-coupling systems. While there is evidence that real polymers are rather characterized by intermediate coupling strengths [3], our results are expected to be (at least qualitatively) relevant for these materials. Moreover, they should stimulate further nonperturbative (numerical) investigations.

THE MODEL AND ITS SOLUTION

We shall consider in the following the Hamiltonian proposed by Kivelson et al. (KSSH) [7,8], supplemented by the kinetic energy of the phonons,

$$\begin{aligned} H &= H_{SSH} + H_{kin}^{(ph)} + H_{int} , & (1) \\ H_{SSH} &= -\sum_{n,s}[t_0 - \alpha(u_{n+1} - u_n)][c_{n+1,s}^{\dagger} c_{n,s} + \text{H.c.}] + \frac{1}{2}K\sum_{n}(u_{n+1} - u_n)^2 , & (2) \\ H_{kin}^{(ph)} &= \sum_n \frac{p_n^2}{2M} , & (3) \\ H_{int} &= \sum_{n,m,l,p,s,s'} V(n,m,l,p) c_{n,s}^{\dagger} c_{m,s'}^{\dagger} c_{l,s'} c_{p,s} + \delta\mu \sum_{n,s} c_{n,s}^{\dagger} c_{n,s} . & (4) \end{aligned}$$

Here, $c_{n,s}^{\dagger}$ creates an electron at site n with spin s, t_0 is the nearest neighbor hopping element and $V(n,m,l,p)$ the matrix element of the electron-electron interaction. $\delta\mu$ is a shift in the chemical potential such that the Fermi energy remains at $E_F = 0$. u_n is the displacement of the ion at site n out of its equilibrium position, p_n its momentum, and K and M are its spring constant and mass, respectively. α is the electron-phonon coupling constant. We only keep electronic couplings involving nearby sites, i.e. the U and V of the extended Hubbard model, and W and X introduced by KSSH. The band is half-filled.

We now go over to the continuum limit and perform a bosonization transformation [11]. For the extended Hubbard model, we arrive at the standard interaction terms denoted by $g_1 \dots g_4$ [13]. As noted by KSSH, the main effect of W is to increase g_1 and decrease g_3. There are, however, two new terms: one is parallel-spin Umklapp scattering

$$H_{3,\parallel} = \frac{2g_{3,\parallel}}{(2\pi\tau)^2}\int dx \cos\left[\sqrt{8}\Phi_\rho(x)\right]\cos\left[\sqrt{8}\Phi_\sigma(x)\right] \qquad \frac{g_{3,\parallel}}{v_F} = -2\frac{V+W}{t_0} . \tag{5}$$

Here, v_F is the Fermi velocity and τ a short distance cutoff. $\Phi_\nu(x)$ is a phase field describing charge ($\nu = \rho$) and spin ($\nu = \sigma$) fluctuations. Moreover, by carefully evaluating the X-term in the boson representation, we find the following nonvanishing Umklapp process

$$H_{3,X} = \frac{2g_{3,X}}{(2\pi\tau)^2}\int dx \sin\left[\sqrt{8}\Phi_\rho(x)\right] , \qquad \frac{g_{3,X}}{v_F} = \frac{X}{t_0} , \tag{6}$$

contrary to the assertions of KSSH that the X-term would not couple Fermi surface states in a half-filled band. Further, the effective interaction coming from (6) in second order perturbation theory has exactly the same form as that due to the ordinary Umklapp scattering indicating that this term must not be neglected. The contribution of X to the coupling constants $g_1 \dots g_4$ vanishes for a half-filled band. X couples charge densities on the bonds and on the sites and breaks the charge conjugation symmetry of (1).

Finally, the electron-phonon Hamiltonian becomes

$$H_{1,ep} = \frac{-8i\alpha}{\pi\tau}\int dx \sin\left[\sqrt{2}\Phi_\rho(x)\right]\cos\left[\sqrt{2}\Phi_\sigma(x)\right]\varphi_{2k_F}(x) . \tag{7}$$

$\varphi_{2k_F}(x)$ is a (real) field describing the $\pm 2k_F$-components of the displacement field $u_{2k_F}(x) \sim \mathcal{R}e\left[\exp(2ik_F x)\varphi_{2k_F}(x)\right]$. Forward scattering by acoustic phonons is negligible [14].

Going over to the Matsubara formalism of imaginary times, the model is mapped onto two XY-models desribing charge and spin degrees of freedom which are both by $g_{3,\|}$ and α. The spin fluctuations scale, for spin-isotropic interactions, along the line $X_\sigma = Y_\sigma$ towards the weak coupling fixed point for repulsive electron-electron interactions and into a strong coupling regime for attractive ones corresponding to a gap in the spin density excitations. For repulsive coupling, the charge density XY-model is placed on the unstable fixed line $(X_\rho < 0, Y_\rho = 0)$ while for attractive electronic coupling, it is on the stable counterpart $X_\rho > 0$. Scaling into the strong-coupling region with positive (negative) Y_ρ implies bond (site) order in the charge degrees of freedom. The equations can practically be guessed from previous work [11] using symmetry arguments. The following general features are noteworthy, however.

(i) The phonon contribution to Umklapp scattering (g_3) has the *opposite* sign of the contribution to backward scattering (g_1) on account of the nonlocal electron-phonon coupling. This is in contrast to previous work by Caron and Bourbonnais [6] and translates into the continuum model the symmetry observed by Hubert [15] yielding an excitation gap independent of the sign of the Hubbard U. Moreover, the effective Umklapp scattering coming from α and U have the *same* sign whereas the sign of their backscattering contributions is different. This observation is not only the basis of the dimerization enhancement by U to be discussed below but also suggests the following very simple picture for the mechanism operating: in weak coupling, only states close to the Fermi surface are relevant. Then, for a fermion operator, $c_{n,s} = \sum_k c_s(k)\exp(ikna) \rightarrow \sum_{\epsilon=\pm} c_s(\epsilon k_F)\exp(i\epsilon k_F na)$ and for a half-filled band $\exp(ik_F na) = i^n$. While the intuitive ideas about electronic interactions being repulsive and electron-phonon coupling being attractive are prefectly valid, one must, in a discrete model, allow for the above phase factors i^n introducing sign changes when intersite and intrasite interactions are compared. It is just their different spatial symmetry which makes electron-electron and electron-phonon interactions cooperate rather than compete.

(ii) It is not possible to derive consistent scaling equations for the SSH-Hubbard model discarding parallel-spin Umklapp scattering ($g_{3,\|}$). It is generated by the renormalization group transformations and therefore must be included from the outset. Its most important property is to couple charge and spin density fluctuations. However, it is irrelevant for weak coupling and only becomes relevant for interactions whose magnitude is a sizable fraction of the bandwidth. A particularly striking example is the extended Hubbard model on the line $U = 2V$ implying $g_{1,\|} = g_{1,\perp} = g_{3,\perp} \equiv 0$ while $g_{2,\|} = g_{2,\perp} \sim U+2V = 4V$, and $g_{3,\|} \sim -2V$. We find that $g_{3,\|}$ goes relevant for $2V \sim 3$. There is a new fixed point ($g_{3,\|} \rightarrow -\infty$, $g_{2,\|} \rightarrow \infty$, $g_{2,\perp}$ finite) in addition to the usual one ($g_{3,\|} = 0$, g_2 finite). It is associated with the first order transition between CDW and SDW at intermediate couplings recently found in numerical work [16]. Our results for the extended Hubbard model suggest a picture where there is long range order in an up-spin CDW and a down-spin CDW with only weak interactions between them. Another interesting consequence of the $g_{3,\|}$-coupling is that vibrational excitations of charged solitons should be accompanied by spin fluctuations [17].

(iii) Also the electron-phonon interaction couples charge and spin density fluctuations at energy scales $E(\ell)$ $[\ell = \ln(\tau/\tau_0)]$ above the phonon frequency $\omega_{2k_F} = 2\sqrt{K/M}$. Contrary to (ii), this coupling is strongly relevant also for weak coupling [11]. Specifically, with $Y_{ph} = 16\alpha^2/\pi v_F M\omega_{2k_F}^2$ the effective electron-phonon coupling, $d\ln Y_{ph}(\ell)/d\ell \cong 1$ for small ℓ, where the equal-sign applies for the case without electron-electron interactions. The latter may become relevant, however with much smaller scaling exponents implying, in practice, significant quantitative corrections to the dimerization but rarely implies qualitative changes in the ground state.

(iv) This difference in scaling behaviour gives a very special role to the phonon frequency, becoming the most important parameter for the control of the interplay of attractive electron-phonon and repulsive Coulomb interaction. This shows up in the scaling equations through an exponential suppression of the contributions of electron-phonon scattering to the effective electron-electron interactions when $E(\ell) < \omega_{2k_F}$. Below this scale, the electron-phonon interaction behaves effectively unretarded.

DIMERIZATION AND ELECTRONIC INTERACTIONS

SSH-model with finite phonon frequencies. **In the abesence of electron-electron interactions, the system dimerizes and opens a gap Δ in its excitation spectrum at any finite coupling constant α and at any phonon frequency. However, our results show a marked crossover between two limiting cases. In a high phonon frequency regime and for relatively weak electron-phonon coupling, we can scale our model consistently onto one containing only instantaneous though renomalized interactions. Then, $\Delta \sim \omega_{2k_F} \exp(-1/\mid Y_\nu^\star \mid)$, where $Y_\rho^\star = -Y_\sigma^\star$ is the effective electron-electron interaction generated once all retardation effects have been scaled out. Consistency requires both electron-phonon and the generated electron-electron interactions to remain small at that energy scale, i.e.**

$$Y_\nu^\star < 1 \, , \qquad Y_{ph}^\star =< E_F/\omega_{2k_F} \, . \tag{8}$$

For frequencies below (or coupling constants above) the values given by this criterion, we determine the gap and the order parameter from the adiabatic limit. Fluctuation effects may then be accounted for by perturbation theory. The consistency condition (8) clearly marks the crossover criterion between (weakly retarded) Gross-Neveu and (strongly retarded) Peierls behaviour. Our results are in good agreement with previous work [10].

SSH-model with electronic interactions. **We first discuss the influence of electronic interactions on dimerization for relatively high phonon frequencies [(8) valid; clearly, this condition now depends *both* on α and $V(n,m,l,p)$]. The renormalized propagator of the $2k_F$-part of the phonon spectrum is given by [18]**

$$\begin{aligned} D(2k_F+q) &= D_0(2k_F+q) + 16\alpha^2 D_0(2k_F+q) R_{BOW}(2k_F+q) D(2k_F+q) \, , & (9) \\ R_{BOW}(2k_F+q) &\approx A_{BOW}(v_F \mid q \mid)^{-\alpha_{BOW}} \, , \quad \mid q \mid \ll 2k_F . & (10) \end{aligned}$$

D_0 is the bare propagator and R_{BOW} the BOW correlation function. The influence of electron-electron interaction on dimerization is assessed quantitatively through its renormalization of both Y_{ph} and α_{BOW}. In the following Figures, we display the scaling flow of these two quantities as high energy fluctuations are integrated out. Fig. 1 shows the influence of a Hubbard U on Y_{ph} and α_{BOW} for fixed electron-phonon coupling and phonon frequency. *Both* Y_{ph} and α_{BOW} (as well as their derivatives) increase with U at any length scale. Moreover, an improved method [19] allows to derive also renormalization equations for the prefactor A_{BOW} in Eq. (10); A_{BOW}, too, increases with U. Thus, although the correlation exponent reaches the value $\alpha_{BOW} = 2$ indicating BOW long range order at $T = 0$ independently of U, it is fair to conclude that dimerization *increases* with U. Notice that at finite temperature, renormalization would stop at $\ell_T = \ln(E_F/T)$ leading to an even increased influence of U on dimerization. We do not find a change in these trends *within* the domain of validity of our approach, i.e. weak coupling, in agreement with the simple picture given above. For $U \to \infty$, spin-Peierls arguments suggest a decrease in dimerization; we thus expect a maximum for $U \approx 4t$ in agreement with earlier work [4,5].

We have also investigated the influence of a finite nearest-neighbour repulsion V on the dimerization and find that the dimerization is increased by nonzero V. Surprisingly, and in contrast to previous work [5], we do not find a decrease in dimerization for $V > U/2$. While both $g_{1,\perp}$ and $g_{3,\perp}$ change sign for $V > U/2$, at the same time the electron-phonon interaction becomes more relevant. However, as demonstrated in Table 1, the value of V where the dimerization starts to decrease, approaches $U/2$ with decreasing Y_{ph} and ω_{2k_F}. The criterion $V > U/2$ given previously [5] is only recovered as $Y_{ph} \to 0$ and/or $\omega_{2k_F} \to 0$ indicating clearly the importance of treating the interactions of electrons among themselves and with dynamic phonons on an equal basis. $V > 0$ favours, in principle, a CDW [18] ground state. In the presence of other interactions, such as e.g. Y_{ph} or U, a weak enough V may cooperate with them actually enhancing the tendency towards ground states different from CDW, and a finite V_{max} is required to enforce the CDW. The point of the present paragraph is that, in a system with both electron-electron and electron-phonon interactions, even the competition between

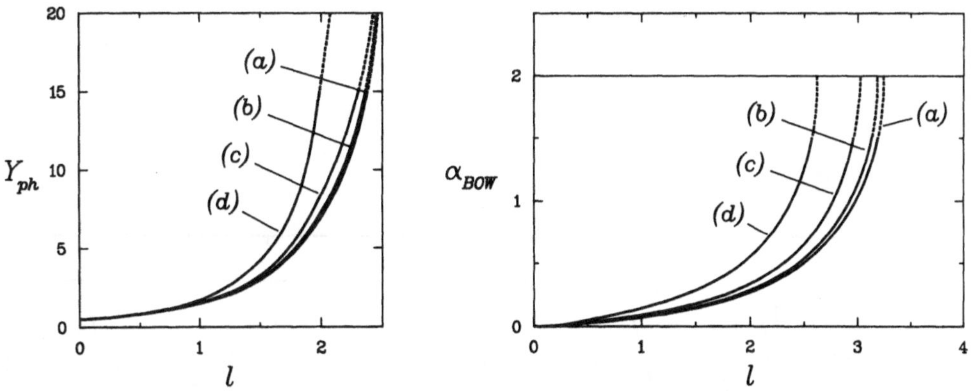

Figure 1. Influence of a Hubbard on-site repulsion on the scaling flow of effective electron-phonon interaction Y_{ph} and BOW correlation exponent α_{BOW}. $Y_{ph}(\ell = 0) = 0.5$, $\omega_{2k_F} = 0.1E_F$, (a): $U = 0$, (b): $U = 0.5$, (c): $U = 1$, (d): $U = 2$ in units of t_0.

Table 1. Dependence of the maximum V enhancing dimerization on (a): phonon frequency ω_{2k_F} and (b): electron-phonon coupling Y_{ph}.

(a):	$\frac{\omega_{2k_F}}{E_F}$	0.1	0.01	10^{-3}	10^{-4}	10^{-5}	...	0
$Y_{ph} = 0.5$	V_{max}	>1.0	0.64	0.41	0.30	0.25	...	0.15
(b):	Y_{ph}	0.5	0.2	0.1	0.05	0.02	...	0
$\frac{\omega_{2k_F}}{E_F} = 0.1$	V_{max}	>1.0	0.75	0.38	0.25	0.20	...	0.15

different electronic interactions depends dramatically on electron-phonon coupling and the frequency of the phonons involved.

Even more interesting in view of the recent debate about possibly missing interactions in the SSH-Hubbard model for a description of conducting polymers is the influence of the new terms W and X [7,8,9]. Fig. 2 displays the change in scaling behaviour with increasing W. Clearly, dimerization strongly decreases and ultimately is completely destroyed. The curve where $\alpha_{BOW} < 0$ for all ℓ describes a system with sizable triplet-superconducting (TS) fluctuations. (This should, however, not be taken as evidence of possible superconductivity in extended SSH-models; actually, a model with $W > U$ and all other $V(n,m,l,p) = 0$ is unphysical in view of recent results on the relative magnitudes of the interaction parameters for physical potentials [8,9].) Our integrating out exactly the phonons shows that W *exactly* opposes Y_{ph}. Interestingly, the effect of W is opposite in the spin-Peierls limit: W decreases the spin exchange integral, thereby increasing the effective electron-phonon coupling and dimerization [20]. Fig. 3 shows that also the interplay of W with U and V depends sensitively on both Y_{ph} and ω_{2k_F}. We keep U and W fixed and try to compensate for the loss in dimerization caused by $W > 0$ by adding V. While dimerization initially increases with V, it is not possible to compensate completely the effect of W. Instead, when V becomes too strong, dimerization breaks down and the system crosses over to a CDW state. A more systematic study [21] seems to indicate that the condition $V > 3W$ given previously [7] for dimerization to be enhanced with respect to the noninteracting case, again is recovered only as $Y_{ph} \to 0$ and/or $\omega_{2k_F} \to 0$.

The X-term produces an Umklapp scattering (6) in a half-filled band. Due its particular symmetry, it is neither renormalized by electron-phonon interaction nor does it renormalize that interaction. X always acts in favour of bond order but could itself become relevant only in the presence of interactions favouring site order. As a consequence, it usually has marginal scaling behaviour. This peculiar behaviour results from the fact that all other interactions

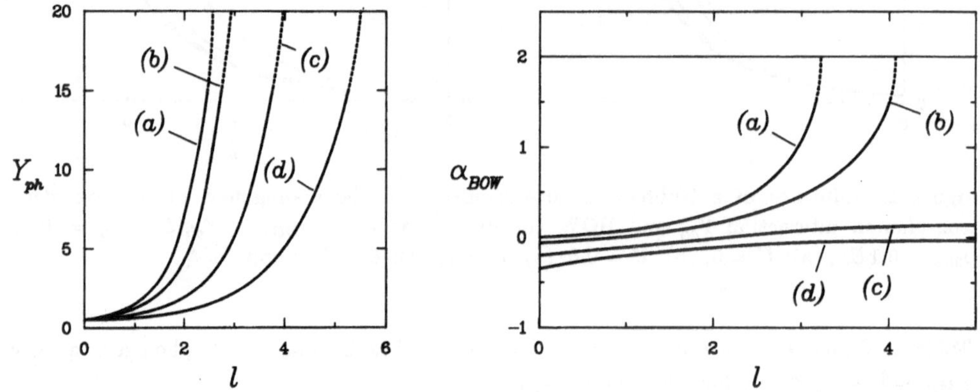

Figure 2. Effect of bond charge repulsion W on dimerization through Y_{ph} and α_{BOW}. $Y_{ph}(\ell = 0) = 0.5$, $\omega_{2k_F} = 0.1E_F$, $U = 0.3$, (a): $W = 0$, (b): $W = 0.1$, (c): $W = 0.3$, (d): $W = 0.5$ in units of t_0.

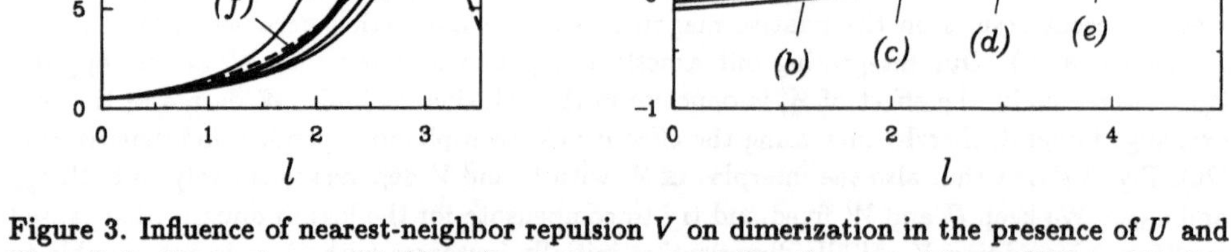

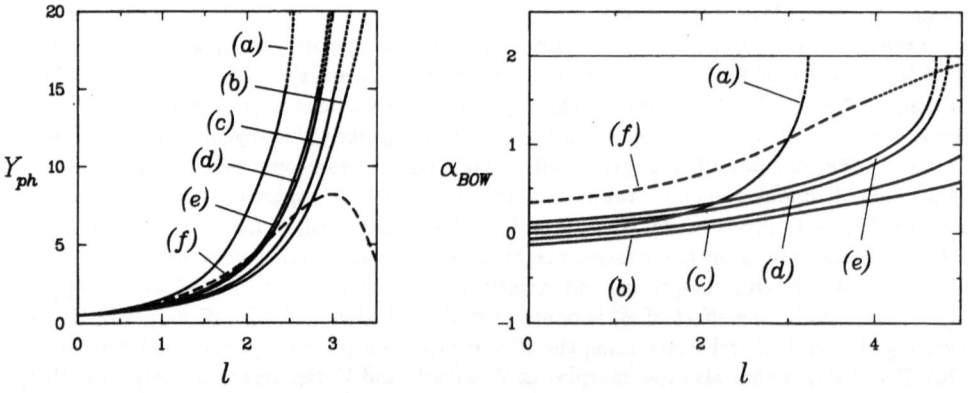

Figure 3. Influence of nearest-neighbor repulsion V on dimerization in the presence of U and W at finite Y_{ph} and ω_{2k_F}. $Y_{ph}(\ell = 0) = 0.5$, $\omega_{2k_F} = 0.1E_F$, $U = 0.3$, W=0.2 [(b) to (f)], (a): $V = 0$ and $W = 0$, (b): $V = 0$, (c): $V = 0.1$, (d): $V = 0.3$, (e): $V = 0.4$, (f): $V = 0.8$ in units of t_0.

drive the system away from the BOW-CDW separation line. From the present continuum model, we suspect the influence of X to be increased considerably on going away from a half-filled band (cf. below).

With decreasing phonon frequency and for fixed bare electron-phonon coupling constant (in units of E_F), dimerization considerably increases. One may then be in a regime where a consistent renormalization group treatment is no longer possible [Eq. (8) violated]. We then start out from the adiabatic limit $[\varphi(x) \to \varphi_0]$ and use standard crossover scaling arguments [22] to determine both the order parameter $\alpha\varphi_0$ and electronic gap Δ [11] as

$$\alpha\varphi_0 \sim \left(\frac{2Y_{ph}}{1+\alpha_{BOW}/2}\right)^{\frac{1+\alpha_{BOW}/2}{\alpha_{BOW}}}, \quad \Delta \sim \frac{v_F}{\tau}(\alpha\varphi_0)^{\frac{2}{2+\alpha_{BOW}}} \tag{11}$$

where α_{BOW} now is the BOW exponent of the electronic system without electron-phonon interaction; e.g. in the Hubbard model at $T = 0$: $\alpha_{BOW} = 1$ for any $U > 0$. In the extended Hubbard model, we have $\alpha_{BOW} = 1$ for $V < U/2$. For $V > U/2$ we have formally $\alpha_{BOW} = 0$ implying $\Delta \sim \exp(-1/2Y_{ph})$ as in the case without interaction, but now, CDW-order is much stronger.

For the SSH-Hubbard model, Δ increases monotonously with α_{BOW}, but $\alpha\varphi_0$ goes through a maximum as a function of α_{BOW} (and, thereby, of U). This behaviour is indeed extremely close to what is observed in numerical work [5]. However, at least at $T = 0$, such a discussion is misleading: then, α_{BOW} has to be determined from the purely electronic *fixed points* which, for the Hubbard model do not depend continuously on U (cf. above). The precise dependence of $\alpha\varphi_0$ and Δ on U depends, however, on the prefactors in Eq. (11). By extending the crossover scaling hypothesis to the full correlation functions [21], the variation of $\alpha\varphi_0$ with U can be related to the dependence on U of the BOW correlation function in the Hubbard model without electron-phonon interaction. Evaluating this correlation function carefully [19], we find

$$R^{Hub}_{BOW}(\vec{r}) = A^{Hub}_{BOW}\, r^{-1} \ln^{-3/2}(r)\,. \qquad [\vec{r} = (x, v_F\tau)] \tag{12}$$

Importantly, for weak U, A^{Hub}_{BOW} increases with U whereas mapping the $U \to \infty$-limit onto a Heisenberg model again leads to a decrease at large U implying a maximum in R^{Hub}_{BOW} for $U \sim 4t$. In the Peierls limit $(\omega_{2k_F} \to 0)$, the dimerization in the SSH-Hubbard model therefore *directly reflects* the behaviour of the BOW correlation function of the pure Hubbard model! Clearly, the discussion can be extended to the presence of further interactions like V, W, and X.

An important effect of the X-term will be mentioned only in passing. It contributes a scale-dependent renormalization to the velocity v_ρ of the collective charge density fluctuations which unlike all other known contributions [11] depends explicitely on the sign of X. Consequently, it should be visible in many thermodynamic properties of the system which generally depend on v_ρ [23]. Moreover, in the dimerized state, the charge conjugation symmetry breaking due to X is expected to influence dramatically the optical absorption of nonlinear excitations such as polarons [24]. Measurements of optical and thermodynamic properties should therefore be extremely valuable to obtain estimates on the interaction parameters discussed above in real systems.

SUPERCONDUCTIVITY

From the above results on the cooperation and competition of electron-phonon and electron-electron interactions in the establishment of a dimerized ground state in extended SSH-models it is clear that, for suitable interaction parameters, dimerization may be destroyed and the system may cross over to a different ground state. It is then natural to wonder if and under what conditions the model considered or modifications thereof exhibit dominant superconducting (SC) fluctuations which, eventually could then be stabilized into a long range ordered SC state by some 3D coupling. In particular, starting from a weak coupling analysis similar to the one presented above, it has been argued recently [12] that doped polyacetylene might be in a resonating valence bond state and therefore a good candidate for high-temperature superconductivity.

1D half-filled bands are, in general, extremely unfavourable to superconductivity. Strong electron-electron and electron-phonon umklapp scattering push the system towards an (insulating) BOW-, CDW- or SDW-state [18], depending on the interactions involved. While it has been shown above that such ground states can be destabilized by suitable interactions of sufficient strength, these situations are generally not realistic. In particular, it has been shown recently [8,9] that a physical electron-electron interaction potential $V(n,m,l,p)$ [cf. Eq. (4)], implies a definite relation $U > X > V \geq W$ among the different interaction parameters used above. Moreover, their relative magnitude depends essentially on the screening length of the potential and on the overlap of the Wannier functions involved in the matrix elements $V(n,m,l,p)$. With these limitations, strong SC fluctuations in a 1D half-filled band are excluded in general.

In incommensurate systems, which model sufficiently highly doped polymers, umklapp processes are absent and consequently density waves are much less stable. We perform the following discussion in the "g-ology"-picture where the interactions are parametrized by the two coupling constants g_1 for large momentum transfer scattering and g_2 for small momentum transfer scattering between the two Fermi points $\pm k_F$. For the extended Hubbard model (t_0, U, V), the mapping is given by $g_1 = U - 2V$ and $g_2 = U + 2V$. Equivalent scattering processes can be defined for electron-phonon interactions. Their relative magnitudes are, however, very different for acoustic phonons [$\omega(q) = v_s \mid q \mid$ for $q \to 0$, v_s being the sound velocity] and dispersionless (intramolecular or optical) phonons [$\omega(q)$ finite for $q \to 0$]: the effective g_1-interactions are comparable in both cases; however, $g_{2,eff} \propto (v_s/v_F)^2$ for acoustic phonons and therefore is negligible compared to $g_{1,eff} \sim Y_{ph}/2$, whereas for dispersionless modes $g_{2,eff} \approx g_{1,eff} \sim Y_{ph}/2$ [14]. Since SC fluctuations require an attractive g_2-interaction it follows immediately that acoustic phonons do not favour SC in 1D. Peierls fluctuations are much stronger here.

The situation is very different for dispersionless phonons where the effects of forward and backward scattering compensate to a large extend so that even weak interactions might lead to a SC state. We have therefore investigated systematically the influence of *both* electron-phonon forward and backward scattering on SC in the weak-coupling "g-ology"-model. In order to ensure a treatment on equal footing of all interaction processes and their scale dependence, we do not make use of an exact solution of the electron-phonon forward scattering problem [14] but include this interaction into our scaling analysis. Fig. 4 shows typical results obtained by searching for the most divergent correlation function in the g_1-g_2-plane. The dashed lines give the phase diagram in the absence of electron-phonon interaction. Adding phonons, we have made the (plausible but arbitrary) assumption of equally strong electron-phonon forward and backward scattering. We summarize the main features, emphasizing their limitation to the *weak-coupling* regime: (i) In the region $g_2 < 0$, $g_1 < 0$, CDW extends at the expense of singlet superconductivity (SS); however, in the region of repulsive electron-electron interactions, adding phonons extends the domain of divergence of SS and TS except in the immediate neighbourhood of the origin. (ii) The electron-phonon-only model $g_1 = g_2 = 0$ displays a CDW divergence independent of Y_{ph} and ω_{2k_F}, albeit only a very weak one. (iii) The Hubbard model (dotted line) is *always outside* the domain of diverging SC fluctuations. Adding a Hubbard-U to the electron-phonon model *suppresses* SC fluctuations. The suppression is at least as strong as that of CDW fluctuations while SDW ones are enhanced. (iv) Increasing Y_{ph} or lowering $\omega_0 = \omega_{2k_F}$ favours CDWs more strongly than SS and TS. (v) Extending the Hubbard model by $V > 0$ corresponds to a displacement from the dotted line in Fig. 4 to the lower right and therefore further decreases SC. (vi) Repulsive off-diagonal interactions like W can be favourable to SC. Their quantitative effect depends sensitively on band-filling. While varying W freely allows easily to enter the SC domain in Fig. 4 for a band not too far from half-filling, physical considerations [8,9] again require $W \leq V$; then, one is limited to the area below the dotted (Hubbard) line. Interestingly, a repulsive X-term in a more than half-filled band generates an attractive forward scattering and possibly favours SC. Moreover, being intermediate between

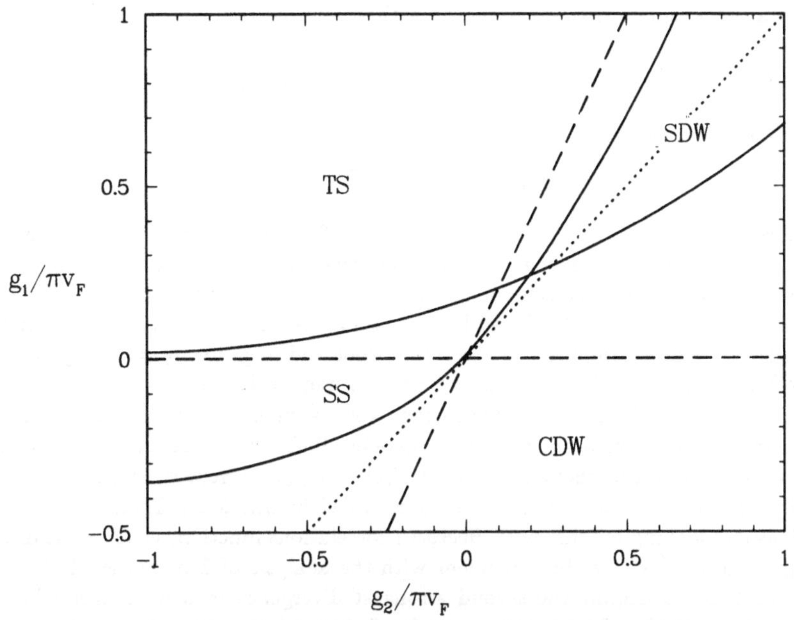

Figure 4. Modified "g-ology" phase diagram including both electron-phonon forward and backward scattering. $g_{2,eff} = g_{1,eff} = 0.25$, $\omega_0 = \omega_{2k_F} = 0.1E_F$. Dashed lines: $g_{2,eff} = g_{1,eff} = 0$, dotted line: locus of the Hubbard model.

U and V, X must not be neglected. Due to the charge conjugation symmetry breaking of X, its contribution to g_2 for less than half-filled band would be repulsive and favour CDWs. On a more speculative level, one might be tempted to relate this observation to the fact that all known organic superconductors (and also the copper oxides) [1] are hole superconductors. A detailed investigation of these issues and a discussion of general conditions for presence or absence of superconductivity in 1D will be given elsewhere [21].

The preceding results sensitively depend on the equality of electron-phonon forward and backward scattering. For $g_{2,eff} > (<) g_{1,eff}$, SC (CDW) is favoured more strongly than indicated in Fig. 4. Further, they are limited to both weak electron-electron and electron-phonon interactions. However, our conclusions are in line with recent work by Tosatti and Yu [25] for a large-U Hubbard model. In particular, they show that adding phonons to such a model rather leads to a spin-Peierls than to a SC ground state. Therefore, if there is no (unknown) intermediate coupling fixed point in the Peierls-Hubbard model, there is strong evidence that the 1D Hubbard model cannot exhibit dominant SC fluctuations - even if coupled in the most favourable way to phonons of arbitrary frequency. A more general model, including both diagonal and off-diagonal electronic interactions, may eventually yield SC fluctuations when coupled to dispersionless phonons. Even then, however, it is not clear to what extend these results are relevant to real systems: acoustic phonons, favouring CDWs, are present in all crystals. Their effect has not been included in the present calculation; it may outweigh any enhancement of SC due to dispersionless modes at higher energies.

These conclusions are in suprising contrast to previous work and do not support the claim that $(CH)_x$ should be a candidate for high-temperature superconductivity [12]. While many features of the analysis are similar to ours, there are important differences. (i) As discussed above, in a half-filled band Umklapp scattering is relevant giving an insulating ground state *without* massless charged excitations. Light doping produces charged solitons which may form lattices. The band formed inside the Peierls gap is completely filled and cannot give rise to SC. (ii) In the incommensurate system, there exist massless excitations of the phase ϕ of the (complex) CDW order parameter φ_0. As discussed elsewhere [11], for very low phonon frequencies, a correlation function exponent for their fluctuations can be obtained as

$$R_\phi(\vec{r}-\vec{r'}) \sim |\,\vec{r}-\vec{r'}\,|^{-\eta} \qquad \eta \approx \left[1 - \frac{2g_2 - g_{1,\|}}{2(\pi v_F + g_4)}\right] / \sqrt{1 + 2\pi v_F \varphi_0^2} \ll 1 \ . \tag{13}$$

Superconducting fluctuations then decay as

$$R_{SS}(\vec{r}-\vec{r'}) \sim |\,\vec{r}-\vec{r'}\,|^{-4/\eta} \ , \tag{14}$$

and therefore are *significantly weakened.* While this formula has been derived for acoustic phonons, it carries over with minor modifications to the case where the coupling to dispersionless phonons cannot be described self-consistently in the high-phonon-frequency limit discussed above. The modification in η introduced by increasing the electron-electron interaction in the square brackets is minor compared to the square-root and, e.g. for $g_1 = g_2 = g_4 = U$ (Hubbard model), even opposes SC. (iii) At higher phonon frequencies, the above analysis, avoiding several important approximations in the previous work, applies directly; in particular, we have used the exact bare phonon propagator and therefore find a renormalization of the electron-electron interactions by the phonons even for $E(\ell) > \omega_{2k_F}$. Moreover, in this regime charge and spin fluctuations are coupled through the retarded interactions. Then, solving the scaling equations and looking for the *most divergent* correlations function, yields immediately the phase diagram of Fig. 4. (iv) In agreement with the analysis of Zimanyi et al. [12] we find the SS fluctuations to exhibit the second strongest divergence in a wide part of the phase diagram and, in particular, for a weakly repulsive Hubbard interaction. We might think about interchain interactions and tunneling stabilizing SC. However, various studies of this problem [26] indicate that interchain tunneling essentially yields an ordered phase of the type showing the strongest fluctuations in the purely 1D model; divergent CDW fluctuations could further take advantage of interchain Coulomb interactions to stabilize an ordered phase. Moreover, experience with inorganic CDW systems [1] shows that the phase degrees of freedom of an incommensurate CDW (and similarly of a SDW) are sufficiently elastic to oppose a destabilization of a CDW by frustration e.g. by impurities etc. on the same or a different chain.

CONCLUSIONS

The SSH-model extended to include electronic interactions shows an extraordinary richness of physical behaviour. The phonon frequency is an important parameter controlling the overall interplay of attractive electron-phonon and repulsive electron-electron interactions and therefore enhancement or suppression of dimerization. However, even the competition between different phases when the relative magnitude of electronic interactions varies, depends sensitively on electron-phonon interaction and phonon frequency. In particular, we have shown that some results given previously for extended SSH-Hubbard models are valid only in the limit $Y_{ph} \to 0$ and/or $\omega_{2k_F} \to 0$.

Concerning dimerization, we have centred our discussion on the influence of various electron-electron interaction terms: while weak U and V increase dimerization, the recently proposed bond charge interactions W (X) depress (enhance) dimerization. In a situation $U > X > V = W$ (but all in the weak-coupling limit) [7], dimerization decreases in general with respect to the noninteracting case. In the spin-Peierls limit $U \gg 4t$ some trends change qualitatively. Magnitudes of the electronic interaction parameters appropriate to conducting polymers have been discussed by other authors [7,8,9]; our theory yields a detailed description of the system once these parameters are known. In particular, it allows to check to what extend the simple dependence of dimerization enhancement or suppression on the screening length of the electron-electron potential [8] also holds when the phonon quantum fluctuations are taken into account. This will be the subject of a forthcoming publication.

We have also searched for superconducting fluctuations in a modified model where the band is not half-filled and electrons are coupled to dispersionless phonons. While we find, for general parameters, some enhancement of superconductivity, this does not occur for physical electron-electron potentials. Within the limits of validity of our analysis, we do not find dominant SC fluctuations in the Hubbard model coupled to phonons nor in the extended Hubbard model. Repulsive off-diagonal interactions are favourable to SC, but a physical W is not enough to overcome CDW and SDW. Preliminary results on the X-term indicate further stabilization of SC due to a strongly attractive forward scattering; due to its sensitive dependence on band-filling, a more definite statement must await, however, further studies [21].

While these results apparently give an explanation for the absence of superconductivity in extremely simple polymers such as $(CH)_x$, together with the current understanding of actual low-dimensional superconductors [1], they might provide some guideline in searching for superconductivity in polymers. While too anisotropic a material in general appears to be unfavourable, the present results point towards a strong enhancement of SC by coupling to dispersionless phonons thereby requiring a minimum of complexity in the monomer structure. Such complexity might also be useful in light of results on a two-band model [27] where SC is generated even from repulsive interactions alone. Moreover, the current theories of organic superconductors [1,28] demonstrate further possibilities to generate SC ground states in quasi-1D model which might be relevant also for polymers.

ACKNOWLEDGMENTS

I wish to acknowledge useful discussions with H. Büttner, D. K. Campbell, K. Fesser, and V. Waas, as well as receipt of ref. [20] prior to publication.

REFERENCES

[1] For reviews on recent progress, see e.g. Proceedings of the International Conference on Synthetic Metals (ICSM'88), Santa Fe, 1988, to be published in Synthetic Metals.

[2] W.-P. Su, J. R. Schrieffer, and A. J. Heeger, Phys. Rev. Lett. **42**:1698 (1979), Phys. Rev. B **22**:2099 (1980), and **28**:1138 (1983).

[3] H. Thomann, L. R. Dalton, Y. Tomkiewicz, N. S. Shiren, and T. C. Clarke, Phys. Rev. Lett. **50**:533 (1983); H. Käss, P. Höfer, A. Grupp, P. Kahol, R. Weizenhöfer, G. Wegner, and M. Mehring, Europhys. Lett. **4**:947 (1987); B. R. Weinberger, C. B. Roxlo, S. Etemad, G. L. Baker, and J. Orenstein, Phys. Rev. Lett. **53**:86 (1984).

[4] P. Horsch, Phys. Rev. B **24**:7351 (1981); D. Baeriswyl and K. Maki, Phys. Rev. B **31**:6633 (1985); S. Kivelson and D. E. Heim, Phys. Rev. B **26**:4278 (1982).

[5] S. Mazumdar and S. N. Dixit, Phys. Rev. Lett. **51**:292 (1983); J. E. Hirsch, Phys. Rev. Lett. **51**:296 (1983); D. K. Campbell, T. A. DeGrand, and S. Mazumdar, Phys. Rev. Lett. **51**:1717 (1983); G. W. Hayden and Z. G. Soos, Phys. Rev. B **38**:6075 (1988). V. Waas, J. Voit, and H. Büttner, Synth. Met. **27**:21 (1988).

[6] L. G. Caron and C. Bourbonnais, Phys. Rev. B **29**:4230 (1984); B. Horovitz and J. Sólyom, Phys. Rev. B **32**:2681 (1985); V. Ya. Krivnov and A. A. Ovchinnikov, these proceedings.

[7] S. Kivelson, W.-P. Su, J. R. Schrieffer, and A. J. Heeger, Phys. Rev. Lett. **58**:1899 (1987), and **60**:72 (1988).

[8] C. Wu, X. Sun, and K. Nasu, Phys. Rev. Lett. **59**:831 (1987). X. Sun, these proceedings.

[9] For a dissenting point of view, see D. Baeriswyl, P. Horsch, and K. Maki, Phys. Rev. Lett. **60**:70 (1988), and J. T. Gammel and D. K. Campbell, ibid. 71; A. Painelli and A. Girlando, Sol. State Comm. **66**:273 (1988).

[10] E. Fradkin and J. E. Hirsch, Phys. Rev. B **27**:1680 (1983); D. Schmeltzer, R. Zeyher, and W. Hanke, Phys. Rev. B **33**:5141 (1986).

[11] J. Voit and H. J. Schulz, Phys. Rev. B **34**:7429 (1986), **36**:968 (1987), and **37**:10068 (1988); J. Voit, Synth. Met. **27**:33 (1989).

[12] G. T. Zimanyi, S. A. Kivelson and A. Luther, Phys. Rev. Lett. **60**:2089 (1988).

[13] J. Sólyom, Adv. Phys. **28**:201 (1979). V. J. Emery, in J. T. Devreese and V. E. van Doren (eds.) *Highly Conducting One-Dimensional Solids,* Plenum Press, 1979, p.247.

[14] J. Voit and H. J. Schulz, Molec. Cryst. Liq. Cryst. **119**:449 (1985) and Phys. Rev. Lett. **57**:370 (1987).

[15] L. Hubert, Phys. Rev. B **36**:6175 (1987).

[16] J. E. Hirsch, Phys. Rev. Lett. **53**:2327 (1984); B. Fourcade, M. Sc. thesis, University of Montreal, 1984, (unpublished); V. Waas, H. Büttner, and J. Voit, to be published.

[17] E. Majerníková and P. Markoš, Phys. Lett. **A 123**:352 (1987).

[18] We reserve the abreviation CDW for a charge density wave centred on the sites and use BOW (=bond order wave) for one centred on the bonds. SDW stands for spin density waves.

[19] J. Voit, J. Phys. C **21**:L-XXXX (1988).

[20] D. K. Campbell, J. T. Gammel and E. Y. Loh, to be published. These authors also find a renormalization of the bare electron-phonon coupling, which is not confirmed by the present work.

[21] J. Voit, to be published.

[22] M. P. M. den Nijs, Phys. Rev. B **23**:6111 (1981).

[23] J. Voit, Synth. Met. **27**:41 (1989).

[24] U. Sum, K. Fesser, and H. Büttner, Phys. Rev. B **38**:6166 (1988).

[25] E. Tosatti and Lu Yu, Physcia **C 153-155**:1253 (1988).

[26] R. A. Klemm and H. Gutfreund, Phys. Rev. B **14**:1086 (1976); Yu. A. Firsov, V. N. Prigodin, and Chr. Seidel, Phys. Rep. 126:245 (1985).

[27] R. Shankar, these proceedings.

[28] C. Bourbonnais, these proceedings.

PHONONS AND QUANTUM FLUCTUATIONS IN A DIMERIZED ELECTRON-PHONON CHAIN

G.C. Psaltakis and N. Papanicolaou*

Research Center of Crete and University of Crete
Physics Department
Heraklion, Crete GR-711 10, Greece

The model Hamiltonian

$$H = \sum_j \{ -\left[t + \alpha(u_j - u_{j+1})\right](c^*_{j+1,\sigma} c_{j,\sigma} + h.c.) +$$

$$\frac{P_j^{\,2}}{2M} + \frac{D}{2}(u_j - u_{j+1})^2 \} \tag{1}$$

was introduced by Su, Schrieffer and Heeger (SSH)[1] to describe electron-phonon interactions in polyacetylene. The index j is summed over the 2K sites of a one-dimensional periodic lattice while σ is summed over the values σ=1,2 to account for the spin multiplicity. In a mean-field adiabatic approximation, where the ionic mass M is assumed to be very large and the phonon degrees of freedom can be treated classically, the system with a half-filled band undergoes a Peierls instability[2] and the ground state is dimerized. More recently Rice et al[3] suggested that the coupling of the two degenerate π-electron half-filled bands of polyyne to lattice vibrations can also be described by Hamiltonian (1) if the degeneracy index σ is allowed to take four distinct values. It is thus useful to study (1) for an arbitrary degeneracy n (σ=1,2,...,n) and eventually calculate quantum fluctuations within a systematic 1/n expansion, setting n=2 or 4 at the end of the calculation to obtain results for polyacetylene or polyyne, respectively.

In this paper we present briefly the results of a study of the SSH model using the 1/n expansion technique developed by the authors[4,5] for tight-binding electron systems. Assuming a half-filled band, i.e., a total of nK electrons, and anticipating a dimerized ground state which leads to the the formation of a sublattice, we introduce Fourier decompositions for the displacement operators u_j at even and odd sites according to

$$u_{2m} = \sqrt{n}u + \frac{1}{\sqrt{2K}} \sum_k \exp(imk) \left[\frac{F_k + F^*_{-k}}{\sqrt{2M\omega_k(+)}} - \frac{G_{-k} + G^*_k}{\sqrt{2M\omega_k(-)}} \right] \tag{2}$$

*Also at the Department of Physics, Washington University, St. Louis, MO 63130, U.S.A.

Interacting Electrons in Reduced Dimensions
Edited by D. Baeriswyl and D.K. Campbell
Plenum Press, New York

$$u_{2m-1} = -\sqrt{n}u - \frac{1}{\sqrt{2K}} \sum_k \exp\left[i(m-\tfrac{1}{2})k\right]\left[\frac{F_k + F^*_{-k}}{\sqrt{2M\omega_k{}^{(+)}}} + \frac{G_{-k} + G^*_k}{\sqrt{2M\omega_k{}^{(-)}}}\right]$$

where m=1,2,...,K labels the sites of the sublattice and k is summed over the K wavevectors of the sublattice Brillouin zone. Operator $F^*_k(G^*_k)$ creates an optical (acoustic) phonon with energy $\omega_k{}^{(+)}(\omega_k{}^{(-)})$ where $\omega_k{}^{(\pm)} = \omega_0\left[1\pm\cos(k/2)\right]^{1/2}$ and $\omega_0 = (2D/M)^{1/2}$. The expressions for the momenta P_{2m} and P_{2m-1} follow easily from (2). Similarly for the electron annihilation operators we set

$$c_{2m,\sigma} = \frac{1}{\sqrt{2K}} \sum_k \exp\left[i(mk + \theta_k/2)\right](-A^\sigma_k + B^{\sigma*}_k)$$

$$c_{2m-1,\sigma} = \frac{1}{\sqrt{2K}} \sum_k \exp\left[i((m-\tfrac{1}{2})k - \theta_k/2)\right](A^\sigma_k + B^{\sigma*}_k) \qquad (3)$$

where A^σ_k and B^σ_k are mutually anticommuting fermion operators. The angle θ_k is determined by a variational calculation optimizing the expected value of (1) in the Fock vacuum associated with the new canonical variables. We find that $\tan\theta_k = (\Delta/2t)\tan(k/2)$ with the gap parameter $\Delta = 4u\bar{\alpha}$ calculated from the self-consistent equation

$$1 = \pi\bar{\lambda}t\,\frac{1}{K}\sum_k \frac{\sin^2(k/2)}{\Omega_k}$$

$$\Omega_k = \{\left[2t\cos(k/2)\right]^2 + \left[\Delta\sin(k/2)\right]^2\}^{\frac{1}{2}} \qquad (4)$$

where $\bar{\alpha} = \alpha\sqrt{n}$ and $\bar{\lambda} = 2\alpha^2/\pi Dt$ is a dimensionless electron-phonon coupling constant. Equation (4) is identical to the gap equation occuring in the mean-field adiabatic approximation[1,3]. However differences result in higher-order corrections which are here organized in inverse powers of the degeneracy n. To obtain such corrections we express the Hamiltonian in terms of bilocal fermion operators which are then bosonized by means of a generalized Holstein-Primakoff realization[4,5]

$$A^{\sigma*}_p A^\sigma_q = \sum_k \xi^*_{pk}\xi_{qk}, \qquad A^{\sigma*}_p B^{\sigma*}_q = \sum_k \xi^*_{pk}R_{qk}$$

$$B^\sigma_p A^\sigma_q = \sum_k R_{kp}\xi_{qk}, \qquad B^\sigma_p B^{\sigma*}_q = n\delta_{pq} - \sum \xi^*_{kq}\xi_{kp} \qquad (5)$$

$$R = (R_{pq}) = (nI - \xi^+\xi)^{\frac{1}{2}}$$

where the entries of the matrix $\xi = (\xi_{pq})$ satisfy the Bose commutation relations $\left[\xi_{pq},\xi_{kl}{}^*\right] = \delta_{pk}\delta_{ql}$. It is understood that on the l.h.s. of (5) the generalized spin index σ is summed over the n values so ξ_{pq} are spin independent. Realization (5) yields a restriction of the Hamiltonian to the sector consisting of nK-electron states that are a singlet under SU(n) transformations.

The square root in R has a simple Taylor expansion for large n leading to a corresponding 1/n expansion of the SSH Hamiltonian. Holding $\alpha = \bar{\alpha}\sqrt{n}$

fixed during the intermediate steps of the calculation, the expansion of H takes the form

$$H = nKE_o + \frac{1}{2}\sum_k \left[\omega_k^{(+)} + \omega_k^{(-)}\right] + H_o + \frac{1}{\sqrt{n}} H_1 + \frac{1}{n} H_2 + \dots \tag{6}$$

For the computation of the first few terms, including H_1, it is sufficient to approximate the square root by $R_{pq} = \sqrt{n}\delta_{pq}$. The constant E_o given by

$$E_o = 4Du^2 - \frac{1}{K}\sum_k \Omega_k \tag{7}$$

is the large-n approximation of the ground state energy which coincides with the mean-field adiabatic approximation of SSH[1]. Linear terms in the Bose operators ξ, F and G do not occur in (6) while small fluctuations are described by the harmonic term H_o given by

$$\begin{aligned} H_o &= \sum_{p,q}(\Omega_p + \Omega_q)\xi^*_{pq}\xi_{pq} + \sum_p\left[\omega_p^{(+)} F_p^* F_p + \omega_p^{(-)} G_p^* G_p\right] \\ &+ \frac{\bar{\alpha}}{4}\sum_{pql}\left[\frac{1}{KM\omega_l^{(+)}}\right]^{\frac{1}{2}} \Delta(p - q + l)\lambda^{(+)}(q,p;l)(F_l + F^*_{-l})(\xi^*_{qp} - \xi_{pq}) \\ &+ \frac{\bar{\alpha}}{4}\sum_{pql}\left[\frac{1}{KM\omega_l^{(-)}}\right]^{\frac{1}{2}} \Delta(p - q + l)\lambda^{(-)}(q,p;l)(G_{-l} + G^*_l)(\xi^*_{qp} - \xi_{pq}) \end{aligned} \tag{8}$$

where $\Delta(p-q+l)$ is equal to unity when $p-q+l = 0 \pmod{2\pi}$ and vanishes otherwise. The quantities $\lambda^{(\pm)}$ involve appropriate phase factors[4]. Diagonalization of the harmonic Hamiltonian H_o yields the normal modes for the singlet sector. The final eigenvalue equation for the quadratic form (8) may be written as

$$\begin{aligned} 1 &- \frac{2\bar{\alpha}^2}{M}\left[\frac{A_l^{(+)}(E)}{(\omega_l^{(+)})^2 - E^2} + \frac{A_l^{(-)}(E)}{(\omega_l^{(-)})^2 - E^2}\right] \\ &+ \left(\frac{2\bar{\alpha}^2}{M}\right)^2 \cdot \frac{A_l^{(+)}(E)A_l^{(-)}(E) - [B_l(E)]^2}{[(\omega_l^{(+)})^2 - E^2][(\omega_l^{(-)})^2 - E^2]} = 0 \ , \end{aligned} \tag{9}$$

$$A_l^{(\pm)}(E) = \frac{1}{K}\sum_k \left[1 \pm \cos(\theta_k + \theta_{k+l})\right]\left[\sin\left(\frac{k}{2}\right) \pm \sin\left(\frac{k+l}{2}\right)\right]^2 \times \frac{\Omega_k + \Omega_{k+l}}{(\Omega_k + \Omega_{k+l})^2 - E^2}$$

$$B_l(E) = \frac{1}{K}\sum \sin(\theta_k + \theta_{k+1})\left[\sin^2\left(\frac{k}{2}\right) - \sin^2\left(\frac{k+1}{2}\right)\right] \times \frac{\Omega_k + \Omega_{k+1}}{(\Omega_k + \Omega_{k+1})^2 - E^2}$$

The roots of this algebraic equation can be labeled as $E = E_s(l)$ where l is the sublattice crystal momentum that runs over K values and s takes K+2 values for each l to account for the expected total of K(K+2) normal modes. The roots of (9) were calculated numerically by reducing the problem to the diagonalization of a suitable (K+2)x(K+2) matrix. Figure 1a shows results for the excitation spectrum using a set of parameters appropriate for polyacetylene[1]. The electron-hole continuum is seen to be separated from the ground state by a gap equal to 2Δ, having an upper cutoff at 4t, as expected. The two low-lying branches are depicted more clearly on the energy scale of Fig.1b and are identified with the renormalized acoustic and optical phonon branches denoted by $\omega_l^{(-)}$ and $\omega_l^{(+)}$. Near the zone center $\omega_l^{(+)}$ has a positive curvature characteristic of the Peierls condensed state. Simple analytic expressions for $\omega_l^{(\pm)}$ can be derived from (9) in the parameter range $\omega_0 \ll \Delta$ which contains both polyacetylene[1] and polyyne[3]

$$\left[\omega_l^{(\pm)}\right]^2 = \frac{1}{2}\,\omega_0^2\{\alpha_l^{(+)} + \alpha_l^{(-)} \pm \left[(\alpha_l^{(+)} - \alpha_l^{(-)})^2 + \beta_l^2\right]^{\frac{1}{2}}\} \tag{10}$$

$$\alpha_l^{(\pm)} = 1 \pm \cos(l/2) - \frac{1}{2}\,\pi\bar{\lambda}tA_l^{(\pm)}(0)\quad , \quad \beta_l = \pi\bar{\lambda}tB_l(0)$$

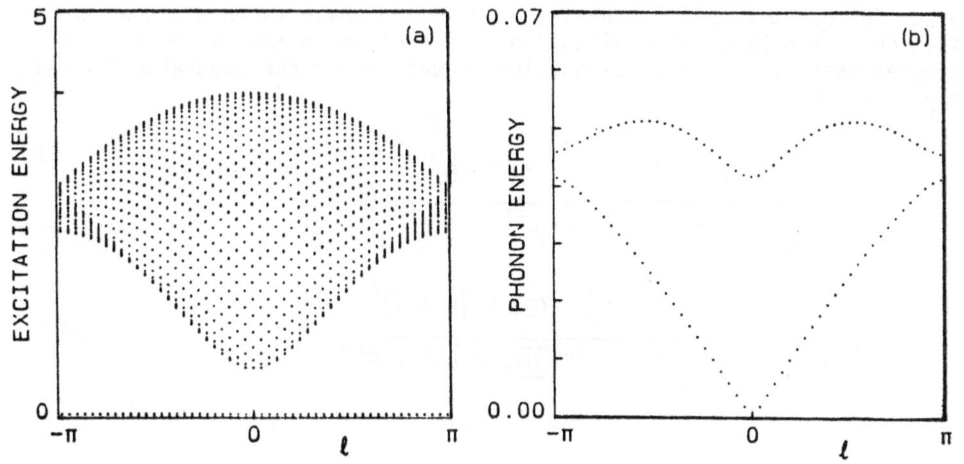

Fig.1 Excitation spectrum of polyacetylene (n=2, $\omega_0/t = 0.0463$, and $\Delta/t = 0.295$ corresponding to $\bar{\lambda} \simeq 0.4$) on a 102-site lattice. Energy is measured in units of t and l is the sublattice crystal momentum. (a) Complete spectrum. (b) Enlargement of the lower part of (a) showing the acoustic and optical phonon branches.

The two phonon branches of (10) are plotted in Fig.2 by a full line for parameters appropriate for polyyne[3]. The dashed lines in the same figure are the corresponding phonon branches calculated using the results of Rice et al[3]. Clearly the phonon dispersions predicted by the 1/n expansion differ appreciably form those of Ref.3, showing a more natural behavior near the zone boundaries. In particular our calculation predicts for the speed of sound $S = \left[d\tilde{\omega}_l^{(-)}/dl \right]_{l\to 0}$ a reduction due to the electron-phonon interaction given by

$$S = S_0 \left[1 - \left(\frac{\Delta}{2t}\right)^2 \frac{\bar{\lambda}F(\bar{\lambda})}{1 - \bar{\lambda}F(\bar{\lambda})} \right]^{\frac{1}{2}}$$

$$F(\bar{\lambda}) = \int_0^{\pi/2} dx \frac{(\cos x \sin x)^2}{\left[\cos^2 x + (\Delta/2t)^2 \sin^2 x\right]^{3/2}} \tag{11}$$

where $S_0 = (D/4M)^{\frac{1}{2}}$ is the speed of sound in the absence of interactions. It should be noted that the condition $\omega_0 << \Delta$ used to derive (10) is not assumed in the derivation of the speed of sound in (11). The dependence of S on $\bar{\lambda}$ is plotted in Fig.3 by a full line while the dashed line is the corresponding result of Rice et al[3]. For small values of $\bar{\lambda}$, e.g., $\bar{\lambda} = 0.8$, our calculation predicts a smaller reduction in the speed of sound ($S = 0.92S_0$) compared to that of Ref.3 ($S = 0.62S_0$). The present calculation also predicts a critical electron-phonon coupling $\bar{\lambda}_c = 3.887$, corresponding to $(\Delta/t)_c = 7.403$, at which the velocity of sound vanishes. For $\bar{\lambda} > \bar{\lambda}_c$ the dimerized ground state is unstable, at least within the 1/n expansion. We believe that this instability is not an effect of quantum fluctuations which in fact decrease with increasing $\bar{\lambda}$. Instead, it is probably due to the fact that this strong coupling region is unphysical in the framework of the SSH model. Indeed for $\Delta/t=2$ (corresponding to $\bar{\lambda}=4/\pi = 1.273$) the ground state corresponds to a collection of dimers and for larger Δ's the effective nearest-neighbor hopping integral, $t-\Delta/2$, changes sign, which is unphysical.

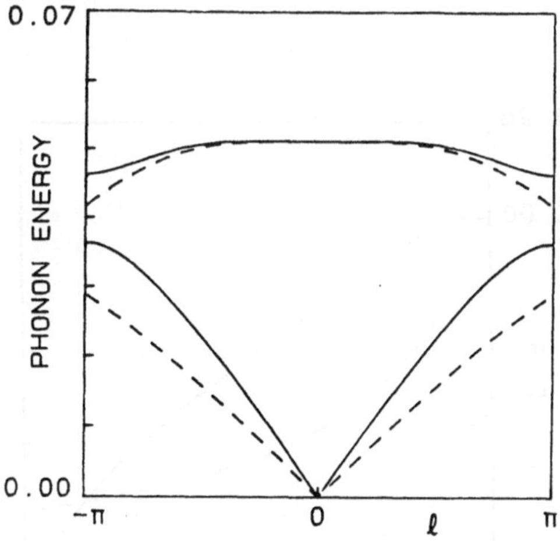

Fig.2 Phonon spectra in units of t for polyyne ($n=4$, $\omega_0/t = 0.0463$, and $\Delta/t = 0.98$ corresponding to $\bar{\lambda} \cong 0.8$). Full lines : present calculation using Eq.(10). Dashed lines : calculation based on the results of Rice et al[3].

In this region an exponential, rather than linear as assumed in (1), dependence of the effective hopping integral on the bond-length changes $(u_j - u_{j+1})$ would be more appropriate. In any case we note that physical systems of current interest are such that $\bar{\lambda} \lesssim 1$.

Concerning the effect of quantum fluctuations on the ground-state energy, the large-n result of Eq.(7) may be improved by including the zero-point fluctuations of the normal modes

$$\frac{E_{gr}}{2K} = \frac{1}{2} nE_0 + \frac{1}{2}\left[\frac{1}{2K}\sum_{s,1} E_s(1) - \sum_k \Omega_k\right] \tag{12}$$

where $E_s(1)$ are the excitation energies discussed in the text. Such a relation is familiar from early calculations of the correlation energy in the high-density electron gas due to Wentzel[6]. Using a 102-site lattice and parameters appropriate for polyacetylene ($n=2$, $\omega_0/t = 0.0463$, $\bar{\lambda} \simeq 0.4$) we can infer from this calculation a small 1.4% increase of the ground state energy, with respect to its large-n value, due to quantum fluctuations. Finally, let us consider briefly the effects of quantum fluctuations on the phonon order parameter

$$m_p = \frac{1}{4K}\sum_{j=1}^{2K} (-1)^{j+1} \langle u_{j+1} - u_j \rangle$$

$$= \sqrt{n}u + \frac{1}{2}\left[\frac{1}{KM\omega_0^{(+)}}\right]^{\frac{1}{2}} \langle F_0 + F_0^* \rangle \tag{13}$$

The term $\sqrt{n}u$ in (13) is the large-n approximation of the order parameter whereas the second term contains the effects of quantum fluctuations. Using Rayleigh-Schrödinger perturbation theory we have calculated[4] the expected value $\langle F_0 + F_0^* \rangle$ to leading order by including the anharmonic term H_1 in the expansion (6). We find that the relative influence of fluctuations decreases

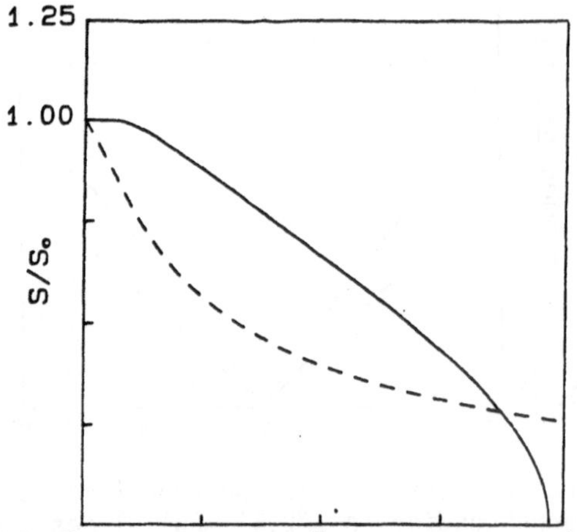

Fig.3 Speed of sound as a function of the dimensionless electron-phonon coupling constant $\bar{\lambda}$. Full line : present calculation using Eq.(11). Dashed line : calculation based on the results of Rice et al[3].

with increasing $\bar{\lambda}$ in general agreement with the behavior found in earlier work[7]. However for the set of parameters describing polyacetylene our calculation predicts[4] a 4% reduction of the order parameter, relative to its large-n value, which is much smaller than the 15% reduction found by a Monte Carlo simulation[7] on a 24-site lattice. We believe that part of this difference is due to finite-size effects involved in the latter calculation[7] and which are probably quite important in this case both due to the Jahn-Teller effect in a 24-site ring and due to the relative large coherence length $2t/\Delta = 6.78$.

Acknowledgement

The work of N.P. was supported in part by the U.S. Department of Energy.

REFERENCES

1. W. P. Su, J. R. Schrieffer and A. J. Heeger, Phys. Rev. B22, 2099 (1980).
2. R. E. Peierls, Quantum Theory of Solids (Clarendon, Oxford, 1955),p.108.
3. M. J. Rice, S. R. Phillpot, A. R. Bishop and D. K. Campbell, Phys. Rev. B34, 4139 (1986).
4. G. C. Psaltakis and N. Papanicolaou, Solid State Commun. 66, 87 (1988).
5. L. R. Mead and N. Papanicolaou, Phys. Rev. B28, 1633 (1983).
6. G. Wentzel, Phys. Rev. 108, 1593 (1957).
7. E. Fradkin and J. E. Hirsch, Phys. Rev. B27, 1680 (1983).

SCALING APPROACH TO ELECTRONIC CORRELATIONS IN ORGANIC CONDUCTORS

Claude Bourbonnais*

Centre de Recherche en Physique des Solides
Université de Sherbrooke, Sherbrooke, Québec, Canada
J1K-2R1

INTRODUCTION

The problem of electronic correlations in organic conductors like the Bechgaard salt series [$(TMTSF)_2X$] is well known to be at the heart of the remarkable richness of phase transitions observed in these materials [1]. The well established quasi-one-dimensionality of these materials makes both the phenomenological and the microscopic descriptions of correlations rather delicate. Correlations are actually characterized by various energy scales. This is mainly imposed for example by strongly anisotropic hopping integrals for the electrons and the sizeable on-chain Coulomb repulsion. The scaling hypothesis [3] and its extended version [3b] for the inclusion of crossover effects between various energy scales then appears as a natural description of correlations in these materials [2]. The microscopic calculations must in turn confirm that this ansatz actually holds. Here, we give a brief survey of the applicability of the scaling hypothesis for the description of itinerant antiferromagnetism in the Bechgaard salts. We will focuss our attention on the recent NMR relaxation rate and EPR line width data. These two observable quantities are directly coupled to magnetic fluctuations and scaling allows a remarkable clarification of several aspects of correlations. Dimensionality and the respective weight associated to uniform and antiferromagnetic correlations are analyzed as function temperature and hydrostatic pressure.

* Also at the Laboratoire de Physique des Solides, Universite de Paris-Sud , Orsay 91405, France

We conclude by discussing the consequences for the mechanisms responsable for magnetic and superconducting ordering.

SCALING HYPOTHESIS AND NUCLEAR RELAXATION

An important property that is usually postulated for antiferromagnetic correlations near a Néel critical point is the one of scaling. The scaling hypothesis is quite useful in the context of critical phenomena [3]. It allows to extract the singular behavior of observable quantities near their critical point uniquely in terms of the correlation length ξ_{AF} and a given set of critical indices. (Here, AF stands for antiferromagnetic correlations). The values of these indices can be calculated for example from the renormalization group method at different levels of approximation [4,5].

Among the experimental probes that are directly coupled to AF correlations, neutron diffraction is well known to give relevant information about fluctuations. For organic conductors like the $(TMTSF)_2X$ series, however, the sample size is too small so that no successfull attemps could be obtained so far. Another probe which is free from these constraints is the nuclear relaxation measured by NMR. Active nucleus (I=1/2) such as ^1H, ^{77}Se and ^{13}C of TMTSF molecules are locally coupled to electronic degrees of freedom through the hyperfine coupling A so that the nuclear relaxation rate T_1^{-1} can probe the dynamic magnetic susceptibility of electrons *for all wave vectors* $\vec{q}$. This is illustrated by the standard Moriya formula [6]:

$$T_1^{-1} = 2\gamma_N^2|A|^2T \int d^dq \; \chi_\perp''(\vec{q},\omega_N)/\omega_N \;, \qquad (1)$$

$(\hbar=k_B=1)$. Here ω_N and γ_N are the Larmor frequency and the gyromagnetic ratio of the nucleus. $\chi_\perp''$ is the imaginary part of the retarded electronic susceptibility.

<u>Antiferromagnetic contribution</u>. Lets starts the scaling analysis with the AF contribution $T_1^{-1}[\vec{q} \approx \vec{Q}_o]$ for $\vec{q}$ near the AF modulation vector $\vec{Q}_o$. According to the scaling hypothesis we can write for the dynamical susceptibility near T_N [3]:

$$\chi_\perp''(\vec{q}'-\vec{Q}_o,\omega_N) \approx \xi_{AF}^{\dot{\gamma}/\dot{\nu}} \; D_{AF}(\vec{q}'\cdot\vec{\xi}_{AF},\omega_N\xi_{AF}^{\dot{z}}) \;. \qquad (2)$$

In the standard notation [3], D_{AF} is a scaling function for $\vec{q}'\cdot\vec{\xi}_{AF}< 1$ and $\omega_N\Gamma\xi_{AF}^{\dot{z}} < 1$. $\dot{\gamma}$, $\dot{\nu}$ and $\dot{z}$ are the critical indices for the correlation length ($\xi_{AF} \propto \dot{r}_{AF}^{-\dot{\nu}}$, $\dot{r}_{AF} = T-T_N/T_N$), the static susceptibility at $\vec{Q}_o$ [$\chi(\vec{Q}_o,T) \propto \dot{r}_{AF}^{-\dot{\gamma}}$] and the fluctuation relaxation time $\tau_{AF} = \Gamma\, \dot{r}_{AF}^{-\dot{z}}$ repectively. Here and in the following, the dots on every quantity refer to the critical domain near T_N ($\dot{r}_{AF} < 1$) [2,3b]. Γ is a microscopic time scale for AF correlations. Now, as q scale like ξ_{AF}^{-1} and ω_N like $\xi_{AF}^{-\dot{z}}$, the relaxation rate profile near $\vec{Q}_o$ and T_N in d dimensions is easily extracted to give:

$$T_1^{-1}[\vec{q} \approx \vec{Q}_o] \approx \dot{C}_1\, \dot{r}_{AF}^{-\dot{\vartheta}} \tag{3}$$

with

$$\dot{\vartheta} = \dot{\gamma} + \dot{z}\dot{\nu} - d\dot{\nu}\ , \tag{4}$$

where $\dot{C}_1$ is a microscopic constant which depends on the anisotropy of the system. Since we deal in the present case with quasi-1D systems, $\dot{C}_1$ will also contain the effects of 1D fluctuations that take place within the 1D energy (temperature) domain. Actually, the form given in (3) is valid whenever the system has already undergone its so-called dimensionality crossover [2]. By making use of the extended scaling hypothesis which precisely deals with crossover effects [3b], one can explicitate the constant $\dot{C}_1$ in the following way:

$$\dot{C}_1 \approx C_\infty [\, g_i(t_\perp)]^{(\dot{\vartheta}-\vartheta)/\phi_{x_i}}\ , \tag{5}$$

where C_∞ is a non-universal constant and $g_i(t_\perp)$ is the small perturbation paramater that gives rise to the dimensionality crossover. The subscript i refers to the different possible mechanisms for the interchain electron hopping $t_\perp$ to lead to the dimensionality crossover [2b,7]. For a given mechanism, one will have a crossover exponent $\phi_x i$ associated to the change $\{\gamma,\nu,\vartheta,\ldots\} \rightarrow \{\dot{\gamma},\dot{\nu},\dot{\vartheta},\ldots\}$ in the set of critical indices. This occurs in a temperature domain centered around the crossover temperature [2]:

$$T_x i \propto E_o\, [g_i(t_\perp)]^{1/\phi_x i}\ . \tag{6}$$

E_o is a characteristic 1D energy scale(e.g, the Fermi energy of the isolated chains). From (5) and (6), purely 1D effects on the relaxation due to the temperature range $T \gg T_x i$ can be easily deduced [2]:

$$T_1^{-1}[\vec{q} \approx \vec{Q}_o] \approx C_\infty\, T\, [\xi_{AF}(T_x i)]^{\gamma/\nu + z - 1}\, \dot{r}_{AF}^{-\dot{\vartheta}}\ . \tag{7}$$

γ, ν, z and $\vartheta=\gamma+z\nu-\nu$ correspond to d=1 indices of the AF susceptibility $\chi(q=2k_F,T)\alpha\ (T/E_o)^{-\gamma}$, the correlation length $\xi_{AF}\alpha\ (T/E_o)^{-\nu}$, the relaxation time $\tau_{AF}\alpha\ (T/E_o)^{-z}$ and the $2k_F$ enhancement of the relaxation rate $(T\ T)^{-1}\alpha\ (T/E\ )^{-\vartheta}$ respectively.

Uniform contribution. Aside from the AF integration domain of the integral in (1), an important fluctuation contribution to T_1^{-1} will come from the $q \approx 0$ magnetic correlations [8,9]. These correlations are uniform and therefore ferromagnetic in origin. Although they are not singular since there is no ferromagnetic phase transition in organic conductors, their amplitude is not small. This is illustrated by the observation of an enhancement for the uniform and static susceptibility χ_s [10]. Furthermore, the enhancement is found to increase as a function of temperature. From the microscopic calculations, it turns out that for a quasi-1D metal it is the backscattering part ($g_1(T)$) of the Coulomb interaction that is mainly responsable for such a temperature dependence [9]. The latter is associated to a characteristic length scale:

$$\xi_F \ \alpha\ [1-g_1(T)\chi_s^o]^{-\bar{\nu}} \equiv r_F^{-\bar{\nu}} \quad , \tag{8}$$

where χ_s^o is the $T\to 0$ static and uniform susceptibility and $\bar{\nu}$ is the related indice. Note that ξ_F has the physical meaning of the ferromagnetic correlation length in the standard paramagnon theory [11] and thus, its relevance is ascribed to the condition $0 \leq r_F \leq 1$. In such a case, a scaling analysis of $T_1^{-1}[q\approx 0]$ in terms of ξ_F can be performed. If we postulate a scaling form for χ'' near $q \approx 0$ namely,

$$\chi''(\vec{q},\omega_N) \approx r_F^{-\bar{\gamma}}\ D_F(\vec{q}\cdot\vec{\xi}_F,\omega_N\xi_F^{\bar{z}}\ ) \ , \tag{9}$$

where D_F is the scaling function for $\vec{q}\cdot\vec{\xi}_F < 1$ and $\omega_N\Gamma_F\xi_F^{\bar{z}} < 1$. $\bar{\gamma}$ and zare the uniform magnetic susceptibility exponent ($\chi_s(T)\ \alpha\ r_F^{-\bar{\gamma}}$) and the dynamical exponent for the ferromagnetic fluctuation time $\tau_F\approx \Gamma_F\ r_F^{-\bar{z}}$, Γ_F being the microscopic unit of time. Following similar steps as those leading to (3) for the AF part, one gets for the uniform contribution in d dimensions [12]:

$$T_1^{-1}[q \approx 0] \approx C_o T\chi_s(T)^{\bar{\vartheta}} \quad , \tag{10}$$

with

$$\bar{\vartheta} = 1+ (\bar{z}\ \bar{\nu}\ -d\bar{\nu}\ )/\bar{\gamma} \quad , \tag{11}$$

where C_o is a constant. Therefore the uniform enhancement of the relaxation rate $(T_1^{-1}T)[q \approx 0]$ is expressed in terms of a power of $\chi_s(T)$ that depends on the dimension d of the system.

SCALING ANALYSIS OF THE EPR LINEWIDTH

Correlation effects in quasi-1D organic conductors can also be analyzed through electron paramagnetic resonance (EPR) [13]. Indeed, the temperature profile of the EPR line width ΔH can be strongly affected by spin fluctuations if there is some magnetic anisotropy in the system [14]. At the approach of an AF transition for example, the critical growth of AF fluctuations leads to a singular behavior of ΔH. There are two important sources of magnetic anisotropy in organic conductors: the dipole-dipole interaction and the spin-orbit coupling. Here, we will focuss on the dipole- dipole interaction for AF fluctuations since it is expected to give the largest singular contribution to ΔH near T_N. For small applied magnetic field, the line width can be expressed in the following way [13-14]:

$$\Delta H(\vec{q} \sim \vec{Q}_o) = \text{cst}\ T\ \chi_s^{-1} \int d^dq \int dt\ G^2(\vec{q},t)\ . \tag{12}$$

Here $G(\vec{q},t)$ is the time dependent correlation function at the wave vector $\vec{q}$. Near T_N and $\vec{Q}_o$, this quantity will scale as:

$$G(\vec{q}-\vec{Q}_o,t) \approx \dot{r}_{AF}^{-\dot{\gamma}}\ f(\vec{q}\cdot\dot{\vec{\xi}}_{AF}, t\dot{\xi}_{AF}^{-\dot{z}}) \tag{13}$$

where f is a scaling function [3]. For the line width, one will have [13]:

$$\Delta H(\vec{q} \sim \vec{Q}_o) \approx \dot{C}\ \dot{r}_{AF}^{-\dot{\mu}}\ , \tag{14}$$

with

$$\dot{\mu} = \dot{z}\dot{\nu}+2\dot{\gamma}-d\dot{\nu}\ , \tag{15}$$

and $\dot{C}$ is a constant. For $\dot{\mu} > 0$, a singularity in ΔH is then expected.

SCALING HYPOTHESIS AND EXPERIMENTS

Nuclear relaxation. Recently, a lot of interest has been devoted to the analysis of electronic correlations in $(TMTSF)_2X$ series via the NMR technique [1,2,7,9]. Nuclear relaxation measurements have been performed on antiferromagnetic and superconducting compounds corresponding to hydrostatic pressures on both sides of the critical pressure P_c for superconductivity. The $(TMTSF)_2PF_6$ compound for example has a critical pressure P_c of 6 kbar or so (fig.1). The low temperature T_1^{-1} profiles measured by Creuzet et al., [16] for this compound at P=1bar, 5.5, 8 and 11kbar are shown in fig.2. For $P < P_c$, a clear divergence of T_1^{-1} is seen at $T_N \simeq 12.2K$ (1bar) and 8.7K(5.5kbar) in agreement with the existence of a magnetic phase transition. In fig.3, the log-log plot of the data in the critical domain are reported, and a singular profile of the form given in (3) with $\dot{\theta} \simeq .5 \pm .03$ is clearly found for both pressures over one decade. It turns out actually that this exponent corresponds to a *d=3 gaussian AF critical behavior* [2,7,16]. Indeed, taking the gaussian static indices [3] $\dot{\gamma}=1$ and $\dot{\nu}=1/2$ together with the AF non-diffusive gaussian dynamic exponent $\dot{z}=2$ [3a], one has according to (4), $\dot{\theta}=1/2$, in d=3. Therefore the analysis of the T_1^{-1} critical domain allows a very clear determination of the type critical ordering, its dimensionality as well as the harmonicity of fluctuations [2,16].

We now focus on the fig.4 for the high temperature domain of the relaxation. The ^{77}Se data were obtained by Wzietek et al., [12] on $(TMTSF)_2PF_6$ (1 bar, circles) and $(TMTSF)_2ClO_4$(1 bar > P_c, triangles). In fig.4, we have also added the low temperature data of fig.2 for $(TMTSF)_2PF_6$(full circles). For $(TMTSF)_2ClO_4$ at 1 bar, we reported the T_1^{-1} data obtained by Creuzet et al., [15,17] at low temperature ($T \leq 40K$, full triangles) which are known to be quite similar to those of $(TMTSF)_2PF_6$ at 8 kbar (fig.2, crosses). From figure 4, the relaxation rate profiles for both compounds are characterized by an upward curvature. The relaxation does not follow a Korringa law for a normal metal ($T_1^{-1} \alpha T$). In fact, this corresponds to an increasing enhancement $(T_1T)^{-1}$with the temperature. This is reminiscent of the one observed for the static magnetic susceptibility $\chi_s(T)$ [10] In fig.5, we have plotted on a logarithmic scale the enhancement $(T_1T)^{-1}$ vs the Miljak et al., [10] $\chi_s^2(T)$ data for both compounds . For $(TMTSF)_2ClO_4$ for example, the figure 5 gives a remarkable support to the following relation

$$T_1^{-1} \alpha T \chi_s^2(T) \qquad (16)$$

between the relaxation rate and the magnetic susceptibility(full line). Therefore, from the scaling analysis that led to (10), uniform magnetic correlations dominates T_1^{-1} above 30K or so with $\theta=2$. This is quite

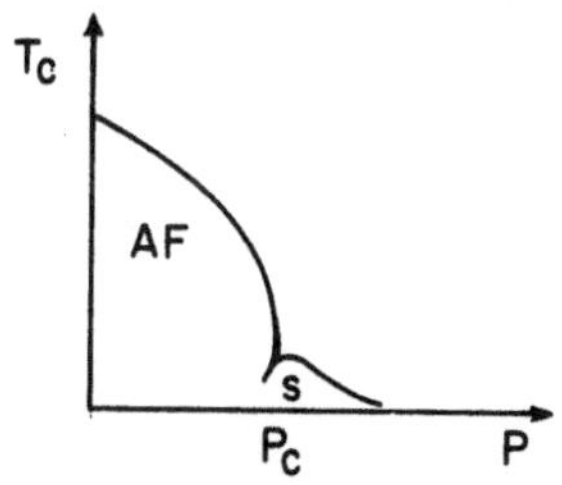

Fig. 1 Typical phase diagram of the (TMTSF)2X compounds. Pc is the critical pressure for superconductivity.

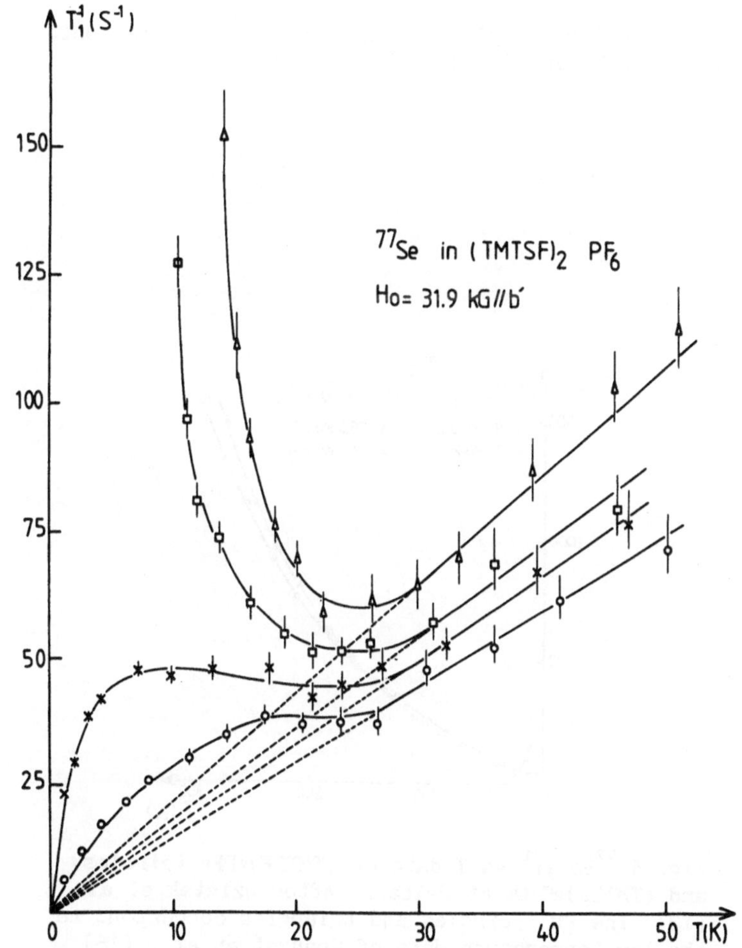

Fig. 2 Low temperature ^{77}Se T_1^{-1} vs T data of Creuzet et al.,[16] at P= 1bar (Δ), 5.5 (□), 8 (×), 11 kbar (o) for $(TMTSF)_2PF_6$.

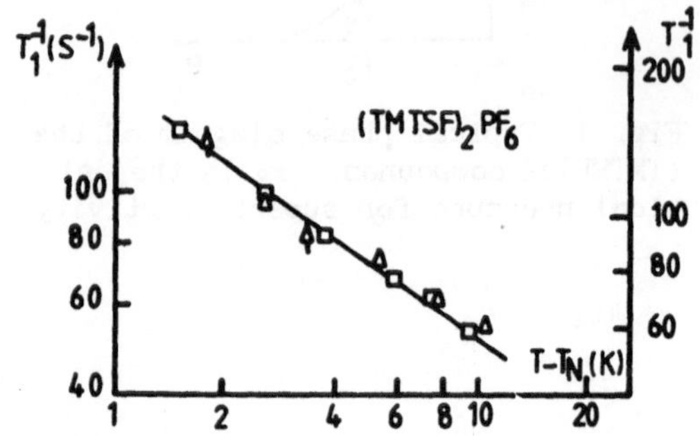

Fig. 3 The log-log plot of the fig.2 T_1^{-1} critical data at 1bar (T_N= 12K) and 5.5 kbar (T_N= 8.7K).

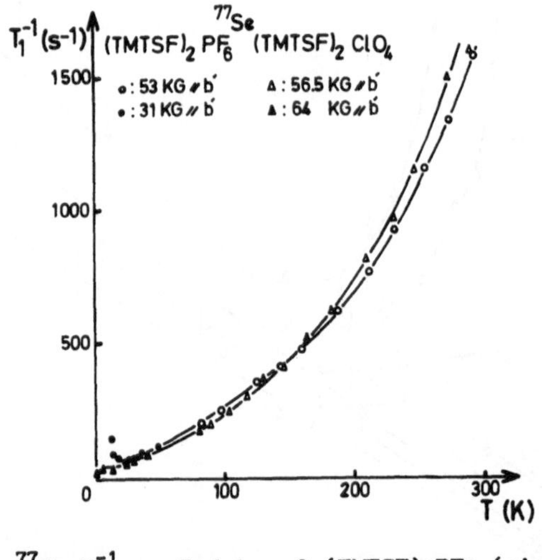

Fig. 4 ^{77}Se T_1^{-1} vs T data of (TMTSF)$_2$PF$_6$ (circles) and (TMTSF)$_2$ClO$_4$ at P=1bar. After Wzietek et al., [12]. The full circles and triangles correspond to the low temperature data of Creuzet et al., [16] (fig.2).

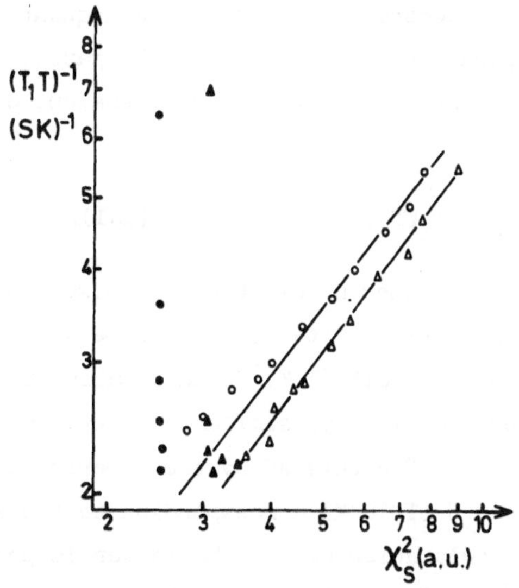

Fig. 5. ^{77}Se $(T_1T)^{-1}$ data of figure 3 vs χ_s^2 values of Miljak et al.,[10] for $(TMTSF)_2PF_6$ (circles)and $(TMTSF)_2ClO_4$ (triangles).

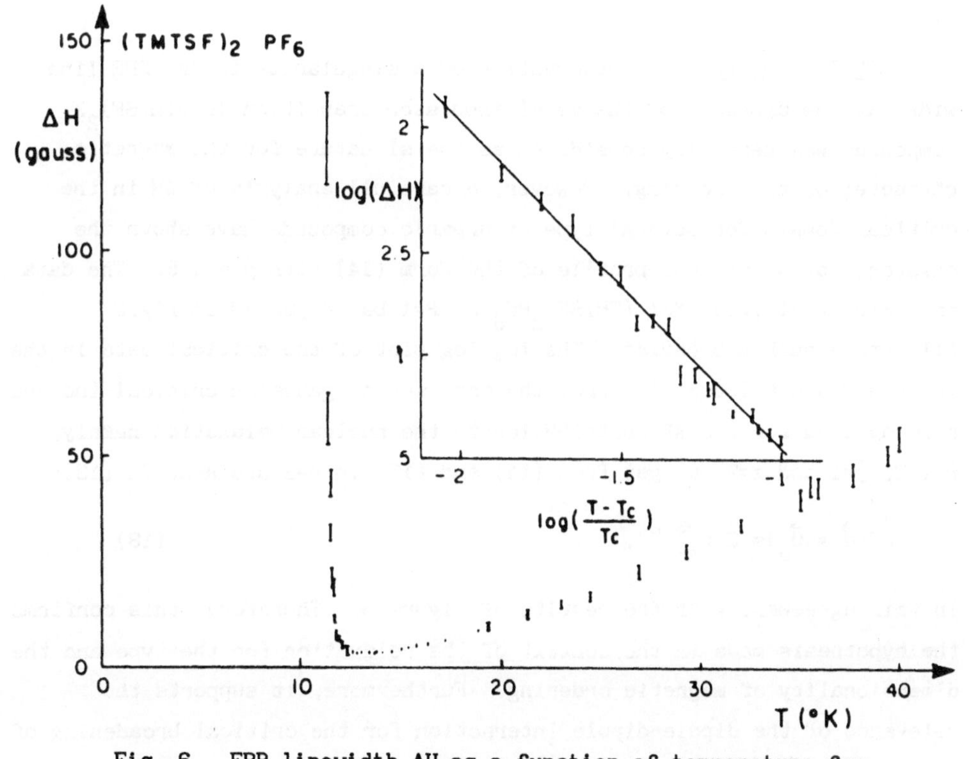

Fig. 6. EPR linewidth ΔH as a function of temperature for $(TMTSF)_2PF_6$ at 1bar. In the inset, the log-log critical profile near T_N. After, Tomic et al.,[13].

interesting. Indeed, according to the standard quantum renormalization group approach to paramagnon systems [11], the gaussian static $\gamma=1$, $\nu=1/2$ and dynamic z=3 exponents can be used for dimensions d ≥ 1. From (10) and (11), this leads to:

$$T_1^{-1}[\vec{q} \approx 0] \approx C_o T\, [\chi_s(T)]^{(5-d)/2} \qquad (d\geq 1). \qquad (17)$$

This result points out rather nicely that *one-dimensional* (ϑ=2) paramagnon type of magnetic excitations would be at the origin of the temperature variations of *both* $(T_1T)^{-1}$ and χ_s above 30K for the perchlorate compound. Obviously, similar conclusions also hold for $(TMTSF)_2PF_6$ at P= 1bar. For this AF system however, deviations are detected below $\chi_s^2 \approx 3.5$ (T<100K) corresponding to the emergence of the AF contribution to the relaxation rate. The latter is probably of 1D origin $[(T_1T)^{-1} \propto (T/E_o)^{-\vartheta}, \vartheta>0]$. It is apparently much weaker at high temperature for the $(TMTSF)_2ClO_4$ salt. The enhancement becomes huge at T < 30K (see fig 5). The same is true for the $(TMTSF)_2PF_6$ salt at P >P_c. This corresponds to the large enhancement commonly seen for these compounds at P > P_c (see for example the crosses and the circles of figure 2). It has been analyzed in terms of the combination of 1D AF and "frustrated" 3D critical ordering effects [2,18,19].

EPR line width. The observation of a singularity in the EPR line width at the approach of the metal-insulator transition in $(TMTSF)_2X$ compounds was generally considered as the signature for the magnetic character of the ordering. However, a carefull analysis of ΔH in the critical domain for several type of organic compounds have shown the existence of a critical profile of the form (14) with $\dot{\mu} \simeq 1.5$. The data of Tomic et al.,[13] for $(TMTSF)_2PF_6$ at P=1 bar reported in fig.6 illustrate such a behavior. The log-log plot of the critical data in the inset gives $\dot{\mu} \approx 1.4 \pm .1$. From the same set of gaussian critical indices already used for the AF contribution to the nuclear relaxation namely, $\dot{\nu}=1/2$, $\dot{\gamma}=1$ and $\dot{z}=2$, we get from (14) and (15) in d=3 and near T_N [13]:

$$\Delta H(\vec{q} \approx \vec{Q}_o) \approx \dot{C}\, \dot{r}_{AF}^{-3/2} , \qquad (18)$$

in fair agreement with the results of figure 5. Therefore, this confirms the hypothesis made in the context of the relaxation for the type and the dimensionality of magnetic ordering. Furthermore, it supports the relevance of the dipole-dipole interaction for the critical broadening of the EPR line. It is worth noting that the spin-orbit coupling in the lowest order would give a smaller value for $\dot{\mu}$.

CONCLUDING REMARKS

From the preceeding discussion, it is clear that the scaling hypothesis together with gaussian exponents give a rather good description of the 3D AF critical behavior of the Bechgaard salts. For both T_1^{-1} and ΔH, the gaussian critical width is found to be large ($\Delta T_{fl.} \approx T_N$, see figs 1 and 2). A reasonable explanation for this would be the small interchain coupling $g_1(t_\perp)$. This leads in turn to small AF coherence lengths in the transverse directions. This enlarges both the gaussian and the non-gaussian (Ginzburg) critical widths. The latter is apparently not observed and it would be detectable, only sufficiently close to T_N. The calculated value of ΔT_{fl} is then strongly dependent on the type of mechanism responsable for AF ordering. There are two proposed mechanisms: the 2D nesting of the Fermi surface [20,21] and the interchain spin exchange interaction [5]. The former mechanism is relevant whenever the 2D band structure or 2D single particle properties are coherent. Owing to the strong anisotropy between the hopping integrals ($t_\perp \ll t_\parallel$) and the effects of Coulomb interaction [17], this is only possible whenever the temperature becomes sufficiently low to reach the crossover region at $T \lesssim T_{x^1} \sim t_\perp^{1/\phi_{x^1}}$. In this case, $g_1(t_\perp)=t_\perp$ corresponding to a *single particle* perturbation and $\phi_{x^1} \leq 1$ [17]. Well below T_{x^1}, the nesting properties at $\vec{Q}_o$ can induce an AF phase transition [20,21]. This mechanism is based on a weak logarithmic singularity of the elementary magnetic response function at $\vec{Q}_o$ and for this reason, it gives rise to a much smaller ΔT_{fl} than observed [7,19] ($\Delta T_{fl} \lesssim 10^{-1}T_N$). Another short comming of this mechanism is despite a calculated value of T_N which drops off to zero as the nesting deviations increases under pressure (see fig.1), the AF correlations that remains at $P \gtrsim P_c$ are so small that the calculated value of T_1^{-1} remains unaffected ($T_1^{-1} \propto T$) over all the temperature domain below T_{x^1}. Therefore one can not from this mechanism explain the huge relaxation rate enhancement seen for $(TMTSF)_2X$ compounds at $P \gtrsim P_c$ (see fig.2). As discussed in details elsewhere [2b,5,18,19,22], in the large temperature domain $[T_{x^1}, E_F]$ where 2D nesting is irrelevant, the spins are thermally localized on each chain. The spins interact in the transverse direction however, through an *interchain exchange interaction* $J_\perp$. This coupling involves the coherent propagation of electron-hole *pairs* perpendicular to the chains and can therefore stabilyzed AF ordering. This can occur despite the metallic character of the isolated chains. This mechanism is mainly active well above T_x1 where it couples to 1D $2k_F$ AF fluctuations. This leads to a different type of crossover $T_x2 \sim g(J_\perp)^{1/\phi_x2}$ which is two-particle like with $\phi_x2 = \gamma$. Whenever T_N becomes sufficiently small so that $T_N \sim T_x1$ the effect of interchain single particle hopping becomes quite relevant and deviations to perfect 1D $2k_F$ nesting frustrate the growth of 1D and 3D AF correlations. The value of T_N is then strongly depressed. This can simulate the effect of pressure. At $P \gtrsim P_c$ however, this mechanism predicts that a huge enhancement of T_1^{-1} remains in agreement with observation [18,19]. It is also consistent with a rather large value of ΔT_{fl}. Of course, whenever $T_N \lesssim T_x1$ the 2D nesting starts to contribute to ordering but the exchange still plays the essential role.

From the observed nuclear relaxation data (e.g. fig.2) at $P \gtrsim P_c$, the existence of strong AF fluctuations above the superconducting transition temperature for the Bechgaard salts indicates that the interactions along the chains are dominantly *repulsive.* Therefore, the electron-phonon interaction is not sufficient to overcome the repulsion and to lead to a BCS type of pairing for electrons. However, an *interchain* pairing would avoid the local repulsion. As proposed in [5], the interchain exchange $J_\perp$ precisely leads to such a pairing. Actually, the exchange invariably opens a new channel of correlations in the transverse direction which is the one of anisotropic superconductivity. As noted by Emery, an interchain pairing would give rise in general to an anisotropic gap with lines of zeros for a 3D Fermi surface. Such an anisotropic gap will be quite sensible to non-magnetic impurities [22], as observed [23]. The absence of a Hebel-Slichter anomaly and the observation of a power law profile near and below T_c, for the nuclear relaxation have also given a strong support to the existence of anisotropic superconductivity in $(TMTSF)_2X$ series [24]. Therefore the interchain exchange can be considered as an essential ingredient for the obtention of an unified description of the phase diagram on both side of the critical pressure for superconductivity.

ACKNOWLEDGMENTS

I would like to thank Drs F. Creuzet, P. Wzietek, P. Baillargeon, S. Tomic, P. Vaca, C. Coulon and Profs D. Jerome, L. G. Caron for their important contribution and fruitfull collaboration on several aspects of the work presented here.

REFERENCES

1. For recent results in the field of organic conductors see: a) Proceeding of the International Conference on Science and Technology of Synthetic metals (ICSM'88), Santa Fe, NM, U.S.A., June-July, 1988, to be published in Synthetic Metals, 27-29; b) Proceedings of a NATO Advanced Study Institute on Low-Dimensional Conductors and Superconductors, ed. by D. Jerome and L. G. Caron, Vol. 155, Plenum Press, New York(1987).
2. a) C. Bourbonnais, Synthetic Metals 19, 57(1987); b) Ibid., in ref. 1b), P:155.
3. a) See for example, S. K. Ma, "Modern Theory of Critical Phenomena," Frontiers in physics, Vol. 46, Bemjamin/Cummings, Reading, Mass.(1976); P. C. Hohenberg and B. I. Halperin, Rev. Mod. Phys. 49, 435 (1977); b) P. Pfeuty, D. Jasnow and Michael Fisher, Phys. Rev. B10, 2088 (1974).
4. J. Solyom, Adv. Phys. 28, 201 (1979); V. J. Emery, in "Hyghly Conducting One-Dimensional Solids" edited by J. T. Devreese, R. P. Evrard and V. E. van Doren , Plenum, N. Y.(1979).
5. C.Bourbonnais and L. G. Caron, Europhys. Lett. 5, 209 (1988); Ibid., Physica 143B, 451 (1986); L. G. Caron and C. Bourbonnais, Physica 143B, 453 (1986).
6. T. Moriya, J. Phys. Soc. Japan 18, 516 (1963)
7. C. Bourbonnais, P. Stein, D. Jerome and A. Moradpour, Phys. Rev. B33, 7608 (1986).
8. G. Soda, D. Jerome, M. Wegner, J. Alizon, J. Gallice, H. Robert, J. M. Fabre and L. Giral, J. Physique 38,931 (1977).
9. C. Bourbonnais, P. Wzietek, D. Jerome, F. Creuzet, L. Valade and P. Cassoux, Europhys. Lett. 6, 177 (1988).
10. M. Miljak and J. R. Cooper, Mol. Cryst. Liq. Cryst. 119, 141 (1985); Ibid., J. Physique (Colloque) 44, C3-893 (1983).
11. J. Hertz, Phys. Rev. B14, 1165 (1985); M. T. Beal-Monod and K. Maki, Phys. Rev. Lett. 34, 1461 (1975).
12. C. Bourbonnais, P. Wzietek, F. Creuzet, D. Jerome, P. Batail and P. Batail, Preprint.
13. P. Baillargeon, C. Bourbonnais, S. Tomic, P. Vaca and C. Coulon, Synthetic Metals, 27B, 73 (1988).
14. H. Mori and K. Kawazaki, Prog. Theor. Phys. 27, 529 (1962).
15. F. Creuzet, Thesis, Université de Paris-Sud, Orsay(1987), unpublished.
16. F. Creuzet, C. Bourbonnais, L. G. Caron, D. Jerome and A. Moradpour, Synthetic Metals 19, 277 (1987).
17. C. Bourbonnais, F. Creuzet, D. Jerome, K. Bechgaard,and A. Moradpour, J. Physique Lett. 45, L-755 (1984).
18. F. Creuzet, C. Bourbonnais, P. Wzietek, H. Nelisse, L. G. Caron, D. Jerome and K. Bechgaard, Synthetic Metals 27, in press.
19. C. Bourbonnais, L. G. Caron, F. Creuzet and D. Jerome, in Poceeding of the International Conference of Magnetism (ICM'88), Paris(1988), to appear in J. Physique.
20. K. Yamaji, J. Phys. Soc. Jpn. 51, 2787 (1982).
21. Chr. Seidel and V. N. Prigodin, J. Phys. Lett. 44, L403 (1983).
22. Y. Suzumura and H. J. Schulz, preprint.
23. R. L. Greene, P. Haen, S. Huang, E. M. Engler, M. Y. Choi and P.M. Chaikin, Mol. Cryst. Liq. Cryst. 79, 183 (1982).
24. M. Takigawa, H. Yasuoka and G. Saito, J. Phys. Soc. Jpn 56, 877 (1987).

PEIERLS INSTABILITY IN WEAKLY NON-IDEAL

ONE-DIMENSIONAL SYSTEMS

V.Ya. Krivnov and A.A. Ovchinnikov

Institute of Chemical Physics
USSR Academy of Sciences
Moscow

INTRODUCTION

According to the known theorem of Peierls[1], a one-dimensional metal is unstable against distortions of the lattice. The dielectric state which then arises possesses a number of unusual physical properties. In particular, in polymers with conjugated bonds, such as polyacetylene, the transition to the Peierls dielectric state leads to alternation of C-C bond lengths and, as a consequence, to the appearance of excited states of special kind, solitons[2].

It should be noted that Peierls' conclusion referred to the system of non-interacting electrons. It is known, however, that electron-electron interactions play an important role in one-dimensional systems and can themselves form a gap in the excitation spectrum. There arises the far from trivial problem of taking simultaneous accounting of two effects leading to instability. There are reasons to expect that the correlation effects can enhance the tendency towards transition to Peierls state[3-8].

In the present paper the problem of the Peierls instability will be considered for the systems with weak electron-electron interaction.

PARQUET APPROXIMATION

We shall start from the Hamiltonian of an electron-phonon system in the adiabatic approximation:

$$\hat{H} = \hat{H}_0 + \hat{H}_1 + \hat{V} \qquad (1)$$

$$\hat{H}_0 = -\sum \cos k \, a^+_{k\sigma} a_{k\sigma}$$

$$\hat{H}_1 = (2N)^{-1} \sum v(q) \, a^+_{k+q\,\sigma} a^+_{k'-q\,\sigma'} a_{k'\sigma'} a_{k\sigma},$$

where $V(q)$ is Fourier transform of the interaction potential, $|V(q)| \ll 1$.

There are two type of the derformation. The first of them:

$$\hat{V} = i\delta \sum \sin k \, a_{k\sigma}^{+} a_{k+\pi\sigma} + N\delta^2/2\pi\lambda \tag{2}$$

leads to the alternation of the bond length and the appearance of a bond charge density wave (CDW). $\delta \ll 1$ is the dimensionless deformation parameter, and λ is the dimensionless electron-phonon coupling. The second term in (2) is the elastic energy of the chain.

The second type of the deformation corresponds to the interaction of intramolecular vibrations with the local electronic density and leads to site CDW:

$$\hat{V} = \delta \sum a_{k\sigma}^{+} a_{k+\pi\sigma} + N\delta^2/2\pi\lambda \tag{3}$$

We shall be interested mainly in the case of a half-filled band ($\rho = 1$), for which a doubling of the period of the chain (dimerization) occurs.

The problem of investigating the instability of the one-dimensional system against the deformation reduces to finding the dependence of the ground state energy E_0 on δ . It is convenient to introduce the quantity $\mathcal{E}(\delta)$

$$\mathcal{E}(\delta) = (E_0(\delta) - E_0(0))/N$$

The deformation is energetically favourable if the quantity

$$\mathcal{E}(\delta) + E_{elast} \tag{4}$$

has a minimum for $\delta \neq 0$.

For the system of noninteracting electrons

$$\mathcal{E}(\delta) = (\delta^2/\pi) \ln \delta$$

and there is a minimum at $\delta = \delta_{eq}$, where

$$\delta_{eq} = \exp(-1/2\lambda)$$

If $V(q) \neq 0$

$$\mathcal{E}(\delta) \sim -\delta^{\mu}$$

and our main task is to calculate a critical index μ.

Since the interaction is weak, it is natural to use perturbation theory in $\hat{H}_1$ (or coupling constant g). This perturbation series is logarithmic, i.e., in the n-th order the leading contribution to $\mathcal{E}(\delta)$ is proportional to

$$\mathcal{E}(\delta) \backsim \delta^2 g^n \ln^{n+1} \delta$$

Therefore it is necessary to sum "parquet" diagrams of $\mathcal{E}(\delta)$. It is convenient to express the sum of these diagrams of $\mathcal{E}(\delta)$ in terms of the vertex $\gamma(K_1\sigma_1, K_2\sigma_2; K_3\sigma_3, K_4\sigma_4)$ (here $K = (k, \omega)$) and σ is the spin index) in parquet approximation[9]:

$$\mathcal{E}(\delta) = -i(8\pi^2)^{-1}\sum_{\sigma_1,\sigma_2}\sum_{K_1,K_2}\int_{-\infty}^{\infty}d\omega_1\int_{-\infty}^{\infty}d\omega_2\, G(K_1,\omega_1)G(K_2,\omega_2)$$

$$\times\, \gamma(K_1\sigma_1, K_2\sigma_2; K_1+\pi, \sigma_1, K_2+\pi, \sigma_2), \tag{5}$$

where $G(K,\omega)$ is a "anomalous" Green function:

$$G(K,\omega) = i\,u_K v_K\left[(\omega-\mathcal{E}(K)+i\lambda)^{-1}-(\omega+\mathcal{E}(K)-i\lambda)^{-1}\right]_{\lambda\to 0}$$

$$u_K^2 = [1+\cos K/\mathcal{E}(K)]/2\,;\quad v_K^2 = 1-u_K^2\,; \tag{6}$$

$$\mathcal{E}(K) = (\cos^2 K + \delta^2 \sin^2 K)^{1/2}$$

(The expression (6) corresponds to the model (2)).
Analogously, for the Peierls gap Δ we have:

$$\Delta = 2\delta - i(2\pi)^{-2}\int_{-\pi}^{\pi}dK'\int_{-\infty}^{\infty}d\omega'\, G(K',\omega')\,\times$$

$$\times\,\gamma\left(\frac{\pi}{2},0;K',\omega';-\frac{\pi}{2},0;K'+\pi,\omega'\right) \tag{7}$$

The solutions of these equations has the form[9]:

$$\mathcal{E}(\delta) = -2\delta^2\int_0^{\Phi} f^2(t)\,dt \tag{8}$$

$$\Delta(\delta) = 2\delta f(\Phi) \tag{9}$$

$$f(t) = \exp\left\{\int_0^t \varphi(t')\,dt'\right\};\quad \Phi = -(2\pi)^{-1}\ln\delta$$

$$\varphi(t) = -\frac{\lambda_4(t)}{2} - \frac{3\lambda_1(t)}{2} \pm \lambda_3(t) \tag{10}$$

The upper sign in Eqn (10) corresponds to the selection of $\hat{V}$ in the form of Eqn (2) and the lower sign to Eqn (3).
The functions λ_1, λ_3 and λ_4 satisfy the known[10] equations in parquet approximations:

$$\lambda_1(t) = g_1 - \int_0^t \lambda_1^2(t')\,dt'$$

$$\lambda_3(t) = g_3 - \int_0^t \lambda_3(t')\lambda_4(t')dt'$$
$$\lambda_4(t) = g_4 - \int_0^t \lambda_3^2(t')dt' \qquad (11)$$

The bare interaction constants g_1, g_3 and g_4, designated as accepted in the so-called "g"-ology[11], are connected with parameters of Hamiltonian (1) in the following way:

$$g_1 = g_3 = V(\pi)\,; \quad g_4 = V(\pi) - 2V(0)$$

It should be noted that in the weakly non-ideal limit the consideration of non-diagonal matrix elements of the interaction potential[12-13] in addition to (1) is reduced to the renormalization of the constants g_1, g_3 and g_4. In particular, confining ourselves to diagonal terms γ_0, γ_1 and non-diagonal terms W,X

$$\gamma_0 = V(n,n,n,n)$$
$$\gamma_1 = V(n,n+1,n+1,n)$$
$$W = V(n,n+1,n,n+1)$$
$$X = V(n,n+1,n+1,n+1)$$

where V(n,m,l,p) are the potential matrix elements in Wannier's representation, we have

$$g_1 = \gamma_0 - 2\gamma_1 + 4W$$
$$g_3 = \gamma_0 - 2\gamma_1 - 4W \qquad (12)$$
$$g_4 = -\gamma_0 - 6\gamma_1 + 4W$$

The solution of the equations for λ_i has the form:[10]

$$\lambda_1(t) = g_1(1+2g_1t)^{-1}$$
$$\lambda_3(t) = C\,sh^{-1}(C_1-2t) \qquad (13)$$
$$\lambda_4(t) = -C\,cth\,C(C_1-2t)$$
$$C = (g_4^2 - g_3^2)^{1/2}$$
$$C_1 = (2C)^{-1}\ln\{(g_4-C)(g_4+C)^{-1}\}$$

We have confined ourselves here to the case $|g_4| \geq |g_3|$.
Substituting (13) into (10), we find

$$f_0(t) = (\operatorname{sh} CC_1)^{1/4} [\operatorname{sh} C(C_1 - 2t)]^{-1/4} (1 + 2g_1 t)^{-3/4} \times$$
$$\times [\operatorname{th}(CC_1/2)/\operatorname{th}(C(C_1-2t)/2)]^{\pm \varkappa/2},$$
$$\varkappa = \operatorname{sign} g_3 \cdot \operatorname{sign} g_4,$$

where the upper sign corresponds to (2), and the lower sign to (3).

The formulas (8) and (9) together (14) express the electron part of the deformation energy and the Peierls gap in terms of the solutions of parquet equations. We consider certain simple particular cases that can be obtained from (8) and (9).

1. The Hubbard model (half-filled band).

$$\varepsilon_1(\delta) = -2\delta^2 \Phi (1 - 4\gamma_0^2 \Phi^2)^{-1/2}$$
$$\Delta_1(\delta) = 2\delta (1 - 4\gamma_0^2 \Phi^2)^{-3/4}$$
$$\varepsilon_2(\delta) = [\gamma_0^{-1} - \gamma_0^{-1}(1 - 2\gamma_0 \Phi)^{1/2}(1 + 2\gamma_0 \Phi)^{-1/2}$$
$$- 2\gamma_0^{-1} \arcsin(2\gamma_0 \Phi)](-2\delta^2)$$
$$\Delta_2(\delta) = 2\delta (1 - 2\gamma_0 \Phi)^{1/4} (1 + 2\gamma_0 \Phi)^{-3/4}$$

The indices 1 and 2 refer to (2) and (3) respectively.

2. The Hubbard model with $\rho \neq 1$ (band is not half-filled).

$$\varepsilon_{1,2}(\delta) = -2\delta^2 \int_0^{\tilde{\Phi}} \exp(4\gamma_0 t)(1 + 2\gamma_0 t)^{-3/2} dt$$

$$\Delta_{1,2}(\delta) = 2\delta \exp(2\gamma_0 \tilde{\Phi})(1 + 2\gamma_0 \tilde{\Phi})^{-3/4}$$
$$\tilde{\Phi} = \left(\sin \frac{\pi \rho}{2}\right)^{-1} \Phi$$

3. The extended Hubbard model with $\gamma_0 = 2\gamma_1$.

$$\varepsilon_{1,2}(\delta) = -\frac{\delta^2}{2\gamma_0} [\exp(4\gamma_0 \Phi) - 1]$$
$$\Delta_{1,2}(\delta) = 2\delta \exp(4\gamma_0 \Phi)$$

4. The extended Hubbard model with $\gamma_0 = -2\gamma_1$.

$$\varepsilon_1(\delta) = -\left(\frac{\delta^2}{2\gamma_0}\right)\ln(1+4\gamma_0\Phi)$$

$$\Delta_1(\delta) = 2\delta(1+4\gamma_0\Phi)^{-1/2}$$

$$\varepsilon_2(\delta) = \left(\frac{\delta^2}{4\gamma_0}\right)\left[(1+4\gamma_0\Phi)^{-2}-1\right]$$

$$\Delta_2(\delta) = 2\delta(1+4\gamma_0\Phi)^{-3/2}$$

5. The model with $\gamma_0 = 2\gamma_1$ and $\gamma_1 = W$.

$$\varepsilon_1(\delta) = [\gamma_0^{-1} - \gamma_0^{-1}(1-2\gamma_0\Phi)^{1/2}(1+2\gamma_0\Phi)^{-1/2} - 2\gamma_0^{-1}\arcsin(2\gamma_0\Phi)](-2\delta^2)$$

$$\Delta_1(\delta) = 2\delta(1-2\gamma_0\Phi)^{1/4}(1+2\gamma_0\Phi)^{-3/4}$$

As can be seen from these expressions, for $\gamma_0\Phi \ll 1$ we have

$$\varepsilon(\delta) = (\delta^2/\pi)\ln\delta\,; \qquad \Delta(\delta) = 2\delta$$

which corresponds to the limit of non-interacting electrons.

However, with decrease of δ two entirely different situations occur. For example, the asymptotics of ε and Δ at $\delta \to 0$ for the models 2, 3, and 4 have the form:

The model 2. ($\gamma_0 > 0$)

$$\varepsilon_{1,2}(\delta) = -\frac{\delta^{\,2-\frac{2\gamma_0}{\pi\sin(\pi\rho/2)}}}{2\gamma_0^{5/2}\left(|\ln\delta|/\pi\sin\frac{\pi\rho}{2}\right)^{3/2}}$$

$$\Delta_{1,2}(\delta) = \frac{2\delta^{\,1-\frac{\gamma_0}{\pi\sin(\pi\rho/2)}}}{\left(\gamma_0|\ln\delta|/\pi\sin\frac{\pi\rho}{2}\right)^{3/4}} \tag{14}$$

The model 3.

$$\varepsilon_{1,2}(\delta) = -\frac{\delta^{2-2\gamma_0/\pi}}{2\gamma_0}\,, \quad \gamma_0 > 0$$

$$\varepsilon_{1,2}(\delta) = \frac{\delta^2}{2\gamma_0}\,, \quad \gamma_0 < 0 \tag{15}$$

$$\Delta_{1,2}(\delta) = 2\delta^{1-2\gamma_0/\pi}$$

The model 4. ($\gamma_0 > 0$)

$$\mathcal{E}_1(\delta) = -\frac{\delta^2}{2\gamma_0} \ln|\ln\delta|; \quad \mathcal{E}_2(\delta) = -\frac{\delta^2}{4\gamma_0}$$

$$\Delta_1(\delta) = \frac{2\delta}{\left|\frac{2\gamma_0}{\pi}\ln\delta\right|^{1/2}}; \quad \Delta_2(\delta) = \frac{2\delta}{\left|\frac{2\gamma_0}{\pi}\ln\delta\right|^{3/2}} \tag{16}$$

As can be seen from (14)-(16) $\mathcal{E}(\delta)$ remain small, though have different asymptotics at $\delta \to 0$. But for the Hubbard model with $\rho = 1$ $\mathcal{E}(\delta)$ diverges as $\delta \to \delta_0$, where

$$\delta_0 = \exp(-\pi/|\gamma_0|)$$

These two situations are typical in the problem of the Peierls instability. The first of them corresponds to the weak-coupling regime for all $\delta \ll 1$. The other is characterized by the increasing of the coupling upon decrease δ. Analysis of the general expression for $\mathcal{E}(\delta)$ shows that the second situation is realized only when there is a correlation gap Δ_{corr} in the excitation spectrum of $\hat{H}_0 + \hat{H}_1$. The values δ_0 at which $\mathcal{E}$ and Δ become singular coincide with the exact[14] values of the correlation gaps of the Hamiltonian $\hat{H}_0 + \hat{H}_1$.

Thus if the spectrum of $\hat{H}_0 + \hat{H}_1$ has no correlation gap, as in models 3 or 2,4 and 5 with $\gamma_0 > 0$, it is legitimate to use perturbation theory in g, i.e. the expressions for $\mathcal{E}$ and Δ which were obtained in parquet approximation are valid for all δ, but the behavior of $\mathcal{E}(\delta)$ as $\delta \to 0$ is not universal.

The presence of a correlation gap in the spectrum implies that Δ_{corr} establishes for the variable δ a scale that divides the range of variation of δ into a weak coupling ($\delta \gg \Delta_{corr}$) and a strong-coupling ($\delta \ll \Delta_{corr}$) region. In the strong-coupling region, methods based on perturbation theory are inapplicable.

Concluding this part, we should point out that equations (8) and (9) have been obtained independently by B.Horowitz and J.Solyom[15] by the renormalization-group method. However, in their work the explicit form of functions λ_i (Eqn (13)) was not utilized, while for λ_i their fixed point values were inserted. As a result the expressions (14) and (16) differ from corresponding expressions of Ref. 15 by logarithmic corrections. Also, in Ref. 15 the regime of the strong interaction was not investigated.

THE STRONG-COUPLING REGIME

We now consider the region of parameters $\delta \ll \Delta_{corr}$, which corresponds to the strong coupling regime. In this case it is natural to use perturbation theory not in $\hat{H}_1$, as was done earlier, but in $\hat{V}$ (or δ). Unfortunately, the direct construction of corresponding series is impossible because the exact wave functions of $\hat{H}_0 + \hat{H}_1$ are unknown. If these had been found, however, in the case of no correlation gap we would have obtained results coinciding with the parquet results as $\delta \to 0$(like the formulas (14)-(16)). It would appear that the presence of a gap simplifies the problem, since $\delta/\Delta_{corr} \ll 1$ and the excited states with $\Delta E \gg \Delta_{corr}$ give small contributions. It is necessary, however, to keep in mind that in the spectrum of $\hat{H}_0 + \hat{H}_1$, besides such excitations there are also gapless excitations, and it is these which lead to dangereous denominators in the perturbation series. It is natural, therefore, to try to separate the excitations of these two types and to reduce the problem to perturbation theory in δ on the gapless excitations.

The indicated separation can be carried out as follows. The initial Hamiltonian $\hat{H}_0 + \hat{H}_1$ can be reduced to the Hamiltonian with a linear spectrum, which, in turn, is reduced to sum of two commuting Hamiltonians $\hat{H}_\sigma + \hat{H}_\rho$ by the bosonization procedure[11].

The spin density Hamiltonian $\hat{H}_\sigma$ has the form:

$$\hat{H}_\sigma = \frac{2\pi v_\sigma}{L} \sum_{K>0} \left[\sigma_1(K)\sigma_1(-K) + \sigma_2(-K)\sigma_2(K)\right] -$$

$$- \frac{g_1}{L} \sum_{K>0} \left[\sigma_1(K)\sigma_2(-K) + \sigma_1(-K)\sigma_2(K)\right] + \tag{17}$$

$$+ g_1 (2\pi d)^{-2} \int_0^L \left[W_2(x) + W_2^+(x)\right] dx,$$

where $\sigma_1(K)$ and $\sigma_2(K)$ are bosonlike operators, d^{-1} is a cutoff parameter, L is the system's length, v_σ is Fermi velocity.

$$W_\alpha(x) = exp\left\{\alpha^{1/2}\left[\varphi^+_{1\sigma}(x) + \varphi_{2\sigma}(x)\right]\right\} \times$$

$$exp\left\{-\alpha^{1/2}\left[\varphi^+_{2\sigma}(x) + \varphi_{1\sigma}(x)\right]\right\}$$

$$\varphi^{+}_{s\sigma}(x) = (2\pi/L) \sum_{K>0} K^{-1} exp(-dK/2 \mp iKx)\sigma_s(K); \quad s=1,2$$

The charge density Hamiltonian $\hat{H}_\rho$ has the form (17) with $\rho_1(K)$, $\rho_2(K)$ replacing $\sigma_1(K)$, $\sigma_2(K)$ and g_4 and g_3 replacing g_1 in the second and the third terms of (17) respectively.

In the case of interest to us (when the spectrum of $\hat{H}_0 + \hat{H}_1$ has a correlation gap and gapless excitations), as follows from Ref.14 one of the Hamiltonian has a spectrum that starts from the gap, while the spectrum of the other is gapless. If, for definitness, we choose as $\hat{H}_0 + \hat{H}_1$ the Hubbard Hamiltonian with $\gamma_0 > 0$, the role of the first Hamiltonian is played by $\hat{H}_\rho$, and that of the second by $\hat{H}_\sigma$. We write $\hat{V}$ in the ρ and σ represantations:

$$\hat{V} = (\delta/2\pi d) \int_0^L dx \, \hat{V}_\rho(x) \hat{V}_\sigma(x),$$

where for $\hat{V}_\sigma(x)$ (and $\hat{V}_\rho(x)$) we have

$$\hat{V}_\sigma(x) = W_{\frac{1}{2}}(x) + W^{+}_{\frac{1}{2}}(x)$$

As a result the initial Hamiltonian takes the form

$$\hat{H} = \hat{H}_\rho + \hat{H}_\sigma + \hat{V}$$

The analysis of the perturbation theory in $\hat{V}$ shows[9] that the most divergent contributions to $\mathcal{E}(\delta)$ correspond to using $\hat{V}$ in the form

$$\hat{V} = (\delta/2\pi d)(V_\rho)_{oo} \int_0^L dx \, \hat{V}_\sigma(x)$$

where $(V_\rho)_{oo}$ is matrix element of $\hat{V}_\rho$ operator with respect to the ground state of $\hat{H}_\rho$ and we suppose that the spectrum of $\hat{H}_\sigma$ is gapless.

As a result full Hamiltonian is the sum of σ and ρ parts. σ part of full Hamiltonian has the form:

$$\hat{H}_\sigma + a \int_0^L [W_{\frac{1}{2}}(x) + W^{+}_{\frac{1}{2}}(x)] dx, \tag{18}$$

where

$$a = (\delta/2\pi d)(V_\rho)_{oo}$$

Thus the calculation of $\mathcal{E}(\delta)$ reduces to finding the dependence of the ground state energy E_0 of (18) on a and the calculation of the average $(V_\rho)_{oo}$. For the quantity

$$\mathcal{E}(a) = (E_0(a) - E_0(0))/L$$

we shall have the expansion

$$\mathcal{E}(a) = \sum_{n=1}^{\infty} a^{2n} C_{2n}(L) \tag{19}$$

where $C_{2n}(L)$ are contributions of the 2n-th order of perturbation series and are, generally speaking, functions on L. The calculation of $C_{2n}(L)$ is based on mapping[16] the model with Hamiltonian (18) onto the two dimensional Coulomb gas. It turns out[9] that

$$C_{2n} = A_{2n} + B_{2n} L^{-2+3n} \tag{20}$$

and A_{2n} and B_{2n} are constants (not depending on L).
It can be seen from (20) that the coefficients C_{2n} diverge when $L \to \infty$.
As a consequence $\mathcal{E}(a)$ has a singular as well a regular part as $a \to 0$, i.e.,

$$\mathcal{E}(a) = \mathcal{E}_{reg}(a) + \mathcal{E}_{sing}(a)$$

As $a \to 0$, $\mathcal{E}_{reg} \sim a^2$. The behavior of $\mathcal{E}_{sing}(a)$ as $a \to 0$ can be established on the basis of scaling arguments proceeding from the type of dependence of perturbation series terms on L. Substituting (20) into (19) we obtain

$$\mathcal{E}_{sing}(a) = |a|^{4/3} R(|a|^{2/3} L),$$

where R(x) is an unknown function. In the thermodynamic limit this function should be a finite negative value, and hence

$$\mathcal{E}_{sing}(a) \sim -|a|^{4/3}$$

As far as the average $(V_\rho)_{00}$ is concerned it is proportional[9] to $\Delta_{corr}^{1/2}$. Hence

$$\mathcal{E}(\delta) = -C \Delta_{corr}^{2/3} \delta^{4/3} \tag{21}$$

The value of C cannot be calculated on the basis of scaling arguments. It is weakly dependent (compared with Δ_{corr}) on interaction constants functions, in particular for the Hubbard model $C \sim |\gamma_0|^{-1}$.

The remarkable property of this dependence is its universality: independent of the specific type of interaction potential (but on condition that there is a correlation gap in the system's spectrum) the $\mathcal{E}(\delta)$ dependence at $\delta \to 0$ characterized by the critical index 4/3. It should also expect that the resulting Peierls gap $\Delta(\delta) \sim \delta^{2/3}$. It is worth mentioning that the same critical index for $\mathcal{E}(\delta)$ asymptotics has been previously found for the spin-Peierls transition[17]. As the Heisenberg model (considered in Ref.17) is equivalent to the Hubbard model with strong interaction that critical indces μ with $\gamma_0 >> 1$ and $\gamma_0 << 1$ coincide. The exact value of μ for intermediate of γ_0 is unknown.

DEPENDENCE OF THE EQUILIBRIUM DEFORMATION PARAMETER OF THE INTERACTION CONSTANT

We now investigate the dependence of the equilibrium value δ_{eq} determined by the full energy (4) minimization on the interaction constant g.
At first we consider the models which have a gapless spectrum(at $\delta = 0$).
Since the behavior of $\mathcal{E}(\delta)$ at $\delta \to 0$ is not universal in this case, the dependence of δ_{eq} on g also is not universal. For example, for model 3

$$\delta_{eq} = \left(1 + \frac{\gamma_0}{\pi}\right)^{-\frac{\pi}{2\gamma_0}} \tag{22}$$

and for model 4 ($\gamma_0 > 0$)

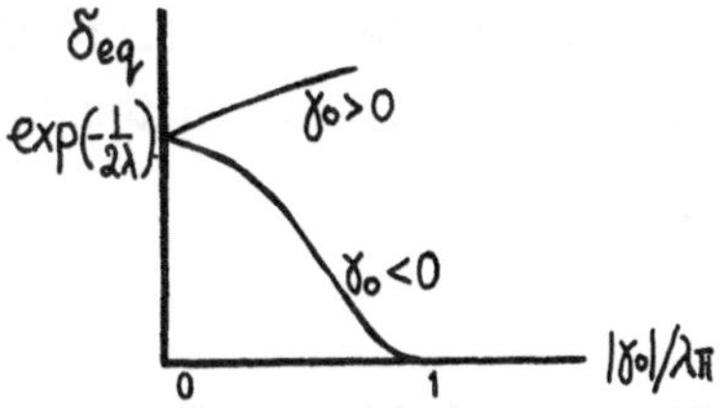

Fig.1. Dependence of δ_{eq} on γ_0/λ for extended Hubbard model with $\gamma_0 = 2\gamma_1$.

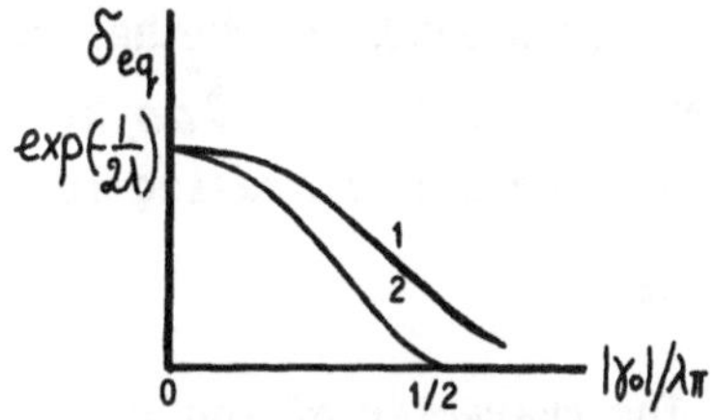

Fig.2. Dependence of δ_{eq} on γ_0/λ for extended Hubbard model with $\gamma_0 = -2\gamma_1$. Curves 1 and 2 correspond to deformations of type (2) and (3).

$$\delta_{eq}^{(1)} = exp\left\{-\frac{\pi}{2\gamma_0}\left[exp\left(\frac{\gamma_0}{\lambda\pi}\right)-1\right]\right\}$$
$$\delta_{eq}^{(2)} = exp\left\{-\frac{\pi}{2\gamma_0}\left[\left(1-\frac{2\gamma_0}{\lambda\pi}\right)^{-1/2}-1\right]\right\} \qquad (23)$$

The superscripts (1) and (2) in (23) refer to Hamiltonians (2) and (3) respectively. Graphs of dependences (22) and (23) are shown in Figs.1 and 2. It follows from (22) and Fig.1 that when $\gamma_0 > 0$ the correlation effects enhance the Peierls instability, whereas at $\gamma_0 < 0$ they decrease it (with $\gamma_0 \geqslant \lambda\pi$ the latter is suppresed). But in model 4 the correlation effects causes a decrease in the instability. In general, δ_{eq} is a monotonical decreasing function of the interaction constant and, consequently, correlation effects reduces the instability if the first order correction for ε is positive, which takes places under the condition

$$\frac{3}{2}g_1 + \frac{g_4}{2} \mp g_3 > 0 \qquad (24)$$

(the upper (lower) sign corresponds to (2) ((3))). This conclusion is an accordance with the results of Ref.12.

Let us consider the most interesting and important case of models with a gap in the excitation spectrum. In this case, as mentioned above, the dependence of $\varepsilon(\delta)$ is known at $\delta >> \Delta_{corr}$ and $\delta << \Delta_{corr}$, which allows the determination of δ_{eq} dependences in the region of correspondingly small and large g/λ parameters values.

When

$$|g|/\lambda << 1 \qquad \delta_{eq} = exp(-1/2\lambda)$$

and when

$$|g|/\lambda >> 1 \qquad \delta_{eq} \sim \Delta_{corr}/(|g|/\lambda)^{3/2} \qquad (25)$$

It is known[14] that Δ_{corr} is sharply increasing function of the interaction constant, hence at $|g|/\lambda >> 1$ correlation effects enhance the Peierls instability. We should emphasize that dependence (25) is universal.

As far as the behavior of δ_{eq} in the region of $|g|/\lambda \approx 1$ is concerned, it largely depends on whether condition (24) is fulfilled or not. In the former case, probably δ_{eq} has a minimum at $|g|/\lambda \approx 1$, and in the latter is a monotonical increasing function of $|g|/\lambda$.

CONCLUSION

The character of the influence of correlation effects on the Peierls instability depends on whether the model spectrum is gapless or has a correlation gap. For the model of the first type the problem can be solved by the summation of parquet diagrams. The behavior of $\mathcal{E}(\delta)$ is not universal. For models of the second type exist two regimes of behavior: weak coupling ($\delta \gg \Delta_{corr}$) and strong coupling ($\delta \ll \Delta_{corr}$). In the strong coupling regime the perturbation theory in g is inapplicable. The critical index μ can be obtained using scaling arguments. The behavior of $\mathcal{E}(\delta)$ is universal. For sufficiently small values of the electron-phonon coupling correlation effects cause an increase in the instability.

REFERENCES

1. R.E.Peierls,"Quantum theory of solids", Oxford University Press, Oxford, (1955).
2. W.P.Su, J.R.Schrieffer and A.J.Heeger, "Soliton excitations in polyacetylene", Phys.Rev.B, 22:2099 (1980).
3. I.I.Ukrainskii, "The influence electron-electron interactions on the Peierls instability", Sov.Phys.JETP, 49:381 (1979).
4. P.Horsch "Correlation effects on bond alternation in polyacetylene", Phys.Rev.B, 24:7351, (1981).
5. H.Fucutome and M.Sasai "Theory of electronic structures and lattice distortions in polyacetylene", Progr.Theor.Phys., 67:41 (1982).
6. D.Baeriswyl and K.Maki, "Electron correlations in polyacetylene", Phys. Rev.B, 31:6633 (1985).
7. J.Hirsch, "Effect of Coulomb interaction on the Peierls instability", Phys.Rev.Lett., 51:296 (1983).
8. S.Mazumdar and S.N.Dixit, "Coulomb effects on 1-d Peierls instability: the Peierls-Hubbard model", Phys.Rev.Lett., 51:292 (1983).
9. V.Ya.Krivnov and A.A.Ovchinnikov, "Peierls instability in weakly nonideal one-dimensional systems", Sov.Phys.JETP, 63:414 (1986).
10. I.E.Dzyaloshinskii and A.I.Larkin "The possible states of quasi-one-dimensional systems", Sov.Phys.JETP, 34:422 (1971).
11. J.Solyom, "The fermi-gas model of one-dimensional conductors", Adv. Phys., 28:201 (1979).
12. S.Kivelson, W.P.Su, J.R.Schrieffer and A.J.Heeger "Missing bond-charge repulsion in the extended Hubbard model:effects in polyacetylene", Phys. Rev.Lett., 58:1899 (1987).
13. C.Wu,X.Sun and K.Nasu "Electron correlation and bond alternation in polymers", Phys.Rev.Lett., 59:831 (1987).
14. G.I.Japaridze, A.A.Nersesjan and P.B.Wiegman, "Exact results in the two-dimensional U(1)-symmetric Thirring model", Nucl.Phys.B, 230:511 (1984).
15. B.Horovitz and J.Solyom, "Charge-density waves with electron-electron interactions", Phys.Rev.B, 32:2681 (1985).
16. S.T.Chui and P.A.Lee, "Equivalence of one-dimensional fermion model and two-dimensional Coulomb gas", Phys.Rev.Lett., 35:315 (1975).
17. M.C.Gross and D.S.Fisher, "A new theory of the spin-Peierls transition", Phys.Rev.B, 19:402 (1979).

USE OF THE SPIN HAMILTONIAN TO STUDY THE SPIN-PEIERLS INSTABILITY AND MAGNETISM IN CONJUGATED POLYMERS

L. N. Bulaevskii

P. N. Lebedev Physical Institute
Moscow, USSR

ABSTRACT

In the framework of Pariser-Parr-Pople π-electron Hamiltonian the spin Hamiltonian (of the Heisenberg type) is obtained for conjugated polymers with half filled π-electron band. Using the spin Hamiltonian the dimerization of molecules like polyacetylene is treated as spin-Peierls instability. The conditions are studied under which the ground state of π-electron system is ferromagnetic.

INTRODUCTION

In the following the spin Hamiltonian will be used to treat the π-electrons in conjugated polymers. The validity of such an approach will be proved in the framework of Pariser-Parr-Pople (PPP) Hamiltonian [1-3]. Actually the spin Hamiltonian treatment is analogous in some sense to the valence bond method (see [4]), and it may be called as modern version of this approach. So the old fashioned valence bond method will be revised in the following and used to study the problem of dimerization in polyacetylene [5]. The possibility to obtain a ferromagnetic ground state in organic molecules will be also considered on the basis of the spin Hamiltonian [6].

The valence bond (VB) method was the first one used in quantum chemistry. It was invented to calculate the ground state energy of chemical bonding. We consider firstly the ansatz of this method for conjugated molecules in the framework of the usual Hamiltonian

$$H = \sum_{i=1}^{N} \frac{\vec{p}_i^2}{2m} + \sum_{i,j}' \frac{e^2}{|\vec{r}_i - \vec{r}_j|} + \sum_{i,k} \frac{e^2}{|\vec{r}_i - \vec{R}_k|} \tag{1}$$

for π-electrons, $\vec{r}_i, \vec{p}_i$ are the coordinates and momenta of electrons, $\vec{R}_k$ are ion coordinates. It is supposed that molecules consist of N conjugated carbon atoms and N electrons of π-type.

A. The function space in the VB treatment is chosen as

$$\theta_i = a^+_{1,\sigma_1} a^+_{2,\sigma_2} \cdots a^+_{N,\sigma_N} \theta_0 \ , \tag{2}$$

Interacting Electrons in Reduced Dimensions
Edited by D. Baeriswyl and D.K. Campbell
Plenum Press, New York

where a^+_{n,σ_n} is the operator which creates the electron on the p_z-orbital of carbon atom n with spin σ_n (up or down), and θ_0 is vacuum. The space $\{\theta_i\}$ consists of 2^N functions with one π-electron on each site with arbitrary orientations of spins $\sigma_1 \cdots \sigma_N$.

B. The matrix elements of H in this space are calculated as

$$V_{ij} = (\theta_i|H|\theta_j) \tag{3}$$

and the secular equation

$$\sum_j (V_{ij} - E\delta_{ij})C_j = 0 \tag{4}$$

has to be solved to obtain the eigenvalues E_i and the eigenfunctions $\chi_i = \sum_k C_k^{(i)}\theta_k$. The minimal value E_i gives the ground state energy. This method may be considered as variational. Here the orbital functions are taken as given (one π-electron for each p_z-orbital) and only the spin configurations are varied.

It is easy to see that the corresponding equation (4) for spin orientations may be represented in the form of a Heisenberg Hamiltonian

$$h_s = \sum_{n,k} T_{nk}(\vec{S}_n\vec{S}_k - \frac{1}{4}) \ , \tag{5}$$

where the $\vec{S}_k$ are 1/2 spin operators. According to (3), all J_{nk} are positive in (5), i.e. the interaction of spins is antiferromagnetic. There are the pair interactions only in this Hamiltonian because the initial Hamiltonian (1) contains only this type of π-electron interaction. For nearest-neighbor sites J_{nk} is large in comparison with others. So in the usual case with two nearest neighboring carbon atoms C in molecule the ground state is a singlet.

C. The parameters J_{nk} can be calculated in the way mentioned above. However, it is possible to choose them phenomenologically to fit the experimental results for small molecules (ethylene) and then use phenomenological parameters to calculate the energies of larger molecules like benzene and so on. The results obtained in this way are surprisingly accurate, although the method seems to be very rough. Actually valence bond method does not take into account the ionic states at all. We know now that the exact ground state of π-electrons contains a large fraction of ionic states because the band width 4t (t is resonance interaction of electrons). Why does ignoring the ionic states in the framework of the VB phenomenological treatment nevertheless produce rather good results?

There is another question. Solving the Heisenberg Hamiltonian we obtain many excited states in addition to the ground state. May we take them seriously as real excited states of π-electrons? We know that variational approaches may be accurate for the ground state and bad for excitations. Is this true for the VB method?

In the following I show that the VB method, i.e. treatment basing on the spin Hamiltonian, is a more correct approach than it seems at first sight. Starting from the PPP Hamiltonian I prove that some eigenstates of PPP Hamiltonian are described very well by the spin Hamiltonian. I give simple expressions for its exchange integrals J as functions of the transfer integral t and the Coulomb parameters U_{nm}. I explain why the spin Hamiltonian takes into account the ionic states. I show also what kind of eigenstates of PPP Hamiltonian are described by the spin Hamiltonian and what kind of states are beyond this consideration.

However the question arises also what is the advantage of using the spin Hamiltonian. The answer is that it is simpler to solve than the initial PPP Hamiltonian.

Numerical methods can be used to solve it exactly for relatively long chain. Besides, it describes directly in a simple manner the magnetic properties of the system. In this respect, the essential point is that the ground state and most of the lowest excited states are described by the spin Hamiltonian. However we pay for this simplification: we do not obtain all the eigenstates of the PPP Hamiltonian. And we cannot study in this way the conjugated molecules with π-band filling different from one-half.

2. FROM THE PPP MODEL TO THE SPIN HAMILTONIAN

The PPP Hamiltonian takes the form

$$H_0 = \frac{1}{2}\sum_{m,k} U_{mk}(n_m - 1)(n_k - 1) \ , \ n_m = \sum_{\sigma} a^+_{m\sigma} a_{m\sigma}$$
$$H' = \sum_{\langle k,n\rangle,\sigma} t_{kn}(a^+_{k\sigma}a_{n\sigma} + a^+_{n\sigma}a_{k\sigma}) \tag{6}$$

Here $\langle k,n\rangle$ are nearest neighboring sites in the molecule. We consider molecules with one π-electron per carbon atom C. Thus the function space consists of $(2N)!/(N!)^2$ functions. It can be reduced to $(N!)^2/[(N/2)!]^2$ (while for the spin Hamiltonian it is reduced to $(N!)/(N/2)!$) by considering the states with $S_z = 0$ only, $\vec{S}$ is the spin of the electron system. This allows to calculate numerically the eigenstates and eigenfunctions of the spin Hamiltonian, while for the PPP Hamiltonian the calculation are much more involved.

Now we consider H_0 as the main term and H' as perturbation. I emphasize that it is not enough to take into account only the lowest order terms of the perturbation series of H' for conjugated molecules. One should sum up all of them, or, at least, the main part of them.

The ground state of H_0 with one electron at every site is 2^N-fold degenerate with respect to spin orientations. We call these states neutral. The excited states of H_0 (with holes or pairs electron on some sites) have a gap $u_1 = U_{00} - U_{01}$, we call them ionic states. Now we turn on the transfer term H' we obtain the secular equation for degenerate states. It is equivalent to the Schrödinger equation with the spin Hamiltonian

$$h_s = PH'\frac{1}{H_0 - E_0}H'P = \sum_{\langle k,n\rangle} J_{kn}(\vec{S}_k\vec{S}_n - \frac{1}{4}) \ ,$$
$$J_{kn} = 4t^2_{kn}/u_1 \tag{7}$$

where $\langle k,n\rangle$ means the summation over the nearest neighboring sites, P is a projection operator onto the space C of neutral states, excluding all the ionic states. Solving the spin Hamiltonian $(h_s - E)C = 0$ we obtain the proper eigenfunctions C_i, eigenvalues E_i and using

$$\chi_i = (1 + \frac{1}{H_0 - E_0}H')C_i \tag{8}$$

the actual wave function χ_i of the state i. Of course χ_i consists of ionic states while C_i does not. Thus turning on H' we obtain the states χ_i which originate from neutral states. We call them as quasineutral. They are the states which are described by the spin Hamiltonian and their eigenvalues are given by h_s. Any physical quantity which is described by the operator A may be found for the quasineutral states by help of the eigenstates C_i using the spin operator a_s which corresponds to A, i.e.

$$(\chi_i|A|\chi_j) = (C_i|a_s|C_j) \,,$$
$$a_s = P(1 + H'\frac{1}{H_0 - E_0})A(1 + \frac{1}{H_0 - E_0}H')P \,. \tag{9}$$

Now, in the second order in H' the quasineutral states are described by the spin operators h_s and a_s which are "projections" of the Hamiltonian H and operator A on the space C. The quasineutral states are those states which become neutral as $t \to 0$. The states which become ionic (I) in the limit $t \to 0$ are beyond our consideration.

Since $J\ 0$, the spins on neighboring sites have antiferromagnetic correlations in the ground state. The physics is simple: due to the Pauli principle transfer can occur if the spins on the neighboring sites are antiparallel. Then we gain energy.

Up to this point we follow the considerations given by Bogoljubov [7] and Anderson [8] in its treatment of the superexchange. Now we should go beyond the perturbation treatment, because the real parameter for this approximation is $4t/u_1$, which may be larger than unity. Thus we should sum up the infinite series in H' to obtain the proper spin Hamiltonian h_s and operators a_s.

The S-matrix has the usual form of T-exponent

$$S(0) = Texp[-i\int_{-\infty}^{0} H'(\tau)d\tau]$$
$$H'(\tau) = e^{iH_0\tau}H'e^{-ih_0\tau+\delta\tau} \,,\ \delta \to 0 \,. \tag{10}$$

However now the matrix $PS(0)P$ is singular as $\delta \to 0$ due to the degeneracy of the ground state of the Hamiltonian H_0. The regular part S_R is obtained with the help of relation

$$S(0)P = S_R P S(0) P \,. \tag{11}$$

The matrix S_R gives the Hamiltonian h_s, the operators a_s and the exact wave function χ_i as

$$h_s = \Gamma^{-1}PH'S_RP\Gamma \,,\quad \Gamma^{-1} = (PS_R^+S_RP)^{1/2} \,,$$
$$a_s = \Gamma PS_R^+AS_RP\Gamma \,,\quad \chi_i = S_R\Gamma C_i \,. \tag{12}$$

The formulas (9)-(12) give the exact operators h_s and a_s. It appears to be very different to find h_s and a_s for long conjugated molecules. Therefore some approximation should be used to obtain h_s and a_s.

3. CLUSTER EXPANSION OF THE SPIN HAMILTONIAN

Now I use cluster expansion to obtain the spin operators h_s and a_s. To clarify its essence I start from two sites (C_2H_4). In this case the PPP Hamiltonian is solved exactly. We obtain three singlet states with energies $E_i = -J$, u_1 and $J + u_1$, and one triplet with energy $E_i = 0$. Knowing E_i and χ_i we write down the spin Hamiltonian

$$h_s = J(\vec{S}_1\vec{S}_2 - \frac{1}{4}) \,,\ J = \frac{u_1}{2}\left(\sqrt{1 + \frac{16t^2}{u_1^2}} - 1\right) \,. \tag{13}$$

The singlet ground states as well as triplet state are described by this Hamiltonian because in the limit $t \to 0$ they belong to neutral C-states.

There is an essential question why the ground state and the lowest triplet are a quasineutral ones. Is this accidental or not? As a rule the ground state is the most symmetric state. The space C contains such a state. Now we use the theorem about the intersection of the energy levels for Hamiltonians which depend on a single parameter (t in our case). The levels of the same symmetry do not intersect as we turn on the term H'. Therefore the ground state with maximal symmetry comes from the space of neutral states and is described by the spin Hamiltonian. The same conclusion is valid for any symmetrical molecule (homogeneous chains, rings like benzene and so on). Thus if a state with some symmetry L is considered and the neutral space is compatible with this symmetry, then the state of symmetry L with lowest energy is quasineutral and is described by spin Hamiltonian.

Knowing the spin Hamiltonian for a two-site molecule we proceed the case of three sites, see Fig. 1. For the symmetrical case (equal distances between atoms 1, 2

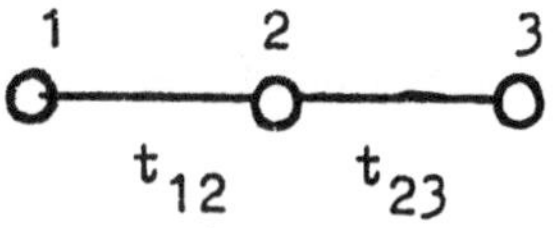

Fig. 1 Three-site linear molecule; t_{12} and t_{23} are the resonance integrals.

and 2, 3) the problem is solved analytically and the corresponding spin Hamiltonian is obtained in a simple form. It may be written as

$$h_s = h_s(2) + h_s(3) \ , \tag{14}$$

where $h_s(2)$ is the pair interaction term obtained above, in Eq. (13):

$$h_s(2) = J_1(2)\sum_k(\vec{S}_k\vec{S}_{k+1} - \frac{1}{4}) \ , J_1(2) = \frac{u_1}{2}\left(\sqrt{1+\frac{16t^2}{u_1^2}} - 1\right) \tag{15}$$

and $h_s(3)$ describes the new triple interaction

$$h_s(3) = [J_1(3) - J_1(2)]\sum_k(\vec{S}_k\vec{S}_{k+1} - \frac{1}{4}) + J_2(3)(\vec{S}_1\vec{S}_3 - \frac{1}{4}) \ , \tag{16}$$

where $J_1(3), J_2(3)$ may be expressed analytically as functions of t_1, u_1 and $u_2 = U(0,0) - U(0,2)$. If we take different transfer integrals t_{12} and t_{23}, then $h_s(2)$ consists of all the terms of perturbation series of the form $t_{12}^2 t_{23}^2$,... contained in $h_s(3)$. If t is small compared with u_1, then $h_s(3) \approx t^4/u_1^2 u_2$ while $h_2(2) \approx t^2/u_1$ and so $h_s(3) \langle\langle h_s(2)$. However, for any parameters t, u_1, u_2 the term $h_s(3)$ is also small in comparison with $h_s(2)$. Thus for $u_k = 0$ we obtain $J_1(2) = 2|t|$, $J_1(3) - J_1(2) = -0.24|t|$, $J_2(3) = 0.47|t|$ while for realistic values $tu_1^{-1} = 0.65, u_2 u_1^{-1} = 1.7$ (see [9]) we get $J_2(3) = 0.13 J_1(2)$ or $h_s(3) \approx 0.13 h_s(2)$.

The analysis of molecules with $N = 4, 5$ and 6 sites and all resonance integrals $t_{k,k+1}$ equal shows that the terms of the series

$$h_s = h_s(2) + h_s(3) + h_s(4) + ... \tag{17}$$

decrease very quickly: $h_s(3) \approx 0.13 h_s(2), h_s(4) \approx 0.05 h_s(2)$. It is worth to remark that $h_s(4)$ is not of the Heisenberg type but consists of the terms with products like

$$(\vec{S}_1\vec{S}_2 - \frac{1}{4})(\vec{S}_3\vec{S}_4 - \frac{1}{4}) \ . \tag{18}$$

I note that we should know the eigenvalues E_i and eigenfunctions χ_i to determine the parameters of the spin Hamiltonian in the case $N \geq 4$, while for $N \leq 3$ the energies E_i of quasineutral states determine unambiguously all the exchange integrals of h_s. For $N = 4$ (see Fig. 2) energies E_i and the functions χ_i were obtained numerically at the values $tu_1^{-1}, u_1 u_2^{-1}$ and $u_1 u_3^{-1}$ which are close to the realistic (see [9]). Knowing χ_i we obtain $P\chi_i$, the matrix $\Gamma_{ik}^2 = (P\chi_i|P\chi_k)$ and then eigenfunctions $C_i = \sum_k (\Gamma^{-1})_{ik} P\chi_k$ of spin Hamiltonian h_s according to expressions (12). The knowledge of E_i and C_i allow to find h_s completely.

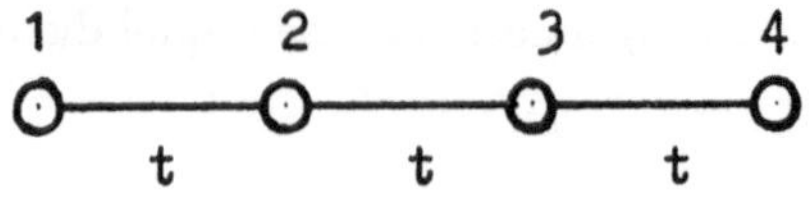

Fig. 2 Four-site symmetrical linear molecule.

The validity of the cluster expansion of other spin operators a_s was also studied. It is found to be about the same as for h_s. For example considering z-component of the spin density, i.e. $A = \sigma_k^a = (a_{k\uparrow}^+ a_{k\uparrow} - a_{k\downarrow}^+ a_{k\downarrow})/2$ we get in the case of three-site molecule for the site K

$$\begin{aligned} &\sigma_{s,k}^z(2) = S_k^z \,, \\ &\sigma_{s,1}^z(3) = \sum_i \alpha_i(3) S_i^z + \beta_i(3) S_i^z (\vec{S}_{i+1}\vec{S}_{i+2} - \frac{1}{4}) \,, \\ &\vec{S}_4 \equiv \vec{S}_1 \,, \ \vec{S}_5 \equiv \vec{S}_2 \,. \end{aligned} \tag{19}$$

The coefficients α_i, β_i for realistic parameters t and u_k are less than 0.11 (for $tu_1^{-1} = 0.65, u_1 u_2^{-1} = 1.7$ we get $\alpha_1 = -0.11, \alpha_2 = 0.09, \alpha_3 = 0.03, \beta_1 = -0.035, \beta_2 = -0.05$ and $\beta_3 = 0.05$).

So we can use the pair interactions in the spin Hamiltonian $h_s(2)$ with accuracy of about 15%. Pair and triple interactions $h_s(2) + h_s(3)$ give the accuracy of about 5%. In both these cases h_s is of the Heisenberg type. Such an approach is equivalent to the spin Hamiltonian (5) of VB method. But now the parameters J_{nk} are quite different from those which are given by ansatz (B) of the old VB method. They depend on the parameters t and u_n of PPP Hamiltonian which take into account the ionic state while (B) accounts the neutral states only. However, if one takes the parameters J_{nk} as phenomenological ones (from the experimental data on small molecules) then there is no difference in old VB method and new spin Hamiltonian treatment in its pair and triple approximation. In this case we obtain the accuracy of about 5%. It is this reason why the ancient phenomenological VB method works well – really it does account for some of the ionic states.

We note that it is easy to include the σ-electrons in this consideration. We obtain then the spin Hamiltonian with extra pair interactions of σ-electrons and the terms in spin Hamiltonian which account the interaction of σ- and π-electrons [2].

4. DIMERIZATION IN POLYACETYLENE AS A SPIN-PEIERLS INSTABILITY

We consider now the first example of using the spin Hamiltonian. Let us study the properties of the long chain (or circle) molecules. The ground state and lowest excitations of homogeneous chains (without dimerization) are described well by the

spin Hamiltonian due to the existence of the energy gap for quasiionic excitations caused by Coulomb interaction of electrons [11,12]. This gap is of order of 1 - 1.5 ev for real conjugated systems. The gap is absent for quasineutral (spin-like) excitations: they are the lowest ones, see Fig. 3. The ground state (^1Ag), the lowest triplet and the lowest singlet excitation (^1Ag) belong to these states and are described by the spin Hamiltonian. Thus all the thermodynamical properties of homogeneous conjugated polymers may be described in the framework of the spin Hamiltonian. The effect of dimerization can be analyzed by using this Hamiltonian also as far as the gap due to the dimerization is less than the Coulomb correlation gap.

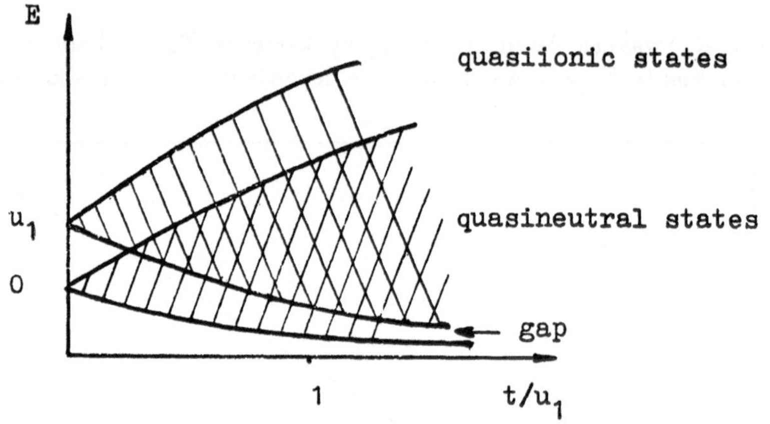

Fig. 3 The lifting of the degeneracy of the eigenstates of PPP Hamiltonian as the resonance integral t turns on.

We consider now the dimerization in polyacetylene. This problem was studied firstly using the model of free electrons without taking into account the Coulomb interaction [13-15]. The role of Coulomb repulsion in long conjugated molecules in connection with correlation gap was stressed out in the review of Ovchinnikov et al. [16]. Later the effect of Coulomb correlations on the dimerization was studied in the framework of PPP Hamiltonian by variational and perturbative methods [17-23] and using the spin Hamiltonian with idea of spin-Peierls instability [5, 24-28].

In so far as dimerization gap is smaller than the Coulomb gap, the ground state of polyacetylene is a quasineutral state, and its structure can be studied by use of spin Hamiltonian. It has the form

$$\begin{aligned} h_s &= \sum_n J(n,v)(\vec{S}_n\vec{S}_{n+1} - \frac{1}{4}) + 2Kv^2 \, , \\ J(n) &= \frac{u_1}{2}\left(\sqrt{1+\frac{16t^2(n)}{u_1^2}} - 1\right) \, , \; t(n) = t_0 + \frac{1}{2}\Delta(-1)^n \, , \\ \Delta &= 4\alpha v \, , \; \delta = \Delta/2t_0 \, , \end{aligned} \tag{20}$$

where v is ion distortion from the homogeneous chain (amplitude of dimerization), i.e. the distance between neighboring atoms is $r(n) = r_0 + v(-1)^n$. The parameters J_n alternates due to the dimerization, $J_n = J_0 + (-1)^n\Delta(\partial J_0/\partial t_0)$. Nakano and Fukujama [5] have used the spin Hamiltonian for the dimerization problem using the

lowest order perturbation term $T(n) = 4t^2(n)/u_1$. This form is inappropriate for polyacetylene because $4t_0/u_1$ is not small ($u_1 = 4 - 6ev$ and $t_0 \approx -2.9ev$ [9]).

Let us introduce the parameters $\lambda = 2\alpha^2/\pi t_0 K$ and $g = (t_0/J_0)(\partial J_0 \partial t_0)$. Then we get

$$h_s/(J_0/2) = 2\sum_k [1 + g\delta(-1)^k](\vec{S}_k\vec{S}_{k+1} - \frac{1}{4}) + \frac{2t_0}{\pi\lambda J_0}\delta^2 . \quad (21)$$

Now the ground state energy of the system $E(y)$ (per site) is

$$2E(y)/J_0 = \varepsilon(y) + ay^2 \ , \ y = \delta g \ , \ a = \frac{2t_0}{\pi\lambda J_0 g^2} \quad (22)$$

The function $\varepsilon(y)$ was calculated numerically by Soos et al. [28] for chains with $N \leq 26$ and extrapolation to $N \to \infty$ was made. The dependence ε on y^2 is shown in Fig. 4.

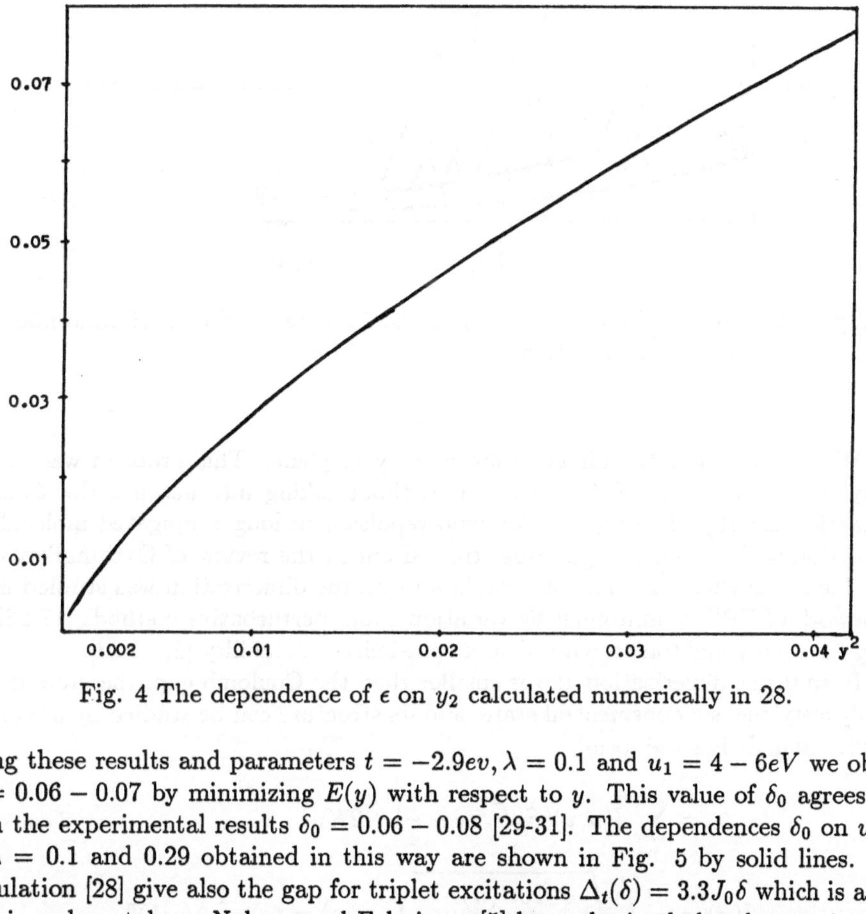

Fig. 4 The dependence of ϵ on y_2 calculated numerically in 28.

Using these results and parameters $t = -2.9ev, \lambda = 0.1$ and $u_1 = 4 - 6eV$ we obtain $\delta_0 = 0.06 - 0.07$ by minimizing $E(y)$ with respect to y. This value of δ_0 agrees well with the experimental results $\delta_0 = 0.06 - 0.08$ [29-31]. The dependences δ_0 on $u_1/4t$ at $\lambda = 0.1$ and 0.29 obtained in this way are shown in Fig. 5 by solid lines. The calculation [28] give also the gap for triplet excitations $\Delta_t(\delta) = 3.3J_0\delta$ which is about $1eV$ in polyacetylene. Nakano and Fukujama [5] have obtained also the quasineutral (magnetic) solitons for dimerized spin-Peierls state using the boson representation for spinless fermion method. The solitons energy is $E_s = (J_0^2/t_0)g^2$, i.e. about $0.45ev$.

These results are valid as long as ground state of the dimerized PPP Hamiltonian belongs to the space of quasineutral states. It does belong to these states (as well as the lowest triplet state) for those values of $4t/u_1$ at which the correlation gap is larger than the dimerization gap and also for large δ independent of $4t/u_1$ (for pairs the lowest states are always quasineutral). In all these regimes δ_0 increases with $4t_0/u_1$.

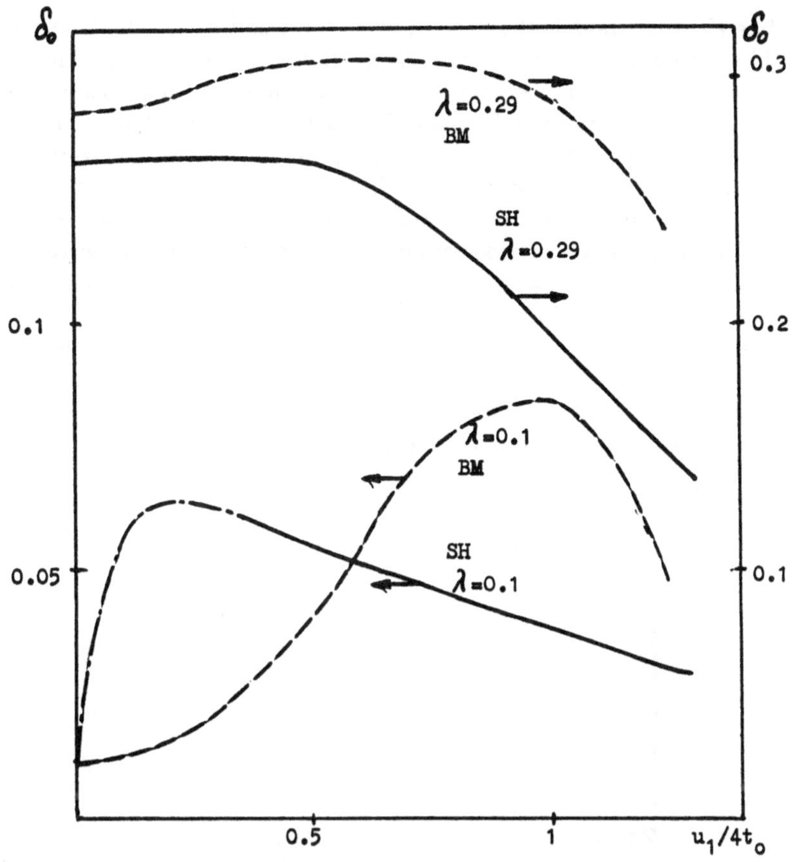

Fig. 5 The dependence of the dimerization parameter on the Coulomb interaction parameter $u_1/4t$. Calculations performed in the framework of spin Hamiltonian for $\lambda = 0.1$ and 0.29 are shown by solid lines. The dotted line for $\lambda = 0.1$ (at small $u_1/4t_0$) is drawn schematically as interpolation between the point $u_1 = 0$ solid line. The dashed lines show the results [19].

However for very large $4t/u_1$, where the dimerization gap is larger than the correlation gap, the ground state may belong to the quasiionic states. Here δ_0 decreases with $4t_0/u_1$. These conclusions follows from the results of Krivnov and Ovchinnikov who use perturbation theory with respect to $u_1/4t$ at given δ and with respect to δ at given $u_1/4t$ [23]. The non-monotonic behavior of δ_0 as function of $u_1/4t_0$ was obtained also by renormalization group numerical calculations done by Hayden and Mele [20] for chains of intermediate length and Monte Carlo calculations of Hirsch [21] at $\lambda = 0.29$. All these results show that a crossover takes place near δ equal to the correlation gap E_c. Thus we have a region in the $\delta - E_c$ plane, where the ground state is quasiionic and the spin Hamiltonian fails to describe the dimerization, see Fig. 6. In this region δ increases with $u_1/4t_0$ while in the other δ_0 decreases as $u_1/4t_0$ grows.

The explanation of this effect was given by Mazumdar and Dixit [22]. They argue that neutral and nearly neutral states (with small numbers of electron pairs and holes) favor the dimerization because the electron transfer from one site to another is restricted here by the Pauli principle only. The ionic states with large number of pairs and holes are unfavorable for dimerization because neighboring pairs and holes depress electron transfer. Quasineutral states are close to the former type while quasiionic states are more of the latter type. As we turn on the Coulomb interaction starting from the free electron model, the ground state for small δ (small λ) becomes more and more like a quasineutral state and thus δ_0 increases with $u_1/4t_0$. The maximum value δ_0 is obtained when ground state becomes quasineutral. Then δ_0 is expected to increase with $4t_0/u_1$ because the Coulomb interaction diminishes the contribution of the ionic states to the wave function and thus suppresses the electron transfer. This is the region on the $\delta - E_c$ plane (see Fig. 6) when the spin Hamiltonian works. Thus δ_0 reaches maximum at some intermediate contribution of the ionic states to the ground state, i.e. at the value $u_1/4t_0$ of order of unity. The same conclusions were reached by Hayden and Mele [20]. For $\lambda = 0.02, 0.3$ and 0.4 they obtained the maximum value δ_0 at $u_1/4t_0 = 1$ for 16-site chain. Monte Carlo calculations done by Hirsch [21] give maximum value δ_0 at $u_1/4t_0 = 0.5$ for $\lambda = 0.29$. For large δ the ground state is a quasineutral at any u_1, so the initial grows of δ_0 at small $u_1/4t_0$ is suppressed and may be absent at all if λ is large enough. In this case we have plateau at small $u_1/4t_0$ and fall for larger values of $u_1/4t_0$, see the curve for $\lambda = 0.29$.

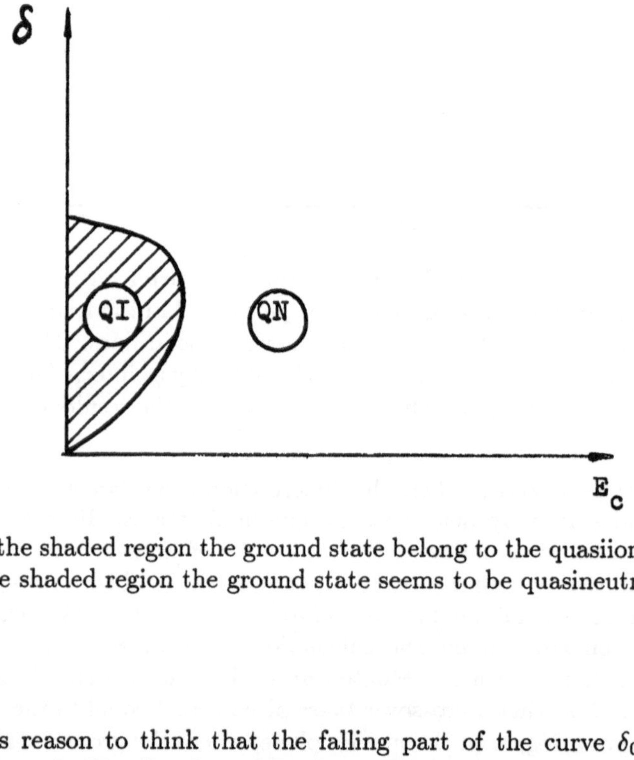

Fig. 6. In the shaded region the ground state belong to the quasiionic states. Outside the shaded region the ground state seems to be quasineutral.

So there is reason to think that the falling part of the curve δ_0 on $u_1/4t_0$ is described by the spin Hamiltonian while the growing part is beyond this approach. Actually the latter part is described better by the approximations which start from H' as main part and take H_0 approximately. For example, the variational method of Baeriswyl and Maki [19] gives more accurately the growing part of the curve δ_0 on $u_1/4t_0$ than its falling part. Their results are shown by dashed lines in Fig. 5. The experimental value of δ_0 in the method of Baeriswyl and Maki at $\lambda = 0.1$ is obtained if u_1 is about $7 - 9eV$. As was mentioned above, the spin Hamiltonian treatment gives for the corresponding value of u_1 about $4 - 6eV$. The latter estimate seems to be more close to the values u_n calculated by Simmons [9].

5. FERROMAGNETISM IN CONJUGATED MOLECULES

As a rule the ground state of conjugated molecules is of a quasineutral type, and it is natural to explore the spin Hamiltonian to study the magnetic properties of such systems.

In the simple homogeneous long chain we have singlet ground state and paramagnetic behavior in the magnetic field. The magnetic susceptibility is $2/\pi^2 J_1(2)$ for infinite chain. Due to the dimerization, the gap appears also for magnetic excitations. As result the dimerized chains are diamagnetic [3].

Is it possible to construct molecules with ferromagnetic ground state? Let us consider the chain with odd (for example, three) carbon atoms in the unit cell, see Fig. 7. Ovchinnikov [6] argued on the basis of the spin Hamiltonian that the ground state

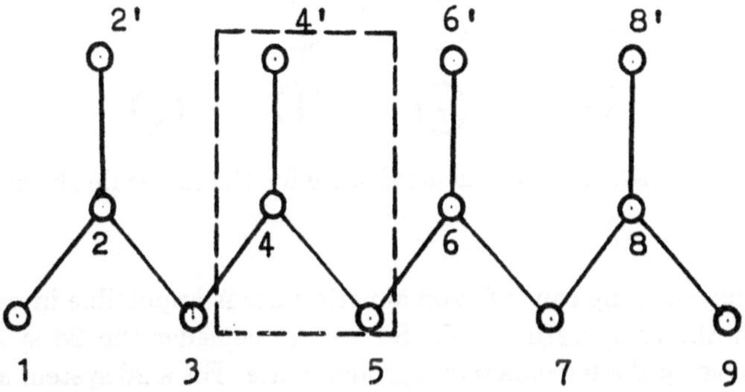

Fig. 7 The polymer with three carbon atoms in the unit cell. The ground state of such hypothetical molecule should be ferromagnetic. The unit cell is shown by dashed lines.

of this chain should be ferromagnetic. Actually, there are ferromagnetic correlations on the even sites 2, 4, 6,... of the chain which leads to the same type of correlations of spins on the side atoms $2', 4', 6'$,... . This fact is obvious from the spin Hamiltonian

$$h_s = \sum_k T_1[(\vec{S}_k \vec{S}_{k+1} - \frac{1}{4}) + (\vec{S}_{2k} \vec{S}_{2k'} - \frac{1}{4})] \ . \tag{23}$$

There is no doubt about the ferromagnetic nature of the ground state in such a system with classical spins. It seems that it is true also for spin 1/2, although a more rigorous analyses is in order. The question is if the ground state of such systems belongs to the quasineutral subspace and is described by the spin Hamiltonian. From this point of view it is interesting to consider the system without Coulomb interactions ($u_n = 0$). There are three bands in the Hückel approximation

$$\begin{aligned} \varepsilon_{1k} &= -t\sqrt{1 + 4sin^2\frac{k}{2}} \ , \ k = \frac{2\pi n}{N} \ , \ n = 0, 1, ...N-1 \ , \\ \varepsilon_{2k} &= 0 \ , \\ \varepsilon_{3k} &= t\sqrt{1 + 4sin^2\frac{k}{2}} \ , \end{aligned} \tag{24}$$

where N is the number of unit cells. The lowest band is filled by $2N$ electrons whereas N electrons should be placed on the $2N$ states of the degenerate middle band. The most symmetrical state belongs to these degenerate states, where most symmetrical means with respect to the translations $r \to r+a$ where a is the period of the structure. This state is with one electron with spin up (or down) at each k-state of middle band, the full spin of the system being $N/2$. The most symmetrical state is presented also in the neutral space C for $t/u_1 \to 0$, such spin configuration is shown in Fig. 8. Basing on the spin Hamiltonian we can argue that the ferromagnetic state with $S = N/2$ is lowest at least for small t. We have seen that it is among the lowest states for $t/u_1 \to \infty$. So we may think that it is lowest at intermediate values of t also.

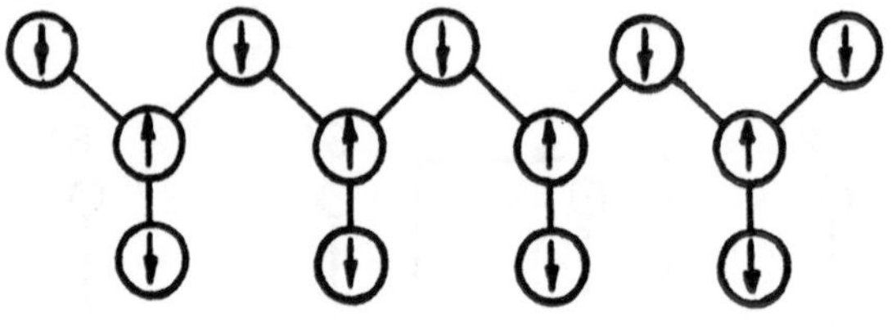

Fig. 8 The most symmetrical neutral state for the molecule shown in Fig. 7.

However the true long range ferromagnetic state is impossible in one-dimensional (1d) system for the temperature T 0. So we can consider the 2d systems with the analogous topology and a ferromagnetic ground state. For a 2d system a small amount of anisotropy gives rise to true ferromagnetic long-range order at T 0.

Let us consider the system which consists of benzene circles with 7 carbon atoms in the unit cell, see Fig. 9. If such a system is infinite in both directions, it should have long range ferromagnetic order.

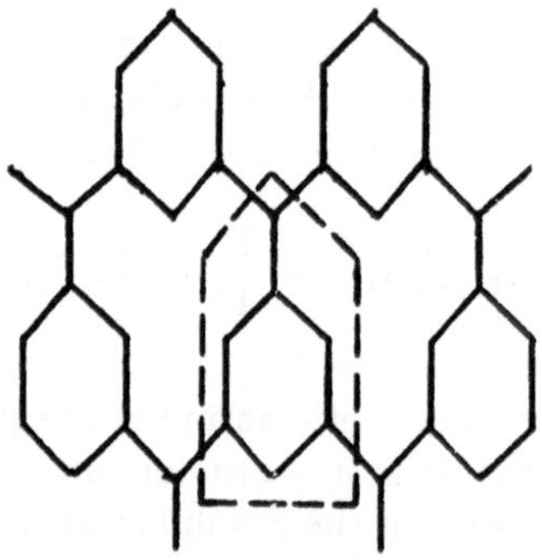

Fig. 9 The 2d infinite sheet of carbon atoms with 7 carbon atoms in the unit cell. Such hypothetical system should have long ferromagnetic order at low temperature.

Up to now the system with $S = 4$ is obtained [32], see Fig. 10. Such a system is described also by the spin Hamiltonian because in π-system we have the equal numbers

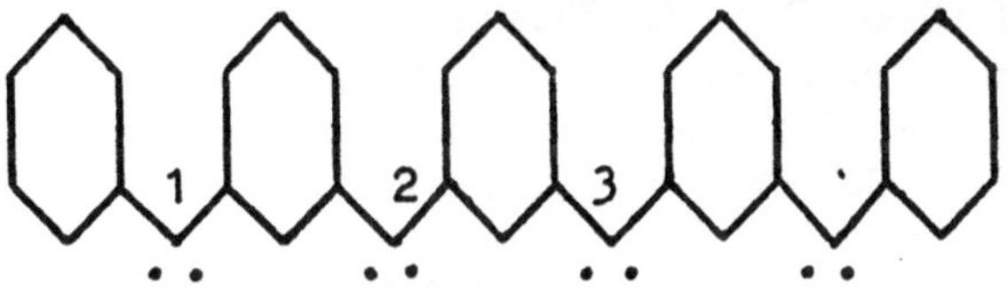

Fig. 10 The molecule with spin $S = 4$ in the ground state [32].

of electron and carbon atoms. The two electrons π and n (2 dots) interact by intra-atomic exchange ferromagnetically and give spin one on carbon atoms 1, 2, 3, 4. On the other hand, π-electrons on different carbon atoms interact antiferromagnetically and give ferromagnetic correlations of π-electrons on atoms 1, 2, 3. As a result we have $S = 4$ in the ground state. Thus the system under consideration may be described by the spin Hamiltonian, all π-electron integrals are positive (as was shown above) and exchange intra-atomic exchange integrals of π- and n-electrons are negative. I note that there are other approaches to study the ferromagnetic ground state. For example, a different approximation was used by Nasu [33], the topology of the studied molecules being the same.

6. CONCLUSIONS

We see that the ancient valence bond method in its modern version is useful to study the ground state of the conjugated molecules with half filled π-band. The magnetic properties of such molecules are described in a very simple form by this method because they can be obtained directly from the spin Hamiltonian.

ACKNOWLEDGMENTS

I thank very much D. Baeriswyl for helpful discussions and reading of manuscript. The discussions with S. Brazovskii were also very useful. I acknowledge the International Centre for Theoretical Physics (Trieste) and Institute for Scientific Interchange (Torino) for the hospitality during the preparation of this paper.

REFERENCES

1. L. N. Bulaevskii, Zh. Eksp. Teor. Fiz. **51**, 230 (1966), Sov. Phys. JETP, **24**, 154 (1967).
2. L. N. Bulaevskii, Teor. i Eksp. Chimija, **4**, 12, 171 (1968), **5**, 267 (1969), in Russian.
3. L. N. Bulaevskii, Izvestija Akademii Nauk SSSR (ser. chim.), **4**, 816 (1972), in Russian.
4. H. Eyring, J. Walter, G. E. Kimbal, Quantum Chemistry.
5. T. Nakano, H. Fukuyama, J. Phys. Soc. Japan, **49**, 1679 (1980), **50**, 2489 (1981).
6. A. A. Ovchinnikov, Dokl. Ak. Nauk SSSR, **236**, 928 (1977), Theor. Chem. Acta **47**, 297 (1978).
7. M. M. Bogoljubov, Lektsii z kwantovoi statistiki, Rad. Shkola, Kiev (1949) (in Russian).
8. P. W. Anderson, Phys. Rev. **115**, 9 (1959).

9. H. Simmons, J. Chem. Phys. **40**, 3554 (1964).
10. T. Morita, Progr. Theor. Phys. **29**, 351 (1963).
11. E. H. Lieb, F. Y. Wu, Phys. Rev. Lett. **20**, 1445 (1968).
12. A. A. Ovchinnikov, Zh. Eksp. Teor. Fiz. **57**, 2137 (1969), Sov. Phys. JETP, **80**, 1160 (1970).
13. S. A. Brazovskii JETP-Lett. **28**, 606 (1978), Zh. Eksp. Teor. Fiz. **78**, 677 (1980), Sov. Phys. JETP, **51**, 342 (1980).
14. W. P. Su, J. R. Schrieffer, A. J. Heeger, Phys. Rev. Lett. **42**, 1698 (1978).
15. D. Baeriswyl, in H. Kamimura ed. Theortical Aspects of Bond Structure and Electronic Properties of Pseudo-One-Dimensional Solids, **1** (1985).
16. A. A. Ovchinnikov, I. I. Ukrainskii, G. F. Kventsel, Usp. Fiz. Nauk **108**, 81 (1972), Sov. Phys. Usp. **15**, 575 (1973).
17. I. I. Ukrainskii, Zh. Eksp. Teor. Fiz. **76**, 760 (1979), Sov. Phys. JETP **49**, 381 (1979).
18. P. Horsch, Phys. Rev. **B34**, 7351 (1981).
19. D. Baeriswyl, K. Maki, Phys. Rev. **B26**, 4278 (1982).
20. G. W. Hayden, E. J. Mele, Phys. Rev. **B34**, 5484 (1986).
21. J. Hirsch, Phys. Rev. Lett. **51**, 296 (1983).
22. S. Mazumdar and S. N. Dixit, Phys. Rev. Lett. **51**, 292 (1983).
23. V. Ja. Krivnov, A. A. Ovchinnikov, Zh. Eksp. Teor. Fiz. **90**, 709 (1986), Sov. Phys. JETP **63**, 414 (1986).
24. E. Pytte, Phys. Rev. **B10**, 4637 (1974).
25. M. C. Cross, D. S. Fisher, Phys. Rev. **B19**, 402 (1979).
26. R. F. Dashen, B. Hasslacher, A. Neveu, Phys. Rev. **D11**, 4114 (1974).
27. W. Duffy, K. P. Barr, Phys. Rev. **165**, 647 (1968).
28. Z. G. Soos, S. Kuwajima, J. E. Michalik, Phys. Rev. **B32**, 3124 (1985).
29. C. J. Bart, C. H. MacGillavry, Acta Crystallogr. Sect. **B24**, 1569 (1968).
30. C. R. Fincher, Jr., C.-E. Chen, A. J. Heeger, A. G. MacDiarmid, J. B. Hastings, Phys. Rev. Lett. **48**, 100 (1982).
31. E. J. Mele, M. J. Rice, Sol. St. Comm. **34**, 339 (1980).
32. H. Iwamura, T. Segawara, K. Itoh, T. Takai, Mol. Cryst. Liq. Cryst. **125**, 251 (1985).
33. K. Nasu, Phys. Rev. **B33**, 330 (1986).

A TWO BAND MODEL FOR HALOGEN-BRIDGED TRANSITION METAL LINEAR CHAIN COMPLEXES

A.R. Bishop, J. Tinka Gammel, and E.Y. Loh, Jr.

Theoretical Division, Los Alamos National Laboratory, Los Alamos, NM

S.R. Phillpot

Materials Science Division, Argonne National Laboratory, Argonne, IL

S.M. Weber-Milbrodt

Physikalisches Institut, Universität Bayreuth, W. Germany

INTRODUCTION

Halogen-bridged transition-metal complexes have been of interest to chemists for many decades as dyes and strongly dichroic materials [1]. However they have only recently begun to receive detailed consideration in the physics community [2-8]. Their potential importance arises because of:

(*i*) The increasing appreciation of strong, competing electron-electron and electron-phonon interactions in low-dimensional materials and the consequent need to expand many-body techniques. The MX materials offer a rapidly expanding, near single-crystal (in contrast to, e.g., polyacetylene), class of quasi-1-D systems which can be "tuned" (by chemistry, pressure, doping, etc.) between various ground state extremes: from strong charge-disproportionation and large lattice distortion (e.g., ~ 20% distortion in PtCl) to weak charge-density-wave and small lattice distortion (e.g., ~ 5% distortion in PtI), to magnetic and undistorted (e.g., NiBr);

(*ii*) The opportunity to probe doping- and photo-induced local defect states (polarons, bipolarons, kinks, excitons) and their interactions in controlled environments and the same large range of ground states; and

(*iii*) The similarities between models and theoretical issues in these materials and the recently discovered oxide superconductors [3]. The MX materials are also closely connected conceptually with mixed-stack charge-transfer salts [9].

The MX class, then, is important in its own right, but also as a template for concepts and electronic structure techniques in strongly interacting (both electron-electron (e-e) and electron-phonon (e-p)), low-dimensional electronic materials.

Interacting Electrons in Reduced Dimensions
Edited by D. Baeriswyl and D.K. Campbell
Plenum Press, New York

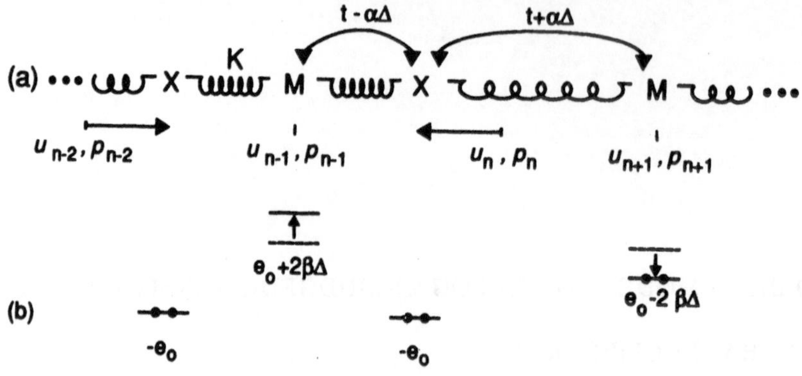

Fig. 1. (a) Schematic of the MX chain showing the model parameters and a CDW distortion; (b) the corresponding $t_0 = \alpha = 0$ energy levels.

THE MODEL

Focusing on a single orbital per site and including only nearest neighbor interactions, we use the following two-band model for the isolated MX chain [8]:

$$H = \sum_{l,\sigma}\{(-t_0+\alpha\Delta_l)(c_{l,\sigma}{}^{\dagger}c_{l+1,\sigma} + c_{l+1,\sigma}{}^{\dagger}c_{l,\sigma}) + [(-1)^l e_0 - \beta_l(\Delta_l + \Delta_{l-1})]c^{\dagger}_{l,\sigma}c_{l,\sigma}\}$$

$$+\sum_l U_l n_{l\uparrow}n_{l\downarrow} + V\sum_l n_l n_{l+1} + V_{MM}\sum_{l\ even} n_l n_{l+2}$$

$$+\frac{K}{2}\sum_l \Delta_l^2 + \frac{K_{MM}}{2}\sum_l(\Delta_{2l}+\Delta_{2l+1})^2 + \frac{1}{2}\sum_l\frac{\hat{p}_l^2}{M_l}$$

with $\beta_l = [\beta_X, \beta_M]$, $U_l = [U_X, U_M]$, relative coordinates $\Delta_l := \hat{u}_{l+1} - \hat{u}_l$, momenta $\hat{p}_l$, and displacements from uniform lattice spacing $\hat{u}_l$. For the discussion here we set $V = V_{MM} = 0$. We have also constrained the sum of the displacements to be zero. This is not necessary, but omitting this constraint just renormalizes t_0, and the parameter to compare with experiment is the renormalized one. The model parameters are shown schematically in Fig. 1. At stochiometry there are 6 electrons per MX unit, and the model has 3/4-filling. We estimate for the dimensionless parameters modeling PtCl in the absence of e-e correlations [8]: $t_0 = .5$, $\alpha = .5$, $e_0 = .6$, $\beta_M = .3$, $\beta_X = -.15$, and $K = 1$. As we have previously discussed [8], if e-e correlations are not too strong, mean-field approximation agrees rather well with exact results, and these numbers should be interpreted as the effective mean field values. For large e_0 this model can be related to the one-band model developed by Baeriswyl and Bishop [3], giving an effective β/α of 3.2. Note that the effective parameters are such that the ground state is always in the CDW part of the one-band phase diagram. As we expect the ligands to strongly constrain the metal-metal distance, we have taken $K_{MM} \to \infty$ when discussing the defects. The values of K_{MM} and β_X do not affect the CDW ground state, and we have set $\beta_X = \beta_M = \beta$ and $K_{MM} = 0$ in the ensuing discussion of the ground states, to reduce the number of parameters and to explore the possibility of a BOW phase. In the discussion of the excitation spectrum below we discuss values for the Hubbard terms. Since we expect this model to be applicable to many other systems it is interesting to explore the full parameter space.

GROUND STATES

A. Long Period

While Peierls' theorem [10] guarantees that a commensurate distortion is lower in energy than no distortion in one-dimensional e-p coupled systems, it does not rule out the possibility that another distortion will be even lower in energy. For this model with $K_{MM} = 0$, the β term drives a long period phase. To see this consider the trivial zero-hopping limit ($t_0 = \alpha = 0$) where the electronic Hamiltonian is site-diagonal. First consider no electron-electron correlations. The lowest energy state phase separates into two regions, one of which has all the sites fully occupied and one of which has all the sites empty. Long range Coulomb effects clearly disfavor long period phases. From our numerical studies we find that this "clumped" long period phase occupies a significant portion of the phase diagram. This phase is not observed experimentally though the inferred parameters for PtCl are in the nominally long period regime. We are currently investigating [11] how inclusion of (site dependent) short- and long-range correlations, allowing β to have a site dependence, quantum phonons, etc., will shift the phase boundary and whether the β term in conjunction with some of these additional interactions can drive "twins" or discommensurations, i.e., long period patterns intermediate between clumping and period 4. Preliminary results indicate that such superlattice orderings are nearby metastable states. For the rest of the discussion we restrict the ground state to be in the experimentally relevant period 4 phase.

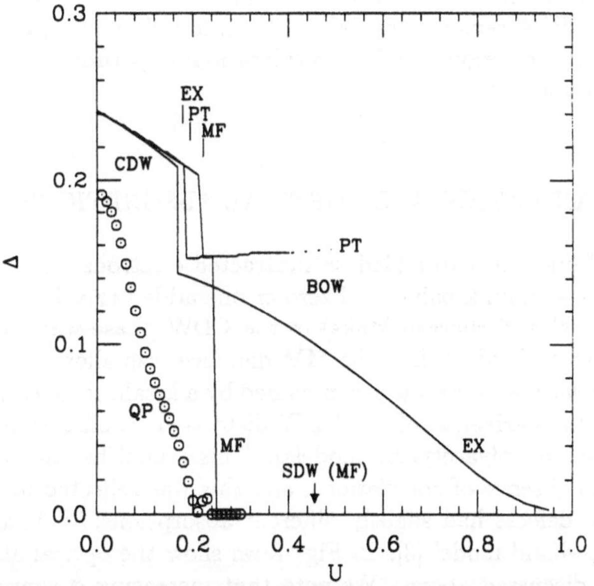

Fig. 2. Lattice distortion amplitude as a function of U from the various approaches. The metastable amplitude of the BOW or CDW in the other phase regime is shown as dotted. The circles are the results from the QPVA.

B. Period 4

In the ground state in the absence of e-e correlations, this problem can be analyzed analytically. The resulting phase diagram is rather featureless: for positive β the lowest energy period 4 phase is nearly always a charge-density-wave on the M sites (CDW) [8]. (There is a bond-order-wave on the X sites, but we label the states by the M behavior in analogy with the one-band model notation [3].) This is in agreement with the fact that all known MX materials with structural distortion are in the CDW phase, and demonstrates one important difference between the two-band model and one-band models [2-7], where the BOW phase is found for a substantial range of parameter values. Using perturbation, Hartree-Fock, and Lanczos exact diagonalization techniques to treat e-e correlations, and treating the phonons in adiabatic approximation, we have mapped out the phase diagram for small β and moderately small to large e_0. (We avoid large β as it drives the long period phase discussed above.) We also set $U_X = U_M = U$. We find [8] for small U that the system has a CDW and for intermediate U the system has a BOW. Since Hartree-Fock contributes in the CDW, explicit correlation corrections are not so striking. This is in contrast to the BOW where, as in polyacetylene [12], Hartree-Fock level terms vanish and exact treatment of the e-e correlations is essential. Fig. 2 shows the amplitude of the CDW or BOW phase as a function of U. We have used a (Gutzwiller-like) variational approach (QPVA) [13], to investigate the effects of quantum phonons. We found [8] that for small values of U, the QPVA always yields a lower bound than the exact diagonalization of the electronic part of our model in combination with a classical treatment of the phonon degrees of freedom (EDCP), which can also be interpreted in terms of a variational Ansatz [13]. The quality of the QPVA, compared to the EDCP bound, increases with e_0. In addition, the relative distance between the two bounds decreases with growing e_0. For reasonably small values of the Hubbard U a CDW-state exists with decreasing amplitude as a function of U, also shown in Fig. 2 (circles). Quantum effects have dramatically reduced the CDW amplitude. The influence of quantum fluctuations contained in the QPVA removes the BOW phase, found above in our adiabatic treatment, in the parameter regime under consideration; experimentally no MX material with a BOW phase is observed.

EXCITATIONS AND ASSOCIATED OPTICAL ABSORPTIONS

The cases of one and two added or subtracted electrons on a 100 site ring (polarons, bipolarons, and kink pairs) and zero or one added or subtracted electron on a 98 site ring (neutral and charged kinks) in the CDW phase were studied. Photoexcitations were also studied. When the MM distance was allowed to vary, the CDW distortion for the various cases was accompanied by a localized BOW distortion, which followed roughly the derivative of the CDW distortion (similar to including acoustic and optic phonons in polyacetylene models). This model has no electron-hole symmetry even in the absence of correlations, and this was reflected in the fact that the electron and hole defects had slightly different absorptions. This asymmetry is not present in the one-band model [3]. In Fig. 3 we show the optical absorption spectra for the polarons discussed above. We note that increasing β sharply decreases the width of the polaron toward the valence-trapped limit [3], whereas increasing e_0 has little effect.

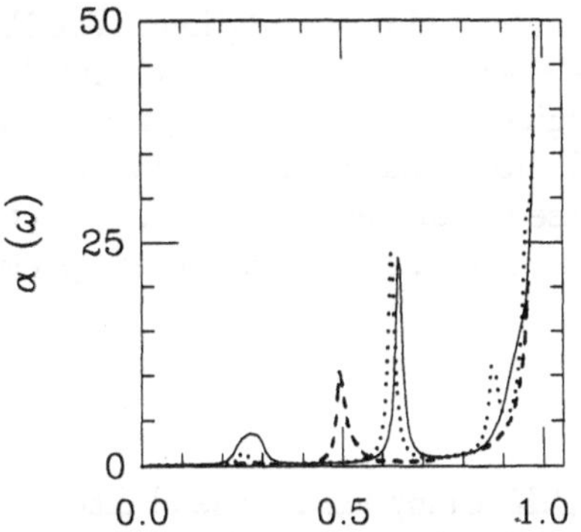

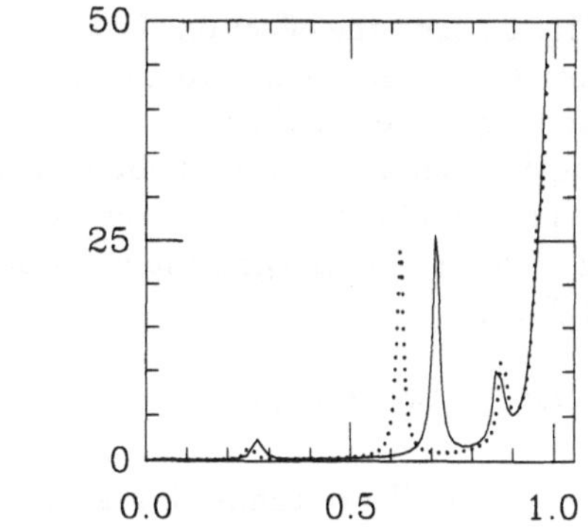

Fig. 3 (left). The optical absorption spectra for a hole (solid) and electron (dotted) polaron in the absence of e-e correlations with $t_0 = 0.5$, $\alpha = 0.5$, $e_0 = 0.6$, $\beta_M = 0.3$, $\beta_X = -0.15$, $K = 1$, and $K_{MM} = \infty$; and the electron-polaron with $K_{MM} = 0$ is also shown (dashed). The lines have been Lorentzian broadened and the inter-valence charge transfer (IVCT) has been scaled to 1.

Fig. 4 (right). The optical absorption spectra for a hole (solid) and electron (dotted) bipolaron with the same parameters as for Fig. 3.

Fig. 4 shows the bipolaron and and Fig. 5 the kink spectra. As might be expected, bipolarons are narrower and deeper than the corresponding polarons, and have associated localized electronic states deeper in the gap. Both the electron polaron and bipolaron are at least as stable as two appropriately charged kinks; their relative stability increases with increasing β, but decreases with increasing e_0. Note however that the relative stability of polarons versus kinks or bipolarons is sensitive to U [3].

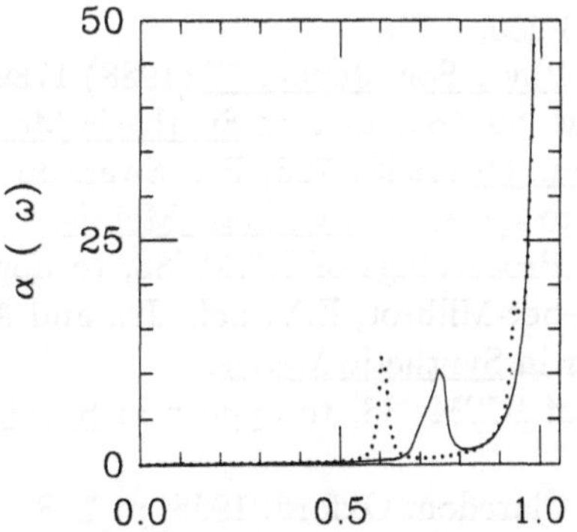

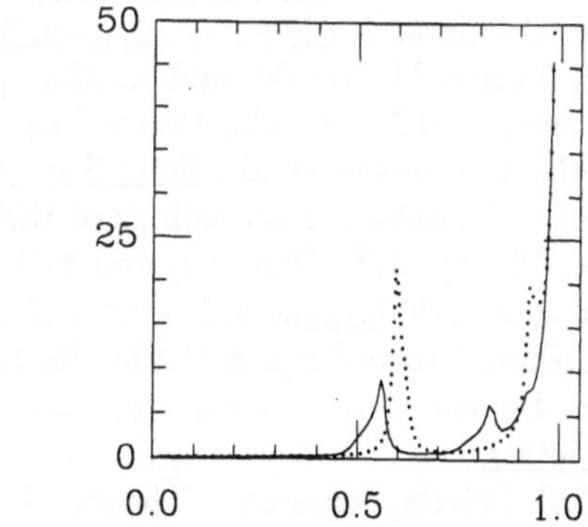

Fig. 5 (left). The optical absorption spectra for a hole (solid) and electron (dotted) kink with the same parameters as for Fig. 3.

Fig. 6 (right). The optical absorption spectra for an electron polaron with $U = 0$ (dotted) and $U = .2$(solid) and other parameters the same as for Fig. 3.

Fig. 6 shows the effect on the optical spectrum of adding correlations. The IVCT absorption has been scaled to 1.0. U_X has little effect, and U_M decreases the IVCT, or effectively increases the polaron gap state energies. We have checked our mean field code against the exact results obtained using a Lanczos diagonalization procedure on 12 sites and find essentially no difference for these parameters values. The spectrum in Fig. 6 with $U = .2$ agrees fairly well with the experimentally measured spectrum for PtCl [14].

ACKNOWLEDGEMENTS

We would like to thank S. Mazumdar and D.K. Campbell for useful discussions. We are happy to acknowledge the Centers for Materials Science and Nonlinear Studies for computational support at Los Alamos National Laboratory, the Computational Sciences Division of the US DOE for computational support at the NMFECC at Livermore, and the Höchstleistungsrechen- zentrum (HLRZ) for computational support at the KFA Jülich, W. Germany. Work of SRP supported by U.S. Department of Energy B.E.S.-Materials Science under Contract W-31-109-Eng-38. We would also like to thank LANL for supporting non-defense basic research.

REFERENCES

1. M.B. Robin and P. Day, in H.J. Emeleus (ed.), "Advances in Inorganic Chemistry and Radiochemistry", Vol. 10, Academic Press, New York, 1967, p. 247; P. Day, in H.J. Keller (ed.), "Low Dimensional Cooperative Phenomena", Plenum Press, New York, 1974, p. 191; H.J. Keller, in J.S. Miller (ed.), "Extended Linear Chain Compounds", Vol. 1, Plenum Press, New York, 1982, p. 357.
2. K. Nasu, J. Phys. Soc. Japan, 50 (1981) 235; 52 (1983) 3865; 53 (1984) 302; 53 (1984) 427; 54 (1985) 1933; J. Luminescence, 38 (1987) 90. K. Nasu and Y. Toyozawa, J. Phys. Soc. Japan, 51 (1982) 2098; 51 (1982) 3111. A. Mishima and K. Nasu, Proceedings of ICSM '88, to appear in Synthetic Metals and unpublished.
3. D. Baeriswyl and A.R. Bishop, Physica Scripta, T19 (1987) 239; J. Phys. C: Solid State Phys., 21 (1988) 339.
4. Y. Ichinose, Solid State Commun., 50 (1984) 137.
5. Y. Onodera, J. Phys. Soc. Japan, 56 (1987) 250.
6. S. Kurita, M. Haruki, and K. Miyagawa, J. Phys. Soc. Japan, 57 (1988) 1789; S. Kurita and M. Haruki, Proceedings of ICSM '88, to appear in Synthetic Metals.
7. S.D. Conradson et al., Solid State Commun., 65 (1988) 723; B.I. Swanson and S.D. Conradson, Proceedings of ICSM '88, to appear in Synthetic Metals.
8. A. Bishop, J.T. Gammel, and S. Phillpot, Proceedings of ICSM '88, to appear in Synthetic Metals; J.T. Gammel, S.M. Weber-Milbrot, E.Y. Loh, Jr., and A.R. Bishop, Proceedings of ICSM '88, to appear in Synthetic Metals.
9. A. Painelli and A. Girlando, Proceedings of ICSM '88, to appear in Synthetic Metals.
10. R.E. Peierls, "Quantum Theory of Solids", Claredon, Oxford, 1955, p. 108.
11. I. Batistic et. al, to be published.
12. D. Baeriswyl and K. Maki, Phys. Rev. B, 31 (1985) 6633.
13. S. Weber and H. Büttner, J. Phys. C: Solid State Phys., 17, 1984, L337-L344; S. Weber and H. Büttner, Solid State Comm. , 56, 1985, 395-398; S. Weber, thesis (Dissertation), Univ. Bayreuth (1985); S. M. Weber-Milbrodt and H. Büttner, to be submitted (1988).
14. R. Donohoe, D. Tait, and B.I. Swanson, private communication.

NUMERICAL SIMULATION OF A ONE DIMENSIONAL TWO BAND MODEL

X.Zotos and W.Lehr*

Institut für Theorie der Kondensierten Materie
Universität Karlsruhe, Karlsruhe, FRG

INTRODUCTION

A great deal of work has been devoted in the study of the Hubbard model as a prototype low dimensional strongly interacting system. The origin of the Hubbard model however is in the two band model describing the magnetism of oxides through the superexchange mechanism[1]. Recently there is renewed interest[2] in the full two band model as describing the physics of the doped high T_c cuprate superconductors and quasi one dimensional metal - halogen materials[3].

In this work we present a report on a study of a one dimensional two band model using the Quantum Monte Carlo method. More details and comparison to analytical results will be presented in an extended presentation elsewhere[4].

HAMILTONIAN

We study the following Hamiltonian in one dimension with periodic boundary conditions:

$$H = -t\sum_{i\sigma}(c^{\dagger}_{i\sigma}c_{i+1\sigma} + H.c.) + (\epsilon_p/2)\sum_{i}(-1)^i n_i + U\sum_{i=odd} n_{i\uparrow}n_{i\downarrow}, \quad i = 1, ..., N$$

where $c_{i\sigma}(c^{\dagger}_{i\sigma})$ are annihilation(creation) operators of a fermion with spin $\sigma =\uparrow,\downarrow$, $n_{i\sigma} = c^{\dagger}_{i\sigma}c_{i\sigma}$ and N the number of sites. We take $t = 1$ as the unit of energy and consider a number of fermions M so that $\rho = M/N$, $1/2 \leq \rho \leq 1$. We used the world line Quantum Monte Carlo method[5] to study lattices up to $N = 64$ sites at a temperature of $T = 0.1, \beta = 10$. This size of lattice and temperature (translating in size of the imaginary time direction) is in the limit of present simulation possibilties. This temperature is low compared to the energies determining the densities and magnetic moments but still relatively high compared to the spin fluctuations energy, proportional to the superexchange.

* Present address: Robert Bosch GmbH, Stuttgart, FRG

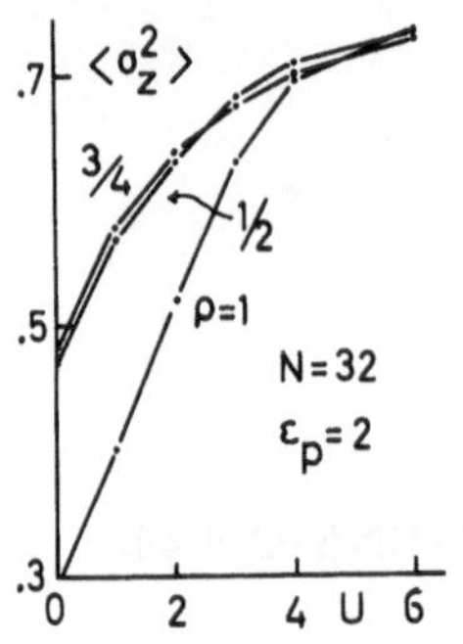

Figure 1.

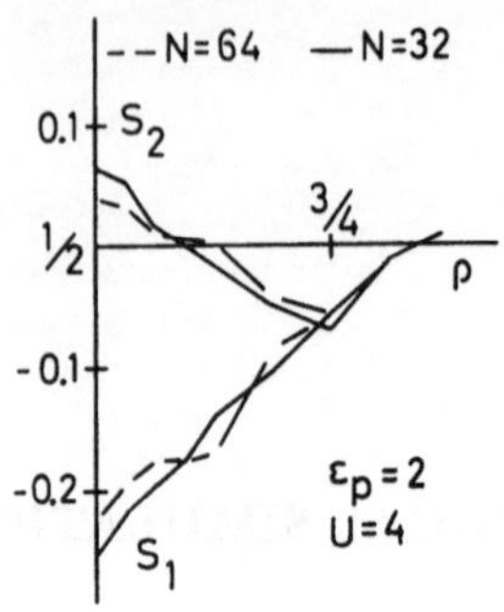

Figure 2.

RESULTS

We are interested in the development of moments as we turn on the on-site interaction U and in the dependence of antiferromagnetic correlations due to superexchange on particle density. In figure 1. we present $< \sigma_z^2 >$ for i=odd as a function of U for different densities ρ. We note that for $\rho \simeq 1/2$ the moment is roughly independent of density as the addition of particles is done mostly on the upper band. In figure 2. we present the $S_l =< \sigma_i^z \sigma_{i+2l}^z >$ spin-spin correlations as a function of density. There is small size dependence compared to the statistical uncertainty. As the density increases we note a transition from antiferromagnetic correlations to paramagnetic ones. The destruction of antiferromagnetism with doping is also a characteristic of the high T_c cuprates.

To conclude, we plan to extend the present study to calculate the superconducting correlations as a function of density and correlate them with the magnetic properties.

ACKNOWLEDGEMENTS

One of the authors (X.Z) would like to acknowledge financial support from the program on "High Temperature Superconductivity" at ISI (Torino-Italy) where this work was completed and the Deutsche Forschungsgemeinschaft. We would like also to thank M. Dzierzawa, D. Baeriswyl and W. Weber for useful discussions.

REFERENCES

1. P.W. Anderson, Solid State Phys. **14**, 99 (1963)
2. J. Müller and J.L Olsen Eds. , Proc. Int. Conf. on High Temperature Superconductors and Materials ans Mechanisms of Superconductivity, Physica **C**152-155 (1988)
3. D. Baeriswyl and A.R. Bishop, Physica Scripta **T19**, 239 (1987)
4. X.Zotos, W.Lehr and W.Weber, Z.Phys. **B** (1989) to be published
5. W. Lehr, Thesis, Univ. Karlsruhe; Z.Phys. **B72**, 65(1988)

DISORDERED QUANTUM SPIN CHAIN AND ITS RELEVANCE TO THE DISORDERED SUPERCONDUCTOR

Naoto Nagaosa

Department of Applied Physics
The University of Tokyo, Hongo
Bunkyo-ku, Tokyo 113, Japan

INTRODUCTION

The interplay between the randomness and superconductivity (SC) has recently gained renewed interest in the light of interest in Anderson localization and metal-insulator transition[1,2]. The randomness enhances the effective Coulomb repulsion and prevents the electron pairing. Even if all the electron form the tightly bound singlet pairs (we shall call it bipolaron in this paper), however, superconductivity as a result of their Bose condensation is disturbed by the randomness[3].

(Quasi) one-dimensional(1D) system have some particular features in this problem. Interactions between electrons and the random potentials both have the essential influence on the electronic states in 1D, and the following conclusions are already known.
(1) In the absence of the electron-electron interactions, all the electronic states are exponentially localized however weak the random potentials are.
(2) The asymptotic behaviour of the various correlation functions of the interacting electron gas in 1D is governed by the fixed line with the continuously varing critical exponent.
(3) The disordered interacting elctron system in 1D shows a metal-insulator transition when the strength of the interaction changes.

In this paper[4], we investigaste the quasi-1D bipolaron system with random potentials treating the interchain Josephson coupling in the mean field approximation. The model is transformed into the S=1/2 XXZ-spin chain with random fields along the z axis and the ordering field along the x axis. This model itself is of interest from the viewpoint of the statistical mechanics of the quantum spin chain, and is closely related to the more general 1D interacting electron systems. Using the quantum transfer matrix method by the Suzuki-Trotter formula[5-7], we calculate the order parameter and the rigidity of the superconductor as functions of the inverse temperature β, the mean strength of the random field σ and the ordering field H^x.

MODEL

We start with the following one-dimensional(1D) Hamiltonian:

$$\mathcal{H} = -\sum_{\ell,\alpha}\left[t\exp\left(-i\frac{eAa}{c\hbar}\right)C^{\dagger}_{\ell\alpha}C_{\ell+1\alpha} + t\exp\left(i\frac{eAa}{c\hbar}\right)C^{\dagger}_{\ell+1\alpha}C_{\ell\alpha}\right] + \sum_{\ell}\varepsilon_{\ell}n_{\ell} - U\sum_{\ell}n_{\ell\uparrow}n_{\ell\downarrow} + V\sum_{\ell}n_{\ell}n_{\ell+1}, \tag{1}$$

where $C_{\ell\alpha}$($C^{\dagger}_{\ell\alpha}$) is the annihilation(creation) operator of the electron at site ℓ with spin α. The occupation number $n_{\ell\alpha}$ is $C^{\dagger}_{\ell\alpha}C_{\ell\alpha}$ and $n_{\ell}=\sum_{\alpha}n_{\ell\alpha}$. The phase of the hopping integral t is modulated by the vector potential A. This vector potential does not represent the physical magnetic field, because the rotation of it vanishes. It is related to the twist of the boundary conditions at the two end points of the sample, and is the test field to measure the rigidity as will be discussed below. The average of the random potential ε_{ℓ} is assumed to be zero, because we consider the case of half-filling. The on-site attractive interaction -U is assumed to be the largest interaction, which leads to the singlet bipolaron formation. We consider also the interaction V between the nearest neighbor sites.

The effective Hamiltonian derived from the above one in the bipolaron limit (large U) is given by[8]

$$\mathcal{H} = -\sum_{\ell}\left\{\frac{J_x}{2}\left(\exp(-i\frac{2eA}{c\hbar})S^{+}_{\ell}S^{-}_{\ell+1} + \exp\left(i\frac{2eA}{c\hbar}\right)S^{+}_{\ell+1}S^{-}_{\ell}\right) + J_z S^{z}_{\ell}S^{z}_{\ell+1}\right\} - \sum_{\ell}H^{z}_{\ell}S^{z}_{\ell} - \sum_{\ell}H^{x}_{\ell}S^{x}_{\ell}, \tag{2}$$

where

$$S^{+}_{\ell} = C^{\dagger}_{\ell\uparrow}C^{\dagger}_{\ell\downarrow},\quad S^{-}_{\ell} = C_{\ell\downarrow}C_{\ell\uparrow},\quad S^{z}_{\ell} = (C^{\dagger}_{\ell\uparrow}C_{\ell\uparrow} + C^{\dagger}_{\ell\downarrow}C_{\ell\downarrow} - 1)/2. \tag{3}$$

The pseudo spin operator S^{-}_{ℓ} (S^{+}_{ℓ}) annihilates (creates) the bipolaron at ℓth site The exchange integrals J^x and J^z are given by $2t^2/U$ and $-2t^2/U$-V, respectively. The random field H^{z}_{ℓ} ($=2\varepsilon_{\ell}$) at each site is generated obeying the Gaussian distribution with the average 0 (half-filled) and the correlation given by

$$\langle H^{z}_{\ell}H^{z}_{\ell'}\rangle = \sigma^2\delta_{\ell,\ell'}, \tag{4}$$

where $\delta_{\ell,\ell'}$ is the kronecker delta. The ordering field H^x is given by

$$H^{x}_{\ell} = H^{x} = 2zJ_{\perp}\langle\langle S^{x}\rangle\rangle \quad \text{(independent of } \ell). \tag{5}$$

where $J_{\perp}$ is the interchain Josephson coupling constant and << >> means the ensemble average with respect to the configuration of the random potentials as well as the quantum statistical average.

BOSONIZATION AND CUMULANT EXPANSION

It is well known that the ground state and the low energy excitations can be described in terms of the bosonized Hamiltonian or the phase Hamiltonian derived from it[9].

$$\mathcal{H} = \mathcal{H}_0 + \mathcal{H}',$$

$$\mathcal{H}_0 = \int dx \left\{ \widehat{A}\left(\frac{d\theta_+(x)}{dx}\right)^2 + \widehat{C}\left(p_+(x) - \frac{eA(x)}{2\pi\hbar c}\right)^2 \right\}, \tag{6}$$

$$\mathcal{H}' = -\int dx \left\{ \frac{H^x}{\pi\alpha}\cos\theta_-(x) + \frac{H^z(x)}{\pi\alpha}\cos\left[\theta_+(x) + 2k_F x\right] \right\},$$

where α and k_F are the momentum cutoff and the Fermi momentum, respectively. The coefficients $\widehat{A}, \widehat{C}$ and the measure of the quantum fluctuation η are given by

$$\widehat{A} = \frac{(1-\Delta^2)^{1/2}(1 - \frac{2}{\pi}\arcsin\Delta)}{16\arccos\Delta},$$

$$\widehat{C} = \frac{\pi^2(1-\Delta^2)^{1/2}}{1 - \frac{2}{\pi}\arccos\Delta}, \tag{7}$$

$$\eta = \frac{1}{2\pi}\sqrt{C/A} = \frac{2}{1 - \frac{2}{\pi}\arcsin\Delta}.$$

$H^z(x)$ is the continuum version of the random field H^z_ℓ , and its average and correlation are given by

$$\langle H^z(x)\rangle = 0,$$

$$\langle H^z(x)H^z(x')\rangle = \sigma^2\delta(x - x'), \tag{8}$$

The phases $\theta_+(x)$ and $\theta_-(x)$ are conjugate to each other quantum mechanically. $p_+(x)$ is the momentum operator conjugate to $\theta_+(x)$.

$$\left[\theta_+(x), p_+(x')\right] = i\delta(x - x'), \tag{9}$$

and $\theta_-(x)$ is related to $p_+(x)$ by the following relation.

$$p_+(x) = -\frac{1}{4\pi}\frac{d\theta_-(x)}{dx}. \tag{10}$$

We discuss in this paper the two physical quantities, i.e., the order parameter $\langle\langle S^x\rangle\rangle$ and the rigidity R. The order parameter $\langle\langle S^x\rangle\rangle$ is roughly given by $\langle\langle\cos\theta_-(x)\rangle\rangle$. The rigidity R is related to the increase in the free energy due to the small vector potential A.

$$F(A) - F(0) \propto RA^2, \tag{11}$$

and this leads to the London equation for the supercurrent J given by

$$J = -\frac{\partial F}{\partial A} \propto -RA, \tag{12}$$

In the language of the pseudo spin, the rigidity is the stiffness constant against the distortion of the phase in the xy-plane.
The free energy up to the second order in A(x) is

$$F(A) - F(0) = \frac{Ce^2}{(2\pi\hbar c)^2}\int dx A(x)^2$$

$$-\frac{C^2e^2}{2(\pi\hbar c)^2\beta}\int_0^\beta d\tau_1\int_0^\beta d\tau_2\int dx_1\int dx_2 A(x_1)A(x_2)\langle\langle T_\tau p_+(x_1,\tau_1)p_+(x_2,\tau_2)\rangle\rangle, \tag{13}$$

where T_τ is the time ordering operator. In the absence of the H^x term in Eqs.(6), the first term cancels the second term in the righthand side of Eq.(13) because the vector potential A(x) does not represent the physical magnetic field. Equation (13) gives the explicit expression of the rigidity R.

At finite temperature, there is no phase transition or discontinuous change in 1D system. All the physical quantities are continuous as functions of temperature $T(=1/\beta)$ and the parameters in the Hamiltonian. Considering this fact, we expand the physical quantities in terms of the perturbative Hamiltonian H' in Eqs.(6). This is done explicitly as follows.

$$\langle\langle \cos\theta_-(x)\rangle\rangle = \sum_{n=0}^{\infty}\sum_{m=0}^{\infty}\frac{1}{(2n+1)!}\frac{1}{(2m)!}$$

$$\times\int_0^\beta d\tau_1\cdots\int_0^\beta d\tau_{2n+1}\int_0^\beta d\tau_1'\cdots\int_0^\beta d\tau_{2m}'\int dx_1\cdots\int dx_{2n+1}\int dx_1'\cdots\int dx_{2m}'$$

$$\times\left(\frac{H^x}{\pi\alpha}\right)^{2n+1}\frac{1}{(\pi\alpha)^{2m}}\langle H^z(x_1')\cdots H^z(x_{2m}')\rangle_{ens}$$

$$\times\langle T_\tau \cos\theta_-(x)\cos\theta_-(x_1,\tau_1)\cdots\cos\theta_-(x_{2n+1},\tau_{2n+1})$$

$$\times\cos[\theta_+(x_1',\tau_1')+2k_F x_1']\cdots\cos[\theta_+(x_{2m}',\tau_{2m}')+2k_F x_{2m}']\rangle_{0C}, \tag{14a}$$

$$\langle\langle p_+(x,\tau)p_+(0,0)\rangle\rangle - \langle\langle p_+(x,\tau)p_+(0,0)\rangle\rangle_{H^x=0} = \sum_{n=0}^{\infty}\sum_{m=0}^{\infty}\frac{1}{(2n+2)!}\frac{1}{(2m)!}$$

$$\times\int_0^\beta d\tau_1\cdots\int_0^\beta d\tau_{2n+2}\int_0^\beta d\tau_1'\cdots\int_0^\beta d\tau_{2m}'\int dx_1\cdots\int dx_{2n+2}\int dx_1'\cdots\int dx_{2m}'$$

$$\times\left(\frac{H^x}{\pi\alpha}\right)^{2n+2}\frac{1}{(\pi\alpha)^{2m}}\langle H^z(x_1')\cdots H^z(x_{2m}')\rangle_{ens}$$

$$\times\langle T_\tau p_+(x,\tau)p_+(0,0)\cos\theta_-(x_1,\tau_1)\cdots\cos\theta_-(x_{2n+2},\tau_{2n+2})$$

$$\times\cos[\theta_+(x_1',\tau_1')+2k_F x_1']\cdots\cos[\theta_+(x_{2m}',\tau_{2m}')+2k_F x_{2m}']\rangle_{0C}, \tag{14b}$$

where $<\ >_{0C}$ and $<\ >_{ens.}$ mean the cumulant averaged over H_0 and the ensemble average, respectively.

Now we carry out the power counting of the integrand of each term. When x, x_i, x_j and t, t_i, t_j are of the order of $\xi(>>a)$, the asymptotic form of the (n,m)-term is $\sigma^m(H^x)^{2n+1}\xi^{-m\eta-(n+1)/\eta}$ in Eq.(14a) and $\sigma^m(H^x)^{2n+2}\xi^{-2-m\eta-(n+1)/\eta}$ in Eq.(14b), respectively. The contribution to $<<S^x>>$ contains (4n+3m)-integrations up to $\xi\sim\beta$ while that to R contains (4n+3m+2)-integrations. Now we discuss the asymptotic behaviour in the limit of large β (low temperature).

When $1/4<\eta<3$, both H^x and $H^z(x)$ are so-called relevant, and the physical quantities remain finite in the "scaling limit" , where β goes to infinity and H^x and σ goes to zero with the products $(H^x)^2\beta^{4-1/\eta}$ and $\sigma^2\beta^{3-\eta}$ being finite. Even for finite but large values of β, and small values of H^x and σ, we expect approximately the following scaling relations.

$$\langle\langle S^x\rangle\rangle = \sum_{n=0}^{\infty}\sum_{m=0}^{\infty} a_{n,m}(H^x)^{2n+1}\sigma^{2m}\beta^{2-1/\eta+n(4-1/\eta)+m(3-\eta)}$$

$$= H^x\beta^{2-1/\eta}\tilde{f}\left((H^x)^2\beta^{4-1/\eta},\sigma^2\beta^{3-\eta}\right)$$

$$= (H^x)^{1/(4\eta-1)}f\left((H^x)^2\beta^{4-1/\eta},\sigma^2\beta^{3-\eta}\right), \tag{15a}$$

$$R = \sum_{n=0}^{\infty}\sum_{m=0}^{\infty} b_{n,m}(H^x)^{2n+2}\sigma^{2m}\beta^{(n+1)(4-1/\eta)+m(3-\eta)}$$

$$= g\Big((H^x)^2\beta^{4-1/\eta}, \sigma^2\beta^{3-\eta}\Big), \tag{15b}$$

where $a_{n,m}$ and $b_{n,m}$ are the expansion coefficients. These relations will be interpreted as follows. We have three characteristic length in the present problem. The first is the thermal cut-off length $\xi_\beta \sim \beta$. The second is the Fukuyama-Lee length ξ_σ, over which the correlation of the charge density phase θ_+ decays exponentially. It is proportional to $\sigma^{2/(\eta-3)}$. The last one is the inverse ξ_S of the gap in the pure superconductor, which is proportional to $(H^x)^{2\eta/(1-4\eta)}$. The radius of the Cooper pair and the lattice constant do not appear in the continuum limit discussed in this section. The physical quantities are determined by the dimensionless ratios of the above three length scales, and the scaling relations Eqs.(15) are rewritten as follows.

$$\langle\langle S^x\rangle\rangle = (H^x)^{1/(4\eta-1)}\bar{f}\Big(\frac{\xi_S}{\xi_\beta}, \frac{\xi_\sigma}{\xi_\beta}\Big), \tag{16a}$$

$$R = \bar{g}\Big(\frac{\xi_S}{\xi_\beta}, \frac{\xi_\sigma}{\xi_\beta}\Big). \tag{16b}$$

In the limit of zero temperature, ξ_β should cancel in Eqs.(16), and we obtain the following.

$$\langle\langle S^x\rangle\rangle = (H^x)^{1/(4\eta-1)}\hat{f}\Big(\frac{\xi_S}{\xi_\sigma}\Big), \tag{17a}$$

$$R = \hat{g}\Big(\frac{\xi_S}{\xi_\sigma}\Big), \tag{17b}$$

These relations are translated into the three dimensional(3D) system by solving the mean field equation Eq.(5). Eliminating the interchain Josephson coupling constant $J_\perp$, we obtain the following conclusions.

(i) The transition temperature as a function of the randomness is expressed by

$$T_c(\sigma) = T_{c0} \times h\Big(\frac{\sigma}{T_{c0}^{(3-\eta)/2}}\Big), \tag{18}$$

where T_{c0} is the transition temperature in the pure system. The scaling function h depends only on η, and $h(0)=1$.

(ii) At zero temperature, the order parameter $\langle\langle S^x\rangle\rangle$ and the rigidity are expressed as function of the randomness σ by

$$\langle\langle S^x\rangle\rangle(\sigma) = \langle\langle S^x\rangle\rangle_0 \times \psi\Big(\frac{\sigma}{T_{c0}^{(3-\eta)/2}}\Big), \tag{19a}$$

$$R(\sigma) = \varphi\Big(\frac{\sigma}{T_{c0}^{(3-\eta)/2}}\Big), \tag{19b}$$

where ψ and φ are the universal scaling functions which depend only on η.

SIMULATION

We used the quantum transfer matrix method proceeding along the spatial direction based on the checker board breakups. The details of the method are found in the literatures[5,6,7]. This method is advantageous in studying the random systems due to the following reasons. [1] There is no statistical errors in contrast to the Monte Carlo method, which enables the numerical differentiation. [2] The theoretical basis of the extrapolation with respect to $1/N_T$ (N_T is the number of the breakups along the Trotter axis) is established. [3] The lattice size L can be easily extended with the CPU-time proportional to L. This is advantageous for the study of the random systems where the large sized system is preferable. (Self-averaging property is assumed.) [4] All the metastable configurations are taken into account exactly.

The lattice size L is taken to be 100 and the extrapolation has been performed using the values with N_T=5,6,7, and 8. We checked that the extrapolation works even at the lowest temperature (β=20.0) by comparing with the result obtained using the data with N_T=4,5,6, and 7. All the simulations in this paper concern the case of XY-model (Δ=0.0 and η=2.0). The order parameter $\langle\langle S^x \rangle\rangle$ and the rigidity R are calculated by the numerical differentiations of the free energy F with respect to the ordering field H^x and the vector potential A, respectively.

Configurations in the Presence of the Random Potentials

We show the expectation values $\langle S^x_\ell \rangle$ and $\langle S^z_\ell \rangle$ around a single impurity in Fig.1. The field H^z_ℓ along the z axis at the impurity site is 240 times larger than the uniform field H^x along the x axis. The expectation value of the bipolaron ocupation number $\langle S^x_\ell \rangle$+1/2 is almost saturated ($\sim$0.9) at the impurity site, and shows the oscillatory behaviour around it. This is due to the tendency toward the CDW ordering. Compared to the drastic change in $\langle S^z_\ell \rangle$, $\langle S^x_\ell \rangle$ is little affected by the potential.

There was a proposal of the 2π-phase soliton formation around the impurity site. The formation of soliton, if it occurs, should be detected by the negative value of $\langle S^x_\ell \rangle$ at the impurity site, because the 2π-rotation of the phase θ_- should results in the change in the sign of $\langle S^x_\ell \rangle$. $\langle S^y_\ell \rangle$ is always zero due to the symmetry between S^y and $-S^y$. We investigated the case of extremely large H^z and extremely small H^x, but we could not observe the negative value of $\langle S^x_\ell \rangle$. This means that the expectation value of the phase $\langle \theta_-(x) \rangle$ is zero, and the soliton formation does not occur.

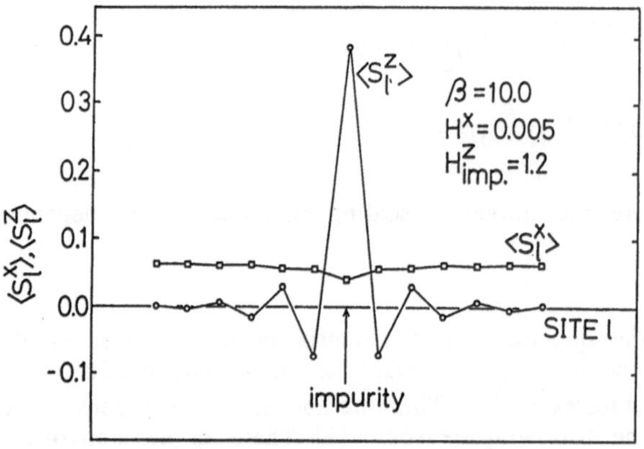

Fig. 1. The effect of a single impurity on the charge density $\langle S^z_\ell \rangle$+1/2 and the order parameter $\langle S^x_\ell \rangle$.

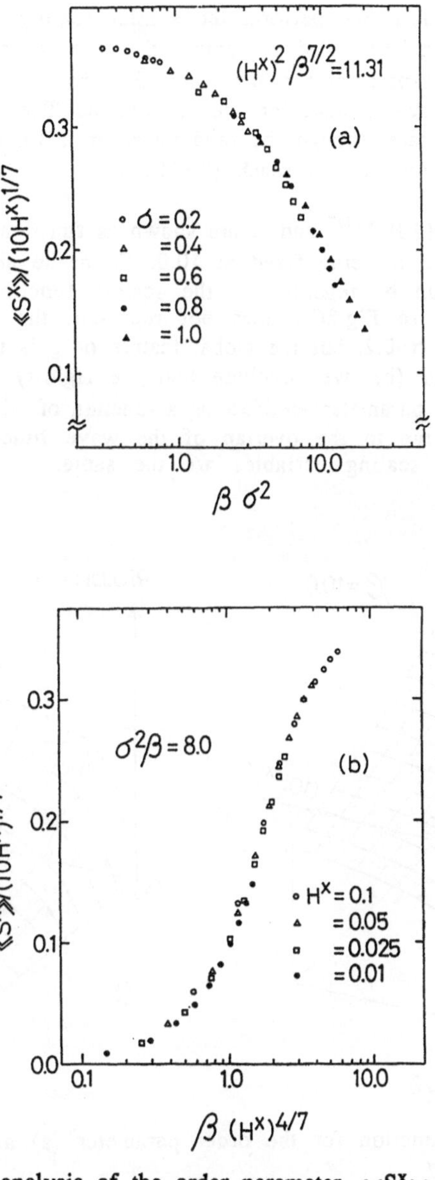

Fig. 2. Scaling analysis of the order parameter $<<S^x>>$ with $(H^x)^2\beta^{7/2}$ being fixed (a) and with $\sigma^2\beta$ being fixed (b).

Scaling Laws for the Order Parameter and the Rigidity

As we discussed in the previous section, we expect the two parameter scaling laws for the order parameter and the rigidity. Then, we fix one of the scaling arguments in Eqs.(15), and examine the scaling relation with respect to the other argument. In Fig.2(a), we plot $<<S^x>>/(10H^x)^{1/7}$ as function of $\sigma^2\beta$ for several values of σ with $(H^x)^2\beta^{7/2}$ being fixed. A single smooth curve is obtained, which shows the scaling relation. We perform the similar scaling analysis for $<<S^x>>$ with $\sigma^2\beta$ being fixed in Fig.2(b). A single universal curve is again obtained. These results confirm the two-parameter scaling law Eq.(15a) with $\eta=2.0$. Similar analysis has been performed also for the rigidity R. The scaling relation Eq.(15b) is obtaied also in this case, though the randomness σ is limited to smal values $\sigma<0.2$ in the temperature region we can reach ($\beta<20.0$).

In Figs.3, $<<S^x>>/(10H^x)^{1/7}$ and R are shown as functions of $(10H^x)^2$ and σ^2 with the inverse temperature β being fixed at 10.0. From the above discussion, the function in Fig.3(a) can be regarded as the scaling function f in Eq.(15a). Strictly speaking, the function in Fig.3(b) does not represent the scaling function g in Eq.(15b) in the region $\sigma>0.2$, but the global feature of g is the same as it. Comparing Fig.3(a) and (b), we conclude that the rigidity R decreases more rapidly than the order parameter $<<S^x>>$ as a function of σ. This is because the rigidity is more sensitive to the overlap of the wave functions than the order parameter, though the scaling variables are the same.

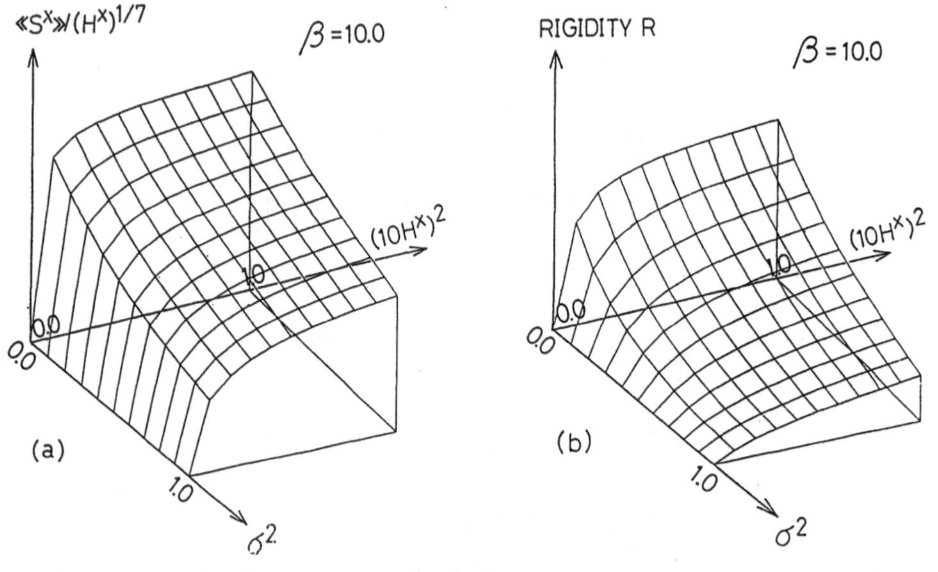

Fig. 3. Scaling function for the order parameter (a) and the rigidity (b), respectively.

In conclusion, we have studied the quasi-one dimensional disordered bipolaronic superconductor. The effective Hamiltonian is that of the S=1/2 XXZ spin chain with the random fields. This model has been investigated by the analytic as well as the numerical methods. Both methods show the two-parameter scaling laws for the order parameter and the rigidity. These scaling relations are interpreted in terms of the thermal cut-off length, Fukuyama-Lee length and the gap in the superconductor.

REFERENCES

1. S. Maekawa and H. Fukuyama, J. Phys. Soc. Jpn. 50, 2516 (1981).
2. M. Ma and P. A. Lee, Phys. Rev. B32, 5658 (1985).
3. M. ma, B. I. Halperin and P. A. Lee, Phys. Rev. B34, 3136 (1986)
4. The results of this paper will be published in
N. Nagaosa, to appear in Phys. Rev. B39, (1989).
5. M. Suzuki, Phys. Rev. B31, 2957 (1985).
6. H. Betsuyaku, Prog. Theor. Phys. 73, 319 (1985).
7. S. Takada and K. Kubo, J. Phys. Soc. Jpn. 55, 1671 (1986).
8. V. J. Emery, Phys. Rev. B14, 2989 (1976).

1/n EXPANSION FOR QUANTUM HEISENBERG ANTIFERROMAGNETS

Gregory C. Psaltakis

Research Center of Crete and University of Crete
Physics Department
Heraklion, Crete GR-71110, Greece

Following the work of Anderson[1], the Hubbard model with large on-site Coulomb repulsion has received considerable attention as a model for the Mott-insulating and high-T_c superconducting phases of the copper oxides.

At the insulating half-filled band limit the model is described by the quantum antiferromagnetic Heisenberg Hamiltonian

$$H = J \sum_{<i,j>} (\vec{S}_i \cdot \vec{S}_j - 1/4) \tag{1}$$

with the spin-1/2 operators expressed in terms of the physical electron operators

$$\vec{S}_i = (1/2)\, c^*_{i\sigma} \; (\vec{\sigma})_{\sigma,\sigma'}\, c_{i\sigma'} \tag{2}$$

and the Mott-Hubbard local constraint of one electron per lattice site $i : c^*_{i\sigma} c_{i\sigma} = 1$ (repeated spin indices such as σ,σ' are summed over the two spin components). Model (1) has been solved exactly by a Bethe ansatz[2] only in one-dimension where rigorous analysis of the low-lying excitation spectrum by Faddeev and Takhtajan (FT)[3] has established its structure as a Fermi liquid continuum formed by spin - 1/2 fermions, or kinks, excited in pairs. On the other hand Anderson and coworkers[4] have developed a mean-field theory for (1), based on the "resonating valence bond" concept, which also leads to a description in terms of a Fermi liquid of neutral fermions, or spinons. Further progress in this approach, however, has been hampered by difficulties in treating the local constraint analytically.

In this paper we derive a perturbation theory that systematizes the quasiparticle Fermi liquid approach for model (1) treating consistently the local constraint. To this end we employ a 1/n expansion technique developed by Papanicolaou and the author[5,6,7] for tight-binding electron systems. The present expansion is built on a translation invariant singlet ground state and therefore differs radically from the usual 1/S expansion (S=1/2) which is based on a broken symmetry Néel state. Our approach should be appropriate for low-dimensional antiferromagnets dominated by strong quantum fluctuations.

Interacting Electrons in Reduced Dimensions
Edited by D. Baeriswyl and D.K. Campbell
Plenum Press, New York

Restricting ourselves in one-dimension we start by rewriting (1) in terms of electron operators as

$$H = -(J/2) \sum_{<i,j>} c^*_{i\sigma} c_{j\sigma} c^*_{j\sigma'} c_{i\sigma'}$$

where following similar work of Affleck and Marston[8] the repeated spin indices are summed now over n components and the local constraint is generalized to : $c^*_{i\sigma} c_{i\sigma} = n/2$. Assuming a 2K-site lattice and a half-filled band, i.e., a total of nK electrons, we introduce a Fourier transform

$$c_{m\sigma} = \frac{1}{\sqrt{2K}} \sum_p \exp(imp/2) \left((-1)^{m+1} A^\sigma_p + B^{\sigma*}_p\right) \tag{3}$$

where m = 1,2,3,....,2K denotes the lattice sites and p is summed over the K wavevectors of the sublattice Brillouin zone. Having expressed H in terms of the mutually anticommuting fermion operators A^σ_p and B^σ_p the bilocal operators are bosonized by means of a generalized Holstein-Primakoff realization[5,6,7]

$$A^{\sigma*}_p A^\sigma_q = \sum_k \xi^*_{p_k} \xi_{qk} \quad , \quad A^{\sigma*}_p B^{\sigma*}_q = \sum_k \xi^*_{p_k} R_{qk}$$

$$B^\sigma_p A^\sigma_q = \sum_k R_{kp} \xi_{qk} \quad , \quad B^\sigma_p B^{\sigma*}_q = n\delta_{pq} - \sum_k \xi^*_{kq} \xi_{kp}$$

$$R = (R_{pq}) = (nI - \xi^\dagger \xi)^{1/2} \tag{4}$$

where the entries of the matrix $\xi = (\xi_{pq})$ satisfy the Bose commutation relations $[\xi_{pq}, \xi^*_{p'q'}] = \delta_{pp'}\delta_{qq'}$. Realization (4) yields a restriction of the Hamiltonian to the sector consisting of states with nK electrons that are a singlet under SU(n) transformations. The square root in R has a simple Taylor expansion for large n leading to a corresponding 1/n expansion of the Heisenberg Hamiltonian. Introducing the rescaled exchange constant $\tilde{J} = nJ$ the expansion of H takes the form

$$H = nKE_0 + H_0 + \frac{1}{\sqrt{n}} H_1 + \ldots$$

For the computation of the first few terms, including H_1, it is sufficient to approximate the square root by $R_{pq} = \sqrt{n}\,\delta_{pq}$. The constant E_0 given by

$$E_0 = -C^2 \tilde{J} \quad , \quad C = \frac{1}{2K} \sum_p \cos(p/2) \tag{5}$$

is the large-n approximation to the ground state energy ($C = 1/\pi$, for an infinite lattice). In this limit, where all fluctuations are ignored, the ground state reduces to the unperturbed Fermi sea satisfying trivially the local constraint as operator $c^*_{m\sigma} c_{m\sigma}$ becomes a c-number equal to n/2. However, strong renormalization effects are present already in the harmonic term H_0 given by

$$H_0 = \sum_{p,q} (\Omega_p + \Omega_q)\xi^*_{pq}\xi_{pq}$$

$$+ \frac{\bar{J}}{4K} \sum_{pp'qq'} \delta(p+q-p'-q')\cos(\frac{p-q'}{2}) (\xi^*_{pp'} - \xi_{p'p}) (\xi^*_{qq'} - \xi_{q'q}) \quad (6)$$

where $\Omega_p = C\,\bar{J}\cos(p/2)$. Using the corresponding large-n expansion of the local number operator $c^*_{m\sigma}c_{m\sigma}$ one can easily verify that

$$[c^*_{m\sigma}c_{m\sigma}, H] = 0 \quad (7)$$

to within harmonic terms. We have also estalished the validity of (7) when leading anharmonic terms are included in H and $c^*_{m\sigma}c_{m\sigma}$. The vanishing commutator (7) implies that eigenstates of the Hamiltonian will also be eigenstates of the local number operator. In the harmonic approximation we have diagonalized the quadratic form H_0 expicitly and verified that the eigenvalues of $c^*_{m\sigma}c_{m\sigma}$ for the resulting ground state and normal modes are in fact all equal to n/2. Therefore fluctuations described by (6) preserve the validity of the local constraint also in this order of the 1/n expansion. This result is implied essentially by the translational invariance of H_0 and the fact that the total number of electrons is fixed to nK in our formalism.

The final eigenvalue equation for (6) reads

$$1 = \frac{\bar{J}}{K} \sum_{p,q} \delta(2\ell-p+q-2\pi\,\mathrm{sgn}(\ell))\cos^2(\frac{p+q}{4}) \frac{\Omega_p + \Omega_q}{(\Omega_p + \Omega_q)^2 - E^2} \quad (8)$$

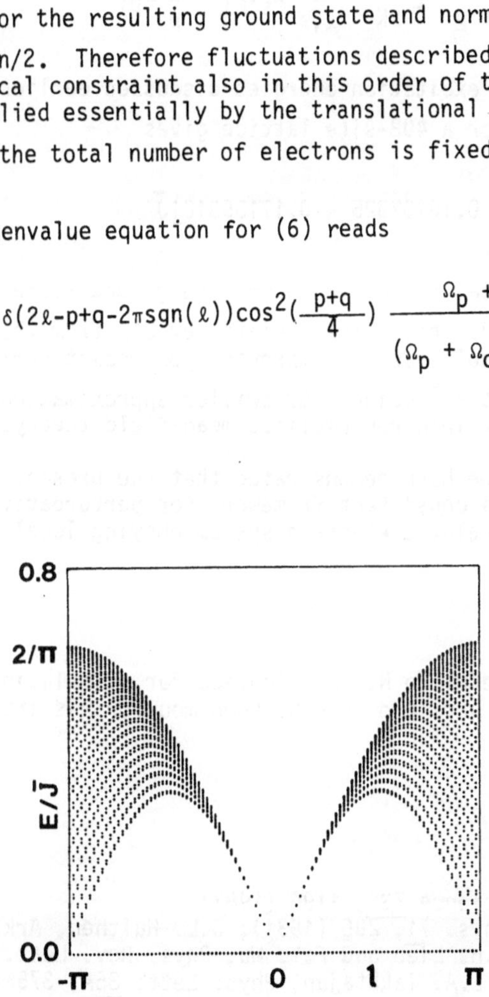

Fig.1. Excitation spectrum for a 102-site lattice. ℓ is the lattice crystal momentum. Note the presence of the zero-energy collective mode throughout the lattice Brillouin zone.

The roots of this algebraic equation can be labeled as $E = E_s(\ell)$, where ℓ is the crystal momentum that runs over 2K-1 values of the lattice Brillouin zone (the value $\ell=0$ is excluded as a limit point), and can be calculated by reducing the problem to the diagonalization of a suitable real symmetric matrix. Figure 1 shows results for the excitation spectrum consisting of a gapless continuum and a featureless zero-energy collective mode extending over the whole Brillouin zone. The renormalized quasiparticles of the continuum are identified with Anderson's[4] neutral fermions, or spinons, while its characteristic Fermi liquid shape is similar to the FT Bethe ansatz results[3]. On the other hand, the bosonic zero-energy collective mode which is tied up to the local conservation law (7) may be considered as a degenerate limit at half-filling of Anderson's[4] holon excitation.

The large-n result of (5) for the ground state energy can be improved by including the zero-point fluctuations of the normal modes[6,7]:

$$\frac{E_{gr}}{2K} = \frac{n}{2} E_0 + \frac{1}{2} \left[\frac{1}{2K} \sum_{\ell,s} E_s(\ell) - \sum_p \Omega_p \right] \tag{9}$$

where $E_s(\ell)$ are the excitation energies discussed earlier. Numerical evaluation of (9) for a 402-site lattice gives

$$\frac{E_{gr}}{2K} = - \left[\left(\frac{n}{2}\right) 0.10132325 + 0.17155310 \right] \bar{J}$$

Hence for the physical flavor value n=2 the ground state energy per lattice site is -0.5457527 J. This value deviates by 21% from the exact result[2] $-(\ell n 2)J$. Yet our result for E_{gr} improves upon recent corresponding result obtained by Zhang et al[9] using a Gutzwiller approximation scheme to treat the local constraint in a renormalized mean-field theory.

In conclusion we have demonstrated that the present 1/n expansion technique provides a consistent framework for perturbative studies of model (1) as strongly correlated electron system obeying local constraints.

ACKNOWLEDGEMENTS

I am very grateful to N. Papanicolaou for stimulating discussions on all aspects of this work and to E.N. Economou for his interest and encouragement.

REFERENCES

1. P.W. Anderson, Science 235, 1196 (1987).
2. H.A. Bethe, Z. Phys. 71, 205 (1931); J.L. Hulthén, Ark. Mat. Astr. Fys. 26A, 1 (1938); E.H. Lieb and F.Y. Wu, Phys. Rev. Lett. 20, 1445 (1968).
3. L.D. Faddeev and L.A. Takhtajan, Phys. Lett. 85A, 375 (1981).
4. P.W. Anderson in "Frontiers and Borderlines in Many Particle Physics", Course C4, edited by R. Broglia and R. Schrieffer (North-Holland, Amsterdam, 1988). This article contains a thorough review of the work of the Princeton group.

5. L.R. Mead and N. Papanicolaou, Phys. Rev. B28, 1633 (1983).
6. G.C. Psaltakis and N. Papanicolaou, Solid State Commun. 66, 87 (1988).
7. G.C. Psaltakis, Phys. Rev. B (1988) to be published.
8. I. Affleck and J.B. Marston, Phys. Rev. B37, 3774 (1988).
9. F.C. Zhang, C. Gros, T.M. Rice and H. Shiba, Supercond. Sci. Technol. 1, 36 (1988).

FRACTIONAL QUANTUM HALL EFFECT: RECENT DEVELOPMENTS

R. Morf

Paul Scherrer Institut
c/o Laboratories RCA Ltd.
Badenerstrasse 569
CH 8048 Zürich, Switzerland

The recent discovery of a Hall plateau at $\nu = \frac{5}{2}$ is reviewed. For a trial wave function recently proposed for this state by Haldane and Rezayi, we present results of Monte Carlo calculations with up to 32 electrons on the surface of a sphere. The Coulomb energy of this trial state is significantly higher than the energy of polarized states at this filling fraction. We also discuss the hierarchical construction of higher order quantized Hall states and evidence in its favor from exact numerical studies of small systems.

1 Introduction

It is by now widely accepted, that the basic physical principles behind the formation of the steps of the Hall conductivity σ_{xy} at rational multiples of e^2/h are well captured by Laughlin's wavefunctions for the ground state at filling fraction $\nu = 1/m$ and its generalizations for excited states which exhibit charged excitations with charge e/m [1], [2],[3]. Denoting the areal electron density by n_S and the magnetic field perpendicular to the two-dimensional electron system by B_0, the Landau level filling fraction ν is defined in terms of the magnetic length $l_0 = (\hbar c/eB_0)^{\frac{1}{2}}$, by $\nu = 2\pi l_0^2 n_S$. While a direct experimental determination of the (fractional) charge of these excitations is still missing, there is experimental evidence that the Laughlin theory predicts the energy gap for neutral excitations quite accurately, provided that the effects of impurities are not too big, which is the case in the highest quality samples with very high electron mobility, produced by means of modulation doping [4], [5],[6].

The theoretical picture of higher order fractional quantum Hall states, e.g at $\nu = \frac{2}{5}$, $\frac{2}{7}$, $\frac{3}{7}$ or $\frac{4}{9}$ is still somewhat controversial. In particular, not everybody is convinced of the correctness and usefulness of the hierarchical construction of higher order states, first introduced by Haldane [7] and by Halperin [8]. In this picture, the $\nu = 2/5$ state is assumed to result from the condensation of the quasiparticles with charge $e/3$ into an incompressible fluid state similar to the Laughlin fluid of electrons in the parent $\nu = 1/3$ state. Invoking this principle iteratively, further daughter states at $\nu = 3/7, 4/9$ are interpreted as condensates of quasiparticle systems with charge $e/5$

Interacting Electrons in Reduced Dimensions
Edited by D. Baeriswyl and D.K. Campbell
Plenum Press, New York

and $e/7$ in the respective parent state at $\nu = 2/5$ and $3/7$. This theory makes definite predictions for the quantum numbers of ground- and excited states at these filling fractions and in their vicinity, which can be tested by means of exact diagonalizations for small systems. Also, in a very strict interpretation of this theory, one would expect that the appearance of a Hall plateau at some higher order hierarchical level would require that all corresponding parent states with their associated Hall plateaux would be present as well [9]. So far, in the lowest Landau level, no exception to this rule has been observed. Also, in the first excited Landau level, there exist experimental results [10], which indicate the existence of a $\nu = 2 + 1/3$ as well as a $\nu = 2 + 2/7$ state. The experimental results at $\nu = 19/7$ are also supported by the results of exact diagonalizations for small systems by Reynolds and d'Ambrumenil [11], who however see no indications in their theoretical results for the existence of a state at $\nu = 7/3$ in agreement with Haldane's conclusions based on his finite system calculations [9].

Exact calculations for small systems have provided a great deal of information about the structure of ground and excited states. Indeed, the discovery by Laughlin of his trial wave function was helped by his exact study of three electrons in a large magnetic field [12]. In this note, I shall discuss recent numerical studies by d'Ambrumenil and myself [13], which show that the hierarchical scheme predicts the quantum numbers for the ground and excited states at filling fractions $\nu = 2/7, 2/5, 3/7$ of finite systems precisely.

Recently, a fractional quantum Hall state at a Landau level filling fraction $\nu = \frac{5}{2}$ has been seen in an experiment [14]. It is the first even denominator fraction, for which not only minima in the longitudinal resistivity ρ_{xx} but also a plateau at a Hall conductance $\sigma_{xy} = \nu e^2/h$ has been observed. This experiment was carried out with a very high electron mobility GaAs-GaAlAs heterojunction. Further experimental studies [10]. have shown that this state disappears rather suddenly, if the magnetic field is tilted away from the direction perpendicular to the semiconductor interface, by adding a field component B_1 parallel to it and keeping the perpendicular component B_0 constant. This tilting allows to increase the Zeeman splitting without affecting the filling fraction ν. The sudden disappearance of the Hall plateau for increasing B_1 has been taken as evidence that a small Zeeman splitting is crucial for this state and thus it appears as though this state is either unpolarized or only little polarized.

An elegant suggestion for a possible trial wavefunction for this state has been proposed by Haldane and Rezayi [15]. In the same way, as the Laughlin wave function at $1/m$ is the exact ground state for a type of hard-core repulsive interaction this state is also the exact ground state for a suitable model interaction, which favors unpolarized over fully polarized states. I have investigated this wave function by means of Monte Carlo calculations and have obtained results for its energy which are significantly higher than results for the fully polarized state obtained by Fano and Ortolani [16]. In this note I shall present these results together with an analysis how properties of a given state can be mapped between different Landau levels. I also include results for an artificial Bose system at $\nu = 1/2$ in the $n = 1$ Landau level, which show that the symmetric Laughlin wave function with $m = 2$ is no good approximation to the ground state for $n = 1$ although for the lowest ($n = 0$) Landau level, it is an excellent trial state for the Bose case.

Finally, I should like to mention a recent experiment by Clark et al. [17] who have investigated the longitudinal resistivity ρ_{xx} of different fractional quantum Hall states at filling fractions $\nu = p, q$, with $q = 3, 5, 7$ and 9. They have analyzed the temperature dependence of ρ_{xx} in the middle of a Hall plateau. Fitting the data to a single exponential

$$\rho_{xx}(\beta) = \rho_{xx}^{c} \exp(-\beta\Delta),$$

the prefactor ρ_{xx}^{c}, which results from extrapolation to $\beta = T^{-1} = 0$, could be used to determine an extrapolated value of the longitudinal conductivity

$$\sigma_{xx}^{c} = \rho_{xx}^{c}/((\rho_{xx}^{c})^{2} + \rho_{xy}^{2}).$$

Using for ρ_{xy} its value at the plateau, $\rho_{xy} = h/e^{2}\nu$, they found that all their experimental results could be fitted with the relation,

$$\sigma_{xx}^{c} = c(e/q)^{2}/h,$$

i.e. completely determined by a fractional charge $e^{*} = e/q$. The constant c was found to be close to unity. The authors interpret their findings as evidence that quasiparticle excitations do indeed have fractional charge e/q, determined solely by the denominator, consistent with Laughlin's theory and its extensions by Haldane [7] and by Halperin [8]

2 The Hierarchy on the Sphere

In the study of fractional quantum Hall states, the spherical geometry has been introduced by Haldane [7]. It has been of great help in the characterization of quantized Hall states, since it is the only translation invariant isotropic finite geometry. As a result, finite size corrections turn out to be small and for isotropic ground states, they show a smooth behavior as a function of the number of electrons N. In this section, we will assume that the magnetic field is so large that the Zeeman energy dominates over the electron interaction energy and thus it suffices to consider fully polarized states only. Following Haldane [7], we put electrons on the surface of a sphere with radius R. In a monopolar magnetic field, $B = \hbar S/eR^{2}$, the single-particle states are solutions of [7]

$$H\psi = \frac{\hbar\omega_c}{2}(\mathbf{L} - S\ \Omega)^{2}\psi = E\psi, \tag{1}$$

where the angular momentum $\mathbf{L}$ is given by

$$\mathbf{L} = \hbar(\mathbf{r} \times \ \pi + S\ \Omega) \tag{2}$$

and Ω is the unit vector in the direction of the electron coordinate $\mathbf{r} = (r, \theta, \phi)$. In the gauge $\mathbf{A} = -(\hbar S/eR) \cot\theta\ \phi$ these may be characterized by the z-component of the angular momentum $\mathbf{L}$ with eigenvalues m_z and written, in the lowest ($n = 0$) Landau level by [7]

$$\phi_{m_z}^{(0)} \sim u^{S+m_z} v^{S-m_z}, \qquad m_z = -S, -S+1, \ldots, S-1, S, \tag{3}$$

where the superscript (0) refers to the $n = 0$ Landau level. The spinor variables, u_i and v_i, are defined in terms of the electron position angles θ_i and ϕ_i as

$$\begin{aligned} u &= \cos\frac{\theta}{2} e^{i\frac{\phi}{2}}, \\ v &= \sin\frac{\theta}{2} e^{-i\frac{\phi}{2}}. \end{aligned} \tag{4}$$

As noted by Haldane [7], the generators of rotations which play the role of angular momentum, can in the lowest Landau level be written as

$$\mathbf{L} = \sum_{i=1}^{N} \mathbf{L}_j, \tag{5}$$

$$L_j^0 = \frac{1}{2}\left(u_j \frac{\partial}{\partial u_j} - v_j \frac{\partial}{\partial v_j}\right), \tag{6}$$

$$L_j^+ = u_j \frac{\partial}{\partial v_j}, \tag{7}$$

$$L_j^- = v_j \frac{\partial}{\partial u_j}. \tag{8}$$

In this geometry Laughlin's wavefunction describing a system at filling fraction $\nu = 1/m$ may be written

$$\psi_m = \prod_{i<j}^{N} (u_i v_j - v_i u_j)^m, \tag{9}$$

which is a $L = 0$ state. The monopolar field strength, $N_\Phi = 2S$, for a system of N particles is given by [7]

$$N_\Phi = 2S = m(N-1). \tag{10}$$

Similarly, Laughlin's states with one quasihole or quasiparticle located at a point with spinor coordinates (u_x, v_x) are given by

$$\psi_m^{(-)} = \left(\prod_{i=1}^{N} (v_x u_i - u_x v_i)\right) \psi_m, \tag{11}$$

and

$$\psi_m^{(+)} = \left(\prod_{i=1}^{N} (v_x^* \frac{\partial}{\partial u_i} - u_x^* \frac{\partial}{\partial v_i})\right) \psi_m. \tag{12}$$

A very convincing test of Laughlin's wave function was performed by Yoshioka [18], who found that for the exact ground state of a system of 8 electrons with periodic boundary conditions, the pair correlation function is characterized by a behavior at short separation R, which up to and including $\nu = 1/3$ has essentially zero coefficients for powers R^2 and R^4, while for filling fractions $\nu > 1/3$ the coeffcient of the R^2 term becomes positive. Precisely, at $\nu = 1/3$ it shows a discontinuity in slope, thus giving rise to the nonanlytic behavior of the energy.

For filling fractions ν other than $1/m$ Haldane's hierarchical classification predicts an equivalent expression to equation (10) relating the number of particles to the monopolar magnetic field strength, N_Φ. Following Haldane [7] we write

$$N_\Phi = m(N_e - 1) + N_{1/m}^{ex} \tag{13}$$

Here N_e is the number of electrons and $N_{1/m}^{ex}$ is given by

$$N_{1/m}^{ex} = \alpha_1 N_1, \tag{14}$$

where N_1 is the number of vortices (additional or missing zeros) nucleated in the Laughlin state at $\nu = 1/m$, and $\alpha_1 = 1$ for quasiholes and $\alpha_1 = -1$ for quasiparticles.

According to eqs. (11) and (12) the angular momentum of a state with one quasihole or quasiparticle is $N_e/2$, [7] and thus there are $N_e + 1$ quasicyclotron orbits for these quasiexcitations. By analogy with the original system of electrons, (equations 13 and 14), in which the number of cyclotron orbits was $N_\Phi + 1$, one now assumes that the quasiparticles will again form a Laughlin condensate with an exponent p_1 and writes for the system of N_1 fractionally-charged excitations:

$$\begin{aligned} N_e &= p_1(N_1 - 1) + N^{ex}_{1/p_1} \\ N^{ex}_{1/p_1} &= \alpha_2 N_2. \end{aligned} \tag{15}$$

Repeating this procedure leads to the general form of equation (14) for the hierarchy in a finite-sized system:

$$N_{i-1} = p_i(N_i - 1) + N^{ex}_{1/p_i}. \tag{16}$$

The requirement of antisymmetry of the electron wave function demands m to be an odd integer, while p_i must be positive even integers. The hierarchy is terminated at level n if

$$N^{ex}_{1/p_n} = 0. \tag{17}$$

As first discovered by Haldane this leads to filling fractions ν which can best be written as continued fractions,

$$\nu = \frac{1}{m + \frac{\alpha_1}{p_1 + \frac{\alpha_2}{p_2 + \cdots}}}, \tag{18}$$

leading to all fractions $0 < \nu \leq 1$ with odd denominator, which results from the conditions m odd and p_i even [19]. As an illustration, we quote the hierarchical prediction for a quantized Hall state at $\nu = 3/7$ where we have to choose $m = 3$, $\alpha_1 = \alpha_2 = -1$, and $p_1 = p_2 = 2$. The resulting magnetic field strength is given by

$$N_\Phi(N_e; 3/7) = \frac{7}{3}N_e - 5. \tag{19}$$

The nontrivial prediction of the hierarchical scheme is the constant term $C(\nu)$ in the expression

$$N_\Phi = N_e/\nu + C(\nu), \tag{20}$$

which is here $C(\frac{3}{7}) = -5$. In the same way, filling fractions, e.g. at $\nu = 2/7$ and $2/5$, which are direct daughter states of the $\nu = 1/3$ state, formed from quasiholes and quasiparticles, respectively, are expected at $C(\frac{2}{7}) = -2$ and $C(\frac{2}{5}) = -4$.

Some of these predictions by the hierarchy have first been verified by Haldane and Rezayi by means of exact diagonalizations using systems with up to 7 electrons [20]. Following their work, further comprehensive numerical studies by Nick d'Ambrumenil and the author [13]. have confirmed this scheme in the following sense: For a sequence of states at a given filling fraction ν, the finite size effects to the energy are smooth functions of system size, provided the number of flux quanta N_Φ and the number of electrons N_e are varied according to the predictions of the hierarchy (20). A change in the number of flux quanta N_Φ by modifying $C(\nu)$ leads to ground states that have angular momentum $L > 0$ and are consequently *not* translationally invariant. These can be identified as excited states of nearby fractional quantum Hall states. From their energies and their dependence on system size the energy of the fractionally charged excitations can be determined reliably.

3 The new state at $\frac{5}{2}$

As already mentioned in the Introduction, the discovery of an even denominator fractional quantum Hall state at $\nu = \frac{5}{2}$ by Willett et al. [14] came as quite a surprise, since theoretical arguments based on Laughlin's theory [1] and its extensions by Haldane [7] and by Halperin [8] generally predict fractional quantization at odd denominators only. It was thus not clear if a completely new explanation has to be found for this state or if slight modifications of the present theory will suffice. An important additional result for this state is the sudden collapse of the excitation gap if the magnetic field is tilted, but keeping the field component perpendicular to the electron sheet constant [10]. This observation is consistent with a strong sensitivity of this state to the strength of the Zeeman splitting and suggests that reversed spins might be important in this state.

The possibility of reversed spins has first been considered by Halperin [21]. In particular, he proposed trial wave functions in the spirit of the Laughlin wave function for systems in which both spin states are equally populated. Denoting the number of electrons by $N = 2M$, his wave functions have the form:

$$\psi_{p,q} = \left(\prod_{i<j}^{M} ((u_i v_j - v_i u_j)(\tilde{u}_i \tilde{v}_j - \tilde{v}_i \tilde{u}_j)) \right)^p \left(\prod_{i,k}^{M} (\tilde{u}_i v_k - \tilde{v}_i u_k) \right)^q , \tag{21}$$

where the spinor coordinates with/without tilde refer to electrons with up/down-spin, respectively. The number of flux quanta N_Φ is thus

$$N_\phi = p(M-1) + qM = \frac{(p+q)}{2} N - p \tag{22}$$

for this state and therefore the filling fraction tends to

$$\nu = \frac{2}{p+q}. \tag{23}$$

Since this wave function must be antisymmetric with respect to interchanges of electrons with the same spin direction, exponent p must be odd. It also has to be $p \geq q$ to prevent the system from phase separating. These conditions are met in particular by an unpolarized $\nu = \frac{2}{5}$ state, with parameters $p = 3$ and $q = 2$, for which Halperin has proposed it [21], and which in the absence of the Zeeman energy is expected to have a lower energy than the fully polarized state at $\nu = \frac{2}{5}$.

Trial wave function (21) has been considered also by Haldane and Rezayi [15] as a trial state for an unpolarized $\nu = \frac{1}{2}$ state, in the $n = 1$ Landau level. There is a unique choice for the parameters p, q meeting the above conditions at $\nu = \frac{1}{2}$: $p = 3$ and $q = 1$. However, it is not an eigenstate of S^2 and does not have a definite symmetry for interchange of electrons with unlike spin. Haldane and Rezayi proposed the following modification to $\psi_{3,1}$ to cure these problems. In the lowest Landau level, it is given by [15]

$$\psi_{HR} = \psi_{3,1} \times perm \left[\frac{1}{\tilde{u}_i v_k - \tilde{v}_i u_k} \right], \tag{24}$$

where the symbol *perm* stands for the permanent of the $M \times M$ matrix in square brackets. The number of flux quanta for this state is thus lowered by one unit,

$$N_\Phi = 2N - 4 = 4(M-1), \tag{25}$$

precisely equal to a Lauglin state (9) of $M = N/2$ Bosons with exponent $m = 4$ formed from pairs of electrons with opposite spin.

It was later observed, that trial state ψ_{HR} could be written in a modified form, convenient for Monte Carlo calculations [15], as

$$\psi_{HR} = \psi_{2,2} \times \det\left[\frac{1}{(\tilde{u}_i v_k - \tilde{v}_i u_k)^2}\right], \tag{26}$$

and in this form, Haldane and Rezayi succeeded to show by induction that this state satisfies the cyclic Fock conditions of a spin singlet state. [22].

Let us now turn to another interesting property of ψ_{HR}: Within a given Landau level n, the electron interaction is fully defined by specifying the energy $V_l^{(n)}$ of two-electron states with relative angular momentum l, for $l = 0, 1, 2, \ldots, N_\Phi$. Haldane and Rezayi [15] discovered that their state ψ_{HR} is a zero energy ground state of a hypothetical 'hollow-core' interaction, defined by $V_0 = 0, V_1 = 1, V_l = 0$ for all $l > 1$, and in fact the one with smallest possible N_Φ for a given number of electrons N. Thus this is the zero energy state with maximum filling fraction for this particular interaction. This property resembles the conditions satisfied by Laughlin's wave functions ψ_m (9), which are the zero energy ground state for a 'hard-core' interaction ($V_l = 1$ for $l \leq m$ and $V_l = 0$ for $l > m$) and again the ones with smallest N_Φ. The important observation was, that the strength of V_0 is significantly reduced in the $n = 1$ compared to its value in the lowest Landau level, and this might then be the reason why such a state would occur only in the higher Landau level.

I have evaluated the Haldane-Rezayi state by the Monte Carlo method, directly computing the wave function ψ_{HR} according to equation (26). The determinant is computed by means of a LU factorization in $O(M^3)$ operations. In Figure 1, the pair correlation function $g^{(0)}(r)$ in this state is shown for a system of 24 electrons in the lowest Landau level $n = 0$ (top line). The separation r is the euclidean distance $r = 2R\sin(\Theta/2)$, measure in units of the ion disk radius R_0, defined in terms of the electron density n_S by

$$R_0 = (\pi n_S)^{-\frac{1}{2}}$$

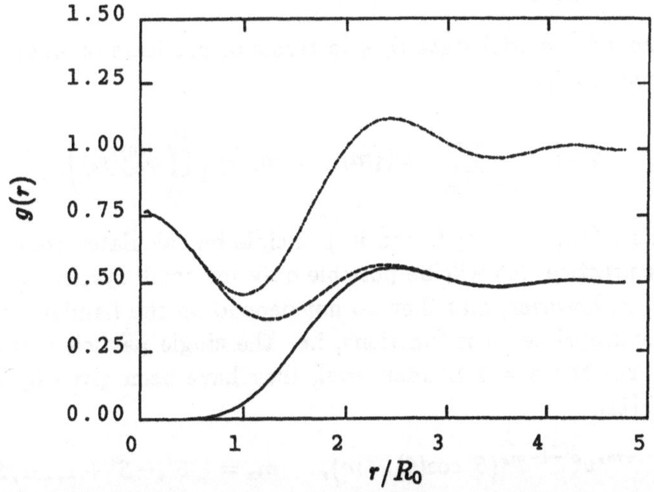

Figure 1. Pair correlation function of the Haldane-Rezayi state in the $n = 0$ Landau level. The full refers to correlation between electrons with like spin, the dashed line to opposite spin orientation. The top curve is the sum. Results are for $N = 24$ electrons.

Shown in Figure 1 are also the individual contributions to $g^{(0)}$, the pair correlation function for like-spins (full line) and for unlike-spins (dashed line). As already noted by Haldane and Rezayi, this state is characterized by the absence of a correlation hole for electrons with unlike spin. Thus the Coulomb energy of this state is rather large. Based on simulations with up to $N = 32$ electrons our result for the energy in the $n = 0$ Landau level is

$$E^{(0)}/N \approx -0.318 \pm 0.002 e^2/\epsilon l_0. \tag{27}$$

This result might be compared with the results for polarized states at $\nu = \frac{1}{2}$ by Fano and Ortolani [23]. They quote a bulk limit for $E/N \approx -0.469 \pm 0.005$, i.e. very significantly below our trial state (26). Of greater interest is of course, what happens in the $n = 1$ Landau level. A direct evaluation of wave function ψ_{HR} (eq. 26) in the $n = 1$ Landau level is at least not easy. Instead, we use a method similar to the one employed by MacDonald [24] and by MacDonald and Girvin [25], in order to derive the pair correlation function $g^{(1)}(r)$ for a given state in the $n = 1$ Landau level from $g^{(0)}(r)$ in the lowest Landau level.

We start by defining the pair correlation function $g(\Theta)$ on the sphere, as

$$g(\Theta_{12}) = c \int d\Omega_3 \dots d\Omega_N |\psi(\mathbf{r}_1, \mathbf{r}_2, \mathbf{r}_3, \dots, \mathbf{r}_N)|^2, \tag{28}$$

which for an isotropic state will only depend on the angle Θ_{12} between electrons at $\mathbf{r}_1$ and $\mathbf{r}_2$. The normalization c is chosen such that integrating $(g(\Theta) - 1)$ over the surface of the sphere yields the usual result -1. We now expand $g(\Theta)$ in terms of spherical harmonics expansion,

$$g(\Theta) = \sum_{\lambda=0}^{N_{max}} g_\lambda Y_{\lambda 0}(\Theta, \Phi). \tag{29}$$

Using the addition theorem of spherical harmonics, the coefficient g_λ can be written as

$$g_\lambda = c\sqrt{\frac{4\pi}{2\lambda+1}} \sum_{\mu=-\lambda}^{\lambda} \left(\int d\Omega_1 \dots d\Omega_N Y^*_{\lambda\mu}(\Omega_1) Y_{\lambda\mu}(\Omega_2) |\psi(\mathbf{r}_1, \mathbf{r}_2, \dots, \mathbf{r}_N)|^2 \right). \tag{30}$$

Let us now expand the trial state ψ_{HR} in terms of products of single particle basis states $\phi^{(0)}_{m_z}$ (equation 3),

$$\psi_{HR} = \sum_{\{m_1,\dots,m_N\}} w(\{m_1, \dots, m_N\}) \left(\prod_{i=1}^{N} \phi^{(0)}_{m_i}(\mathbf{r}_i) \right). \tag{31}$$

The coefficients $w(\{m_1, \dots, m_N\})$ can in principle be calculated from the definition of ψ_{HR} but in practice this will be possible only for small systems ($N < 10$). The important fact is, however, that they do not depend on the Landau level. It suffices to insert the appropriate basis functions, i.e. the single particle states of the n-th Landau level. For the $n = 1$ Landau level, they have been given by Reynolds and d'Ambrumenil [11],

$$\phi^{(1)}_{m_z} \sim u^{S'-1+m_z} v^{S'-1-m_z} (S' \cos(\theta) - m), \qquad m_z = -S', -S'+1, \dots, S'-1, S', \tag{32}$$

where the modified magnetic strength $S' = S + 1$ has been introduced. This leads to a one-to-one correspondence between $\phi^{(1)}_{m_z}$ and $\phi^{(0)}_{m_z}$ if the magnetic strength in the $n = 1$ Landau level is reduced by one unit relative to its strength at $n = 0$. In this case there is a very useful relation for matrix elements of the spherical harmonics between these single particle states,

$$\langle\phi^{(1)}_{m+\mu}|Y_{\lambda\mu}|\phi^{(1)}_{m}\rangle = N_{\Phi}(1-\frac{\lambda(\lambda+1)}{N_{\Phi}})\langle\phi^{(0)}_{m+\mu}|Y_{\lambda\mu}|\phi^{(0)}_{m}\rangle, \tag{33}$$

with $|\mu| \leq \lambda \leq N_{\Phi} = 2S$. Making use of this expression and inserting the expansion (equation 31) into equation (30), we ca derive a simple relation between the coefficients g_λ for different Landau levels:

$$g^{(1)}_{\lambda} = \left(1-\frac{\lambda(\lambda+1)}{N_{\Phi}}\right)^2 g^{(0)}_{\lambda}, \tag{34}$$

which may be inserted to compute the pair correlation function $g^{(1)}(\Theta)$ using equation 29, where the upper limit is given by $N_{max} = N_{\Phi}$.

The computation by Monte Carlo of $g^{(0)}_{\lambda}$ in the lowest Landau level poses no difficulty, either by using equation 30 or by using instead of the μ-sum the appropriate $Y_{\lambda 0}(\Theta_{12})$. Clearly, for the minimization of statistical errors all pairs of electrons i and k are used for the computation of $g^{(0)}_{\lambda}$. The statistical error in $g^{(1)}_{\lambda}$ is amplified a great deal for λ close to N_{Φ}. This leads to oscillations in $g^{(1)}(\Theta)$.

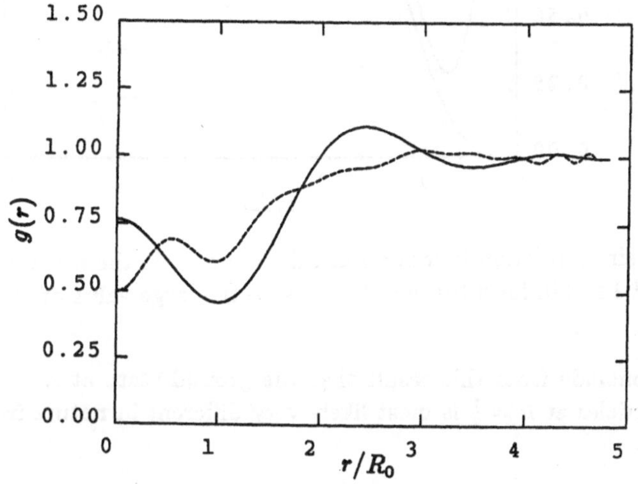

Figure 2. Pair correlation function of the Haldane-Rezayi state in the $n = 0$ (full line) and in the $n = 1$ Landau level (dashed line).

In Figure 2, I illustrate again the pair correlation function for the Haldane-Rezayi state 26, but this time in the $n = 0$ (full line) and in the $n = 1$ Landau level (dashed line). Based on calculations with between 6 and 32 electrons, I obtain a value for the Coulomb energy

$$E^{(1)}/N \approx -0.300 \pm 0.008 e^2/\epsilon l_0, n = 1. \tag{35}$$

This result must be compared with the exact result for a polarized state at $\nu = \frac{1}{2}$ in the $n = 1$ Landau level, which has been obtained by Fano and Ortolani. For $N = 12$ electrons, their value is very significantly lower [16],

$$\begin{aligned} E^{(1)}/N &= -0.3793 e^2/\epsilon l_0, \quad N = 8, \\ E^{(1)}/N &= -0.3743 e^2/\epsilon l_0, \quad N = 12, \end{aligned} \tag{36}$$

while the ground state at $N = 10$ has $L = 2$ and an energy of $E/N = -0.3752$.

Finally, I should like to present a result for the symmetric Laughlin state with $m = 2$ (equation 9) at $\nu = \frac{1}{2}$ in both the $n = 0$ and $n = 1$ Landau levels. In Figure 3, the pair correlation function is depicted for this state for $N = 24$ electrons. The full line, having a r^4 behavior as r tends to zero, represents the $n = 0$ result, while $g^{(1)}(r)$ (dashed line) has a large non-zero value at $r = 0$, which is responsible for its large Coulomb energy at $n = 1$,

$$E^{(0)}/N = -0.486 \pm 0.002e^2/\epsilon l_0, \qquad N = 24,$$
$$E^{(1)}/N = -0.250 \pm 0.005e^2/\epsilon l_0, \qquad N = 24. \qquad (37)$$

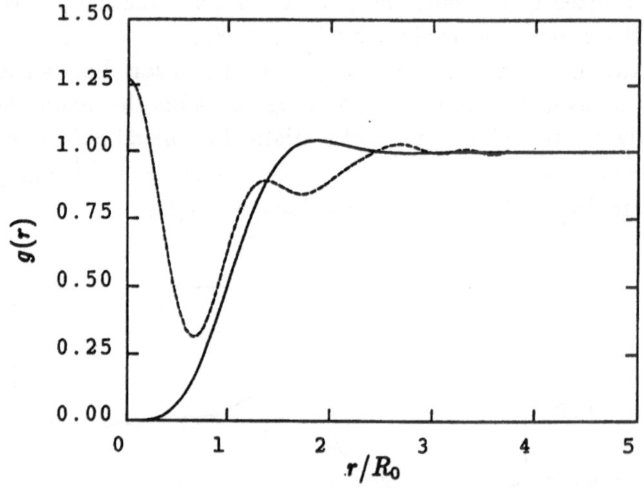

Figure 3. Pair correlation function Laughlin state ψ_2 in the $n = 0$ (full line) and in the $n = 1$ Landau level (dashed line). Note the large value of $g^{(1)}(r = 0)$.

We have to conclude from this result that the ground state at $n = 1$ for a system of charged particles at $\nu = \frac{1}{2}$ is most likely very different in nature from the ground state at $n = 0$.

ACKNOWLEDGEMENTS

I have benefitted from discussions with N. d'Ambrumenil, P. Béran, G. Fano, F. Ortolani and B.I. Halperin.

References

[1] R.B. Laughlin, Phys. Rev. Lett. **50** (1983) 1395.

[2] The Quantum Hall Effect, eds. R.E. Prange and S.M. Girvin, Springer 1987.

[3] R.B. Laughlin, in [2].

[4] R.L. Willett, H.L. Störmer, D.C. Tsui, A.C. Gossard and J.H. English, Phys. Rev. **B37**,(1988), 8476.

[5] A.M. Chang, in [2].

[6] S.M. Girvin, A.H. MacDonald and P.M. Platzman, Phys. Rev, **B33** 2481. cf. also S.M. Girvin in [2], p377.

[7] F.D.M. Haldane, Phys. Rev. Lett. **51** (1983) 605.

[8] B.I. Halperin, Phys. Rev. Lett. **52** (1984) 1583 and *ibid.* 2390(E).
For the discussion of fractional statistics of quasiparticles, see also: D. Arovas, J.R. Schrieffer and F. Wilczek, Phys. Rev. Lett. **53** (1984) 722.

[9] F.D.M. Haldane in [2] p303.

[10] J.P. Eisenstein, R. Willett, H.L. Störmer, D.C. Tsui, A.C. Gossard and J.H. English, Phys. Rev. Lett. **61**(1988), 997.

[11] A.M. Reynolds and N. d'Ambrumenil, J.Phys.**C21** (1988),119.

[12] R.B. Laughlin, Phys. Rev. **B27**(1983),3383.

[13] N. d'Ambrumenil and R. Morf, to be published in Phys. Rev. B (1989).

[14] R. Willett, J.P. Eisenstein, H.L. Störmer, D.C. Tsui, A.C. Gossard and J.H. English, Phys. Rev. Lett. **59**(1987), 1776.

[15] F.D.M. Haldane and E.H. Rezayi, Phys. Rev. Lett. **60**(1988),956 and *ibid.* **60**(1988),1886.

[16] G. Fano and F. Ortolani, private communication.

[17] R.G. Clark, J.R. Mallett, S.R. Haynes, J.J. Harris and C.T. Foxon, Phys. Rev. Lett.**60** (1988) 1747.

[18] D. Yoshioka, Phys. Rev. **B29**,(1984),6833.

[19] If a state at such fractional filling ν is a stable quantized Hall state will depend on the electron interaction as well as on the strength of impurities and cannot be generally answered without detailed microscopic calculations.

[20] F.D.M. Haldane and E.H. Rezayi, Phys. Rev. Lett. **54** (1985) 237.

[21] B.I. Halperin, Helv. Phys. Acta **56** (1983) 75.

[22] M. Hamermesh, Group Theory (Addison-Wesley, Reading, MA, 1962),p.249.

[23] G. Fano, F. Ortolani and E. Tosatti, Nuovo Cimento **9D**(1987),1337.

[24] A.H. MacDonald, Phys. Rev. **B30**(1984),3550.

[25] A.H. MacDonald and S.M. Girvin, Phys. Rev. **B33**(1986),4009.

FRÖHLICH CONDUCTION OF SPIN DENSITY WAVES

IN QUASI-TWO DIMENSIONAL SYSTEMS

Kazumi Maki and Attila Virosztek*

Department of Physics
University of Southern California
Los Angeles, CA 90089-0484

INTRODUCTION

Since the discovery of nonlinear dc conductivity in charge density waves[1] in $NbSe_3$ a large class of quasi-one dimensional compounds which exhibit the Fröhlich conduction are synthesized and studied.[2,3]

More recently we have shown theoretically[4] that spin density waves (SDW) in Bechgaard salts like $(TMTSF)_2PF_6$ and $(TMTSF)_2ClO_4$ exhibit the Fröhlich conduction. In this review we shall first summarize mean field theoretical results of the Hubbard model in a quasi-two dimensional system. Then we shall describe the transport properties of spin density waves related to the Fröhlich conduction. At this moment only two experimental reports[5,6] on possible Fröhlich conduction of SDW in (TMTSF)-salts are available. However, we believe that not only SDWs in (TMTSF)-salts but also SDWs in recently synthesized (DMET)-salts should exhibt the Fröhlich conduction, since (DMET)-salts behave similarly to (TMTSF)-salts in a number of ways (e.g. the phase-diagram, appearence of superconductivity in high pressures).

MODEL HAMILTONIAN

The electron spectrum of (TMTSF)-salts is described approximately by[7].

$$\varepsilon(\vec{p}) = -2t_a\cos ap_1 - 2t_b\cos bp_2 - 2t_c\cos cp_3 \tag{1}$$

with

$$t_a/\ t_b/\ t_c \approx 10/\ 1/\ 0.03$$

where a is the chain direction. Further in most of compounds the electron band is 3/4-filled, which implies $p_F = \frac{3}{4}\pi a^{-1}$ since the Fermi surface in these systems consists of a pair of separate sheets near $p_1 = \pm p_F$. In the following we shall neglect completely the t_c term in

* Present Address: Department of Physics, University of Virginia Charlottesville VA 22901, Permanent Address: Central Research Institute for Physics H-1525 Budapest 114., P.O. Box 49 Hungary

Interacting Electrons in Reduced Dimensions
Edited by D. Baeriswyl and D.K. Campbell
Plenum Press, New York

Eq(1), though we still assume that mean field theory is reliable, since this system is essentially three dimensional. Then appearance of a SDW rather than a CDW at low temperatures in these systems indicates that the Coulomb interaction is more important than the electron phonon interaction, though the latter may not be completely negligible. First, the appearance of the singlet superconductor under high pressure ($P \approx$ 7-10k bar) requires the electron-phonon interaction. Second, at high temperatures the electron-phonon scattering should be the dominant quasi-particle scattering mechanism. However, for simplicity we neglect the electron-phonon interaction and we approximate the Coulomb term by the Hubbard term

$$H_C = U \sum_i n_{i\uparrow} n_{i\downarrow} \tag{2}$$

where $n_{i\alpha}$ is the electron density at the site i and spin α.

Following Yamaji[7] let us consider the Hamiltonian given by

$$H = \sum_{\alpha,p} (\varepsilon(p) - \mu)\, a^+_{p\alpha} a_{p\alpha} + H_C \tag{3}$$

and solve the Hamiltonian within mean field theory.
Here

$$\mu \approx -2t_a \cos ap_F \tag{4}$$

is the chemical potential.

Assuming that the ground state develops a SDW with spin

$$\langle S_z(\vec{x})\rangle = 4\Delta\, U^{-1} \cos\left(\vec{Q}\cdot\vec{x} + \phi(\vec{x})\right) \tag{5}$$

we construct the Green's function[4] as

$$G^{-1}(\omega, \vec{p}) = i\omega_n - \eta(\vec{p}) - \xi(\vec{p})\rho_3 - \Delta\rho_1\sigma_3 \tag{6}$$

where

$$\xi(\vec{p}) = \frac{1}{2}\left[\varepsilon(\vec{p}) - \varepsilon(\vec{p} + \vec{Q})\right]$$

$$= 2t_a \sin ap_F \sin a(p_1 - p_F) - 2t_b \cos bp_2 \tag{7}$$

$$\eta(p) = \frac{1}{2}\left[\varepsilon(\vec{p}) + \varepsilon(\vec{p} + \vec{Q}) - 2\mu\right]$$

$$= -2t_a \cos ap_F \left(1 - \cos a(p_1 - p_F)\right) - \delta\mu$$

$$\approx \varepsilon_0 \cos(2bp_2) \tag{8}$$

with

$$\varepsilon_0 = -\frac{1}{4} t_b^2 \cos ap_F \left(t_a \sin^2 ap_F\right)^{-1} \tag{9}$$

and $Q = (2p_F\ ,\ \pi/b\ ,\ \pi/c)$

Here ω_n is the Matsubara frequency, Δ is the SDW order parameter and ρ_i

and σ_i are Pauli matrices operating on the spinor space consisting of the right-going electron and the left-going electron and on the ordinary spin space respectively. The self-consistent equation for $\langle S_z \rangle$ gives the gap equation

$$1 = \bar{U} T \sum_n \int d\xi \frac{1}{2\pi} \int_0^{2\pi} d\phi \ \left((\omega_n - i\eta(\phi))^2 + \xi^2 + \Delta^2\right)^{-1}$$

$$= \bar{U} T \sum_n \frac{1}{2} \int_0^{2\pi} d\phi \ \left((\omega_n - i\eta(\phi))^2 + \Delta^2\right)^{-\frac{1}{2}} \quad (10)$$

where $\bar{U} = UN_0$ and N_0 is the density of states at the Fermi surface per spin.

The frequency sum in Eq(10) has to be cut off around $\omega_n = \mu$. Here

$$\eta(\phi) = \varepsilon_0 \cos\phi \quad (11)$$

describes deviation from the prefect nesting.

As ε_0 increases the nesting between two Fermi sheets are spoiled and the SDW transition is completely suppressed[8] when $\varepsilon_0 \approx \Delta_0$ where Δ_0 is the order parameter at T = OK. The phase diagram is schematically shown in Fig. 1. If we assume that the pressure increases ε_0, the pressure dependence of the SDW transition temperature is understood semi-quantitatively[7].

Futher the present model describes Δ_0 the order parameter at T = OK independent of ε_0, which is also consistent with NMR experiments on $(TMTSF)_2PF_6$ by Kawamura et al[9].

In the presence of a magnetic field perpendicular to the ab plane the effect of the magnetic field is incorporated into the theory by replacing p_2 by p_2-eHx, where x is the space coordinate along the chain direction. We don't go into the theory [10-13] of field induced spin density waves but limit ourselves to a few comments. The theory appears to describe semi-quantitatively a cascade of SDW transitions [14-17] observed in $(TMTSF)_2ClO_4$ for H(5 ~ 15T). On the other hand the theory is unable to describe the suppression of FISDW in extremely high magnetic field [18,19] (H ~ 25T).

Further the plateaus in the Hall resistance is currently interpreted in analogy to the quantum Hall effect in a two dimensional electron system. However unlike the two dimensional system the magnetoresistance is still substantial and the Hall resistance of the last FISDW (H > 10T) is almost three times of the one before.[19]

PHASON DYNAMICS AND THRESHOLD FIELDS

The Fröhlich conduction[20] is understood as an electric conduction associated with the bodily motion of a CDW or a SDW. However as pointed out by Lee, Rice and Anderson[21] the CDW is in general pinned by impurities or by crystalline defects and the CDW does not superconduct. The relevance of the Fröhlich conduction to the nonlinear conduction observed in CDWs of $NbSe_3$ is first recognized by Bardeen[22].

Here following Fukuyama and Lee[23] and Lee and Rice[24] we shall write down the Hamiltonian[25] for phase $\phi(x,t)$ of the order parameter defined in Eq(5)*

$$H(\phi) = f \int d^P x \ \{ \tfrac{1}{4} N_0 [(\frac{\partial\phi}{\partial t})^2 + v^2 (\frac{\partial\phi}{\partial x})^2] - en\, Q^{-1} \phi\, \varepsilon \} + V_{pin}(\phi) \tag{12}$$

where $Q = 2p_F$, $V_{pin}(\phi)$ is the pinning potential given by

$$V_{pin}(\phi) = -2N_0 \lambda^{-1} \Delta(T)\, V_2 \sum_i \cos (\vec{Q} \cdot \vec{x}_i + \phi(\vec{x}_i))$$

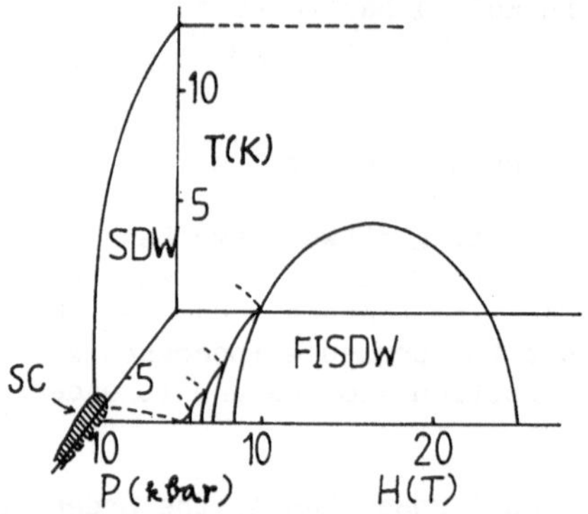

Fig. 1. The phase diagram of $(TMTSF)_2PF_6$ under pressure P and magnetic field H is sketched. At P = H = 0 $(TMTSF)_2PF_6$ undergoes a transition into the SDW around T = 13K. When pressure P is applied, the SDW transition temperature is suppressed with pressure and for P ≈ 7k bar the ground state becomes the superconductor (shaded area). When a magnetic field is applied, the magnetic field suppresses the superconductor and in higher magnetic fields the SDW is resurected (FISDW). The Phase diagram of $(TMTSF)_2C\ell O_4$ at ambient pressure is very similar to the one for $(TMTSF)_2PF_6$ at P = 7k bar.

for a CDW and

$$V_{pin}(\phi) = - (\frac{\pi}{2} N_0 V_2)^2 \Delta(T) \tanh (\frac{\Delta(T)}{2T}) \times \sum_i \cos (2\vec{Q} \cdot \vec{x}_i + 2\phi(x_i))$$

for a SDW (13)

* For a CDW $(\frac{\partial\phi}{\partial t})^2$ in Eq(12) has to be multiplied by m*/m where m* is the temperature dependent phason mass.[26]

Here V_2 is the momentum Q component of V, which gives rise to the back scattering, v is the Fermi velocity along the chain direction and f is a complicated function as defined later (see Eq (28)). In particular f takes different values depending whether $\omega/vq \to 0$ or $\to \infty$, where ω and q are the frequency and the wave vector associated with $\phi(\vec{x}, t)$. In particular we have

$$f_0 = \lim_{\omega\to 0} \lim_{q\to 0} f = n_c(T)/n$$

$$f_1 = \lim_{q\to 0} \lim_{\omega\to 0} f = \rho_s(T)/\rho \tag{14}$$

In particular $f_0 = f_1 = 1$ at T = OK, while in the vicinity of $T = T_c$, we have

$$f_0 \approx 2{,}406 \left(1 - \frac{T}{T_c}\right)^{1/2}, \; f_1 \approx 2\left(1 - \frac{T}{T_c}\right) \tag{15}$$

Further the spatio-temporal variation of $\phi(x,t)$ generates the excess electric charge and current given by

$$\rho_{CDW}(\vec{x},t) = - enf\, Q^{-1} \frac{\partial\phi}{\partial x} \tag{16}$$

and

$$J_{CDW}(\vec{x},t) = enf\, Q^{-1} \frac{\partial\phi}{\partial t} \tag{17}$$

These equations satisfy automatically the current conservation

$$\frac{\partial}{\partial x} J_{CDW} + \frac{\partial}{\partial t} \rho_{CDW} = 0 \tag{18}$$

In order to consider the threshold electric field E_T, which depins the SDW or CDW it is crucial to take the correct limiting value of f. Since $\phi(\vec{x},t)$ is dominated by the spatial variation near $E = E_T$ we have to take $f = f_1$.

After this identification we follow the classical analysis by Fukuyama, Lee and Rice[23,24] (FLR). In general it is necessary to discriminate the strong pinning limit and the weak pinning limit.[24] For a CDW in the strong pinning limit the threshold field at T = OK is given by

$$E_T^S(0) = \frac{2Q}{e\lambda} \left(\frac{n_i}{n}\right) N_0 V_2 \Delta(0) \tag{19}$$

which is proportional to n_i the impurity concentration while for $T \neq OK$ we obtain

$$\frac{E_T^S(T)}{E_T^S(0)} = e^{-\frac{T}{T_0}} \frac{\Delta(T)}{\Delta(0)} \left(\frac{\rho}{\rho_s(T)}\right) \tag{20}$$

where we include the effect of the thermal fluctuation[27] in ϕ. We note Eq(20) gives a diverging expression when T approaches T_c

$$\frac{E_T^S(T)}{E_T^S} \approx 0.868 \left(1 - \frac{T}{T_c}\right)^{-\frac{1}{2}} \tag{21}$$

Further for not too large T_0, Eq(20) predicts the minimum in $E_T^S(T)$ at

$T = T_c - \frac{1}{2}T_0$. We compare Eq(20) with an early experiment on $NbSe_3$ by Fleming.[28] As easily seen from Fig. 2, we obtain an excellent agreement with the observed $E_T(T)$ for the second CDW below $T_{c2} = 59$ K. For the present fit we need $T_0 = 14.6$K, which is again consistent with T_0 deduced from $E_T(T)$ in $NbSe_3$ in a strong transverse magnetic field (~23T) by Coleman et al.[29] On the other hand Eq(20) cannot describe the threshold field of the first CDW below $T_{c1} = 144$ K.

In the weak pinning regime on the other hand we obtain

$$E_T^W(0) \propto n_i^{2/4-D} \tag{22}$$

and

$$E_T^W(T)/E_T^W(0) = \left(E_T^S(T)/E_T^S(0)\right)^{4/4-D} \tag{23}$$

where D is the effective dimension of the CDW. Since the divergence of $E_T(T)$ at $T = T_c$ is stronger in the weak pinning limit, we obtain an excellent fit for $E_T(T)$ of the first CDW by choosing D = 2. This implies that the Fermi velocity in the third direction is extremely small compared with Fermi velocities in other directions. We note also D = 2 give $E_T^W(0) \propto n_i$, which is again consistent with the observed n_i dependence[30] of E_T.

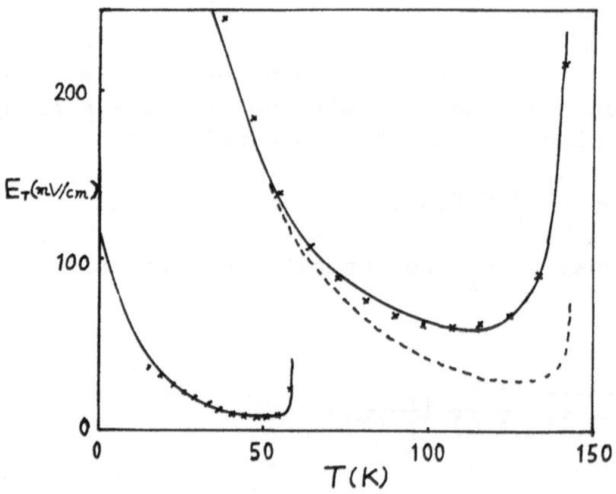

Fig. 2 The threshold electric field E_T of $NbSe_3$ are shown as function of temperature. (x) are the experimental data taken from Ref. 28, while solid curves are theoretical results; for the first CDW with $T_c = 144$K we used the D = 2 weak pinning model with $E_T(0) = 765$mVcm^{-1} and $T_0 = 60.6$K, while for the second CDW with $T_c = 59$K we used the strong pinning model with $E_T(0) = 177$mVcm^{-1} and $T_0 = 14.6$K. The broken curve is a tentative comparison of the strong pinning model with E_T for the first CDW.

We can repeat a similar analysis for a SDW. For the strong pinning regime, we have

$$E_T^S(0) = \frac{Q}{e}\left(\frac{n_i}{n}\right)(\pi N_0 V_2)^2 \Delta(0) \tag{24}$$

and

$$E_T^S(T) / E_T^S(0) = \frac{\Delta(T)}{\Delta(0)} \tanh\left(\Delta(T)/2T\right)\left(\rho/\rho_s(T)\right) \tag{25}$$

Here we neglect the effect of the thermal fluctuation of ϕ, since T_c in the SDW of organic salts is rather low ($T_c \approx 10K$) in general. We plot Eq (25) as a function of temperature in Fig. 3. Unlike $E_T(T)$ in a CDW, $E_T(T)$ in a SDW does not diverge at $T = T_c$ but depends only weakly on T. This is consistent with a recent observation of $E_T(T)$ in $(TMTSF)_2NO_3$ by Tomic et al[6].

ELECTRIC CONDUCTIVITY

The electric conductivity is expressed in terms of the current-current correlation function as

$$\sigma(\omega,q) = \langle [j, j] \rangle (\omega, q)(-i\omega)^{-1} \tag{26}$$

where ω and q are the frequency and the momentum of the electromagnetic field. Since the electric current couples to the sliding motion of a SDW, it is necessary to include the contribution from the phase fluctuation. In the collisionless limit we obtain*.

$$\sigma(\omega,q) = \frac{e^2 n}{m} \frac{i\omega\left(\omega^2 - \omega_p^2(1 - f) - \zeta^2\right)}{(\omega^2 - (1+\bar{U})\zeta^2)(\omega^2 - \omega_p^2 - \zeta^2) - \bar{U}\zeta^2\omega_p^2 f} \tag{27}$$

where

$$f = (2\Delta)^2(\zeta^2 - \omega^2)\frac{1}{2\pi}\int_0^{2\pi} d\phi \int_\Delta^\infty dz\,(z^2 - \Delta^2)^{\frac{1}{2}} \tanh\left(\frac{z - \eta}{2T}\right)\frac{N}{D} \tag{28}$$

and

$$N = (\zeta^2 - \omega^2)^2 - 4z^2(\omega^2 + \zeta^2) + 4\Delta^2 z^2$$

$$D = N^2 + 64\, z^2\omega^2\zeta^2(\Delta^2 - z^2) \tag{29}$$

and $\zeta = vq$, m is the band mass and ω_p is the pinning frequency given by

$$\omega_p^2 = 2\, n_i\, N_0^{-1}(\pi N_0 V_2)^2\, \Delta(T) \tanh\left(\frac{\Delta(T)}{2T}\right) f^{-1} \tag{30}$$

in the strong pinning regime.[26] It is perphaps of interest to note that the pinning frequency for a SDW as given in Eq (30) vanishes like $\left(1 - \frac{T}{T_c}\right)^{1/2}$ as the temperature approaches T_c, while the corresponding expression for a CDW (in the strong coupling regime) as given by

$$\omega_p^2 = 4\, n_i\, V_2 \lambda^{-1}\left(m/m^*(T)\right)\Delta(T)\, f^{-1} \tag{31}$$

*For a corresponding expression for a CDW see Maki and Virosztek.[31]

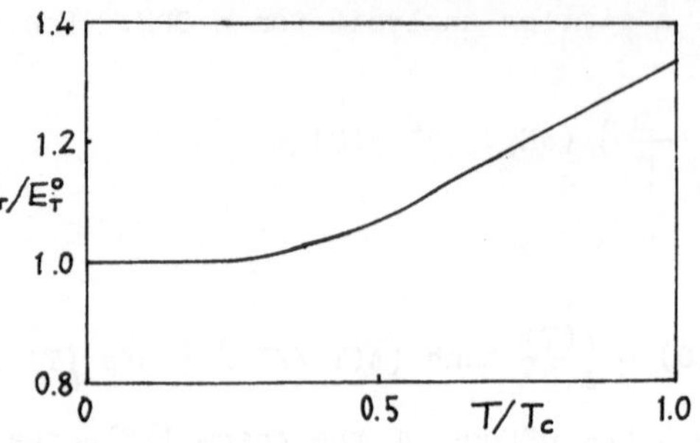

Fig. 3 The predicted temperature dependence of $E_T(T)$ of a SDW in the strong pinnning limit. $E_T(T)$ depends only weakly on T.

where

$$m^*(T)/m = 1 + \lambda^{-1} \left(2\Delta(T)/\omega_Q\right)^2 f^{-1} \tag{32}$$

diverges like $\left(1 - \frac{T}{Tc}\right]^{\frac{1}{2}}$.

We note that Eq(27) contains two poles

$$\omega^2 = \omega_p^2 + (1 + \bar{U}f)\,\zeta^2$$

and

$$\omega^2 = \left[1 + \bar{U}\,(1 - f)\right]\zeta^2 \tag{33}$$

the phason and the zero sound mode. In deriving the phason Hamiltonian (12) we neglected the coupling to the zero sounds. Although this neglect is fully justified for conducting CDWs like in $NbSe_3$, this coupling may become very important in semiconducting CDW's and SDWs. Especially at low temperatures when a CDW (or a SDW) is insulating the zero sound mode becomes a plasmon with an energy gap. Then if the spatial distorsion of ϕ is mostly transverse to the chain direction the energy gap is given by

$$\omega^2 = \omega_{p\ell}^2\,(1-f)\left(\frac{v_\perp}{v}\right)^2 \tag{34}$$

and $\omega_{p\ell}$ is the bulk plasma frequency with q parallel to the chain and $v_\perp$ is the Fermi velocity transverse to the chain direction.

The microwave conductivity is obtained from Eq (26) by putting q = 0

$$\sigma(\omega) = \frac{e^2 n}{m}\left\{\Gamma_n^{-1}\,(1-f_0) + i\omega f_0\left[\omega\,(\omega + i\Gamma_p) - \omega_p^2\right]^{-1}\right\} \tag{35}$$

where Γ_n and Γ_p are the quasi-particle and the phason relaxation rate.[26] These relaxation rates are calculated within a model where the quasi-particle scattering is mostly due to impurities. In such a model we can show

$$2\Gamma_2 \geq \Gamma_n \geq \Gamma_p \tag{36}$$

where Γ_2 is the back scattering rate (i.e. $\Gamma_{tr} = 2\Gamma_2$) and the equality holds only at $T = T_c$. Further though both Γ_n and T_p decreases with temperature, Γ_n depends only weakly on T, while Γ_p vanishes exponentially at low temperatures.[26] Especially in the vicinity of $T = T_c$, we obtain[26]

$$\Gamma_n = 2\Gamma_2 \{ 1 - 4 (1 - \frac{T}{T_c})\}$$

and

$$\Gamma = 2\Gamma_2 \{ 1 - \frac{1}{28\zeta(3)} \quad (\frac{2\pi T_c}{\Gamma_1 + {}^1/_2\, \Gamma_2})^2 (1 - \frac{T}{Tc} \quad) \} \tag{37}$$

where Γ_1 and Γ_2 are the forward and the backward scattering rate. This means that in the microwave range the SDW response is described by a damped harmonic oscillator.[32,33]

The real part of $\sigma(\omega)$ takes the maximal value when $\omega = \omega_p$

$$\sigma_{max} = \frac{e^2N}{m} \quad (\Gamma_n^{-1} \;(1 - f_0) + \Gamma_p^{-1} f_0 \;) \tag{38}$$

while the dc conductivity in the Ohmic regime is given by

$$\sigma_{dc} = \frac{e^2n}{m} \quad \Gamma_n^{-1} \;(1 - f_0) \tag{39}$$

Since the threshold electric field E_T scales with ω^2_p, it is natural to take σ_{max} as the dc conductivity of the unpinned SDW (or CDW). If these indentifications are made Eqs (38) and (39) in the gapless region agree with an earlier result by Gor'kov and Dolgov[34], though their derivation is limited in the vicinity of $T = T_c$ and further they assumed $\Gamma_1 = \Gamma_2$.

ELECTROMECHANICAL EFFECTS

In elegant experiments Brill et al.[35] and Mozurkewichr et al.[36] have shown that the elastic constant like Young's modulus depends sensitively on the electric field in CDWs of TaS_3 and $NbSe_3$. We succeeded in interpreting the anomaly in Young's modulus in TaS_3 with help of the concept of pinned portion[31] of the CDW.

Roughly speaking in a CDW both the quasi-particles and the CDW participate in the screening of the ion-ion interaction. When the CDW is unpinned, the sum of these screenings are the same as the quasi-particle screening in the normal state. Therefore when the pinning is absent, the longitudinal sound velocity is unaffected by the CDW transition.[37] However a CDW in real systems is always pinned. Then the screening is reduced below $T = T_c$, since the CDW portion is not available for the screening. This implies an increase in the sound velocity in a CDW as observed in TaS_3.

When an electric field is applied, the electric field can depin the CDW when it exceeds $E_T(T)$. Then the depinned portion of the CDW starts contributing in the screening resulting in a decrease in the elastic constant. We have extended such an analysis for a SDW.

The longitudinal sound velocity in a SDW is given by

$$C^2 = C_0^2 \quad \{1 - \lambda \;(1 - f_1 + \frac{\zeta^2}{\zeta^2 + \omega_p^{*2}} \; f_1 \;) \;\} \tag{40}$$

where

$$\omega_p^{*2} = \omega_p^{\,2}\,(f_0/f_1) \tag{41}$$

and C_0 is the bare sound velocity, λ is the electron-phonon coupling constant and the pinning frequency ω_p^2 has been already define in Eq(30). It is more useful to rewrite Eq(40) as

$$C^2 = C_0^2 \left\{ 1 - \lambda \left(1 - f_1\, P(E)\right) \right\} \tag{42}$$

where P(E) is the pinned portion. In the absence of an electric field, all the SDW is pinned and we have P(0) = 1. Then Eq(42) predict an increase in the sound velocity below the SDW transition as shown in Fig. 4. We compare the theoretical curve with an observation made in $(TMTSF)_2PF_6$ by Chaikin et al.[38] As seen from Fig. 4 we obtain a fair agreement. In the presence of an electric field, the sound velocity should decrease according to Eq(42), when the electric field exceeds E_T. Although no such an experiment is reported for a SDW, it will provide a test of generality of the concept of the pinned portion. The function P(E) may be deduced from the nonlinear conductivity, since the same concept predicts

$$\sigma\,(E) = \frac{e^2 n}{m} \left\{ \Gamma_n^{-1}\,(1-f_0) + \Gamma_p^{-1}\, f_0 \left(1 - P(E)\right) \right\} \tag{43}$$

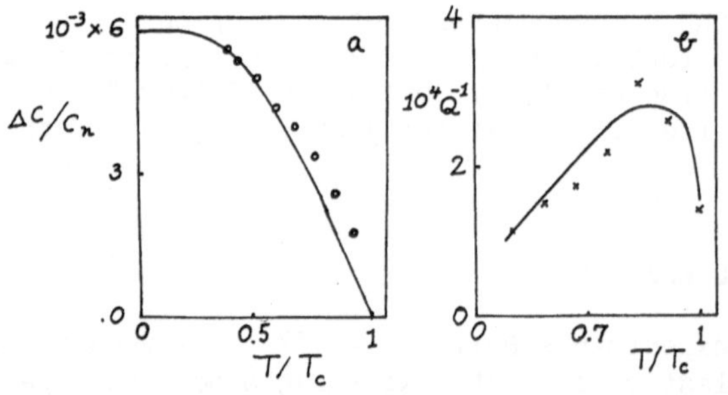

Fig. 4 The temperature dependence of the sound velocity (a) and the attenuation coefficient (b) in a pinned SDW. Circles and crosses are taken from Ref. 38. The theory describes reasonably well both the observed sound velocity and the attenuation coefficient in a SDW in $(TMTSF)_2PF_6$.

We study also the effect of the quasi-particle scattering on the sound propagation. We find the attenuation coefficient in a SDW is given by[39]

$$\alpha/\alpha_N = 2\, f(\Delta)\left(1 + \frac{\Delta(T)}{T}\,(1 - f(\Delta))\delta\right) - (1 - P^2)\,\frac{\pi\Delta^2}{4T\Gamma_2}\,\left(1-\frac{1}{8}\left(\frac{\Delta}{T}\right)^2\right) \tag{44}$$

with $\delta \approx \frac{2}{3}\,\ln\left(\frac{\Delta}{\Gamma}\right)$ for $Dq^2 >> \omega$

where $D \approx v^2\,(2\Gamma_2)^{-1}$ is the diffusion constant, $f(\Delta) = (1 + e^{\Delta/T})^{-1}$ is the Fermi distribution functon $\Delta = \Delta(T)$ is the temperature dependent order parameter. The first term in Eq (44) has a rather broad peak below $T \approx T_c$, while the second term gives a similar peak but with negative sign.

This implies that the attenuation coefficient decreases when the SDW is unpinned. This anomalous behaviour appears to be related to the presence of the diffusion pole in the sound wave propagator. On the other limit $Dq^2 \ll \omega$, the attenuation constant decreases in a SDW. Further there will be only negligible contributions from sliding SDW.

CONCLUDING REMARKS

We reviewed our theoretical results on Fröhlich conduction in SDWs in organic conductors. We believe that SDWs in organic conductors like (TMTSF)-salts and (DMET)-salts behave similarly to CDWs in quasi-one dimensional inorganic conductors. Perhaps the most important differences are the temperature dependence of $E_T(T)$ and ω_p^2. Further SDWs possess the spin wave modes, which can be detected by ESR. We predict the presence of the longitudinal resonance in addition to the ordinary transverse resonance.[40,41]

Another open problem is the topological defects in a SDW. Most common defects are phase vortices and magnetic solitons. But implication of these defects deserves further studies.

ACKNOWLEDGEMENTS

This work is supported in part by the National Science Foundation undergrant No. DMR86-11829

REFERENCES

1. P. Monceau, N.P. Ong, A.M. Portis, A. Meerschaut, J. Rouxel, Phys. Rev. Lett 37 602 (1977)
2. G. Grüner and A. Zettl, Phys. Rep. 119 117 (1985)
3. P. Monceau, in "Electronic Properties of Inorganic Quasi-One Dimensional Materials II", P. Monceau, ed. Reidel, Dordrecht (1985)
4. A. Virosztek and K. Maki, Phys. Rev. B 37 2028 (1988); K. Maki and A. Virosztek, Phys. Rev. B (submitted)
5. T. Osada, N. Miura, I. Ogura, and G. Saito, Phys. Rev. Lett. 58 1563(1987)
6. S. Tomić, J.R. Cooper, D. Jérome, and K. Bechgaard, Phys. Rev Lett. (Submitted)
7. K. Yamaji, J. Phys. Soc. Jpn 51 2787 (1982); ibid. 52 1361 (1983)
8. Y. Hasegawa and H. Fukuyamm, J. Phys. Soc. Jpn 55 3978 (1986)
9. H. Kawamura, T. Ohyama, Y. Maniwa, T. Takahashi, K. Murata and G. Saito, Jpn J. Appl Phys 26 583 (1986)
10. K. Yamaji, J. Phys. Soc. Jpn 54 1034 (1985); Synth. Metals 13 29 (1986)
11. L.P. Gor'kov and A.G. Lebed, J. Phys (Paris) 45L 433 (1984)
12. M. Héritier, G. Montambaux and P. Lederer, J. Phys (Paris) 45 L943 (1984); D. Poilblanc, M. Héritier, G. Montambaux and P. Lederer, J. Phys. C 19 L321 (1986)
13. A. Virosztek, L. Chen and K. Maki, Phys. Rev. B 34 3371 (1986)
14. J.F. Kwak, J.E. Shirber, R.L. Greene, and E.M. Engler, Phys. Rev. Lett. 46 1296 (1981)
15. K. Kajimura, H. Tokumoto, M. Tokumoto, K. Murata, T. Ukachi, H. Anzai, T. Ishiguro and G. Saito, Solid State commun 44 1573 (1982)
16. M. Ribault, J. Cooper, D. Jérome. D. Mailly, A. Moradpour, and K. Bechgaard, J. Phys (Paris) 45 L 953 (1983)
17. M.J. Naughton, J.S. Brooks, L.Y. Chiang, R.V. Chamberlin, and P.M. Chaikin, Phys. Rev. Lett. 55 969 (1985)

18. T. Osada, N. Miura, and G. Saito, Solid State Commun 60 441 (1986)
19. R. V. Chamberlin et al, Phys Rev. Lett. 60 1189 (1988); M.J. Naughton et al, Phys Rev. Lett. 61 621 (1988)
20. H. Fröhlich, Proc. Roy. Soc. A223 296 (1954)
21. P.A. Lee, T.M. Rice and P.W. Anderson, Sol. State. Commun. 14 703 (1974)
22. J. Bardeen in "Quasi-One Dimensional Conductors I", Lecture Notes in Physics Vol. 95. S. Barisic, A. Bjelis, J.R. Cooper and B. Leontic, eds. Springer-Verlag. Berlin (1979)
23. H. Fukuyama and P.A. Lee, Phys. Rev. B17 535 (1987)
24. P.A. Lee and T.M. Rice, Phys. Rev. B 19 3970 (1979)
25. K. Maki and A. Virosztek, Phys. Rev B (in press)
26. K. Maki and A. Virosztek, Phys Rev. B 39 (in press)
27. K. Maki, Phys. Rev. B 33 2852 (1986)
28. R.M. Fleming, Phys. Rev. B 22 5606 (1980)
29. R.V. Coleman, M.P. Everson, G. Eiserman, A. Johnson, Phys. Rev. B32 537 (1985)
30. M. Underweiser, M. Maki, B. Alavi and G. Grüner, Solid. State. Commun 64 181 (1987)
31. K. Maki and A. Virosztek, Phys. Rev B 36 2910 (1987)
32. M.J. Rice, S. Strassler and W.R. Schneider in "One Dimensional Conductors" H.G. Schuster, ed. Springer-verlag. Berlin 34 282 (1975)
33. G. Grüner, A. Zawadowski and P.M. Chaikin, Phys. Rev. Lett, 46 511 (1981)
34. L.P. Gor'kov and E.N. Dolgov, Zh, Eksp. Teor. Fig. 77 396 (1979); Soviet Physics JETP 50 203 (1979); J. Low Temp. Phys. 42 101 (1981)
35. J.W. Brill and W. Roark; Phys. Rev. Lett. 53 846 (1984); J.W. Brill, W. Roark and G. Minton, Phys. Rev. B 33 6831 (1986)
36. G. Mozurkewich, P.M. Chaikin, W.G. Clark and G. Grüner, Solid State Commun 56 421 (1985)
37. Y. Nakane and S. Takada, J. Phys. Soc. Jpn 54 977 (1985)
38. P.M. Chaikin, T. Tiedje and A.N. Bloch, Solid, State, Commun 41 739 (1982)
39. A. Virosztek and K. Maki, unpublished
40. K. Maki and A. Virosztak, Phys. Rev B 36 511 (1987)
41. K. Maki and A. Virosztek, Phys. Rev. B38 2691 (1988)

A UNIFIED THEORETICAL APPROACH TO SUPERCONDUCTORS WITH STRONG COULOMB CORRELATIONS: THE ORGANICS, $LiTi_2O_4$, ELECTRON- AND HOLE-DOPED COPPER OXIDES AND DOPED $BaBiO_3$

Sumit Mazumdar

Department of Physics
University of Arizona
Tucson, AZ 85721

ABSTRACT

A dopant-induced first-order valence transition is proposed in the copper oxides and $BaBiO_3$ that makes them exactly analogous to the organic superconductors and $LiTi_2O_4$, other materials with strong Coulomb correlations. The number of carriers is different from the number of chemical holes, and superconductivity occurs within a narrow carrier concentration range. The low T_c LaSrCuO is inhomogenous, and the 60 K and the 90 K YBaCuO are different materials. The discovery of the new electron-doped materials only reinforces the above conclusions. Diverse experiments in the oxides are explained, and specific testable predictions about Nd(Ce)CuO are made.

I. INTRODUCTION

In the rush to develop theoretical models for high temperature superconductors, we believe that three factors are being overlooked. First, there are several other families of superconducting materials in which Coulomb correlations are strong compared to the one-electron bandwidth, and if the electron-phonon coupled BCS mechanism or its strong coupling version did suffice until now for all the "normal" superconductors, we should perhaps similarly expect a single (though possibly different from BCS) mechanism for all such systems with strong Coulomb interactions. Second, even with the copper oxides alone, apparently conflicting experimental data (antiferromagnetism, structural distortion, charge-transfer, etc.) have led to conflicting theories, and either some of these experimental findings are incorrect, or theory should reconcile them. Third, there are numerous "chemical" anomalies associated with the normal states, which we believe would have more than just raised eyebrows under ordinary circumstances. We elaborate further on these themes below.

<u>Superconductors with Strong Coulomb Correlations</u>

Among superconductors with strong Coulomb interactions, the organics have been known for ten years.[1] A very large number of experiments indicate strong Coulomb interaction between holes.[1,2] Merely lower T_c is not a criterion for considering them different, as, (i) in 1987, a $T_c \sim 11$ K was reached[3] for $(BEDT\text{-}TTF)_2$ $[Cu(NCS)_2]$, and,

(ii) in nearly all "high T_c" theories presented until now, T_c scales as some power of the hopping integral t, which is only ~0.1 eV in the organics,[1,2] as compared to 0.5-1 eV in the copper oxides. Aside from the organics (there are about 30 of them), we would want to include $LiTi_2O_4$, T_c = 13 K, in this list. This material has long been mysterious, mostly because of a well-separated narrow d-band that lies at the Fermi surface (see below), and, again, speculations that it is "high T_c" already exist.[5] Finally, because of structural similarity to the copper oxides, as well as a countercation (Bi) with the same electronegativity as Cu, we would include $BaPb_{1-x}Bi_xO_3$[6] and $Ba_{1-x}K_xBiO_3$[7] in this list. We do not include the heavy fermion superconductors, since the behavior there is intimately related to the existence of more than one orbital (as opposed to one in all the above materials) per site. We believe that there is a "global" mechanism for SC in all Coulomb correlated superconductors (whatever the mechanism finally is), and within this, for some as-yet-unspecified but eventually understandable reason, copper oxides have the highest T_c. Theoretical challenge is not merely to find an exotic mechanism for SC, but, (a) to understand the common theme between these materials, and (b) to explain the absence of SC in related systems (see discussion on organics in Section II). Certainly a correct theory should not only explain (ideally, predict) occurrence of SC, but also its absence.

Conflicting Experiments and Theoretical Models

We limit the discussion in this subsection to the copper oxides only. Very broadly speaking, three classes of exotic pairing mechanisms have been proposed: (a) magnetic pairing, based on the antiferromagnetism (AF)[8,9] in the semiconducting La_2CuO_4 and $YBa_2Cu_3O_6$; (b) pairing driven by charge-density wave (CDW), based on a structural distortion that occurs within the orthorhombic phase at superconducting compositions;[10] and (c) pairing mediated by a charge transfer (CT) exciton.[11] There are then many variations of each theme, but what is important is that these mechanisms are all based on experimental claims and/or observations, and that such observations are seemingly conflicting. For example, AF in the semiconductor would seem to preclude CDW near this composition.[12] Probably because of this very reason, lattice distortions are not being considered seriously any longer by theorists. Similarly, a low lying $Cu^{2+}O^{1-} \leftrightarrow Cu^{3+}O^{2-}$ exciton would preclude the magnetic mehcanisms, and so on. The approach taken in the present paper is that all the observations are for real, and a single theory should explain the seemingly contradictory experiments. Clearly, this would require interpretations of the CDW and the CT absorption that are different from the standard interpretations (there are, of course, several "standard" interpretations). We will see, however, that the present interpretations also explain the related dopant concentration dependences and other anomalies.

Chemical Anomalies

There are too many, and most have not been discussed at all. One could, for example, ask the following simple questions:

(i) Considering that the dopant Sr in LaSrCuO or the chain oxygen (O_{ch}) in YBaCuO are three and four bonds away from the layer oxygen (O_ℓ), how does the charge reach the layer at all? Why should not the bridging oxygen (O_{br}) instead of O_ℓ become O^{1-} (we refer to the apical oxygen in both LaSrCuO and YBaCuO as O_{br})?

(ii) Why are there plateaus[13] in T_c and carrier concentrations in $YBa_2Cu_3O_{6+\delta}$ for $0.5 < \delta < 0.9$?

(iii) With seven oxygens in $YBa_2Cu_3O_7$, why is O_{ch} so easy to remove, particularly considering that there is more crowding in the layers?

(iv) T_c in LaSrCuO supposedly decreases for $x > 0.2$ because of oxygen vacancies. Why is T_c still only 40 K even when creation of such vacancies is prevented?[14] Why does SC disappear even here for large x? Similarly, why is T_c near 40 K for all La_2CuO_{4+y}, and there is no 10 K superconductor here?[15]

(v) Why are the recent Bi and Tℓ-based materials nearly stoichiometrically Cu^{2+}?

There are many similar chemical anomalies, but with the exception of (ii) and (v), none of them are even being asked. However, all of the above are intimately related to the question of the location and magnitude of the dopant-induced charge, which in its turn determines the normal state of these materials.

In the present paper, we deal with all of the above relevant features of the normal state. We show how there is a common feature between the organics and $LiTi_2O_4$, and then indicate how the same is attained in the copper oxides and doped $BaBiO_3$. In the process, we explain the seemingly conflicting physical data and explain the chemical anomalies. The highly nontrivial roles of the bridging, chain and layer oxygens are made clear. All these require only two realizations. First, because of their positions in the periodic table, large ionization energies of the ions favor lower charge in both Cu and Bi. Second, exactly as mixed valence is possible in wide-band systems, integral valence can occur in narrow-band ionic solids, and a first-order "jump" between different valence states occurs under appropriate conditions. In conducting (but not semiconducting) copper and bismuth oxides, we argue that the cations are Cu^{1+} and Bi^{3+}. Before discussing the current oxides, we briefly discuss the organics and $LiTi_2O_4$ to indicate that SC in correlated systems might require specific carrier concentrations, which has not been recognized until now.

Only as this manuscript was nearing completion did we learn about the new materials in which electrons are carriers. These only reinforce our conclusions. The present model can successfully explain the electron-hole symmetry and the absence of Cu^{3+} in the hole-doped materials. The new materials are discussed at the end.

II. ORGANIC SUPERCONDUCTIVITY

The reason we believe that SC in the organics gives us a hint about the SC mechanism in correlated systems is because there are so few of them, only about thirty out of hundreds (perhaps greater than 500) of nonsuperconducting conductors. It is only reasonable to assume that the correct mechanism would explain the absence of SC in the majority of the materials.

The reasons it had been difficult to arrive at a theory even for the normal state of the organic conductors were, (a) strong Coulomb correlations that made band theory inapplicable, and (b) the extreme variety in physical behavior -- even among materials that were obviously structurally related.[16-18] It is impossible to go into detail here, but solids with the same crystal structures and nearly identical chemical components exhibit very different behavior. The two commonly accepted signatures of strong Coulomb correlations are (a) enhanced static magnetic susceptibility (compared to the theoretical Pauli susceptibility), and (b) a "$4k_F$" instability, whereby a phase transition occurs at low temperatures due to the formation of density waves with periodicity $4k_F$ (where $k_F = \rho\pi/2a$, ρ the number of carriers per molecular site and a the lattice constant), instead of the normal $2k_F$. While both enhanced susceptibility and the $4k_F$ instability are common, so are unenhanced susceptibility and only a $2k_F$ instability (i.e., signatures of "weak correlation"). Following the principle that a correct theory must explain both kinds of behavior, we demonstrated the following: if in addition to the on-site Coulomb correlation (the Hubbard U) a nearest neighbor repulsion (the "extended" Hubbard parameter) is explicitly included, and if the discreteness of the lattice is maintained (i.e., no transformation to a continuum model is made), physical properties become strongly dependent on ρ. Within the single-band extended Hubbard model,

$$H = U \sum_{i} n_{i\uparrow} n_{i\downarrow} + \frac{1}{2} V \sum_{\langle ij \rangle} n_i n_j + t \sum_{\substack{\langle ij \rangle \\ \sigma}} c^{+}_{i\sigma} c_{j\sigma} \quad , \tag{1}$$

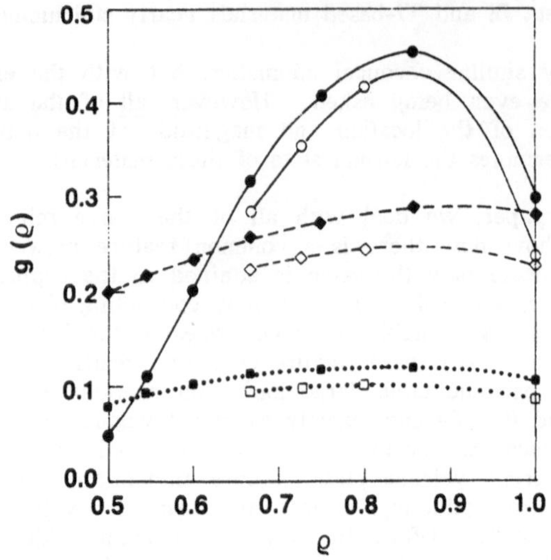

Fig. 1. The normalized probability of double occupancy $g(\rho)$ vs. ρ. The solid circles, solid diamonds and solid squares correspond to the number of electrons $N_e = 4n + 2 = 6$, while the open circles, open diamonds, and open squares correspond to $N_e = 4n = 8$. The $N_e \to \infty$ curve is bounded by $N_e = 4n + 2$ and $N_e = 4n$. Solid line: $U = 7\sqrt{2}t$, $V = 3\sqrt{2}t$; dashed line: $U = 4\sqrt{2}t$, $V = 0$; dotted line $U = 7\sqrt{2}t$, $V = 0$.

the ground state normalized probability of double occupancy $g(\rho) = \langle n_{i\uparrow}n_{i\downarrow}\rangle/\langle n_{i\uparrow}\rangle\langle n_{i\downarrow}\rangle$ is nearly independent of ρ for $V = 0$, but becomes strongly ρ-dependent for $V > 0$, as shown in Fig. 1. From Fig. 1, one would predict strongly enhanced susceptibility and $4k_F$ instability for $\rho \to 0.5$, gradual disappearance of these signatures with increasing ρ, followed by a reappearance as $\rho \to 1$ is reached.

Experimentally, this is exactly what is observed.[16-18] All materials that exhibit "weak-correlation" behavior are in the $0.6 < \rho < 0.8$ regime, and the complete family of CT solids can be described within Eq. (1) with virtually the same explicit parameters. Given the variety in crystal and molecular structures, interchain coupling and electron-molecular vibration interactions, the "fit" to the experimental data[18] is actually amazing. Given the extreme importance of ρ in the above theory, and the fact that the effect of V is most important at $\rho = 0.5$, it is absolutely intriguing that organic SC is limited strictly to $\rho = 0.5$. Of the thirty known superconductors, in four different cationic families, not one is known in which ρ is different.

Aside from $\rho = 0.5$, organic superconductors are distinquished by quasi two-dimensionality (2D), as opposed to the more common quasi one-dimensionality (1D) in the conductors. Since the question of dimensionality has been a controversial issue, we explain how we arrive at this conclusion. We do not attempt to calculate the interchain hopping integral $t_\perp$. Rather, our argument is based on the observed spatial broken symmetry as well as the lack of it. In the organics, $(TMTSF)_2X$ exhibit a spin density wave (SDW) at low temperatures[1] (which give way to SC under pressure). We have shown that in 1D, the spin-Peierls (SP) phase is unconditional for all realistic (i.e., convex) Coulomb correlations and the SDW never occurs.[19] In contrast, for $t_\perp > t_{\perp c}$, the SP phase is not possible and SDW can appear.[12] Thus the very appearance of a SDW in the TMTSF-based superconductors indicates 2D.[20]

The question often asked is why the newer family of superconductors, $(BEDT\text{-}TTF)_2X$, does not exhibit SDW. Indeed, these are much more isotropic within the 2D layer, with $t_{\|} \sim 2 - 3t_\perp$, in contrast to $t_{\|} \sim 10t_\perp$ in the TMTSF! The answer lies in the fact that while $t_\perp \neq 0$ is necessary for SDW, in $\rho = 0.5$ (as opposed to $\rho = 1$) systems a new feature appears. For $t_\perp$ greater than an upper critical value, spin fraustrations will destroy the SDW[20] in $\rho = 0.5$ within the 2D version of Eq. (1).

Recent observations of a pressure-induced SP-to-SDW transition[21] in the sulphur containing $(TMTTF)_2X$, as well as the magnetic-field-induced SDW[22] in $(TMTSF)_2X$, can all be explained within the same model. To summarize then: (i) the extended Hubbard model, Eq. (1), explains the extreme variety of behavior in organic solids in terms of the implicit parameter ρ; (ii) SC is limited to $\rho = 0.5$; (iii) the variety within the superconductors can be explained with Eq. (1), provided due consideration is given to the importance of ρ and the dimensionality. It is then natural to expect that SC also should be explained within Eq. (1), and that the 2D and $\rho \to 0.5$ are important. Aside from this, we believe that weak incommensurability effects[1] are essential. How such incommensurability arises is another matter altogether. We discuss below the case of $LiTi_2O_4$, which we show is amazingly similar to the superconducting organics. This has not been recognized until now.

III. $LiTi_2O_4$

Superconducting $LiTi_2O_4$ shares the following features with the organics: (i) $LiTi_2O_4$ also has $\rho = 0.5$, (ii) theoretical description of at least the normal state is the single-band extended Hubbard model, Eq. (1). Let us elaborate.

Ti is to the left of the periodic table, implying that true and formal valences in the Ti oxides are the same (unlike that in Cu and Bi oxides, see below). This is confirmed in recent band calculations,[23a,b] which find the Fermi energy lies in a well-separated (by 2.4 eV) Ti d-band. This implies complete CT, or $Li^{1+}(Ti^{3.5+})_2\ (O^{2-})_4$, which in its turn implies one d-electron per two Ti atoms, i.e., $\rho = 0.5$!

Now we come to the second similarity. $LiTi_2O_4$ is an inverse spinel with the Li atoms in the tetrahedral hole, and both Ti atoms occupying octahedral holes. Structurally it is related to Fe_3O_4 (magnetite), which should be written as $Fe^{3+}(Fe^{2+}Fe^{3+})O_4$, where an Fe^{3+} occupies the tetrahedral hole, and an Fe^{2+} and a second Fe^{3+} occupy the octahedral holes. This system is again $\rho = 0.5$, with an extra electron being shared by the equivalent metal sites. The material is a conductor at high temperatures, but at 115 K a sharp metal-insulator (MI) transition (the Verwey transition[24]) is observed that has been ascribed to[25] charge segregation between Fe^{2+} and Fe^{3+}. Two facts are relevant now: (i) this is exactly the same mechanism for the $4k_F$ instability in the nonsuperconducting organic conductors[17] with $\rho = 0.5$, and (ii) as with the $4k_F$ instability in the organics, the Verwey transition has been explained within Eq. (1)[25] (rather a "spinless" version of it). Therefore, the organic superconductors and $LiTi_2O_4$ are both (i) $\rho = 0.5$, (ii) related to nonsuperconducting materials that show a MI transition driven by Coulomb correlations, which (iii) is explained with the single-band extended Hubbard model. We find the similarities simply too strong to be coincidences.

One last point. Anderson has discussed how in the inverse spinel lattice one can have perfect short-range order but still finite entropy.[26] We suspect that this situation in $LiTi_2O_4$ and incommensurability in the organic superconductors, or at least the consequence thereof, may be related. We postpone further discussion until the very end, where we discuss a possible mechanism for the SC.

IV. THE COPPER OXIDES

We now come to the copper oxides, which we will attempt to show are also $\rho = 0.5$ and are described within Eq. (1) at the superconducting composition. We believe that in each copper oxide, SC occurs at a specific carrier concentration at nearly one T_c. The T_c vs. x "phase diagram" seen in $La_{2-x}Sr_xCuO_4$, which theorists are attempting to explain, is misleading -- the number of carriers is different from x. Indeed, this phase diagram is not seen in any other system, and experimentalists have claimed that the 90 K and the 60 K YBaCuO are different, a conclusion that we will agree with. We believe that the low T_c $x < 0.15$ LaSrCuO is simply inhomogenous, and only the $x > 0.15$ 40 K material is truly superconducting.

Within a CuO_2 layer, we believe that a dopant-induced valence transition, $Cu^{2+} \rightarrow Cu^{1+}$, occurs that makes half the oxygens singly charged (O^{1-}). With the Cu in the closed-shell d^{10} configuration, we now have a $\rho = 0.5$ oxygen band in which SC occurs. In the rest of the discussion in this subsection, we will indicate why this transition occurs and how this mechanism can explain the apparently conflicting physical data and the chemical anomalies mentioned in the Introduction.

Molecular CuO is a linear combination of $Cu^{1+}O^{1-}$ and $Cu^{2+}O^{2-}$ according to $\psi_{CuO} = \alpha|Cu^{1+}O^{1-}\rangle + \beta|Cu^{2+}O^{2-}\rangle$, where α and β depend on the second ionization potential (I.P.) I_d of Cu, the second electron affinity (E.A.) A_p of O and the hopping intergral t_{pd}. For $t_{pd} << I_d$, valencies are integral, as opposed to mixed. Note that I_d and A_p are "bare" quantities that are different from the "site energies" that go into the standard Hamiltonian describing a CuO_2 layer: such site energies are combinations of I_d, A_p and Madelung energy and can be small. In copper oxides, $t_{dp} \sim 0.5$-1 eV, while the atomic bare $I_d \sim 20$ eV. I_d in Cu, because of the closed shell nature of Cu^{1+}, is unusually large, so that we expect $\alpha >> \beta$. This is supported by our recent diatomic molecular calculation.[27] Therefore, the only quantity that drives the system towards Cu^{2+} is the large Madelung energy ΔE_M in a solid that is gained from the attraction between large charges. Ionicity in narrow-band ($t_{pd} << I_d$, ΔE_M) systems is given by the McConnell inequality,[28]

$$I_d + A_p - \Delta E_M \lesseqgtr 0 \quad , \tag{2}$$

where ΔE_M is the difference in the Madelung energies between the Cu^{1+} and the Cu^{2+} configuration. A large ΔE_M can compensate for large I_d, but inequality (2) also implies that for a system close to the border, a change in ΔE_M can cause a first-order transition. To put it differently, even if a large ΔE_M favors the higher ionic charges, such a system will revert to the lower charges if for any reason ΔE_M becomes smaller. This is different from mixed valence. Such a transition (a "neutral-ionic" transition) has been seen in the organic mixed-stack donor-acceptor insulators:[29] decreasing temperature or increasing pressure increases ΔE_M, and there is a resultant jump in the ionicity.[29] A similar transition occurs in LaSrCuO and YBaCuO because of the 3D nature of the Madelung energy, even though fermion hopping is limited to 2D.

We describe the Hamiltonian in terms of holes (as opposed to electrons in Eq. (2)), in accordance with the current convention. With Cu^{1+} (d^{10}) and O^{2-}(p^6) as the vacuum, the Hamiltonian is written as,

$$H = U_d \sum_i n_{i\uparrow}n_{i\downarrow} + U_p \sum_j n_{j\uparrow}n_{j\downarrow} + \epsilon_d \sum_i n_i + \epsilon_p \sum_j n_j + {}^1\!/_2\, V_{dp} \sum_{\langle ij\rangle}{}' n_i n_j$$
$$+ {}^1\!/_2\, V_{pp} \sum_{j,j'}{}' n_j n_{j'} + t_{dp} \sum_{\substack{\langle ij\rangle \\ \sigma}} (a^+_{i\sigma}c_{j\sigma} + h.c.) + t_{pp} \sum_{\substack{j,j' \\ \sigma}} c^+_{j\sigma}c_{j'\sigma} \quad . \tag{3}$$

Here $a^+_{i\sigma}$ creates d holes of spin σ on $i \in$ Cu sites, $c^+_{j\sigma}$ created p holes on $j \in$ O sites, U_d, U_p are the corresponding on-site repulsions and ϵ_d, ϵ_p the site energies. The number operators $n_{i\sigma} = a^+_{i\sigma}a_{i\sigma}$, $n_{j\sigma} = c^+_{j\sigma}c_{j\sigma}$, $n_i = \Sigma_\sigma n_{i\sigma}$, $n_j = \Sigma_\sigma n_{j\sigma}$. I_d and A_p in Eq. (2) are contained in the hole site energies ϵ_d, ϵ_p, and the Madelung energy terms involving repulsion V_{dp} and V_{pp} have been limited to nearest neighbor for convenience. The primes on the summations indicate 3D Madelung sums, even though the hopping is limited to 2D. This will be important for our discussion below.

The magnitudes of ϵ_d, ϵ_p, V_{dp}, and V_{pp} are such that they obey Eq. (2), and undoped La_2CuO_4 and $YBa_2Cu_3O_6$ are semiconductors. We discuss LaSrCuO first. With the replacement of La^{3+} with Sr^{2+}, initially O_{br} becomes charged (O^{1-}). This leads to a repulsion V_{dp} between the holes occupying O_{br} and the nearest Cu, or rather, a substantial decrease in $|\Delta E_M|$ in Eq. (2). In fact, if in Eq. (3) the Madelung energy terms are not limited to nearest neighbors, the decrease in $|\Delta E_M|$ due to substitution of La^{3+} by Sr^{2+} is even larger. For 2D transport within the CuO_2 layer, however, now some of the O_ℓ have to be charged, and there are only two routes after O_{br} has a hole in it. This same hole can migrate to the layer through CT or the Coulomb repulsion V_{dp} can "push" a hole occupying a Cu^{2+} site onto an O_ℓ. We show the two possibilities in Fig. 2. The former requires 3D t_{dp} and/or t_{pp}, in contradiction to all experiments. Therefore, O_ℓ necessarily becomes charged due to V_{dp} and the Cu involved becomes Cu^{1+} (route 2 in Fig. 3). In reality, after sufficient doping, we expect $|\Delta E_M|$ to be small enough that the left-hand side (LHS) in Eq. (2) becomes larger (as even the undoped system was close to the border, see above), and the same valence transition seen in the organics[29] will occur here. The CuO_2 layer now consists of only Cu^{1+}, and half the oxygens are O^{1-}. In the semiconductor, only the Cu orbitals were effective with the O-band empty, so that Eq. (3) reduced to a ρ = 1 Cu-based Hubbard model. Now the reverse is true, the Cu-band is empty and we have a 2D ρ = 0.5 O-band described by

$$H_O = U_p \sum_j n_{j\uparrow}n_{j\downarrow} + \frac{1}{2} V_{pp} \sum_{j,j'} n_j n_{j'} + t_j \sum_{j,j',\sigma} c^{+}_{j\sigma} c_{j'\sigma} \quad , \tag{4}$$

where j,j' are nearest neighbors in the O sublattice, and $t_j = (t^2_{dp}/\Delta E) + t_{pp}$, with ΔE the energy of the CT $Cu^{1+}O^{1-} \leftrightarrow Cu^{2+}O^{2-}$.

Before discussing YBaCuO, we point out what has been achieved. First, we have explained the role of O_{br}. Second, we have made the relationship between the organics and $LiTi_2O_4$ on the one hand, and superconducting copper oxides on the other, clear. Third, since we ascribe SC to Eq. (4) at a specific ρ, it is no longer mysterious why $La_{2-x}Sr_xCuO_4$ under oxygen pressure still shows a maximum T_c of 40 K[14] or why La_2CuO_{4+y} exhibits only this high T_c.[15] The "low T_c" $La_{2-x}Sr_xCuO_4$ at x < 0.15 is inhomogenously doped and contains superconducting islands within this scenario. Indeed, many experimentalists have suggested this, and near x = 0.1, both semiconducting and superconducting materials are known. Similarly for x much larger than 0.15-0.2, ρ is much larger than 0.5, and even though the system is conducting, it is nonsuperconducting (as with the organics). Finally, the valence transition implies a concomitant structural transition. This has been claimed in LaSrCuO, but even more clearly in $LaCuO_{4+y}$, in which distinct semiconducting La_2CuO_4 and structurally different La_2CuO_{4+y} have been seen.[15]

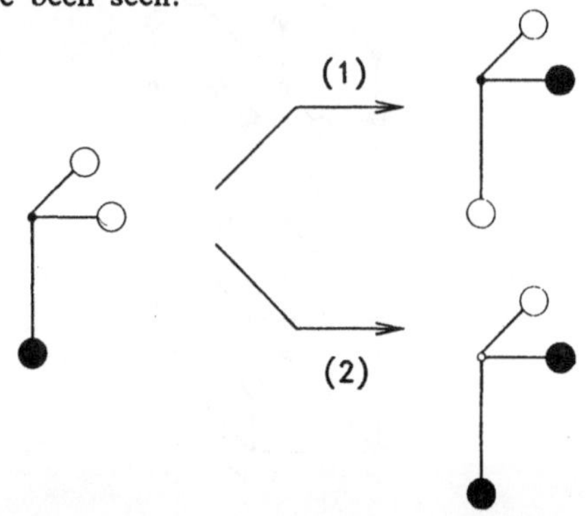

Fig. 2. Two possible mechanisms by which the layer oxygen becomes charged (see text). Here the small circle denotes Cu, the large circles O, and black and white circles are filled and empty sites, respectively.

In YBaCuO the scenario is more dramatic. Note that there are four O_ℓ, two O_{br} and one O_{ch}. Semiconducting $YBa_2Cu_3O_6$ must necessarily have charges +2, +2 and +1 for the layer and chain Cu. Introducing one O_{ch} increases the total negative charge by 2 units, and the most obvious way to balance this is to make the chain copper Cu^{3+}. Because of the large U_d, this is forbidden, and the next possibility is to charge the two O_{br} to O^{1-}. Each of these "pushes" the hole occupying the layer Cu which is bonded to it onto an O_ℓ, so that we have two layered $\rho = 0.5$ O-bands in $YBa_2Cu_3O_7$ with all Cu as Cu^{1+}. We have thus not only explained the roles of O_{br} and O_{ch}, but have also explained why O_{ch} is lost so easily in spite of greater crowding in the CuO_2 layer. In the superconducting layer described by Eq. (4), the charges on the O_ℓ are $-(1.5-\delta)$ and $-(1.5+\delta)$, respectively, while charge separation between O_{ch} (2-) and O_{br} (1-) is nearly complete. Since Cu^{1+} and O^{2-} are both closed shell, there can be no bonding between them, so that O_{ch}^{2-} is easily separated.

This mechanism explains the occurrence of a distinct 60 K material. Notice that each O_{ch}^{2-} creates holes on two bridging O on both sides of the chain. For δ increasing from 0 in $YBa_2Cu_3O_{6+\delta}$ then, there are two possibilities: holes are introduced into the two CuO_2 layers randomly or one-by-one: first one layer becomes completely Cu^{1+}, and only then does the other one begin to be affected. The former scenario requires neighboring Cu^{2+} and O^{1-}, which we believe is severely forbidden by Madelung energy considerations. In the latter case only for $\delta \geq 0.5$, one layer is completely Cu^{1+} (since here each bridging O hole pushes out one Cu hole), while only for $\delta \to 1$, we have two Cu^{1+} layers. Thus within the valence transition mechanism, the 60 K and the 90 K materials are one-layer and two-layer materials, as shown in Fig. 3.

The present scenario explains the plateaus in T_c and carrier concentration (see below), but more importantly for "doping" to occur layer-by-layer, it is essential that for large δ, especially near $\delta = 0.5$, the occupancy of O_{ch} is periodic. The particular periodicity that would be most favorable is shown in Fig. 4. Experimentally, all the three periodicities ($^1/_2$,0,0), (0,$^1/_2$,0) and ($^1/_2$,$^1/_2$,0) have been claimed,[30] although the most recent x-ray data claim to find a ($^1/_2$,0,0) periodicity only.[31] However, this requires alternation among a nearly intact and a nearly completely fragmented chain near $\delta = 0.5$, which would have a highly disruptive effect on the overall stability of the crystal, while the alternation shown in Fig. 4 does not affect this stability adversely. At the same time, we believe that it is very significant that the authors of Ref. 31 find that even for $\delta = 0.67$, a periodicity of ($^1/_2$,0,0) is "strongly preferred" over ($^1/_3$,0,0). Again, if we read this as ($^1/_2$,$^1/_2$,0), this is simply related to the fact that at $\delta = 0.67$ only one layer is completely Cu^{1+}, which requires tendency to alternate occupancy.

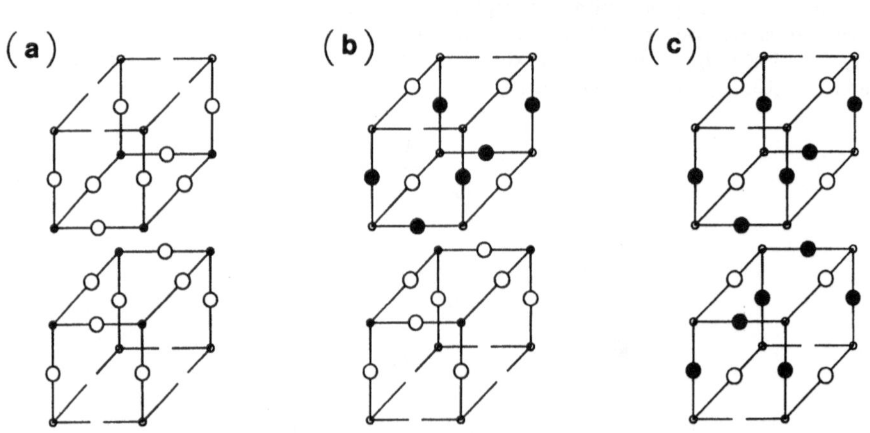

Fig. 3. Hole occupancies in $YBa_2Cu_3O_{6+\delta}$ for (a) $\delta = 0$, (b) $\delta = 0.5$, (c) $\delta = 1$. Small circles are Cu, larger ones O; black and white circles hole-occupied and empty sites. In (a) layer coppers are Cu^{2+}, in (b) one layer has Cu^{1+}, in (c) both layers have Cu^{1+}. *For oxygen occupancy of chain, see Fig. 4.*

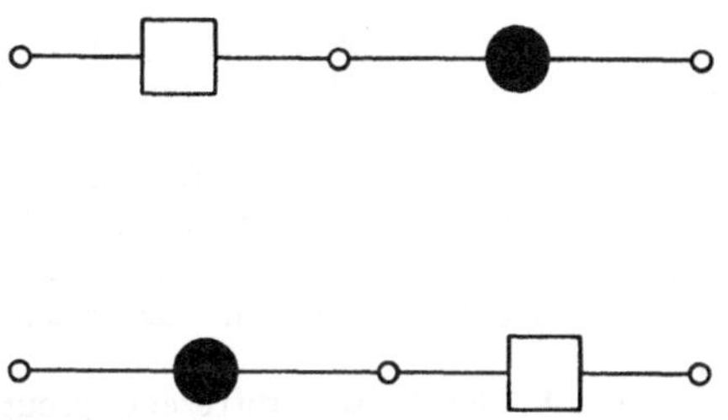

Fig. 4. Chain oxygen (black circle) and vacancy (white square) occupancy scheme in neighboring chains that would be favored in the 60 K YBaCuO.

The above scenario is qualitative, but whether it is correct depends entirely on whether or not it explains experiments. We emphasize that the assumption of Cu^{2+} is not based on a microscopic calculation, but rather on the observation that the semiconductors are antiferromagnetic. There is no *a priori* reason to assume that this situation persists in the doped state. We have already indicated how the valence-transition mechanism succesfully explains the doping, the role of the various oxygens, and the occurrence of 90 K and a 60 K YBaCuO. Disappearance of SC in strongly doped LaSrCuO, even under oxygen pressure,[14] is expected. This corresponds to $\rho >> 0.5$, a region where the organics are highly conducting but nonsuperconducting. That the chemical formulas of the Tℓ-Cu-O or Bi-Cu-O compounds suggest formal Cu valency of 2.0 is explained, since a valence transition will still give $\rho \rightarrow 0.5$ O-sublattices. With increase in the number of CuO_2 layers, T_c increases initially, but then reaches a peak. This is expected, since the intermediate layers do not have the bridging oxygens which we believe are necessary for the hole and the valence transition. We predict that the infinite layer semiconductor $Ca_{0.86}Sr_{0.14}CuO_2$ cannot be doped at all (note that this prediction does not follow from other doping mechanisms). We are unaware of any other mechanism explaining all of the above. We now show how all the other seemingly conflicting experiments can be explained within this picture.

Structural Distortions

The structural distortion, like the AF, is an intrinsic feature of the materials.[10] Distortion and lattice softening are seen in both LaSrCuO and YBaCuO at temperatures $T_s > T_c$ (but $T_s < T_0$, where T_0 is the tetragonal-orthorhombic transition temperature) and only at the superconducting compositions. Sound velocity, ultrasound, x-ray and neutron scattering measurements have separately claimed distortion, while the very recent studies[10] have shown that the distortions occur within the CuO_2 layers. Such experimental results continue to be reproduced,[32] but after initial theoretical interest, these results are being largely neglected by theorists, presumably because CDW within a Cu^{2+} lattice is not expected. Within the present scenario, these distortions are expected: after the valence transition, the $\rho = 0.5$ O-sublattice is described by H_O in Eq. (4), within which a $4k_F$ O-based CDW will occur. The oxygens with charges $-(1.5-\delta)$ and $-(1.5+\delta)$ align along orthogonal axes due to V_j. Since a Cu^{1+}-O^{1-} bond is stronger than a Cu^{1+}-O^{2-} bond, a lattice distortion follows. Electron-phonon interactions are an effect, rather than a cause.

Optical Absorptions

The 0.4 eV absorption. A broad electronic absorption at 0.4 eV is a characteristic feature of the superconductors. Within the CT exciton model(s), this corresponds to a low-frequency CT, $Cu^{2+}O^{1-} \leftrightarrow Cu^{3+}O^{2-}$. However, such a CT should grow in oscillator strength monotonically with doping, in complete disagreement with what is observed: in $La_{2-x}Sr_xCuO_4$, the oscillator strength peaks at $x = 0.15$-0.2 and then decreases[33] for larger x. Furthermore, this electronic absorption is coupled to a phonon mode,[33,34] suggesting once again a CDW!

Within the present picture, this absorption corresponds to a nearest-neighbor CT within the single-band extended Hubbard model, Eq. (4), that gives rise to an electronic absorption.[18] In the present case, this CT would correspond to $O^{2-}O^{1-} \leftrightarrow O^{1-}O^{2-}$, which near $\rho = 0.5$ has a finite energy barrier for nonzero V_{pp}. The oscillator strength is now expected to decrease at large x ($\rho >> 0.5$). Two additional points are to be noted. The absorption is across a $4k_F$ CDW gap and should be coupled to a vibrational mode, and a similar absorption is seen in the organics.[18]

The 1.7 eV absorption. At least two different groups have found a new electronic absorption in YBaCuO at 1.7 eV that is distinct from the 0.4 eV and the 2.5-2.8 eV absorptions.[35,36] Both groups associate this absorption with Cu^{2+}, and experimentally, the absorption occurs only at the semiconducting compositions. We point out that in their original work on $La_{2-x}Sr_xCuO_4$, an absorption near this energy (~2.0 eV) and near the semiconducting limit ($x \to 0$) was claimed by Etemad et al.[33] The vanishing of this absorption at the superconducting composition is perplexing, since within standard two-band models, it should continue to persist.

Within the present model, this absorption corresponds to the CT $Cu^{2+}O^{2-} \to Cu^{1+}O^{1-}$. The disappearance of the absorption at the superconducting composition is related to the valence transition proposed here -- there is no Cu^{2+} for large doping.

The 2.5-2.8 eV absorption. We ascribe this to the reverse CT $Cu^{1+}O^{1-} \to Cu^{2+}O^{2-}$. Note that is is not at all necessary that the forward and the reverse CT processes cost the same energy. In the case of the neutral-ionic transition in the organic mixed-stack compounds, these energies are quite different, and we expect the same here. Etemad et al.[33] have claimed a continuous increase in the oscillator strength of this absorption wtih doping in $La_{2-x}Sr_xCuO_4$ (in contrast to the 2.0 eV absorption), and this agrees with our interpretation.

Magnetic Moment

Search for Cu^{2+} magnetic moment in the superconducting compositions has involved epr, nmr, nqr, Raman scattering, neutron scattering, and muon spin rotation. LaSrCuO and (RE)BaCuO (where RE = rare earth) have been studied at both semiconducting and superconducting compositions, while we are at present aware of only a limited number of studies involving the newer materials. In no case has the Cu^{2+} moment been detected in superconducting (RE) BaCuO or in the Bi-Sr-Cu-O compounds, although, in general, this is easily detected in the semiconducting region. While epr silence may not necessarily rule out Cu^{2+}, it is harder to argue against the nmr[37] and nqr[38] results. A neutron scattering study of $YBa_2Cu_3O_7$ claims that at 110 K, less than 2% of the Cu have spin.[39] While an earlier Raman scattering data claimed spin fluctuations[40] in $YBa_2Cu_3O_7$, the strength of the Raman peak related to the magnon scattering was vanishingly weak in the superconductor.[41] A more careful study[42] has claimed that the Raman peak in question is absent in the superconductor. The only studies we are aware of that claim Cu^{2+} moments in the superconducting composition all involve LaSrCuO.[43,44] Neutron scattering[43] and μsr[44] have claimed Cu^{2+} here, but in both studies, an appreciable Cu^{2+} moment is found only in the "low T_c" region. At the "high T_c" compositions, the moments are again vanishingly small.

Static Magnetic Susceptibility

The static magnetic susceptibility χ exhibits a peculiar behavior as a function of x in $La_{2-x}Sr_xCuO_4$. $\chi(x)$ at 150 K initially decreases with x until a minimum is reached near x ~ 0.06, beyond which $\chi(x)$ actually increases again.[45] $\chi(x)$ peaks near x = 0.25 ($\chi(0.25) > \chi(0)$), beyond which it again decreases.

The initial decrease in χ is expected, but we do not believe it is possible to explain the increase beyond x = 0.06 without invoking the valence transition we have proposed. Immediately after the valence transition, ρ is exactly 0.5, where within a single-band extended Hubbard model, χ reaches a peak.[16,18] Following this, for larger ρ, χ is expected to decrease theoretically,[16,18] in agreement with the observation.

Chemical Substitution

As expected, the strongest effects are seen when Cu is substituted by another transition metal. What stand out from all these studies are that, (a) the most drastic effects on SC occur when Cu is substituted by Zn, and (b) substitution by Ag has the least destructive effect on SC. The effects of Zn-substitution have been widely discussed,[46] and there are several speculations based on the fact that Zn^{2+} is non-magnetic (d^{10}). However, we do not believe that this alone can explain why even a few percent of Zn can lead to semiconduction (in LaSrCuO, the transition to semiconductor can be masked by inhomogeneities). On the other hand, Ag substitution has a very weak effect on the SC, and there are at least two published reports which claim that even complete replacement of Cu with Ag leaves the 123 material superconducting (albeit with $T_c \leq 20$ K).[47] We believe that both of these experiments indicate the role of Cu^{1+} in the superconductors, as explained below.

Elemental Zn has the configuration $3d^{10}4s^2$, indicating that now Zn^{2+} is more stable relative to Zn^{1+}. For charge balance, each Zn^{2+} will convert a neighboring O^{1-} to O^{2-}. This in its turn increases the $|\Delta E_M|$ in Eq. (2) such that each Zn^{2+} surrounds itself with large numbers of O^{2-} and Cu^{2+}, until at a small critical concentration there is no more Cu^{1+} or O^{1-} at all. We believe that this is the only way to explain semiconduction in the weakly doped system.

In contrast to Zn, Ag is always monovalent. Thus, if indeed YBaAgO is a superconductor, as claimed in Refs. 47 (a) and (b), we see no alternative to Cu^{1+} in the parent compound. In this connection, it is absolutely intriguing that a recently developed procedure by a different group to make the 90 K superconductor malleable involves making a composite[48] with Ag. We strongly believe that Ag-substitution should be studied in much greater detail.

Hall Coefficient

The plateau in the Hall coefficient in YBaCuO for YBaCuO for $0.5 < \delta < 0.9$ is easily understood from Fig. 3. The 60 K material is a one-layer superconductor in which one layer has only Cu^{1+}, the other still has Cu^{2+}. In the first layer, the CT O^{1-}-$O^{2-} \leftrightarrow O^{2-}$-$O^{1-}$ is a low energy process. In the second layer, this CT would lead to neighboring $Cu^{2+}O^{1-}$, which has a high energy barrier due to Madelung energy considerations. Therefore, in the region $0.5 < \delta < 0.9$, the extra holes go to the second layer in accordance with the process in Fig. 2, but they virtually sit in their respective positions. The number of carriers and holes are simply different. Only for ρ very close to 1 we have two active layers, explaining why the Hall coefficient here is twice that of the 60 K material.[13] All other explanations of this phenomenon have requirements[49] that would be highly detrimental to crystal stability.

We now come to the Hall coefficient of $La_{2-x}Sr_xCuO_4$. Experimentally,[50,51] it is claimed that until $x \sim 0.16$, the number of carriers n_c increases linearly with x. Beyond this point n_c shows a drastic increase (there is an increase by nearly a factor of 30 between $x = 0.16$ and $x = 0.2$). There are two separate points that are of interest to us regarding this measurement. First, the sudden increase in n_c is a signature of the valence transition. In the original papers, it is assumed that bulk SC occurs at $x < 0.15$, instead of just beyond this point, as claimed here. Even more significantly, the magnitudes of the Hall coefficients of $x \sim 0.16$ LaSrCuO and the 60 K YBaCuO are virtually identical, at least from the Princeton-Bell Core data (note, however, that the former corresponds to measurements at 77 K and the latter at 95 K). This strongly supports our idea of both being one-layer superconductors (as opposed to the 90 K 123 material). A second point that we would like to emphasize is that even though the original authors claim that n_c in LaSrCuO increases linearly with x up to $x = 0.16$, there seem to be only two data points in this region (including the $x = 0.16$ point), while most of the experimental points are for $x > 0.16$, a region where $n_c >> x$.

Photoemission. In addition to magnetic measurements in the semiconductors, photoemission data have been repeatedly cited as the evidence for Cu^{2+}. The following points are to be noted.

First, the motivation of many studies mainly seems to be to preclude Cu^{3+}, thereby precluding the one-band magnetic mechanisms of pairing. Second, sizeable amounts of O^{1-} (although often thought of as O_2^{2-}, O_2^{1-}, etc.) have been determined in O(1s) XPS by several groups[52-54] (most notably by the Bangalore group[52]), and to satisfy the existence of so many O^{1-}, the bulk of the Cu must be Cu^{1+}. Angle-resolved photoemission finds an O-band with small dispersion that crosses the Fermi surface,[55] thus supporting the present idea. Indeed, in a number of publications, this is exactly what the Bangalore group has claimed -- singly charged Cu.[52] Finally, Bianconi et al.[56] find that the bridging oxygens in superconducting YBaCuO, but not in semiconducting YBaCuO, are O^{1-}. This is not only in agreement with what has been said (see Fig. 3), but gives us (we believe) only two choices, Cu^{1+} or 3D transport. We do not believe experiments indicate the latter. We believe that while it may be relatively simple to detect the presence or absence of Cu^{3+}, it is simply much more difficult to spectroscopically distinguish between Cu^{1+}, Cu^{2+}, and true mixed valence. Mixed valence has been suggested by many investigators, but within this it is difficult to explain the other observations, viz., vanishing of SC for large x in $La_{2-x}Sr_xCuO_4$, the decrease in the oscillator strength of the 0.4 eV absorption even as the number of carriers increases, and so on.

Because of space restrictions, we have listed only a limited number of experiments, but many seemingly unrelated observations are directly related to the valence transition. For example, the rather detailed analysis of the role of O in substituted (RE)BaCuO by Tokura et al.[57] can be very precisely explained within our model: the chain oxygen does not create carriers along the chain, but charges the bridging oxygens into O^{1-} (which are noncarriers), which in their turn create carriers in the layers. Indeed, the observation by Tokura et al. that there is a narrow region of charge within which high T_c occurs is in total agreement with our viewpoint. These and other experiments will be discussed elsewhere.

V. DOPED $BaBiO_3$

We now discuss $BaPb_{1-x}Bi_xO_3$ (superconducting at x = 0.25) and $Ba_{1-x}K_xBiO_3$ (superconducting at x = 0.25) in which we argue that the situation is similar to that in copper oxides, viz., true valence and formal valence are different, so that the number of O^{1-} ions is again much larger than what is superficially obvious. Once again the ionization processes $Bi^{3+} \rightarrow Bi^{5+}$ or $Pb^{2+} \rightarrow Pb^{4+}$ require huge amounts of energy inputs, but in a commensurate insulating crystal, the higher ionicities can be stabilized by the Madelung energy gain $|\Delta E_M|$. In this context, it is extremely relevant that Cu, Bi and Pb have virtually the same electronegativities. In a doped, conducting situation, the equilibrium is again disturbed and a sharp transition to the lower charge is expected. This results in one-third of the oxygen in $BaBiO_3$ as O^{1-}.

Much more experimental work is needed, but magnetic experiments preclude Bi^{4+}. Similarly, it has been claimed that SC is not due to electron-phonon coupling.[58] However, a much clearer experimental discussion is found in the recent XPS and UPS work by Hegde and Gauguly.[59] These authors show that in conducting Pb and Bi oxides, there is no signature at all of Pb^{4+}, Bi^{4+} or Bi^{5+}, evidently because the 6s band is way below the O 2p band (as opposed to the 6p band), so that any ionization has to occur from the O. This is in total agreement with our conclusion. In fact, we believe that the authors do not go far enough: if indeed Bi is totally in the Bi^{3+} state in the superconductors, then charge balance requires that one out of three oxygens is O^{1-}.

In the next section we will suggest a mechanism of SC that requires tendency towards a $4k_F$ periodicity. In a cubic perovskiite this periodicity is obtained at ρ = $^1/_3$ within Eq. (4). In Fig. 5, we show schematic structures that describe doped

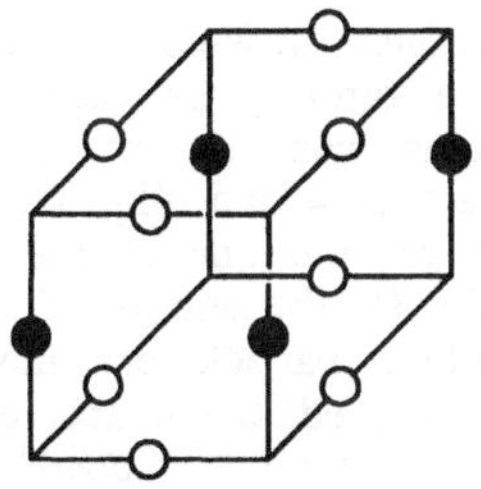

Fig. 5. Hole-occupancy scheme in doped $BaBiO_3$. Only the oxygen atoms are shown. Hole occupancies alternate in 3D along any O-O axis with only one-third of the oxygens as O^{1-}.

$BaBiO_3$. Notice that even with $\rho = {}^1/_3$, sites along any O-O axis are alternately occupied and unoccupied due to V_{pp}, a situation that in 2D requires $\rho = {}^1/_2$.

VI. THE N-TYPE COPPER OXIDES

Very recently SC has been found[60] in the n-type material $Nd_{2-x}Ce_xCuO_4$ between x = 0.15 and 0.18. This is a major theoretical challenge,[61] as O^{1-} instead of Cu^{3+} in hole-doped materials has been convincingly demonstrated, while electron doping leads only to Cu^{1+} and not O^{1-}! The present picture explains this: due to the repulsive interaction between the dopant Ce^{4+} and Cu^{2+}, $|\Delta E_m|$ in Eg. (2) becomes smaller and valence transition occurs at the same dopant concentration. We expect Nd(Ce) CuO_4 to have bulk amounts of Cu^{1+} and O^{1-}. The actual sign of the carrier within Eq. (4) depends on whether ρ is slightly greater or less than 0.5 (charge balance requires former in hole-doped and latter in electron-doped systems). The narrowness of the superconducting region, the magnitude of the Hall coefficients ($n_c >> x$) and the impossibility of hole doping (Nd_2CuO_4 does not have O_{br}) all[60] agree with our model.

The following predictions are made for these systems: O^{1-} in electron spectroscopy (note that without the valence transition this is not possible), a sharp 0.4 eV absorption for $0.15 \leq x \leq 0.18$ drastic effect of Zn-substitution (in spite of the fact that Zn^{2+} is isoelectronic with Cu^{1+} and should not severely affect magnetic pairing now), and absence of Cu^{2+} moments within the above concentration range.

VII. SUPERCONDUCTIVITY MECHANISM

We have, in the above, consistently claimed that SC can occur close to a correlated $4k_F$ CDW without actually describing how. The reason is that our motivation here was to suggest that experiments indicate that a common thread does seem to connect all of the materials described above. Perhaps the most important observation is that SC seems to occur within a narrow range of dopant or carrier concentration in the copper oxides and the two bismuth oxides, exactly as in the organics and in $LiTi_2O_4$. For example, to the best of our knowledge, it is agreed that bulk SC in $BaPb_{i-x}Bi_xO_3$ and $Ba_{1-x}K_xBiO_3$ is limited to narrow regions of x. If we recognize that the 90 K and 60 K YBaCuO are different materials, a similar conclusion is reached. Only in LaSrCuO SC occurs over a broad range of (extrinsic) hole concentrations, but as we have tried to show, the observed phase diagram is misleading. LaSrCuO is an extremely inhomogeneous material, and the "low T_c" materials can have superconducting islands. Limitation of SC to narrow carrier concentration range strongly suggests proximity to commensurate SDW or CDW. the former does not occur in most of the systems described above, while in $(TMTSF)_2X$ the $2k_F$ SDW has to coexist at least with a short-range $4k_F$ CDW order at the specific bandfilling.[17] The oxygens in the Cu and Bi oxides have just the right number and positions for a $4k_F$ periodicity, while the Ti are the active sites in $LiTi_2O_4$.

The actual mechanism of SC near a weak $4k_F$ CDW is still being investigated, but we have been able to find a binding between pairs of defects for ρ slightly different from 0.5 in 2D. This can be physically explained within strong coupling.[20,41] Consider the perfect $\rho = 0.5$ $4k_F$ CDW in 2D with alternate sites occupied. An extra particle will go into one of the unoccupied sites and has four nearest neighbors (so that within Eq. (1) the energy of this configuration relative to the original one is 4V). A second added particle can now occupy any unoccupied site, and independent of which site is occupied, the matrix element of the Coulomb term in Eq. (1) or (4) is 8V, relative to the original "undoped" configuration. However, there is a polarization-induced net gain in energy if the added particles occupy neighboring plaquettes: the energy of CT from the occupied site(s) that is (are) common to both plaquettes is reduced from 2V to V if neighboring plaquettes are occupied. This results in an attraction between the added particles, even though in first order the Coulomb energy is the same even if the particles are well-separated. Notice that exactly similarly, if particles are removed from the $4k_F$ CDW, the new unoccupied sites would tend to be on neighboring sites. We have numerically demonstrated this[20] in a 4×4 lattice for a spinless version of Eq. (1). We calculated the energies E(N) for N particles on such a lattice and showed that the binding energy of the pair, defined as $2\Delta = E(8) + E(10) - 2E(9)$, is negative. In contrast, 2Δ in 1D is positive, indicating the lack of binding there. The calculation is for spinless fermions, but finite U will strongly favor singlet pairing, as opposed to triplet pairing.

While this result is suggestive, the problem with Coulomb-induced binding is that there is always tendency of more than two particles binding (except in a "negative U" Hubbard model). This is true in several of the recent mechanisms of SC proposed for the copper and bismuth oxides, and is true here too. It has been suggested that longer-range Coulomb interactions and/or finite bandwidth can lead to "windows" within parameter space where binding leads to pairs, but clusters of larger size are avoided. This will be true in our case even more strongly, as all binding here is in second order already. All the same, it is at this point not clear that this is the only way to have SC near a weakly incommensurate $4k_F$ CDW, and further studies are going on. What is interesting is that there is finite binding even for small V,[20,41] so that a true $4k_F$ CDW with long-range order is not actually necessary.

Note added in proof: As this manuscript was nearing completion, I received a preprint by Goodenough and Manthiran that essentially comes to the same conclusion of an abrupt transition from the antiferromagnetic to the superconducting state. The authors show that this transition manifests itself in sharp structural changes as functions of compositions in four different materials. Such a sharp structural change is in total agreement with the valence transition proposed here.

REFERENCES

1. See, for instance, Proc. Int. Conf. Synth. Met. (ICSM88), Santa Fe, 1988. Synth. Met., in press.
2. See, for instance, D. Jérome and F. Creuzet in Novel Mechanisms of Superconductivity, eds. V. Krezin and S. Wolf (Plenum, N.Y., 1987).
3. H. Urayama et al., Chem. Lett., 463 (1988).
4. D. C. Johnston et al., Mater. Res. Bull. 8, 777 (1973).
5. P. W. Anderson et al., Phys. Rev. Lett. 58, 2790 (1987).
6. S. Uchida et al., Phase Transitions, 8, 95 (1987).
7. R. J. Cava et al., Nature 332, 814 (1988).
8. D. Vaknin et al., Phys. Rev. Lett. 58, 2802 (1987).
9. J. M. Tranquada et al., Phys. Rev. Lett. 60, 156 (1988).
10. S. Bhattacharya et al., Phys. Rev. Lett. 60, 1181 (1988), and references therein.
11. C. M. Varma et al., Sol. St. Commun. 62, 681 (1987).
12. S. Mazumdar, Phys. Rev. B 36, 7190 (1987).
13. Z. Z. Wang et al., Phys. Rev. B 36, 7222 (1987).
14. J. B. Torrance et al., Phys. Rev. Lett. 61, 1127 (1988).
15. J. D. Jorgensen et al., Phys. Rev. B 38, 11337 (1988), and references therein.
16. S. Mazumdar and A. N. Bloch, Phys. Rev. Lett. 50, 207 (1983).

17. S. Mazumdar, S. N. Dixit, and A. N. Bloch, Phys. Rev. B 30, 4842 (1984).
18. S. Mazumdar and S. N. Dixit, Phys. Rev. B 34, 3683 (1986).
19. S. Mazumdar and D. K, Campbell, Phys. Rev. Lett. 55, 2067 (1985).
20. S. Mazumdar and S. Ramasesha, Synth. Met. 27, 103 (1988).
21. L. G. Caron et al., in Ref. 1 and references therein.
22. P. Chaikin et al., in Ref. 1 and references therein.
23 (a) S. Satpathy and R. M. Martin, Phys. Rev. B 36, 7269 (1987); (b) S. Massidda, Jaejun Yu, and A. J. Freeman, Phys. Rev. B 38, 11352 (1988).
24. E. J. W. Verwey and P. W. Haayman, Physica VIII, 979 (1941).
25. J. R. Cullen and E. Callen, J. Appl. Phys. 41, 879 (1970).
26. P. W. Anderson, Phys. Rev. 102, 1008 (1956).
27. A. Redondo, S. Mazumdar, and D. K. Campbell, in preparation.
28. H. M. McConnell et al., Proc. Natl. Acad. Sci. 53, 46 (1965).
29. J. B. Torrance et al., Phys. Rev. Lett. 47, 1747 (1981).
30. H. W. Zandbergen et al., Phys. Stat. Sol. (a) 103, 45 (1987).
31. R. M. Fleming et al., Phys. Rev. B 37, 7920 (1988).
32. X. D. Shi et al., Phys. Rev. B 39, 827 (1989).
33. J. Orenstein et al., Phys. Rev. B 36, 8892 (1987); S. Etemad et al., Phys. Rev. B 36, 733 (1987).
34. S. L. Herr et al., Phys. Rev. B 36, 733 (1987).
35. M. K. Kelly et al., Phys. Rev. B 38, 870 (1988).
36. J. Humlicek, M. Garriga, and M. Cardona, Sol. St. Commun. 67, 589 (1988); see also, M. Garriga et al., JOSA B, to be published.
37. G. J. Bowden et al., J. Phys. C20, L545 (1987).
38. I. Furo et al., Phys. Rev B 36, 5690 (1987).
39. T. Brückel et al., Europhys. Lett. 4, 1189 (1987).
40. K. B. Lyons et al., Phys. Rev. Lett. 60, 732 (1988).
41. S. Mazumdar, Sol. St. Commun, 69, 527 (1989).
42. D. M. Krol et al., Phys. Rev. B 38, 11346 (1988).
43. R. J. Birgeneau et al., Phys. Rev. B 38, 6614 (1988).
44. A. Weidinger et al., Phys. Rev. Lett. 62, 102 (1989).
45. L. F. Schneemeyer et al., Phys. Rev. B 35, 8421 (1987).
46. G. Hilsher et al., Z Phys. Lett. B 72, 461 (1988), and references therein.
47. (a) K. K. Pan et al., Phys. Lett. A 125, 147 (1987); (b) Y. D. Yao et al., Proc. MRS meeting, 1988.
48. See, for instance, Nature, 336, 607 (1988).
49. J. Zaanen et al., Phys. Rev. Lett. 60, 2685 (1988).
50. N. P. Ong et al., Phys. Rev. B 35, 8807 (1987).
51. C. Uher et al., Phys. Rev. B 35, 5676 (1987).
52. D. D. Sarma and C. N. R. Rao, J. Phys. C20, L659 (1987).
53. J. H. Weaver et al., Phys Rev. B 38, 4668 (1988).
54. J. W. Rogers et al., Phys. Rev. B 38, 5021 (1988).
55. T. Takahashi et al. Nature 334, 691 (1988).
56. A. Bianconi et al., Phys. Rev. B 38, 7196 (1988).
57. Y. Tokura et al., Phys. Rev. B 38, 7156 (1988).
58. B. Batlogg et al., Phys. Rev. Lett. 61, 1670 (1988).
59. M. S. Hegde and P. Ganguly, Phys. Rev. B 38, 4557 (1988).
60. Y. Tokura, H. Takagi and S. Uchida, Nature 337, 345 (1989).
61. V. J. Emery, Nature 337, 306 (1989).

SPIN FLUCTUATIONS AND QUASIPARTICLES IN Cu-O PLANES

George Reiter

Physics Department and
Texas Center for Superconductivity
University of Houston
Houston, TX 77204-5504

There is general agreement that the Cu-O planes of the High Temperature superconductors may be described by a nearly isotropic Heisenberg model of localized spins when the doping is such that there is one hole per copper site. There has been considerable controversy over the nature of the ground state,[1-3] although the consensus now is that it is Neel ordered, as spin wave theory suggests. I present here results for the dynamics of the two dimensional Heisenberg model with arbitrary spin based upon the spin wave expansion away from the Neel state that are asymptotically exact for large S and low temperatures, and that are accurate for spin 1/2. They give a satisfactory explanation of the neutron scattering data, at least at near zero temperatures, and, although the relevant correlation functions have not been calculated yet, a qualitative explanation for the differences in the two magnon light scattering from previous theoretical predictions.[4] This work was done in conjunction with T. Becher[3] and K. Stuart.[5] It predicts that dynamical scaling is violated in the classical system, a result that is consistent with recent simulations of Wysin, Bishop and Gouvea[6] and with a similar theory for the ferromagnet published some years ago.[7]

There is considerably less agreement about the proper Hamiltonian to describe the system when additional holes are added. An extended Hubbard model introduced by Emery[8] presumably contains all of the effects one might wish to include, but is regarded by many as unnecessarily complicated. Zhang and Rice[9] have argued that a simplified version of this model, neglecting oxygen-oxygen and copper-oxygen Coulomb interactions, is in fact equivalent in its low energy physics, to the one band model, now called the t-J model. I will argue, based upon an exact solution for the extended Hubbard model in a ferromagnetic background (obtained by Emery and myself),[10] that there are, in fact significant differences with the physics of the t-J model.

The theory for the dynamics of the Heisenberg model that we will develop is valid at finite but low temperatures, where there is no long range order, even in the classical system, and we will use a spin wave expansion to calculate with. It is worth asking then, why there are spin waves in the system at all, in the absence of long range order. The answer is understood to be that within a region the size of a coherence length, the system is locally ordered, with all spins pointing in nearly the same direction, this direction differing from region to region. Spin waves with wavelengths much less than this coherence length can be regarded as propagating in an ordered region, and hence can be expected to be well defined excitations. This answer is correct, as far as it goes, but is insufficient to determine the lifetime of the excitations. To make contact between this picture and a quantitative calculation, we will use an argument akin to that used in the

Interacting Electrons in Reduced Dimensions
Edited by D. Baeriswyl and D.K. Campbell
Plenum Press, New York

ordered phase of three dimensional systems to demonstrate the existence of spin waves and calculate their frequency.

From the equations of motion for the Hamiltonian

$$H = J \sum_{\vec{i},\vec{\Delta}} \vec{S}_i \cdot \vec{S}_{i+\Delta} \quad , \qquad \vec{\Delta} = (0,\pm1), (\pm1,0) \tag{1}$$

it can be shown that

$$\frac{\partial^2}{\partial t^2} S_q^\alpha(t) = N^{-1} \sum \Gamma(q_1 q_2 q_3)[S_{q_1}^\alpha \vec{S}_{q_2} \cdot \vec{S}_{q_3} + \vec{S}_{q_2} \vec{S}_{q_3} S_{q_1}^\alpha] \tag{2}$$

where

$$\Gamma(q_1 q_2 q_3) = 1/2(J_{q_1} - J_{q_2})(J_{q_3} - J_{q_1+q_2})\delta_{\vec{q},\vec{q}_1+\vec{q}_2+\vec{q}_3}$$

and

$$J_q = 2J(\cos q_x + \cos q_y)$$

The quantity $\vec{S}_{q_2} \cdot \vec{S}_{q_3}$ describes the local order in the system, and can be written as

$$\vec{S}_{q_2} \cdot \vec{S}_{q_3} = \delta_{q_2+q_3,0} \langle \vec{S}_{q_2} \cdot \vec{S}_{-q_2} \rangle + \Delta \vec{S}_{q_2} \cdot \vec{S}_{q_3} \tag{3}$$

If we neglect the fluctuations in the local order for the moment, we find that

$$\frac{\partial^2}{\partial t^2} S_q^\alpha(t) = -\omega_{q,T}^2 S_q^\alpha(t) \tag{4}$$

where

$$\omega_{q,T}^2 = N^{-1} \sum_{q'} (J_{\vec{q}'} - J_{\vec{q}-\vec{q}'})(J_{\vec{q}'} - J_{\vec{q}}) \langle \vec{S}_{q'} \cdot \vec{S}_{-q'} \rangle \tag{5}$$

As the temperature T approaches zero for the classical system, $\langle \vec{S}_q \cdot \vec{S}_{-q} \rangle \rightarrow \delta_{\vec{q},\vec{q}_0} S^2$ where $\vec{q}_0 = (0,0)$ for a ferromagnet, and (π,π) for the antiferromagnet,in units in which the lattice constant is one, and hence

$$\omega_{q,T}^2 \rightarrow \begin{cases} S^2(J_o - J_q)^2 & \text{(ferromagnet)} \\ S^2(J_o^2 - J_q^2) & \text{(antiferromagnet)} \end{cases} \tag{6}$$

which are the classical spin wave frequencies. The lifetimes of the spin waves are determined by the fluctuations that have been neglected. These vanish as the temperature approaches zero for the classical system. To see this, observe that the equations of motion couple spins that are at most two lattice sites apart, for nearest neighbor interactions, so that we need only consider $\langle (\Delta \vec{S}_i \cdot \vec{S}_j)^2 \rangle$ for i and j a short distance apart to estimate the magnitude of the fluctuation terms. Rotationally invariant quantities involving spins at any fixed distance apart, such as this one, can be calculated perturbatively from spin wave theory, with no divergences, even in the absence of long range order. This was conjectured in Ref. (7) and, independently, by S. Elitzur,[12] and proven by F. David.[13]

We can, therefore, use the standard Holstein Primakoff theory to evaluate rotationally invariant averages, even though the symmetry is not broken. We find that for the classical system, and large separations,

$$S^{-4}\langle(\Delta\vec{S}_i\cdot\vec{S}_{i+n})^2\rangle = 4\left(\frac{KT}{2\pi JS^2}\right)^2 \ln^2|n| \tag{7}$$

Thus, if one makes the temperature sufficiently small, the fluctuations will always be negligible compared to the average of $\langle\vec{S}_i\cdot\vec{S}_j\rangle$ at T = 0, and the spin waves will be well defined. For the quantum case, the fluctuations do not vanish at T = 0, for the antiferromagnet, and for spin 1/2 the fluctuations are not small compared to S^2, but, as we shall see, spin waves are still well defined.

To calculate the lifetimes of the spin waves, and more generally the lineshapes, it is necessary to include the effects of the fluctuations away from local order. This is done naturally in the projection operator formalism.[7,14] Defining an inner product on the space of operators on the Hilbert space of the system, for any two operators A and B

$$\langle A|B\rangle \equiv \left\langle\int_0^\beta e^{\tau H}A^+e^{-\tau H}B\,d\tau\right\rangle \tag{8}$$

we find that

$$R_q(z) \equiv \int_0^\infty e^{izt}\langle\vec{S}_q|\vec{S}_q(t)\rangle dt = i\chi_q/(z - \omega_q^2/(z + \gamma_q(z)) \tag{9}$$

where ω_q^2 is the exact second moment

$$\omega_q^2 = \langle\dot{\vec{S}}_q|\dot{\vec{S}}_q\rangle/\langle\vec{S}_q|\vec{S}_q\rangle \tag{10}$$

and $\langle\vec{S}_q|\vec{S}_q\rangle$ is the static, isothermal susceptibility χ_q. The quantity $\gamma_q(z)$ has an exact definition in terms of $\ddot{\vec{S}}_q$, and projection operators. It can be shown,that to leading order in the spin wave expansion one can omit projection operators appearing in the definition[14] and obtain for γ_q in the time domain

$$\gamma_q(t) = iN^{-2}\sum \Gamma(q_1q_2q_3)(\vec{S}_{q_1}\Delta\vec{S}_{q_2}\cdot\vec{S}_{q_3}|\Delta\vec{S}_{q'_2}\cdot\vec{S}_{q'_3}\vec{S}_{q'_1}(t))\Gamma(q'_1q'_2q'_3)/\langle\dot{\vec{S}}_q|\dot{\vec{S}}_q\rangle \tag{11}$$

so that γ_q explicitly incorporates the effects of the fluctuations neglected in obtaining Eq. (4).

It can be shown that $\langle\dot{\vec{S}}_q|\dot{\vec{S}}_q\rangle = (J_o - J_q)\langle\vec{S}_i\cdot\vec{S}_{i+\Delta}\rangle$ and $\langle\vec{S}_i\cdot\vec{S}_{i+\Delta}\rangle$ can be calculated accurately from the spin wave theory. To lowest order

$$\chi_q = \frac{2}{(J_o + J_q)} \frac{(1 - .195/S)}{(1 + .079/S)} \equiv z_\chi \frac{2}{(J_o + J_q)} \tag{12}$$

This is accurate for $q >> \kappa$, where κ is the inverse coherence length, but one must use a renormalization group argument, such as that of Chakravarty, Halperin and Nelson[15] to obtain the correct longwavelength behavior.

In calculating (11), the two operators $\vec{S}_{q_1}$ and $\vec{S}_{q'_1}(t)$ appearing in the average would be replaced to lowest order, by $NS^2\delta_{\vec{q},\vec{q}_1}\delta_{-\vec{q},\vec{q}'_1}$. A better approximation, that takes us beyond a straightforward pertur-bation expansion, would be to replace this pair of operators by their average, and neglect the fluctuations in the product. We obtain then

$$\gamma_q(t) = \frac{1}{S^2} \sum_{q'} \langle \vec{S}_{q'} \cdot \vec{S}_{-q'} \rangle \gamma^o_{q-q'}(t) \tag{13}$$

where $\gamma^o_q(t)$ is the leading term in the spin wave expansion for $\gamma_q(t)$. This may be calculated readily from (12) and is

$$\gamma^o_q(t) = i \frac{S^4}{\langle \vec{S}_q | \vec{S}_q \rangle} \sum_{q_1 q_2} \{ |\Gamma^+(q_1 q_2)|^2 (\eta_{q_1} + \eta_{q_2} + 1)(\varepsilon_{q_1} + \varepsilon_{q_2})^{-1} \cos(\varepsilon_{q_1} + \varepsilon_{q_2})t)$$

$$+ |\Gamma^-(q_1 q_2)|^2 (\eta_{q_2} - \eta_{q_1})(\varepsilon_{q_2} - \varepsilon_{q_1})^{-1} \cos(\varepsilon_{q_2} - \varepsilon_{q_1})t)\} \delta(q - q_1 - q_2) \tag{14}$$

where

$$\varepsilon_q = 2\bar{J}S[4 - (\cos q_x + \cos q_y)^2]^{1/2}, \quad \bar{J} = z_c J, \quad z_c = 1 + .079/S,$$
$$\eta_q = [e^{\beta \varepsilon_q} - 1]^{-1} \tag{15}$$

and

$$\Gamma^\pm(q_1 q_2) = \{[(J_o - J_{q_1})(J_o - J_{q_2})]^{1/2}[(2J_o - J_{q_2} + J_{q_1})(J_{q_2} - J_{q_1})]$$
$$\pm [(J_o - J_{q_1})(J_o - J_{q_2})]^{1/2}[(2J_o + J_{q_2} - J_{q_1})(J_{q_2} - J_{q_1})] \tag{16}$$
$$+ (J_o^2 - J_q^2)\left([(J_o - J_{q_1})(J_o + J_{q_2})]^{1/2} \pm [(J_o + J_{q_1})(J_o - J_{q_2})]^{1/2}\right)\}$$
$$/(J_o^2 - J_{q_1}^2)^{1/4}(J_o^2 - J_{q_2}^2)^{1/4}$$

Note that there are two distinct processes that enter into Eq. (14). One involves the decay of two spin waves into a pair of others. This exists at T = 0 for any finite S. The other involves the absorption by the spin wave of a preexisting, thermally generated spin wave, and vanishes at T = 0 for any value of S. It is the sum process that dominates the low temperature region in the High Tc materials, and most of the surprising features of the neutron scattering data can be understood by looking at the T = 0 results.

Neutron scattering actually measures $S(q,\omega)$, which is related to $R_q(\omega+i\varepsilon)$ by a detail balance factor. Explicitly

$$S(q,\omega) = \frac{1}{2\pi}\int_{-\infty}^{\infty} e^{i\omega t}\langle\vec{S}_{-q}\vec{S}_q(t)\rangle dt$$

$$= \frac{\omega}{\pi}(1-e^{-\beta\omega})\chi_q\, \omega_q^2\gamma_q''(\omega)/[(\omega^2-\omega\gamma_q'(\omega)-\omega_q^2)^2 + (\omega\gamma_q''(\omega))^2] \qquad (17)$$

where $\gamma_q(\omega) = \gamma_q'(\omega) + i\gamma_q''(\omega)$.

For small ω, $\vec{q}^* = \vec{q} - (\pi,\pi)$. From Eq. (14), at $T = 0$, with q^* denoted by q.

$$\gamma_q''(\omega) = (4\sqrt{2}JS^2|\omega|)^{-1}(\omega^2 - c^2q^2)^{3/2}\theta(\omega^2 - c^2q^2)z_c^{-6} \qquad (18)$$

where $c = 2\sqrt{2}\,(JS)$. For small ω, $\gamma_q'(\omega) \propto \omega$, and we find numerically

$$\gamma_q'(\omega) = -.279\ \omega/S \qquad (19)$$

At $T = 0$, in the vicinity of the spinwave peak, and for small q^*, we find using Eqs. (17)-(19), (9) and (12)

$$S(q,\omega) = \frac{8^{1/2}}{\pi}\ \frac{(z_c)^{-4}\theta(\omega^2-c^2q^2)(\omega^2-c^2q^2)^{3/2}}{\left(\omega^2(1+.279/S) - \dfrac{c^2q^2}{z_x}\right)^2 + \dfrac{(\omega^2-c^2q^2)^3}{(2cSz_c^5)^2}} \qquad (20)$$

The linewidth is very small compared to the frequency despite the fact that the total fluctuation $\langle(\delta\vec{S}_i\cdot\vec{S}_j)^2\rangle$ is not small. This is due to the fact that the fluctuations are distributed over a broad band of frequencies, and only the spectrum at the spin wave frequency is effective. There is also a shift of the spin wave frequency away from the value cq due to the real part of γ_q and the quantum renormalization of χ_q. It is this shift of the resonance frequency of the spin propagator that is responsible for there being any linewidth at all, since energy conservation and the convexity of the spectrum imply that no decay of a bare spin wave into a pair of others is possible. However, $(1 + .279/S)z_x = 1 + O(1/S^2)$, so that this shift is of order $1/S^2$ at least, and is not calculated accurately by the present work. We nevertheless, show the result in Fig. 1 of the calculation (20) as it demonstrates that the linewidths are very small at $T = 0$. The value of J deduced from Aeppli et al.'s[16] measurement of the spin wave velocity (.85 ± .03 eV Å) in La_2CuO_4 and this renormalization ($c' = 1.29\ (2\sqrt{2}JS)$) is $J = .125 \pm .007$ eV, in agreement with the result of Singh et al., .128 ± .006 eV, from the light scattering measurements. The agreement may be fortuitous because of the inaccurate treatment of the $1/S^2$ terms.

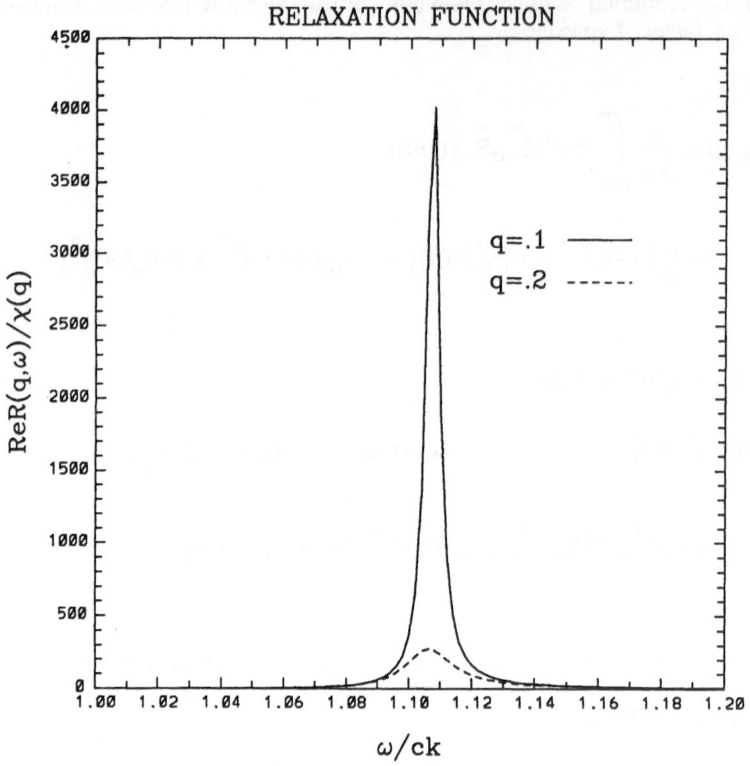

Fig. 1. $R(q,\omega)$ at $T = 0$, using the long wavelength approximation to $\gamma_q(\omega)$, Eq. (18), (19). $S = 1/2$.

The classical limit of Eq. (16) is easily taken by letting $S \to \infty$, $J \to 0$ in such a way that JS is fixed. Again, for q and ω small, we find that

$$\gamma_q^0(\omega) = 2KT(cS|\omega|)^{-1}[(\omega^2-(cq)^2)\theta(\omega^2-c^2q^2) + (c^2q^2-\omega^2)\theta(c^2q^2-\omega^2)(2\ \tan^{-1}(|\omega|/(c^2q^2-\omega^2)^{1/2})] \quad (21)$$

The spectrum below the spin wave frequency is due to the difference process, the spectrum above to the sum. When this is convolved with the static pair correlation function as in Eq. (5), which can be calculated perturbatively from the spin wave theory, we find that

$$\gamma_q^{\cdot}(\omega) = A(KT/JS^2)^2 JS \quad (22)$$

for small q and ω. That is, nearly independent of q and ω. The coefficient A is approximately 1. The spectral function that results is in strong violation of the dynamical scaling hypothesis, which would predict $\gamma_q \propto C\kappa f(q/\kappa)$, since for low temperatures $KT/JS^2 \gg \kappa$. There is preliminary numerical evidence that this is correct in the simulations of Wysin, Bishop and Gouvea[6] where linewidths do not appear to vary in a manner consistent with scaling.

The failure of dynamical scaling occurs because the short wavelength spin waves give a significant contribution to the damping of the longest wavelength mode. As a physical picture, observe that there are short wavelength fluctuations in the locally ordered regions, with an amplitude proportional to KT and a characteristic spread in frequencies of order J. As a consequence, the spin waves are subject to a rapid random fluctuation that leads to a motional narrowed linewidth of order $(KT)^2/J$. This is a much larger decay frequency than $c\kappa$, the reciprocal time it would take for a spinwave packet to cross a locally ordered region. In one dimension, this is not the case, since the the coherence length is proportional to kT, and the short wavelengths are ineffective. The same sort of analysis as above was used to predict scaling in the one dimensional antiferromagnet and a violation of scaling in the two dimensional ferromagnet some years ago.[7]

There have been several published works that conclude that dynamical scaling holds. Chakravarty, Halperin and Nelson have given a plausibility argument, based on the idea that the rapid short wavelength fluctuations serve only to renormalize the spin wave stiffness constant, and that the modes effective for the damping at any particular wavelength are only those at approximately the same wavelength, concluding that the linewidth for $q >> \kappa$ is proportional to $cq(KT)^{\nu}$, with ν some low power. While this may be a plausible picture, it explicitly ignores the possibility of motional narrowed lines resulting from the short wavelength fluctuations, and is not a calculation. We note, however, that if one uses the lowest order expression (21) in the exact result (17), one can make contact with the results of Chakravarty, Halperin and Nelson. The second moment is shifted to smaller values than the square of the $T = 0$ spin wave value, by an amount proportional to T, and as a consequence, the damping function in (21) is non zero at the actual resonance frequency. The lifetime of the spinwaves is then $\propto cq(KT)^2$, and the characteristic time proportional to $c\kappa T^{-1/2}$. In order for this to be correct the approximation (13), which is excellent in one dimension,[14] must fail. A more accurate calculation to higher orders in T would have to vanish, term by term, at the spin wave frequency in order for the leading term at small q and ω to be given by Eq. (21). We think this unlikely, but are calculating $\gamma_0(0)$ exactly to second order in KT to check the possibility. A preliminary analysis indicates that the linewidths observed in the simulation of Wysin, Bishop and Gouvea[6] are in agreement with Eq. (22).

Grempel[18] has done a calculation based upon the mode coupling approximation that obtains the same characteristic relaxation time as Chakravarty, Halperin and Nelson, but, as he intimates, vertex corrections are essential for discussing the dynamics when the fixed length of the spins is a relevant constraint. In fact, the mode coupling theory is the first term in an expansion of the dynamics of the dynamical spherical model, where this constraint is relaxed, and that model has very different dynamics from the Heisenberg model in one or two dimensions.[19] In particular, it does not have well defined spin waves at $T = 0$. The work of Auerbach and Arovas[20] is essentially a mean field calculation that neglects the short wavelength fluctuations that appear to dominate the dynamics. It gives results equivalent to the non interacting spin wave theory at $T = 0$, which are less sophisticated than the present results. We find, therefore, no compelling theoretical reason to doubt the conclusion that dynamical scaling does not hold in the two dimensional antiferromagnet.

We turn now to the issue of the nature of the quasiparticles introduced by adding holes to the Cu-O planes. There is general agreement that this can be described by the extended Hubbard model Hamiltonian introduced by Emery,[8] with perhaps an extension to include oxygen-oxygen hopping. When the oxygen-oxygen and copper-oxygen coulomb interactions are neglected, it has been argued by Zhang and Rice[9] that the Hamiltonian that one obtains by treating the hopping to second order is equivalent, in its low energy physics, to the one-band t-J model.The second order Hamiltonian from which we and they start describes the motion of oxygen holes on a background of fixed copper holes. (There is some hybridization that we ignore for simplicity of exposition although it is included in the calculation.)

$$H = 2(t_1 + t_2)N^{-1} \sum_{q,q'} \alpha_q \alpha_{q'} \vec{s}_{q-q'} \cdot \vec{S}_{\sigma\sigma'} b^{+}_{q'\sigma} b_{q'\sigma} + 1/2(t_2 - t_1) \sum_{q,\sigma} \alpha_q^2 b^{+}_{q\sigma} b_{q\sigma} \tag{23}$$

where $\vec{S}_{\sigma\sigma'}$ are the matrix elements of the spin operator and $\vec{s}_q$ is the fourier transform of the copper hole spin operators. $\alpha_q^2 = 2(2 + \cos q_x + \cos q_y)$.

The states that are created by $b_{q,\sigma}^+$ are the antibonding combination of the two oxygen sites in each cell, with the non-bonding states playing no role in the discussion, as they do not couple to the states being considered. There is then one Wannier oxygen state/cell. Zhang and Rice consider the action of this Hamiltonian in a subspace consisting of a singlet combination of the oxygen Wannier states with the copper hole spin on the same site. As a consequence they obtain a mapping to the "t-J model" in which the singlet combination maps onto a vacancy. This mapping is supposed to work for the Hamiltonian (23) whatever the state of the background copper spins. Emery and I[10] have solved for the eigenstates of (23) when there is a single hole in a ferromagnetic background. The problem can be solved exactly when one includes a ferromagnetic Heisenberg term to produce this background as well, but it adds nothing to the present discussion to include such a term. With the vacuum being the state with all copper hole spins up, and no holes on the oxygen, the state

$$|\psi_K\rangle = b_{K,\downarrow}^+|0\rangle + N^{-1}\sum W_{K'}^K\, b_{K-K',\downarrow}^+\, s_{K'}^+ \tag{24}$$

is an eigenstate of the system with an eigenvalue λ_K that satisfies

$$\frac{\alpha_K^2}{\lambda_K - t_1\alpha_K^2} + N^{-1}\sum_q \frac{\alpha_q^2}{\lambda_K - t_1\alpha_q^2} = \frac{-1}{t_1+t_2} \tag{25}$$

The wavefunction is then given by

$$W_q^K = -\frac{\alpha_q}{\alpha_K}\left[\frac{\lambda_K - t_1\alpha_K^2}{\lambda_K - t_1\alpha_q^2}\right] \tag{26}$$

The state (24) describes the exact quasiparticle of the system. We find that the energy surface is rather accurately given by the Zhang and Rice mobile singlets, which are the states obtained from (24) by setting $W_{K'}^K$ to -1, and that the overlap of the singlet states with the exact quasiparticle is over 95% throughout most of the band. On this basis, one would certainly say theirs was an excellent approximation. However, the Zhang and Rice singlet is, of necessity, a spin zero object, whereas we find that the spin on the oxygen site is not zero. We show in Fig. (2) the average spin on an occupied oxygen site as a function of the wavevector of the quasiparticle.

$$\sigma_K = \frac{1}{2}\,\frac{1 - N^{-1}\sum_q |W_q^K|^2}{1 + N^{-1}\sum_q |W_q^K|^2} \tag{27}$$

While the spin at the bottom of the band is not large, it is close to the value 1/6 that one would expect if the quasiparticle were described locally by the ground state of the Hamiltonian formed by a single oxygen hole and its two neighbors, when the hopping of the oxygen to another site is not included

$$H = J(\vec{s}_i + \vec{s}_{i+\Delta})\cdot \vec{S}_{i+\Delta/2} \tag{28}$$

where $\vec{S}_{i+\Delta/2}$ is the spin on the oxygen hole.

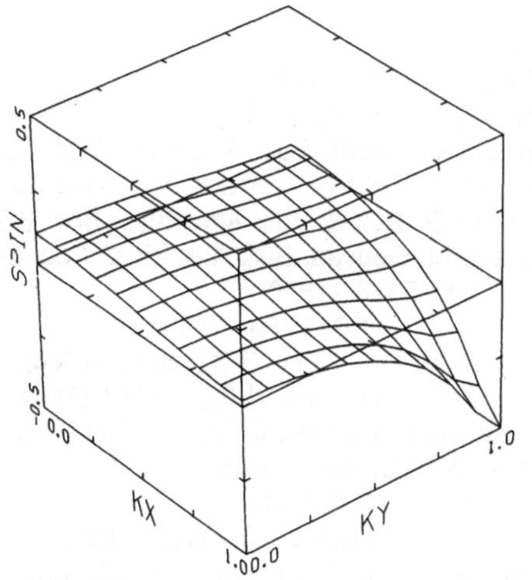

Fig. 2. The average spin on an oxygen site for $t_1 = t_2$, for arbitrary wavevectors. The value of 1/6 corresponds to the polaron of (29).

This state, which we will call a polaron for lack of a better name, has the configuration

$$|\phi_o\rangle = 2|\uparrow\downarrow\uparrow\rangle - |\uparrow\uparrow\downarrow\rangle - |\downarrow\uparrow\uparrow\rangle \tag{29}$$

In fact, the ground state at K = 0 for $t_1 = 0$ is exactly a superposition of such states and the overlap of such a state, for any t_1 with the exact ground state is higher than for the Zhang and Rice approximation. As one moves away from the bottom of the band, some of the first excited state of the Hamiltonian (28) are mixed in, and the overlap of the straightforward superposition of polarons with the exact quasiparticle diminishes. One can say then, that the low energy quasiparticle looks locally like the polaron state above, hopping from site to site with a slight distortion that increases as the wavevector increases. It is clear that the polaron is a spin 1/2 object, whereas the Zhang and Rice singlet is spinless. In fact the difference is moot in a ferromagnetic background, since in either case, if one considers a region of a few lattice sites surrounding the oxygen hole, one finds a total spin deviation of -1/2. This would not necessarily be the case in a disordered background, which is the situation of interest. In the absence of long range order, one could not tell from the location of a vacancy what the spin in the region was, whereas the polaron would carry a definite spin with it. This spin might be neutralized by the adjustment of the background,but in that case the spinlessness of the quasiparticle would not be a consequence of the second order Hamiltonian. Clearly, whether the quasiparticle does or does not carry a spin with it is an essential distinction of the physics. Near the middle of the band, the spin on the oxygen is nearly zero and the quasiparticle is very close to a moving singlet. Should it be the case that the minimum of the quasiparticle band is in this region for the appropriate disordered background, then the mapping suggested by Zhang and Rice may indeed capture the essential physics of the Hamiltonian (26), but this has not been shown, and may well not be true.

Acknowledgement: This work was supported in part by DARPA Grant #MDA972-88-G-0002.

REFERENCES

1. G. Shirane et al., PRL 59, 1613 (1987).
2. P. W. Anderson, PRL 59, 2497(c) (1987).
3. T. Becher, K. Stuart and G. Reiter, Bull. Am. Phys. Soc. 33, 1362 (1988).
4. K. B. Lyons, P. Fleury, L. F. Schneemyer and J. V. Waszcak, PRL 60, 732 (1988).
5. K. Stuart, T. Becher and G. Reiter, Bull. Am. Phys. Soc. 33, 1363 (1988).
6. G. Wysin, A. Bishop and M. Gouvea, Bull. Am. Phys. Soc. 34, 847 (1989).
7. G. Reiter, Phys. Rev. 21, 5356 (1980).
8. V. J. Emery, Phys. Rev. Letts 58, 2794 (1987).
9. F. C. Zhang and T. M. Rice, Phys. Rev. B 37, 3759 (1988).
10. V. J. Emery and G. Reiter, Phys. Rev. B 38, 11938 (1988).
11. V. J. Emery and G. Reiter, Phys. Rev. B 38, 3759 (1988).
12. S. Elitzur, Nucl. Phys. B 212, 501 (1983).
13. F. David, Phys. Lett 69 B, 371 (1980).
14. G. Reiter and A. Sjolander, J. Phys C 13, 3027 (1988).
15. S. Chakravarty, B. Halperin and D. Nelson, PRL 60, 1057 (1988).
16. G. Aeppli, S. M. Hayden, A. Mook, Z. Fisk, S. W. Cheonut, D. Rytz, J. R. Remeika, G. P. Espinosa and A. S. Cooper (preprint).
17. R. R. P. Singh, P. A. Fleury, K. B. Lyons and P. E. Sulewski (preprint).
18. D. R. Grempel, PRL 61, 1042(c) (1988).
19. G. Reiter, Phys. Rev. B 19, 1582 (1979).
20. A. Auerback and D. P. Arovas, PRL 61, 617 (1988).

ELECTRONIC PROPERTIES AND SPIN-CORRELATIONS OF CuO_2-PLANES IN HIGH TEMPERATURE SUPERCONDUCTORS

P. Horsch and W.H. Stephan

Max-Planck-Institut für Festkörperforschung
D-7000 Stuttgart 1, Federal Republic of Germany

INTRODUCTION

Considerable insight into the electronic structure and the nature of the charge carriers in high-T_c superconductors (HTSC's) comes from various types of photoemission and inverse photoemission experiments [1]. Such experiments showed that the states close to the Fermi level in the metallic samples have strong oxygen character [2], i.e. additional holes go essentially on oxygen. By angle resolved photoemission in superconducting $Bi_2CaSr_2Cu_2O_8$ even a band crossing the Fermi level could be resolved [3]. A Fermi edge has been seen by several groups, which suggests that there is a Fermi liquid which becomes superconducting. A further success of this class of spectroscopies was the observation of the superconducting gap by high-resolution UV-photoemission [4]. The discussion about the precise character of the carriers, however, is still on a qualitative level and controversial. Recent investigations of the O 1s absorption edge [5,6] showed that the relevant oxygen orbitals have 95% $p_{x,y}$-symmetry, with x and y in the plane [6].

The theoretical difficulty with HTSC's is their closeness to Mott-Hubbard insulators. The undoped reference materials such as La_2CuO_4 and $YBa_2Cu_3O_6$ are antiferromagnetic insulators. Standard bandstructure calculations predict metallic behavior instead, with no tendency towards antiferromagnetism [7]. Bandstructure theory also fails to explain photoemission or optical absorption in this situation, because the single particle picture no longer provides a valid description. This is well known for the transition metal oxides [8], and suggests that the underlying physics

Interacting Electrons in Reduced Dimensions
Edited by D. Baeriswyl and D.K. Campbell
Plenum Press, New York

has certain similarities. Under these circumstances it is evident that electronic correlation effects must be taken into account properly to achieve an understanding of the electronic properties of these materials. Present theoretical studies of HTSC's are still limited to simple models for the electronic structure, due to the complexity of the many-body problem at hand, and mainly focus on ground state properties. Nevertheless such a model should be able to explain these experiments at least on a qualitative level.

Excitation spectra of strongly-correlated electronic systems are even more difficult to calculate than ground state properties. In such cases the problem is often first simplified by a further reduction of the model Hamiltonian, and (or) by searching for reasonable approximations. The method of exact diagonalization of finite clusters can be of great value in this process, as approximations may be more rigorously tested than is otherwise possible. We present here first calculations of the single particle (hole) excitation spectrum of extended Hubbard models. This provides a direct comparison of the physical content of these models and actual photoemission and inverse photoemission spectra (PES) of HTSC's.

Concerning the mechanism responsible for superconductivity [9] there has been considerable controversy over whether the essential physics is describable in terms of a single band Hubbard model with large U, which is the basis of e.g. the resonating valence bond (RVB) model [10] , or by a multi-band Hubbard model [11] where Cu and O degrees of freedom are explicitly allowed. Zhang and Rice [12] suggested that a local singlet formed by a hole on Cu and a dopant hole distributed over 4 neighboring $O(p_\sigma)$ orbitals leads for certain parameter choices again to an effective single band model. Numerical results for the pure Hubbard models indicate that attraction via spin correlations is either not present or not strong enough [13,14] .

The question whether holes introduced upon doping feel an attractive force and the particular role of the near neighbor Coulomb repulsion is discussed in the following paragraph. New physics is introduced into the Hubbard model for the CuO_2-planes by addition of the Coulomb repulsion $V = U_{pd}$ between Cu and O neighbors. This interaction, if sufficiently strong, may lead to a considerable attraction between carriers. Varma et al [15] have emphasized the importance of charge transfer excitations induced by a Cu-O Coulomb interaction as a crucial polarization mechanism. The analogy of HTSC's with quasi-one-dimensional halogen-bridged metal complexes where V plays an important role has been stressed by Baeriswyl and Bishop [16] . Recent studies have shown that the interplay of a large U_d, i.e. strong correlations at Cu-sites, and a sufficiently large V is the source of strong attractive interactions between charge carriers [17,18,19] .

The final section gives an analysis of the effect of charge transfer excitations on spin- correlations upon doping. These results are discussed in connection with a spin-fermion Hamiltonian which excludes charge transfer excitations.

BANDSTRUCTURE AND MODEL HAMILTONIAN

The hybridized Cu(d) and O(p) band complex looks rather complicated, nevertheless the essential features can be understood in a simple 3-band tight-binding model [20] . The problem is thought to simplify firstly because there is only one electron per CuO_2 unit cell missing to fill all bands. Secondly the largest overlap is between Cu($d_{x^2-y^2}$) and O(p_σ)- orbitals in the Cu-O planes. These orbitals have a relatively small energy separation $\epsilon = \epsilon_p - \epsilon_d$, leading to strong covalent splitting of the corresponding bands. The resulting bands form the top and the bottom of the band complex, and the topmost anti-bonding band is half-filled. Apart from the bonding and antibonding band there is a nonbonding band which is dispersionless as long as the direct hopping matrix element t_{pp} between oxygens is zero.

The Coulomb interaction is characterized by 3 terms, a Hubbard repulsion $U_d(U_p)$ for two holes on Cu(O)-sites, respectively, and a Coulomb repulsion $V = U_{pd}$ between nearest neighbor Cu-O pairs. This model, which is now also known as Emery's model [21,22] , is able to describe a variety of situations, e.g. insulators with Mott-Hubbard and charge-transfer gaps.

$$H = \sum_{i,j,\sigma} \epsilon_{i,j} a^+_{i,\sigma} a_{j,\sigma} + \frac{1}{2} \sum_{i,j,\sigma,\sigma'} U_{i,j} n_{i,\sigma} n_{j,\sigma'}. \tag{1}$$

Throughout this work we use the hole picture, i.e. the $a^+_{i,\sigma}$ are creation operators for holes in copper $3d_{x^2-y^2}$ or oxygen $2p_x(2p_y)$ orbitals, respectively, and $n_{i,\sigma}$ are the corresponding particle number operators. We use particle with the same meaning as holes. The vacuum state has no holes, and corresponds to Cu^+ and O^{2-}, i.e. all 3d and 2p states occupied. Besides the site diagonal terms $(\epsilon_{i,i}, U_{i,i})$, i.e. (ϵ_d, U_d) and (ϵ_p, U_p), we include the nearest neighbor repulsion $V = U_{pd}$ which represents the long range Coulomb interaction. The hopping matrix element $\epsilon_{ij} = t_{pd} = t = 1$ is taken as unit of energy, and the appropriate phases are taken into account according to the symmetry of the wave functions. In certain cases also the hopping matrix element t_{pp} between neighboring oxygen sites is taken into account.

SINGLE-PARTICLE EXCITATIONS OF EXTENDED HUBBARD MODELS

There are a wide variety of experimental probes of condensed systems (XPS, UPS, BIS, EELS, etc.) [1] for which a theoretical description may be formulated simply in terms of a single-particle (-hole) propagator [23] . In the particular case of transition metal oxides, these measured excitation spectra cannot be understood fully using conventionally calculated band structures. These discrepancies may however be understood to arise from correlation effects which are not properly treated in single particle band structure calculations. These features may most simply and successfully be handled within an impurity model [24] , where a single metal ion coordinated by the appropriate ligands models the crystal. This approach has the advantage of being simple enough to allow for a realistic treatment of many atomic orbitals with crystal-field effects etc. On the other hand, there are interesting questions concerning for example quasi-particle dispersion which cannot readily be considered within the impurity approach.

The model calculation presented here may be considered as a step from the impurity problem toward the crystal. The advantage of increasing the range of allowed questions to include dispersion comes however, at the cost of increasing (numerical) difficulty. Even with the restriction of the model to only a single orbital per site, the practical limit is at present reached with a cluster of only 4 CuO_2 units. This is a rather severe limitation, as one may not be able to extrapolate to the thermodynamic limit. The related problem of the discrete nature of the spectra of finite systems may also cause difficulties in the interpretation of the calculated spectra. This problem is at least partially overcome through the use of modified periodic boundary conditions (MPBC) [25,26] . This consists of the inclusion of complex phase factors $\exp(i\phi)$ when a particle crosses the boundary of the cluster and enters again at the other side. Through this device it is possible by varying the angle ϕ to alter the set of allowed single particle momenta continuously. The accompanying energy shifts result in a filling in of the calculated spectra.

The spectral density of the single-hole excitations may be defined as

$$A_{k\sigma}(\omega) = \frac{1}{N_c} \sum_m |\langle \psi_m(N-1)|a_{k\sigma}|\psi_0(N)\rangle|^2 \delta(\omega - E_0(N) + E_m(N-1)). \quad (2)$$

Here the operator $a_{k\sigma}$ annihilates a particle with momentum k and spin σ, $|\psi_0(N)\rangle$ is the ground state eigenfunction of an N-particle system, and $|\psi_m(N-1)\rangle$ is an eigenstate of the (N-1)-particle system, which have energies $E_0(N)$ and $E_m(N-1)$, respectively. An analogous definition applies for the particle spectral function. Naturally momentum conservation restricts the possible final states $|\psi_m(N-1)\rangle$ which may contribute to the spectral function. The hole spectral function may be rewritten as

$$A_{k\sigma}(\omega) = \frac{1}{\pi} Im\langle u_0(N-1)|[E + H - i\delta]^{-1}|u_0(N-1)\rangle, \quad (3)$$

where $|u_0(N-1)\rangle = a_{k\sigma}|\psi_0(N)\rangle$, and $E = \omega - E_0(N)$ is the excitation energy measured from the ground state energy of the N-particle system, and $\delta \to 0^+$. In practice a finite value of δ leads to a useful smoothing of the spectra, which would otherwise consist of a set of spikes due to the relatively small number of k-points used.

The desired expectation value is readily calculated iteratively by using the Lanczos algorithm [27] to generate a basis in which the Hamiltonian is tri-diagonal. To begin with the N-particle ground state eigenvector $|\psi_0(N)\rangle$ is calculated using the standard Lanczos technique. Then a particle is annihilated in this state in a fashion which may depend on the experiment which is being mimicked when there is more than a single site per unit cell. This new $N-1$ -particle state is then expressed in the appropriate basis, and the problem of calculating the propogator has been reduced to the evaluation of the expectation value of the matrix $(H+z)^{-1}$ in this state $|u_0(N-1)\rangle$. The desired expectation value may be found by renewed application of the Lanczos algorithm, with the starting vector given by $|u_0(N-1)\rangle$. After M Lanczos iterations, an M dimensional tridiagonal representation of the Hamiltonian is generated. The coefficients of this matrix may then be used in a continued-fraction expansion [28] of the inverse $(H+z)^{-1}$, or equivalently this expectation value may be expressed in terms of the eigenvalues and eigenvectors of the (small) M-dimensional tridiagonal matrix. The latter approach is to be preferred if frequency integrals over the density of states are desired.

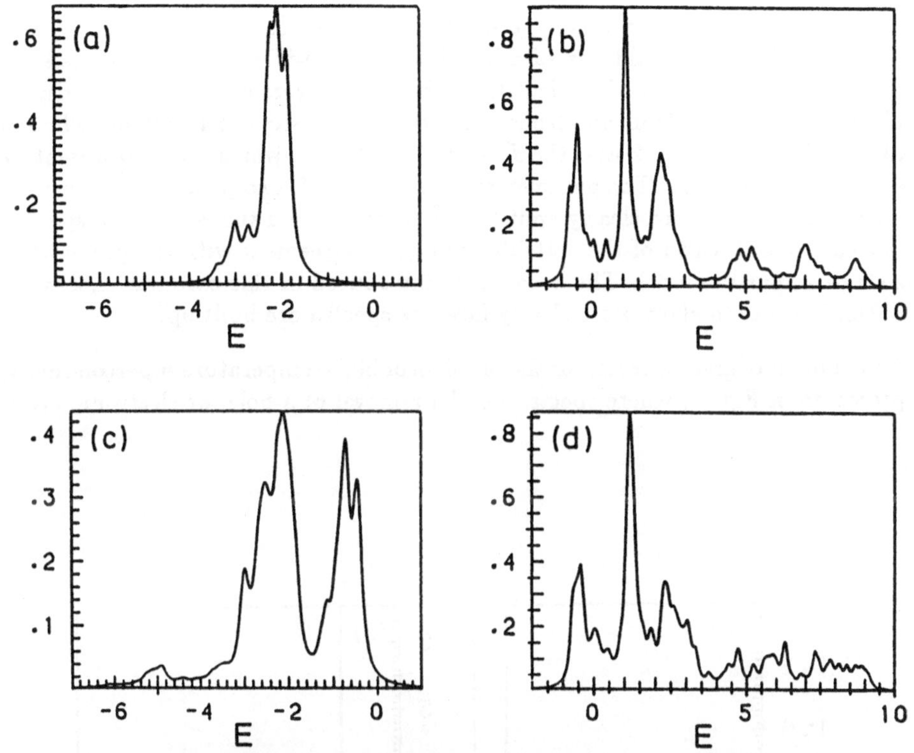

Fig. 1. **Photoemission spectra for the parameters $\epsilon = 2.0$, $U_d = 6.0$, $U_p = 3.0$, $U_{pd} = 0.0$, with $t = 1.0$ taken as the unit of energy. Five different boundary conditions were used, along with a broadening of $\delta = 0.1$. In figs. 1(a) and (b) the initial state is half-filled, i.e. 4 holes in 4 unit cells. The final state in Fig. 1(a) has one hole less, corresponding to inverse photoemission of electrons, while in Fig. 1(b) a hole is added, corresponding to photoemission of electrons. Figures 1(c) and 1(d) represent a repetition of the same experiments starting from an initial state which is doped with one extra hole, which implies a doping concentration of 25%.**

Figure 1(a,b) shows typical photoemission and inverse photoemission spectra starting from half-filling. The existence of a gap between the top of Fig. 1(a) and the bottom of Fig. 1(b) is consistent with an insulating ground state at half-filling, as expected. The most notable change in spectra (1c) and (1d) of the doped system is the appearance of states within the energy region corresponding to the gap at half-filling shown in figs. 1(a) and (b). Another feature to note is the sharp peak at $E = 1$ in figs. 1(b) and (d), which may be attributed to the addition of holes into the dispersionless non-bonding oxygen band.

A further possibility for this type of calculation is illustrated in Fig. 2, where another example of the addition of a hole into the half-filled ground state is shown. Fig. 2(a) shows the result of a calculation where holes were created only on Cu sites, while Fig. 2(b) results from an incoherent sum of the O site contributions. The sharp peak near $E = 13$ in Fig. 2(a) is the $d^9 \rightarrow d^8$ Cu sattelite, which is clearly absent from the O spectrum in 2(b). The parameters are as in Fig. 1, except for an increase of U_d from 6.0 to 12.0 to more clearly split the d^8 sattelite from the rest of the spectrum. The location and width of the sattelite is in good agreement with the predictions of the single impurity model. The broadening introduced in this case was chosen to be $\delta = 0.02$ in order to show more clearly how the spectra are built up.

A final point of great interest for the problem of high-temperature superconductivity is presented in Fig. 3, where spectra for the removal of a hole, or electronic inverse-

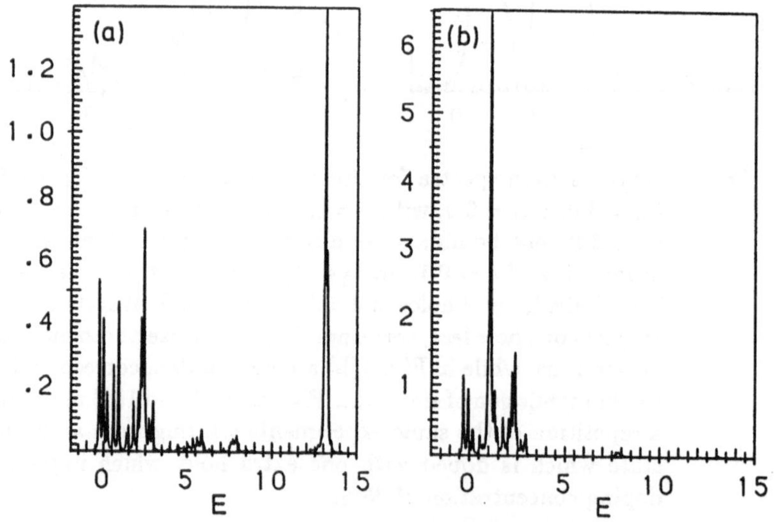

Fig. 2. Creation of a hole in the insulating state (a) on a Cu- and (b) on a O-site (corresponding to photoemission of electrons). Parameters as in Fig.1, except for $U_d = 12$. The Cu-spectrum (a) reveals the $d^9 \rightarrow d^8$ sattelite which is not present in the O-spectrum.

photoemission, are shown for two different parameter sets. In both cases the initial state was doped with one extra hole (25%), and the system parameters were $\epsilon = 1.0$, $U_d = 6.0$ and $U_p = 3.0$. For Fig. 3(a) the nearest neighbor Coulomb repulsion U_{pd} was 0, and for 3(b) $U_{pd} = 2.0$ was used. In Fig. 3(b) a "pseudo-gap" is clearly visible between the states introduced by doping (energy region $E \geq 0$) and the hole states which are occupied at half-filling. In Fig. 3(a) for $U_{pd} = 0$ this "gap" is absent. These results may also be compared to Fig. 1(c), where a similar "gap" also appears, however in this case due to larger charge-transfer energy $\epsilon = 2.0$ rather than due to the Coulomb interaction. This point is of particular interest due to a number of experiments which may be consistent with the existence of a similar gap in HTSC's in inverse-photoemission and EELS [29] .

ATTRACTION BETWEEN CARRIERS

Electronic polarization can be the source of an effective attractive interaction between carriers (holes) introduced upon doping. As the relevant electronic excitation energies are much larger than typical phonon energies this might provide an explanation of the high superconducting transition temperatures. The short coherence length of HTSC's, i.e. the small extend of Cooper pairs, suggests that we can learn about the pairing mechanisms from small systems calculations, without making further unnecessary approximations. The results may provide insight into appropriate approximation schemes.

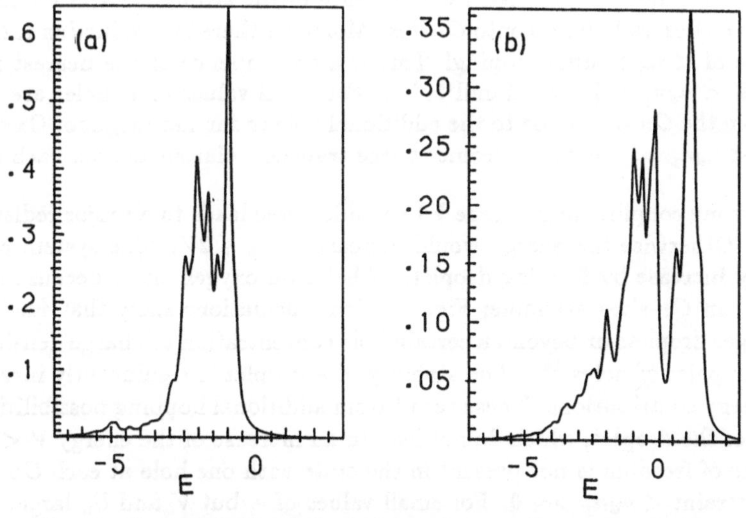

Fig. 3. Removal of a hole (electronic inverse photoemission) for two parameter sets: $t = 1$, $\epsilon = 1$, $U_d = 6$, $U_p = 3$ and (a)$U_{pd} = 0$ and (b) $U_{pd} = 2$, respectively. In both cases the initial state was doped with one extra hole per 4 CuO_2-unit cells (25%).

The binding energy between two carriers added to the undoped system is defined by

$$E_B = (E_2 - E_0) - 2(E_1 - E_0). \quad (4)$$

Here E_0 is the ground state energy of the undoped system, i.e. with 1 hole per CuO_2-unit cell, and $E_{1(2)}$ refer to the ground state energies of systems with 1(2) more holes. Negative binding energy indicates an instability of the system, which in the large system limit is related to the energy of a Cooper pair unless e.g. a charge density wave state turns out to be more favorable. The low-lying eigenvalues and eigenvectors of this many-body problem are found for particular numbers of spin-up and -down holes by generating the complete set of atomic limit basis states, finding all non zero matrix elements of the Hamiltonian (1), and then applying the Lanzcos algorithm for diagonalization [19] . The size of the matrices to be diagonalized limits the number N_C of CuO_2-unit cells which can be handled. We use here periodic boundary conditions, and typical sizes are up to $N_C = 6$ in the spin-polarized case and $N_C = 4$ otherwise.

To understand the physical mechanism leading to negative binding energy in the presence of V, it is helpful to consider some limiting cases. We note that the binding energy saturates as a function of U_d. Figure 4 gives the binding energy as function of V in the limit $U_d \to \infty$ for 3 different cases: (i) $U_p = 0$, (ii) $U_p = \infty$ and (iii) all holes with same spin. The binding energy does not essentially depend on these choices, which clearly indicates that the spin degrees of freedom only play a minor role. A clear crossover between weak and strong binding is seen for V comparable to ϵ. Direct insight into the nature of the ground states with 0,1 and 2 additional holes follows from the dependence of the hole density $< n_d >$ on Cu and the density correlation function $< n_d n_p >$ between neighboring Cu and O sites, Fig. 4(b) and (c). In the undoped ground state holes are preferentially located on Cu-sites ($< n_d > \cong 0.7$ for $\epsilon = 2$ and $V = 0$). As a consequence of the strong correlation on the Cu sites (large U_d) extra holes go onto O-sites. Moreover there is a reduction rather than an increase of $< n_d >$ upon doping! This is a consequence of the nearest neighbor repulsion $V < n_d n_p >$ in the Hamiltonian. For small values of V holes are virtually excited from the Cu sites close to the additional hole to further neighbor Oxygen sites to reduce $< n_d n_p >$. These electronic charge transfer polarons attract each other.

In the strong coupling limit (large V) an added hole leads to a major redistribution of charge. Otherwise the energy would increase by $\epsilon_p + 2V$. The system can avoid this energy increase by forming droplets of holes on oxygen sites, because there are more O- than Cu-sites available; Fig. 5. Our calculations show that further holes lead to larger droplets or beyond a certain hole concentration to charge density waves, rather than pairs of holes [19] . The stability of a droplet is significantly increased by kinetic energy contributions. These result from additional hopping possibilities at the boundary of the droplet, which do not lead to an increase of the energy $V < n_d n_p >$. This degree of freedom is not present in the state with one hole at each Cu, because of the constraint $< n_d n_p > \sim 0$. For small values of ϵ, but V and U_d large, even the undoped ground state becomes unstable against the formation of droplets, leading to mixed phase of $Cu^{2+}O^{2-}$ and $Cu^{+}O^{-}$.

The character of the different ground states with 1 extra hole is summarized in a phase diagram of the (ϵ, V)-plane (Fig. 6). The parameters chosen here are typical for $HTSC's$, as found from various LDA-calculations [30] (Energies in eV are obtained by multiplying these parameters with $t \sim 1.5eV$). For $\epsilon \geq 2$ the extra hole will either

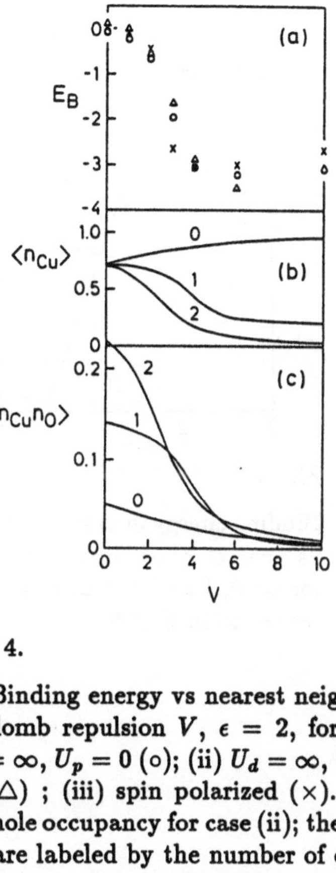

Fig. 4.

(a) Binding energy vs nearest neighbor Coulomb repulsion V, $\epsilon = 2$, for: (i) $U_d = \infty$, $U_p = 0$ (o); (ii) $U_d = \infty$, $U_p = \infty$ ($\triangle$) ; (iii) spin polarized ($\times$). (b) Cu-hole occupancy for case (ii); the curves are labeled by the number of extra holes (0,1, and 2). (c) Density-density correlation function between Cu-O neighbors.

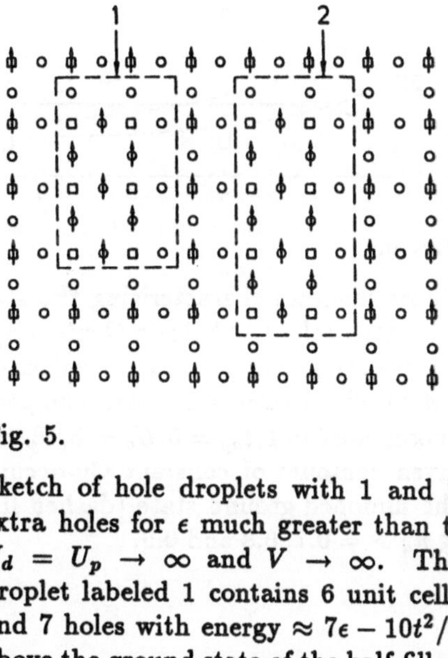

Fig. 5.

Sketch of hole droplets with 1 and 2 extra holes for ϵ much greater than t, $U_d = U_p \to \infty$ and $V \to \infty$. The droplet labeled 1 contains 6 unit cells and 7 holes with energy $\approx 7\epsilon - 10t^2/\epsilon$ above the ground state of the half-filled case. The box labeled by 2 contains 8 unit cells and 10 holes, and has energy $\approx 10\epsilon - 12t^2/\epsilon$ above the undoped system. The estimated binding energy is $E_2 - 2E_1 \approx -4\epsilon + 8t^2/\epsilon$ [19] .

go (A) on an O-site (charge transfer regime) or (B) on a Cu-site (Mott-Hubbard regime), with energies $E_1^{CT} = \epsilon + 2V$ and $E_1^{MH} = U_d$, depending on the magnitude of V. The droplet energy $E_1 = 7\epsilon + 10(\epsilon/2 - \sqrt{\epsilon^2/4 + t^2})$ (in the limit discussed in Fig. 5) gives an estimate of the phase boundary of these regimes.

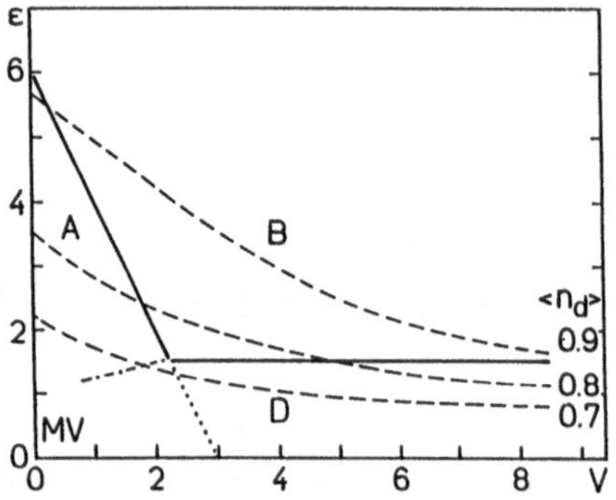

Fig. 6.

Phase diagram characterizing the state with one extra hole: (A) hole on O-site, (B) hole on Cu-site, (D) droplets for large values of V and (MV) mixed valence regime. The parameters chosen are $t = 1, t_{pp} = 0, U_d = 6, U_p = 3$. Also given contours of constant Cu-occupancy in the undoped ground state (dashed lines), for $< n_d > = 0.7$, 0.8 and 0.9.

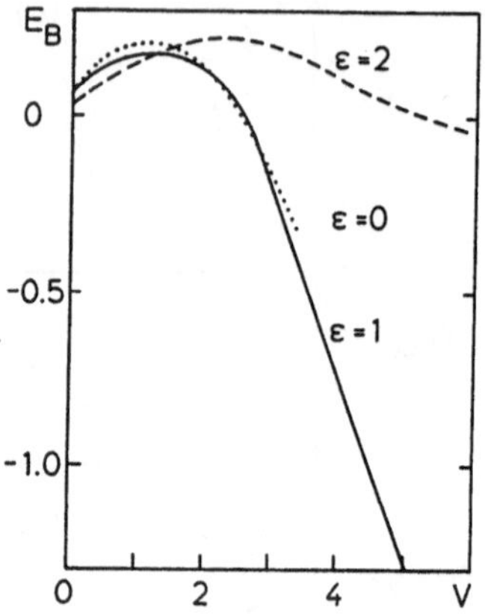

Fig. 7.

Binding energy in units of t (t=1) vs nearest neighbor repulsion V for ϵ= 0, 1 and 2. Same parameter set as in Fig. 6.

Figure 7 shows the binding energy as function of V, and for various values for ϵ. Negative E_B is found for $V \geq 2.5$ and $\epsilon = 0$ and 1, for slightly larger values of ϵ the critical value V becomes large. The proximity of ϵ_p and ϵ_d and also of the ionization levels of Cu and O appears to be essential for pairing. The stability of a pair is found from $\Delta E_B^{(3)} = (E_3 - E_0) - ((E_2 - E_0) + (E_1 - E_0))$, where E_3 is the ground state with 3 extra holes. If $\Delta E_B^{(3)}$ becomes negative droplets will form. There is only a small region at the borderline of this instability where pairs are stable; e.g. $2.5 < V < 3.7$ for $\epsilon = 1$. These values as the binding energy [19] are expected to depend on the size of the cluster.

The values required for pairing seem to be larger than those estimated from local density calculations [30]. On the other hand, the real substances are much more polarizable than the 3-band model. There are many bands near the Fermi level, not taken into account in the model, which would effectively contribute to charge transfer excitations. This can be modeled to a certain extent in the 3-band model by choosing $t_{pp} \neq 0$, such that the nonbonding band comes close to the Fermi level. This already lowers the critical V from 2.5 to 1.8 for $\epsilon = 1$.

SPIN-CORRELATIONS IN THE MULTI-BAND MODEL

Charge-transfer excitations modify the spin-correlation functions (CF) particularly in the vicinity of the extra holes. It is instructive to compare this to a spin-fermion Hamiltonian which leaves out charge excitations and includes only the coupling of mobile extra holes on the O-sublattice with spins on Cu-sites.

$$H = t' \sum_{<i,j>\sigma} (c^{+}_{i,\sigma} c_{j,\sigma} + h.c.) + U_p \sum_i n_{i\uparrow} n_{i\downarrow} + 2J_K \sum_{<i,l>} \vec{S}_l \cdot \vec{\sigma}_i + 2J_S \sum_{<l,m>} \vec{S}_l \cdot \vec{S}_m, \quad (5)$$

where $\vec{\sigma}_i = \frac{1}{2} c^{+}_{i\sigma} \vec{\sigma}_{\sigma\sigma'} c_{i\sigma'}$ describes the spin of a hole on a O-site, and t' is the effective hopping matrix element of these holes. The spin-fermion Hamiltonian is closely related to the original Hamiltonian Eq.(1) for $U_d \gg \epsilon \gg t$, and has been investigated in great detail by Imada [31]. The spin-fermion coupling J_K

$$J_K = \frac{2t^2}{U_p + \epsilon} + \frac{2t^2}{U_d - \epsilon - 2V} - J_D \ , \ J_S = \frac{2t^4}{(\epsilon + V)} \left[\frac{1}{U_d} + \frac{2}{U_p + 2\epsilon} \right], \quad (6)$$

frustrates the antiferromagnetic superexchange interaction J_S [22,31]. The sign of J_K depends on the magnitude of the direct exchange J_D. Particularly the case of localized holes on oxygen $t' = 0$ has been discussed in considerable detail [32,22]. The frustration leads to a ferromagnetic alignment of the two Cu-spins close to the O-hole. It has been argued that the frustration is a significant pairing mechanism [32].

The frustration is measured directly by the conditioned spin-CF (Fig.8(b) and 9), i.e. the correlation function between two n.n. Cu-spins under the condition that there is a hole on the O-site in between. While the averaged spin-CF between nearest neighbors, Fig.8(a), is always antiferromagnetic, the conditioned CF shows ferromagnetic frustration as expected from the simple considerations. The size of the conditioned CF, however, is an order of a magnitude smaller in the region where the binding energy becomes negative, because the charge transfer excitations induced by the extra holes weaken the spin-correlations. Hence the dominant mechanism determining the binding energy is due to charge-transfer excitations.

These results in the regime, where the level difference ϵ is comparable to t, differ from those obtained by Imada in the region $t' \sim J_K \gg J_S$. In this region the spin-spin correlations between the hole's spin and the localized spins are always antiferromagnetic, which is interpreted as an extended singlet cloud [31].

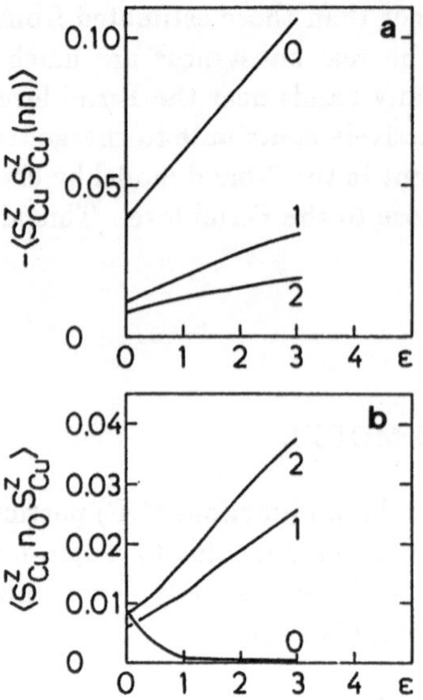

Fig. 8.

The nearest neighbor spin-CF (a) for V=1 vs ϵ is antiferromagnetic, while the conditioned n.n. spin-CF (b) is ferromagnetic.

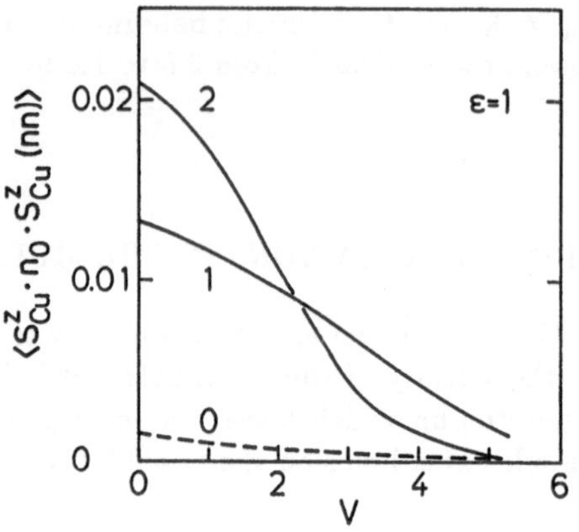

Fig. 9.

The conditioned CF between Cu-spins vs. V for $\epsilon = 1$. The number of extra holes is labeled by 0,1, and 2.

SUMMARY

To provide a basis for comparison with the wealth of spectroscopic data, we have presented first results of photoemission and inverse photoemission spectra for the strongly correlated Cu-O layers. The results presented here focus in particular on the change of these spectra upon doping.

In conclusion we have shown that the combination of strong correlations on Cu (large U_d) and a sufficiently large Coulomb repulsion between Cu and O leads to an attraction between charge carriers (extra holes). For realistic choice of parameters (t,U_d,U_p) pairing is found for small level distance ϵ and intermediate values of V, indicating the closeness of the ionization levels of Cu and O. This proximity of the ionization levels has been argued to be characteristic for HTSC's [33] . In this regime quasiparticles, i.e. holes on oxygen, are dressed by charge transfer excitations from neighboring Cu-sites to further neighbor O-sites (electronic CT polarons). For larger values of V pairs become unstable, and the Coulomb repulsion favors the formation of droplets of holes on oxygen atoms and at higher doping concentration of charge density waves.

The 3-band model is, of course, a crude simplification leaving out many excitation channels, which all contribute to the polarizability and further modify the effective interaction between charge carriers. A further extension in this direction is the d-d excitation model investigated by Weber et al [34] , where in addition the d($3z^2 - r^2$)-orbitals of Cu are taken into account. Also the lattice degrees of freedom not considered here couple to charge transfer excitations and are expected to contribute to binding.

ACKNOWLEDGEMENTS

The authors acknowledge helpful discussions with O. Gunnarsson and J. Zaanen. They are particularly grateful to W.v.d. Linden for his continuous advice. One of us (P.H.) would like to thank the Institute for Scientific Exchange, Torino, were part of this manuscript was prepared.

REFERENCES

1. J. C. Fuggle, J. Fink, and N. Nücker, Int. J. Mod. Phys. B1:1185 (1988) and references therein.
2. H. Rietschel et al., Physica C153-155:1067 (1988).
3. T. Takahashi et al., Nature 334:691 (1988).
4. J. M. Imer et al., (preprint).
5. N. Nücker et al., (preprint).
6. F. J. Himpsel et al., Phys. Rev. B38:11946 (1988).
7. J. Zaanen, O. Jepsen, O. Gunnarsson, A.T. Paxton, and O. K. Andersen, Physica C153-155:1636 (1988).
8. J. Zaanen, G.A. Sawatzky, and J.W. Allen, Phys.Rev.Lett. 55:418 (1985).
9. P. Fulde, Physica C153-155:1769 (1988).
10. P. W. Anderson, Science 235:1196 (1987).
11. A. Muramatsu, (this volume).
12. F. C. Zhang and T. M. Rice, Phys.Rev. B37:3759 (1988).
13. J. E. Hirsch, E. Loh, D. J. Scalapino, and S. Tang, Physica C153-155: 549 (1988).
14. M. Ogata and H. Shiba, J.Phys.Soc.Jpn. 57:3074 (1988).
15. C. M. Varma, S. Schmitt-Rink, and E. Abrahams, Physica C153:1622 (1988).
16. D. Baeriswyl and A. R. Bishop, Phys. Scr. T19:239(1987).
17. J. E. Hirsch, S. Tang, E. Loh, and D. J. Scalapino, Phys.Rev.Lett. 60: 1668 (1988).
18. C. A. Balseiro et al., Phys. Rev. B38:9315 (1988).
19. W. H. Stephan, W. v.d. Linden, and P. Horsch, Int. J. Mod. Phys. B1: 1005 (1988), and Phys. Rev. B39,Feb.(1989).
20. M. S. Hybertsen and L. F. Mattheiss, Phys. Rev. Lett. 60:1661 (1988).
21. V. J. Emery, Phys. Rev. Lett. 58:2794 (1987).
22. V. J. Emery and G. Reiter, Phys. Rev. B38:4547 (1988).
23. G. D. Mahan, Many-Particle Physics, (Plenum, New York, 1981).
24. J. Zaanen, C. Westra, and G. A. Sawatzky, Phys.Rev. B33:8060 (1986).
25. R. Jullien and R. M. Martin, Phys. Rev. B26:6173 (1982).
26. A. M. Oles, G. Treglia, D. Spanjard, and R. Jullien, Phys. Rev. B32:2167 (1985).

27. B. N. Parlett, The symmetric Eigenvalue Problem, (Prentice Hall, Englewood Cliffs,1980).

28. R. Haydock, V. Heine, and M. J. Kelly in Solid State Physics, Vol.35, edited by H. Ehrenreich, F. Seitz, and D. Turnbull (Academic, New York, 1980).

29. J. Fink, N. Nücker, H. Romberg, and J. C. Fuggle, IBM Journal of Research and Development (in print).

30. M. Schlüter, M. S. Hybertsen, and N. E. Christensen, Physica C153-155: 1217 (1988); M. S. Hybertsen et al. (preprint) and references therein.

31. M. Imada, N. Nagaosa, and Y. Hatsugai, J.Phys.Soc.Jpn. 57:2901 (1988); M. Imada, Proceedings of the 2nd NEC Symposium on Mechanisms of HTSC'y. (Springer).

32. A. Aharony, R. J. Birgeneau, and M. A. Kastner, Int.J.Mod.Phys. B1:649 (1988).

33. Z. Tesanovic, A. R. Bishop, R. L. Martin, and K. A. Müller, (submitted to Nature).

34. W. Weber, A. L. Shelankov, and X. Zotos, Proceedings of the 2nd NEC Symposium on Mechanisms of HTSC'y. (Springer,Heidelberg, in print).

NUMERICAL APPROACH TO MAGNETIC MECHANISM OF SUPERCONDUCTIVITY

Masatoshi Imada

Department of Physics, College of Liberal Arts
University of Saitama, Shimo-Okubo, Urawa 338

ABSTRACT

Since the discovery of high-Tc superconductors, various models have been proposed and investigated from different point of view to clarify the mechanism of high-Tc superconductivity. We start with a two-band d-p model with copper and oxygen orbitals in the CuO_2 plane of the high-Tc superconductors. The models investigated in the numerical analysis include the original two-band d-p model, the coupled spin-fermion model and the t-J model.

The d-p model given by

$$\begin{aligned} H = &-t_1 \sum_{<ij>,\sigma} (d_{i\sigma}^{\dagger} p_{j\sigma} + d_{i\sigma}^{\dagger} q_{j\sigma} + h.c.) \\ &- t_2 \sum_{<ij>\sigma} (p_{i\sigma}^{\dagger} q_{j\sigma} + h.c.) \\ &+ U_d \sum_i n_{di\uparrow} n_{di\downarrow} + U_p \sum_j (n_{pj\uparrow} n_{pj\downarrow} + n_{qj\uparrow} n_{qj\downarrow}) \\ &+ V \sum_{<ij>} n_{di}(n_{pj} + n_{qj}) \\ &+ \epsilon_d \sum_i n_{di} + \epsilon_p \sum_j (n_{pj} + n_{qj}), \end{aligned}$$

has been investigated by the quantum simulation method [1], where fermion operators d, p and q represent holes in Cu-$3d_{x^2-y^2}$, p_x and p_y orbitals, respectively. The results show small but finite enhancement of the superconducting susceptibilities as compared to the noninteracting case.

The coupled spin-fermion model given by

$$\begin{aligned} H = &-t \sum_{<i,j>\sigma} (c_{i\sigma}^{\dagger} c_{j\sigma} + c_{j\sigma}^{\dagger} c_{i\sigma}) + U_h \sum_i n_{i\uparrow} n_{i\downarrow} \\ &- 2J_K \sum_{<i,l>} \vec{S}_l \cdot \vec{\sigma}_i - 2J_S \sum_{<l,m>} \vec{S}_l \cdot \vec{S}_m \qquad (J_S < 0) \end{aligned}$$

Interacting Electrons in Reduced Dimensions
Edited by D. Baeriswyl and D.K. Campbell
Plenum Press, New York

$$\vec{\sigma}_i = \frac{1}{2} c^\dagger_{i\sigma} (\vec{\sigma})_{\sigma\sigma'} c_{i\sigma'}$$

$$n_{i\sigma} = c^\dagger_{i\sigma} c_{i\sigma}$$

is derived as the Kondo limit of the d-p model, where the charge fluctuation at the Copper sites is suppressed and the local moment is preserved. The ground state of this model has been examined [2] by the exact diagonalization of finite lattices. The system with one itinerant fermion shows the existence of the extended cloud around the fermion, when the transfer of the itinerant fermion and the Kondo coupling are large. In the cloud, the antiferromagnetic correlation in the substrate of the Heisenberg spins is reduced from the pure Heisenberg system. The fermion's spin couples antiferromagnetically to the localized spins in the cloud. It shows the developement of totally singlet structure constructed from the fermion's spin and the localized Heisenberg spins in the cloud. The interaction of two fermions has also been calculated. It seems to show attractive behavior irrespective of the detailed fermion lattice structure and the dimensionality. The origin of the attractive interaction is attributed to the fact that the total area of the reduced antiferromagnetic correlation is smaller if two clouds overlap. The developement of the singlet cloud and the attractive interaction seem to be the characteristic feature only in the case of the antiferromagnetic Kondo coupling. In fact, the distortion of the localized spins around the fermion is not remarkable in the case of ferromagnetic Kondo coupling.

REFERENCES

1) M. Imada: J. Phys. Soc. Jpn. 56, 3793 (1987).
M. Imada: J. Phys. Soc. Jpn. 57, 3128 (1988).
2) M. Imada, N. Nagaosa and Y. Hatsugai: J. Phys. Soc. Jpn. 57, 2901 (1988).
M. Imada, Y. Hatsugai and N. Nagaosa: Proceedings of the Adriatico Conference "*Towards the Theoretical Understanding of High* − T_c *Superconductors*" ed. by S. Lundquist et al. (World Scientific, Singapore, 1988) p423.
Y. Hatsugai, M. Imada and N. Nagaosa: J. Phys. Soc. Jpn. 58 No.4 (1989).
M. Imada: Proceedings of the 2nd.NEC Symposium "Mechanism of High-Tc Superconductivity" (Springer Verlag, 1989).
N. Nagaosa, Y. Hatsugai and M. Imada: J. Phys. Soc. Jpn. 58 No.3 (1989).

PARASTATISTICS FOR HIGHLY CORRELATED TWO DIMENSIONAL FERMI SYSTEMS

E. J. Mele and D. Morse
Department of Physics
University of Pennsylvania
Philadelphia, PA 19104
USA

Abstract

A parastatistics representation for the quasiparticles of the large U Hubbard model for fermions in two dimensions is discussed. Our representation replaces the dynamical Gutzwiller constraint in the large U Hubbard model by a kinematical constraint obeyed by the quasiparticles. The resulting parastatistical objects interact through a U(1) gauge field, which can be interpreted in a mean field theory as a long range interaction of the quasiparticles with an effective boundary current on a macroscopic (finite) system, with the size of this current specified by the net spin of the enclosed many body system. Illustrative examples of calculations employing this statistical representation are discussed for the strongly correlated Hubbard problem both at the half filled band and away from the half filled band.

I. Introduction

There is a renewed interest in electron correlation effects in the low temperature properties of the two dimensional Hubbard model. This interest is stimulated in part by the possibility that the high temperature superconductivity of the doped CuO_2 superconductors may be associated with novel quantum behavior in the "large U" limit of this model [1]. In this paper I discuss an unconventional but computationally tractable approach to this problem, which exploits a well known property of the dynamics of inpenetrable quantum objects in two dimensions. Specifically, I propose that the formidable problem of accurately describing correlation effects off the half filled band in the strong positive U limit, may be addressed easily by adopting a parastatistics (nonfermionic) representation for the underlying quasiparticles. This idea provides a physically sensible picture of the many body dynamics in the strongly correlated state whose essential qualitative features will be summarized here; we are continuing to study a number of other unusual and intriguing aspects of this parastatistics representation.

Our parastatistics formulation makes use of a general observation about the dynamics of strongly correlated and highly quantum mechanical two dimensional systems, and ultimately provides a picture of the many body dynamics which is very closely related to the RVB model for the highly correlated Hubbard model as described by Anderson and colleagues [1-7] over the last several years. Importantly, this approach leads us to the idea that the dynamics of the underlying quasiparticles is dominated by an interaction with an effective internally generated U(1) gauge field [3,5,6,7] . In our formulation this field may be interpreted as resulting from a boundary current around the two dimensional system prepared in a spin eigenstate , where the magnitude of the current is specified by the *spin* of the enclosed many body system. The resulting picture of the quantum fluid ground state is quite closely related to present ideas about a gauge representation of the RVB state in two dimensions at the half filled band, and the modification of these phases as one moves off the half filled band. In applications of this idea to physical systems, we have found that this approach also provides a suprisingly simple and physically appealing explanation for some very perplexing low temperture properties exhibited by the Bechgaard salts, which have been observed for some time, and which have remained poorly understood [9].

II. Parastatistics Representations in Two Dimensions

The quantum dynamics of hard core objects in two dimensions is special, in the sense that the specification of the particle statistics is ambiguous and may be formally replaced by an effective interaction between the particles [8]. The replacement of the particle statistics by an interaction is accomplished by fixing a fictititous flux tube to each particle of the theory each threaded by ν flux quanta. For two particles, with complex coordinates in the plane denoted by z_1 and z_2, the effect of the flux tube is to make a Wilczek singular gauge transformation of the two particle wavefunction:

$$\Psi(z_1,z_2) = \exp\left(i\nu \, \mathrm{Im} \log (z_1-z_2) \right) \Phi (z_1,z_2) \tag{1}$$

so that for $\nu = 1$, the gauge factor is antisymmetric under exchange of the coordinates, and thus the dynamics of a two particle Fermi system in the Ψ space, may be described by the dynamics of two fictituous (interacting) bosons in the transformed Φ space.

A minor modification of this idea provides a powerful scheme for representing strongly correlated many *fermion* states in two dimensions. Consider fermions in a lattice model, where $a^+_{i\sigma}$ creates a particle with spin σ on site i. Two fermions coupled to a singlet state may occupy the same lattice site of course, so that configurations containing $a^+_{i\sigma} a^+_{i-\sigma}$ may appear in a general many body state. For Hubbard fermions in the large positive U limit the two particle repulsive interaction dynamically excludes these configurations from the low energy spectrum. Conventionally, one may attempt to describe this situation by constructing a correlated function, by projecting out configurations of the form (1) with the Gutzwiller projection:

$$\psi_G = \prod_i \left(1 - n_{i\uparrow} n_{i\downarrow}\right) \psi_o \tag{2}$$

However, configurations with double site occupancy can be excluded in another, more natural way. Note that an SU(2) singlet for two fermions on different sites is represented as the sum of the determinants:

$$a^+_{i\uparrow} a^+_{j\downarrow} - a^+_{i\downarrow} a^+_{j\uparrow} \tag{3}$$

This function is invariant under rotations of the spin coordinates, and is symmetric upon exchange of the spatical indices, so that doubly occupied configurations are of course kinematicaly admissible.

We now consider a spin dependent Wilczek transformation of the underlying fermion fields:

$$s^{+}_{i\sigma} = a^{+}_{i\sigma} \exp\left(i\sum_{j} n_{j-\sigma} \ \mathrm{Im}\log(\tau_i - \tau_j)\right) \tag{4}$$

where the τ_i denotes the complex lattice coordinate of the ith site in the system. The interpretation of this transformation is that particles with spin σ experience a unit flux attached to particles with spin $-\sigma$. The resulting field operators now exhibit mixed statistics: they are anticommuting for pairs with the same spin, but commuting for pairs with opposite spin. This leads to an important kinematical constraint on the dynamics of these objects; namely the SU(2) singlet combination:

$$s^{+}_{i\uparrow} s^{+}_{j\downarrow} - s^{+}_{i\downarrow} s^{+}_{j\uparrow} \tag{5}$$

is now ***antisymmetric*** upon exchange of just the spatial coordinates, and thus singlet configurations with i=j are kinematically excluded from consideration. In essence, the particles of opposite spin avoid each other due to a repulsive centrifugal barrier between opposite spin pairs. The Wilczek gauge transformation is a singular transformation which will thus exclude the doubly occupied configurations of the s's from the accessible Hilbert space. Note however, that the particle density for the s's is identically the particle density for the original fermions, so the Wilczek tranformation provides a gauge representation in which the original fermions are forced to avoid each other. In addition to the kinematical constraint provided by the Pauli principle which requires the coordinate wavefunction for particles of like spin to vanish when the coordinates coincide, this representation enforces an identical "exchange hole" for pairs of particles with opposite spin.

This construction is straightforward for two particles; the generalization of this result to an N-particle system is conceptually similar. Let us return to the case of an N-fermion wavefunction with N an even integer, and examine the spatial symmetry of a spin singlet state constructed from the N identical spin 1/2 particles. The Schrodinger operator acts only on the spatial coordinates so the wavefunction may be factored into a space part and a spin part, $\Psi = \psi(z_i)\,\chi\,(\sigma_i)$. We represent the spin singlet factor in this function by the elementary Young tableau consisting of two rows each with N/2 columns. [In the tableau, the spin function is constructed by partitioning the particles into two groups P and Q, and distributing the set P in the first row, with set Q in the second. For the spin singlet we assign the upper row the spin "up" magnetic state, and the lower row to spin "down". The desired singlet function is then constructed by first symetrizing the function over permutations within a row, and then antisymmetrizing with respect to the exchange of pairs in each of the columns. The resulting spin function is both a spin eigenfunction and a basis function of an irreducible representation of the symmetric group S_N]. The product state Ψ must project into the fully antisymmetric product space, which is possible if and only if the ψ factor transforms according to the conjugate (associated) tableau, i.e. a spatial function obtained by symmetrizing and antisymmetrizing the particles coordinates according to the tableau obtained by reversing the rows and columns describing the spin part. The resulting spatial part is now symmetrized over N/2 pairs of particles, and thus it is possible, in principle, to construct a fermionic function in which particles are paired and these pairs occupy $n_d < N/2$ of the available lattice sites.

We now inquire whether it is possible to make a further factorization of the spatial part of this wavefunction. We are interested in factoring a pure "fermionic" wavefunction out of the spatial tableau, i.e. extracting a function which belongs to the antisymmetric one dimensional representation of the permutation group (i.e. a function which is symetric under even permutations of the spatial coordinates and antisymmetric under the odd permutations) . Noting that the fermi statistics requires that the full spatial wavefunction must have the permutational symmetry of the conjugate tableau, we will attempt to absorb the remaining permutational symmetry in a pure gauge function, i.e a complex function of the particle coordinates with unit modulus. Thus we wish to factorize $\psi = \psi' g$ where g is a pure gauge function, and ψ' is a function describing a fluid of underlying "spinless" lattice fermions. Is such a factorization possible? Generally it is not, the difficulty being that fermions with spin 1/2 *may* doubly occupy lattice sites, a possibility which is manifestly not possible in our factored function ψ'. However, if the many body dynamics excludes double site occupancy, as occurs in the *infinite U* limit of the Hubbard model, then a gauge function may be defined, and one may verify that it is simply the Wilczek factor:

$$g(z_i) = \prod_{\substack{i \in P \\ j \in Q}} \exp(i \, \phi_{ij})$$

where $\phi_{ij} = \operatorname{Im} \log (z_i - z_j)$. This corresponds to an effective gauge interaction between pairs of particles from the partitions P and Q. For two particles, this trivially reduces to the two particle transformation in equation (4).

Note that in the spin singlet state, the number of flux quanta seen by a single particle in the system is the number of particles in the opposite partition (namely N/2). One may generalize this argument to derive a similar relationship for any pure spin state of the many body system, with the result that the number of quanta $N_\phi = N/2 - S$ where S is the total spin. This is a very interesting result. Note for example, that in a "fluid" state, for which the particle density in the plane is uniform, a Hartree average over the resulting gauge field describes a vector potential produced by a uniform magnetic field piercing the area of the sample. This in turn may be formally replaced by a surface current circulating around the sample boundary. Thus to maintain the system in a pure spin state, quantum fluctuations in our state must be correlated over a long range and interestingly these long range (algebraic) correlations are enforcible by introducing an *effective vacuum current circulating around the sample boundary*. It is interesting that a similar observation has been given by Kalmeyer and Laughlin [9], who, starting from a very different argument (studying the Holstein Primakoff bosons for the spin 1/2 Heisenberg antiferromagnet), find that analyticity of the many body state for these bosons constrains the system to a global singlet.

The parastatistics operators described by these s's thus appear to provide a very natural set field operators for describing the dynamics in the large U Hubbard system; in essence they describe a basis for the effective spinless fermions in the theory. These particles bear some resemblance to "half rotons" in that they consist of a bare particle (the original fermions a) surrounded by a vortex of circulating particles in the opposite spin sector. We refer to these excitations which consist of a bound electron-vortex pair as "vortons." Double site occupancy is kinematically excluded for the quantum vortons, although the price one pays for this simplification is that there is a long range gauge interaction between the particles.

Can one calculate within this parastatistics representation? The effective gauge field resulting from the Wilczek transformation is a complicated object. Nevertheless, there is an important simplifying feature; namely, this vector potential is very long ranged falling off only as 1/r between pairs of particles. Thus as one takes the thermodynamic limit in this problem, the gauge field "seen" by a particle should be dominated not by local fluctuations in the particle density, but instead by the overall gross features of the distribution of the particle density. As a consequence it is plausible that a mean field theory (in the sense of a Hartree average over the gauge field) may capture the essential features of the dynamics of the gauge transformed state by properly describing this boundary effect. If the particle density is uniform, the vector potential seen by a particle on the interior is thus the vector potential produced by a surface current circulating around the sample boundary. As noted above the nonlocal nature of this interaction physically originates in our constraint which confines the system to a spin eigenstate (in this case a global singlet), and thus the underlying quantum fluctuations of the spins exhibit long range (algebraic) correlations in this "fluid" state.

III. Illustrative Examples

(a) Half Filled Band

We first consider an application of this representation to describe Hubbard fermions exactly at the half filled band, i.e. with one fermion per lattice site. Here the mean field theory of this parastatistics representation correctly recovers a (particularly trivial) limiting case of the interacting problem.

We consider a finite segment (say a circular disc) of the square lattice, containing N sites and N fermions, with N even. Our Hamiltonian for the fermions is the Hubbard form:

$$H = \sum_{\substack{\langle ij \rangle \\ \sigma}} t(a^+_{i\sigma} a_{j\sigma} + a^+_{j\sigma} a_{i\sigma}) \tag{6}$$

$$+\sum_i U n_{i\uparrow} n_{i\downarrow} + E_o (n_{i\uparrow} + n_{i\downarrow})$$

and we assume that U/8t >>1. We focus on the singlet subspace, and transforming to the vorton representation, the transformed Hamiltonian reads:

(7)

$$H' = \sum_{\substack{<ij> \\ \sigma}} t' s^+_{i\sigma} s_{j\sigma} + t'^* s^+_{j\sigma} s_{i\sigma}) + E_o (n_{i\uparrow} + n_{i\downarrow})$$

with

$$t' = t \exp \left(\frac{2\pi i}{\phi_o} \int_i^j A \cdot dl\right)$$

Note the interaction term is now excluded by the vorton kinematics. Nevertheless the bare hopping terms are dressed by an phase factor resulting from the gauge interaction. In the mean field limit, the net vector potential is $\frac{r \times H^*}{2}$ so that the particles on the disc appear to be propagating on a cross section of a long solenoid with a surface current (particle flux per unit length) given by $\lambda = \frac{nc}{4\alpha}$ where n is the two dimensional particle density and $\alpha = e^2/\hbar c$.

For the half filled band we have N vortons in the system, and these vortons behave as "spinless" fermions, i.e. the wavefunction for the vorton spatial coordinates changes sign under any odd permutation of the coordinates. With only N sites in our system the only possible antisymmetric spatial function for the vortons is then a determinant which saturates the spatial degrees of freedom and it may thus be written in the form:

$$\psi' = s^+_1 s^+_2 s^+_3 \ldots | 0 >$$

The free particle energy for this system is just E_s = NEo. Note that the singlet state is "locked" against density fluctuations: there is a gap of order U for the lowest lying charge excitation in the system which require creation of a particle-hole pair (i.e. a doubly occupied and an empty site somewhere in the system) .

Where are the magnetic excitations? If we were to couple our many body system to a higher spin state, say a triplet (S=1) we must reduce the surface current since the number of enclosed quanta is fixed by the spin according to $\lambda' = \frac{c}{2\alpha} \frac{N/2 - S}{A}$ Nevertheless, at the half filled band, the resulting antisymmetric vorton state saturates the spatial determinant, and the energy is then identically that of the singlet state (E_t = NE_o) given above.

Thus the ground state in the problem is an extensively degenerate, correlated state with one particle per site, and completely "soft" with respect to local fluctuations of the orientations of the spins. This is of course the correct asymptotic behavior of the Hubbard model in the infinite U limit. Our model has reduced to the Heisenberg spin 1/2 antiferromagnet with J-->0. In this limit, we find that the parastatistics representation is correct, but not very useful. To break the ground state degeneracy one needs to include effects to leading order in t/U, i.e. include virtual excitations into the doubly occupied subspace which mix the various low lying spin configurations, just as in the usual canonical projection of the Hubbard Hamiltonian into the singly occupied subspace. Nevertheless this limit is reassuring, since it demonstrates that our "mean field" over the gauge interaction recovers the correct limiting behavior of our model at half filling.

(b) Non Half Filled Band

As soon as we dilute the system described in section (a) the ground state degeneracy is relieved, and the low energy states are dispersed over a range of energy given by t in our model. The stable states of our system are spinless fermionic configurations which minimize the residual kinetic energy in the gauge transformed Hamiltonian (7) for the correlated system. Note that the system exactly satisfies the Gutzwiller constraint with only singly occupied configurations, so the only remaining energy in the problem is fixed by the gauge transformed kinetic energy term.

We consider first how we may characterize the mean field states of this system. Our correlated fluid states are states with uniform particle density. Simple off diagonal correlation functions of the form $\langle s_i^+ s_j \rangle$ are not gauge invariant quantities, and thus do not describe fields which are ordering in the ground state. Instead we are led to consider a class of loop correlation functions of the form $Z_p = \prod_{\text{plaquette}} \langle s_i^+ s_j s_j^+ s_k \ldots s_i \rangle$ which are gauge invariant. The first notice of the importance of loop correlation functions of this form in the context of the RVB theory at half filling is given by Baskaran and Anderson [3]. The plaquette state is defined as the Im log Zp with the product taken following a counterclockwise loop around the plaquette; a possible mean field state of the system will be represented by defining the plaquette states of all the cells in our system. There is a sum rule, stating that the integrated flux through the system yield N_ϕ =N/2 for the singlet state, and these fluxes may be distributed through the system in various patterns, some of which are illustrated (for a quarter filled band) in Figure 1. The kinetic energy of the vortons, and thus the total energy of the system, will be optimized for a particular ordered epitaxial arrangement of the flux configuration on the lattice. Thus the stable mean field states of the system correspond to an ordering of an array of ring currents on the two dimensional lattice.

The effect of the ordered "epitaxial" flux lattice on the underlying fermionic spectrum is itself very interesting. For illustration we consider the case of the 1/4 filled band with one bare fermion for every two lattice sites. In the absence of the gauge field the quasiparticle spectrum for the spinless fermions is just the free particle dispersion:

$$E(k_x,k_y) = 2t\,(\cos(k_x a) + \cos(k_y a))$$

with the negative energy states occupied. This yields a "square" fermi surface inscribed within the original Brillouin zone. For the N vortons, our sum rule requires the presence of N/2 quanta in the system; the lowest energy mean field state for this system corresponds to a "uniform" state, with half each of the primitive cells enclosing a net flux $\pi/2$ in the arrangement shown in Figure 2. For this case the quasiparticle spectrum is given by:

$$E(k_x,k_y) = \pm \{ 2 \pm \sqrt{4 - \sin^2(2k_x a) - \sin^2(2k_y a)} \}$$

with negative energy states occupied. Thus the fermi surface shrinks to a Fermi point, similar to that described by Kotliar for the s + id phase of the RVB state *at half filling* on the square lattice [6]. The Fermi points occur in the Brillouin zone at the points (0.,0) and $(\frac{\pi}{2a}, 0)$ with opposite chirality. As a consequence the long wavelength excitation spectrum of such a system is "bosonic" with $\omega = cq$. These excitations are spinless, charged, (and gapless) excitations. We suspect (but can not prove) that the stable mean field vorton states will always exhibit a point Fermi surface in the quasiparticle spectrum. The suppression of the quasiparticle spectral density at the effective "Fermi energy" is an essential feature of the low lying states in this problem, since it is precisely this depression which lowers the vorton kinetic energy in the system.

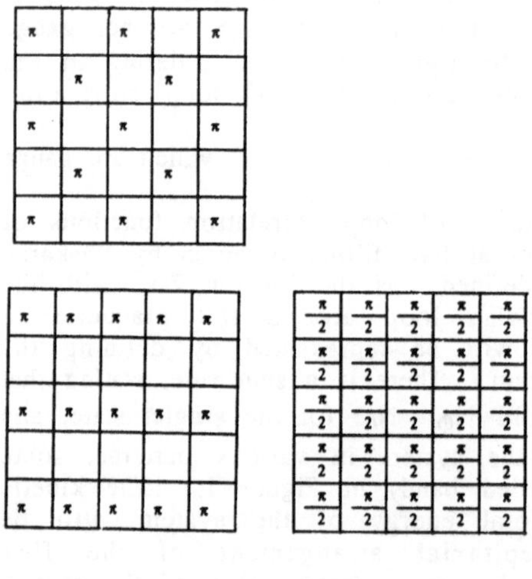

Figure 1. Three possible flux phases for the strongly correlated Hubbard model for quarter filling of the square lattice. In each state the particle density is uniform and the average flux distribution is one quantum for four plaquettes. The uniform state in the loweset panel is the lowest energy configuration for this system.

There are also magnetic excitations in this system. The simplest of these may be constructed by returning to the spin tableau and constructing a symmetrized spin function corresponding to a triplet excitation of the N particle system. According to our counting rules given above, this corresponds to removing a single flux quantum from the system (i.e. the boundary current is reduced by the amount $\frac{c}{2\alpha A}$). An interesting construction to describe this situation is the addition of a single "external" flux quantum (with a sign opposing the internally generated U(1) gauge field) to the quasiparticle Hamiltonian for the singlet state. The net effect of this addition is to remove four quarter quanta from the ground state flux pattern in Figure 1, i.e. to create a quartet of vacancies in the flux crystal. The elementary spin 1 excitation can thus be constructed from a a quartet of neutral elementary excitations which fractionalize the "spin" of the excitation. The analog to the spinons of the RVB state at half filling are striking.

Our picture the ground state for the diluted fluid of strongly interacting Hubbard fermions is that of a correlated state in which the electrons exactly satisfy the Gutzwiller constraint while minimizing the residual kinetic energy by ordering their zero point fluctuations into a pattern of ring currents circulating on the two dimensional lattice. It would be quite interesting to construct a realizable scattering experiment which would be sensitive to the spatial distribution of these vacuum currents, to test the presence of RVB correlations of this form for diluted two dimensional strongly interacting Hubbard systems.

The picture further suggests that there are "special" particle densities for this problem for which the flux lattice in the singlet state is commensurate with the underlying coordinate space lattice. In these configurations, the stabilization provided by the reduction of the vorton kinetic energy in the mean field ground state should be particularly large. It has been suggested that these commensurate "epitaxial" states describe particularly stable singlet states for this problem [see for example ref. 11], although the question of whether these would describe Mott insulators or some other exotic many body state of the system is presently unresolved.

IV. Outlook

We are continuing to investigate a number of formal questions related to the vorton representation discussed above. For example, at present we are uncertain how realistic to regard the mean field Hartree average over the internal gauge field. There is some reason to trust the qualitative properties of the mean field ground state, based on the long range nature of the gauge interaction, as argued above, but a systematic treatment of the fluctuations, to examine the validity of this assumption, would of course be desirable. Similarly, one is concerned about the stability of these quantum fluid states against lower energy ordered broken symmetry states. These may occur in response to terms of order t/U in the effective "vorton" Hamiltonian (as in the case of magnetically ordered states at the half filled band), occur in response to residual repulsive interactions not included in the on site Hubbard Hamiltonian, or simply involve a disordering (melting) of the flux crystals described above. At low temperature, broken symmetry configurations involving spatial ordering of the charge density, possible magnetic ordering, or the stabilization of a pair amplitude in a superconducting state are all of interest. Preliminary work suggests that unlike the one dimensional problem, the singlet "fluid" state persists over a finite range of coupling constants at zero temperature for the two dimensional system on the square lattice. Finally, it would be useful to generalize the model discussed above for the infinite U limit, to finite U (i.e. the intermediate couling regime) which is almost certainly the relevant parameter range for most physical systems. While one may argue that the physics in this regime is governed by a "large U" fixed point, the parastatistics described above may only strictly be defined in the infinite U limit. It will be interesting to examine how this thinking may be generalized to the intermediate coupling regime relevant to most physical systems.

V. Acknowledgement

I thank the I.S.I. for their hospitality during this workshop, and to B. Altshuler, L. Bulaevskii, S. Ostlund and J. Voit for a number of useful discussions. This work was supported in part by the NSF under grant number DMR 87-03551. DM is supported by the NSF MRL program under grant DMR 85-19059.

References

1. P.W. Anderson Science 235, 1196 (1987)

2. P.W. Anderson, G. Baskaran, Z. Zou and T. Hsu Physical Review Letters 58, 2790 (1987)

3. G. Baskaran and P. W. Anderson Physical Review B37 580 (1988)

4. S. Kivelson, D. Rokhsar and J. Sethna Physical Review B35, 8865 (1987)

5. R. B. Laughlin Science 242, 525 (1988)

6. G. Kotliar Physical Review B 37, xxx (1988)

7. I. Affleck and B. Marston Physical Review B 37, 3774 (1988)

8. F. Wilczek Physical Review Letters 49, 957 (1982)

9. R. V. Chamberlin, M.J. Naughton, X. Yan and P.M. Chaikin, Physical Review Letters 60, 1189 (1988)

10. V. Kalmeyer and R. Laughlin (private communication)

11. P. Wiegman (preprint, 1988)

COMPOSITE VS CONSTITUENT PARTICLE ASPECTS OF HOLE MOTION IN THE CuO_2 PLANE

Zhao-bin Su and Lu Yu*

Institute of Theoretical Physics, Academia Sinica
Beijing, China

1. INTRODUCTION

There is accumulating evidence that the CuO_2 plane is responsible for both magnetism and superconductivity in several classes of oxide superconductors/1/. However, the question of what is the simplest model to describe the essential physics for the hole motion in this quasi-two-dimensional system, is still under active discussion. Anderson /2/ proposed in the early days of the Hi-Tc heat wave to use the one-band Hubbard model, according to which there is only one state per each unit cell. On the other hand, the experiments have shown clearly that the antiferromagnetic (AF) spins are located on Cu sites , while the additional holes reside on O sites/3/. Emery and several other authors/4/ have considered an extended Hubbard model which includes both Cu and O bands. Later, Zhang and Rice/5/ have shown that one can derive under certain assumptions an effective one-band model, starting from the two-band scheme. Their main observation is that the Cu-O hybridization binds a hole on each square of O atoms to the central Cu^{2+} ion to form a local singlet. Nevertheless, there is still no consensus on this issue/6/.

We agree with Zhang and Rice that there is a strong hybridization of the Cu-d and the O-$p\sigma$ orbits and that the singlet state formed by the symmetrized square O wave function (of d symmetry) with the central Cu function has much lower energy compared with other possible configurations (e.g., a singlet formed by an O orbit in the middle with a symmetrized Cu orbit on the two ends). In our opinion, the character of the hole motion in the CuO_2 plane depends strongly on the energy scale involved. For "low" energy phenomena, when the characteristic energy $\hbar\omega$ is much less than the binding energy E_b , the singlet state appears as a composite particle and its internal structure can be ignored. DC

---

* Present Address: Intern. Centre for Theor. Physics, Trieste, Italy

transport and infrared absorption belong to this category. In the opposite limit, when $\hbar\omega >> E_b$, the constituent particles themselves should show up. A typical example is photoemission and other "high" energy experiments. The virtual transitions between constituent and composite particles may play a role even in "low" energy phenomena.

In this paper we derive an effective Hamiltonian up to the second order of the Cu-O hopping and it can be written in two different forms, either in terms of singlet-singlet and triplet-triplet interactions (Sec.2), or in terms of a Kondo lattice Hamiltonian, i.e., a spin-hole interaction (Sec.3). The composite particles are introduced via a Hubbard-Stratonovich transformation (Sec.2). The weak coupling effects like effective hole attraction induced by AF spin fluctuations and the strong coupling effects, like the mass renormalization and the self-consistent spin-polaron state, are briefly discussed (Sec. 2&3). Furthermore, as an example of the applications of the proposed scheme, the linear temperature dependence of the resistivity is agrued (Sec.4). Finally, an alternative scenario of superconductivity is speculated (Sec.5).

2. EFFECTIVE HAMILTONIAN IN SINGLET AND TRIPLET STATES

Our starting Hamiltonian for the hole motion is the same as in Ref. 5, i.e.,

$$H = H_0 + V , \tag{1}$$

$$H_0 = \varepsilon_d \sum_{i,\sigma} d^\dagger_{i\sigma} d_{i\sigma} + \varepsilon_p \sum_{l,\sigma} p^\dagger_{l\sigma} p_{l\sigma} + U \sum_i n_{i\downarrow} n_{i\uparrow} , \tag{2}$$

$$V = \sum_{i,l,\sigma} (V_{il}\, d^\dagger_{i\sigma} p_{l\sigma} + h.c.), \tag{3}$$

where index i runs over Cu sites, while index l runs over O sites surrounding Cu in each square and the hopping integral

$$V_{il} = t_0 \zeta_{il} \tag{4}$$

with

$$\zeta_{il} = \pm 1 \tag{5}$$

depending on the relative sign of O-p and Cu-d wave functions. The rest of the notation is obvious.

Using the perturbation theory with degenerate ground states/7/, we can obtain the following effective Hamiltonian up to the second order of the Cu-O hopping integral t_0 as

$$H = H'_0 + V', \tag{6}$$

$$H'_0 = \sum_{i,\sigma} (\varepsilon_d - \frac{1}{\Delta} \sum_l |V_{il}|^2) d^\dagger_{i\sigma} d_{i\sigma} + \varepsilon_p \sum_{l,\sigma} p^\dagger_{l\sigma} p_{l\sigma} , \tag{7}$$

$$V' = \left(\frac{1}{\Delta} + \frac{1}{U-\Delta}\right)\sum_{i,l,l',\sigma} V_{il'}^{*} V_{il} \left[p_{l'-\sigma}^{\dagger} d_{i\sigma}^{\dagger} d_{i-\sigma} p_{l\sigma} - p_{l'\sigma}^{\dagger} p_{l\sigma} d_{i-\sigma}^{\dagger} d_{i-\sigma}\right]$$

$$+ \frac{1}{\Delta}\sum_{i,l,l',\sigma} V_{il'}^{*} V_{il} p_{l'\sigma}^{\dagger} p_{l\sigma} \left(\sum_{\sigma'} d_{i\sigma'}^{\dagger} d_{i\sigma'}\right), \tag{8}$$

where

$$\Delta \equiv \varepsilon_p - \varepsilon_d > 0. \tag{9}$$

One can easily check that if the single occupation constraint

$$\sum_{\sigma} d_{i\sigma}^{\dagger} d_{i\sigma} = 1 \tag{10}$$

is imposed, (6)-(8) is the same as that used by Ogata and Shiba/8/ in their numerical studies.

Using the symmetrized O orbits

$$p_i^s = \frac{1}{2}\sum_{l} \zeta_{il} p_l, \tag{11}$$

the interacting part of the effective Hamiltonian (8) can be rewritten as

$$V' = J_s \sum_{i} \phi_i^{s\dagger} \phi_i^s + J_t \sum_{i,\alpha} \phi_{\alpha,i}^{t\,\dagger} \phi_{\alpha,i}^t, \tag{12}$$

where

$$J_s = -4t_0^2\left(\frac{2}{U-\Delta} + \frac{1}{\Delta}\right), \tag{13}$$

$$J_t = 4t_0^2/\Delta, \tag{14}$$

$$\phi_i^s = \frac{1}{\sqrt{2}}\left(d_{i\uparrow} p_{i\downarrow}^s - d_{i\downarrow} p_{i\uparrow}^s\right), \tag{15}$$

$$\phi_{+i}^t = d_{i\uparrow} p_{i\downarrow}^s,$$

$$\phi_{0i}^t = \frac{1}{\sqrt{2}}\left(d_{i\uparrow} p_{i\downarrow}^s + d_{i\downarrow} p_{i\uparrow}^s\right), \tag{16}$$

$$\phi_{-i}^t = d_{i\downarrow} p_{i\downarrow}^s.$$

As follows from (12), the energy difference between the triplet and the singlet states is given by

$$\Delta E = J_t - J_s = 8t_0^2\left(\frac{1}{U-\Delta} + \frac{1}{\Delta}\right) \tag{17}$$

in agreement with Ref. 5. It is obvious that the binding energy E_b is of the same order as (17).

The effective interaction (12) appears very simple, but we have to keep in mind that the symmetrized functions p_i^s in the neighboring cells are not orthogonal to each other, which gives rise to finite band width of singlet and triplet states. The whole discussion makes sense only if the band width is smaller than the energy difference ΔE, which still remains an assumption. To avoid the difficulty due to non-orthogonality, Zhang and Rice/5/ have introduced the Wannier function for the O state defined as

$$p^{w}_{i\sigma} = N^{-1}\sum_{k} e^{i\,k\cdot R_i}\beta_k \sum_{j} p^{s}_{j\sigma}\, e^{-i\,k\cdot R_j} \tag{18}$$

with

$$\beta_k = [\,1-\frac{1}{2}(\,\cos k_x + \cos k_y\,)]^{-1/2} \tag{19}$$

and then considered the effective hopping of the singlet states in this representation. Here we will adopt an alternative approach of introducing the composite particles of singlet (e_i) and triplet ($f_{a,i}$) symmetries via the Hubbard-Stratonovich transformation

$$\exp\frac{i}{\hbar}[-J_s\sum_i \phi_i^{s\dagger}\phi_i^{s} - J_t\sum_{\alpha,i}\phi^{t\dagger}_{\alpha,i}\phi^{t}_{\alpha i}\,] = \int [de\,][de^{\dagger}][df_\alpha][df^{\dagger}_\alpha]\times$$

$$\exp\frac{i}{\hbar}\{J_s\sum_i e_i^{\dagger}e_i + J_t\sum_{\alpha,i} f^{\dagger}_{\alpha,i}f_{\alpha,i} - J_s\sum_i [e_i^{\dagger}\frac{1}{\sqrt{2}}(d_{i\uparrow}p_{i\downarrow} - d_{i\downarrow}p_{i\uparrow}) + h.c.\,]$$

$$-J_t\sum_i[(f^{\dagger}_{+i}d_{i\uparrow}p_{i\uparrow} + h.c.\,) + (f^{\dagger}_{0i}\frac{1}{\sqrt{2}}(d_{i\uparrow}p_{i\downarrow} + d_{i\downarrow}p_{i\uparrow}) + h.c.) + (f^{\dagger}_{-i}d_{i\downarrow}p_{i\downarrow} + h.c.)]\}. \tag{20}$$

The single occupation constraint then becomes

$$e_i^{\dagger}e_i + \sum_{\alpha} f^{\dagger}_{\alpha,i}f_{\alpha,i} + \sum_{\sigma} d^{\dagger}_{i\sigma}d_{i\sigma} = 1. \tag{21}$$

For "low" energy phenomena we can limit ourselves to the singlet states and use the vertices of the type $e^{\dagger}pd$ as the basic ingredient in describing various processes. The oxygen propagation effect is easy to incorporate by including the inverse oxygen propagator

$$(S_0^{-1})_{ll'} = i\hbar\frac{\partial}{\partial t}\,\delta_{ll'} - t_{ll'} \tag{22}$$

in the functional integral. We can integrate out the oxygen variables and obtain an effective Hamiltonian in terms of copper and singlet variables as done in Ref.5, but we prefer to leave the present form untouched in our further discussions.

3. SPIN-HOLE COUPLING HAMILTONIAN

The effective interaction (8) can be rewritten in another form to emphasize the coupling of the hole to the spin system, namely,

$$V' = -t_{eff}\sum_i p_i^{s\dagger}p_i^{s} + J_k\sum_i s_i\cdot S_i\,, \tag{23}$$

where

$$s_i = p_i^{s\dagger} \frac{\sigma}{2} p_i^s , \tag{24}$$

$$S_i = d_i^\dagger \frac{\sigma}{2} d_i , \tag{25}$$

$$t_{eff} = 2t_0^2 \left(\frac{1}{U - \Delta} - \frac{1}{\Delta}\right), \tag{26}$$

$$J_k = 8t_0^2 \left(\frac{1}{U - \Delta} + \frac{1}{\Delta}\right). \tag{27}$$

Combined with the Heisenberg part of the Cu-Cu superexchange

$$H_h = J_h \sum_{<i,j>} S_i \cdot S_j \tag{28}$$

the interaction Hamiltonian (23) is nothing but the Kondo lattice Hamiltonian which has been discussed by several authors/9-11/ in the context of Hi-Tc superconductivity. In the present form the coupling of moving holes with strongly correlated AF background is emphasized. In heavy-fermion systems the band width proportional to t_{eff} is much larger than the coupling constant J_k , and the Fermi-liquid behavior is expected below the characteristic Kondo temperature/12/. Here both t_{eff} and J_k are determined by similar combinations of parameters and, apparently, the effective hopping t_{eff} is smaller, hence the behavior may be quite different. In fact, Prelovsek et al./9/ have found some critical value for J_k , beyond which the localized spin polaron picture holds, while a Fermi-liquid type behavior is expected below that value.

Our main concern in this paper is the strong coupling regime where the localized spin polaron picture holds. The interaction vertices of the type $e^\dagger pd$ introduced in the previous section to materialize the composite particle picture, provide a possibly new way for describing this localized state. The singlet state and a strong spin-hole coupling are the two sides of the same coin. In fact, the singlet state considered in the previous section is a propagating state, and it becomes localized only if the strong AF fluctuations are taken into account. We should also remember that the symmetrized p_i^s orbits appearing in (23) are not orthoganal, so the spin-hole coupling in the Wannier representation is nonlocal/13/. The effect of the on-site Cu-O superexchange has been discussed by Andrei and Coleman/11/ in the mean field approximation, whereas the effect of the nearest-neighbor Cu-O superexchange was considered earlier by us/14/ to study the possible triplet pairing of holes due to AF spin fluctuations, although at that time the effective Hamiltonian was written down mainly on the physical grounds. Now it can be derived in a more systematic way.

Generally speaking, if one treats the two-band model in a mean-field, or slave-boson appraoch, a typical Fermi-liquid behavior should be expected/15/. In order to find non-Fermi-liquid behavior one has to go beyond the mean-field approximations. One of the outstanding problems in this regard is the character of the quasiparticle states in the large J_k limit, which depends on the AF background in a crucial way. The similar problem in the one-band Hubbard model (or equivalently, the so-called t-J

model), is still quite controversial, in spite of the great efforts dedicated to this issue/16/. It seems that the low energy part of the spectrum can be described approximately in terms of quasiparticles (or coherent states), but the effective mass of these particles is strongly renormalized. The band width is of the order J_h, instead of the order of hole hopping t. Here the situation should be similar, although it is further complicated by the presence of J_k which should also be renormalized. The propagation of the singlet state should be equivalent to the propagation of a hole on an AF background, as discussed in Refs. 16. One may expect that in the low energy part of the spectrum, there should also be coherent, quasiparticle states. In treating this important issue of hole motion on an AF background many powerful techniques have been used/16/, but it seems to us that the self-consistency of the spin configuration with the hole states materialized in terms of the Bogoliubov-de Gennes equation, is rather crucial/17/. Similar approach should also be useful in the two-band case/17/.

4. TRANSPORT PROPERTIES

The linear temperature dependence of the resistivity is one of the characteristic features of the oxide superconductors/1/. There have been many proposals to explain this peculiar phenomenon /1,18,19/. In particular, Anderson and Zou/18/ have suggested to attribute this dependence to the scattering of holons on spinons within the RVB framework. However, their result was debated by Kallin and Berlinsky/19/ who showed that a more careful calculation using the same model would give $T^{3/2}$ instead of T. The previous authors further argued/18/ that the linear dependence might be connected with a special distribution of holons in the momentum space due to a "hard core" repulsion in that space.

Here we would like to point out that the scattering of composite particles on constituent partners as discussed in this paper may give rise to such linear temperature dependence. In fact, the change of the boson distribution n_k due to scattering can be written as

$$\left(\frac{\partial n_k}{\partial t}\right)_{col} = -A \sum_{k'} [n_k(1-f_d(k-k'))(1-g_p(k')) - (1+n_k)f_d(k-k')g_p(k')]$$

$$\times \beta_k \delta(\omega_k - \varepsilon_p(k') - \varepsilon_d(k-k')), \tag{29}$$

where f_d and g_p are Fermi distribution functions for Cu and O holes, respectively, β_k is given by (19) and the coupling constant A is proportional to J_s^2.

If we assume d electrons dispersionless and take its energy as zero, it is easy to see that the average scattering rate is given by

$$\left\langle \frac{1}{\tau_k} \right\rangle = \sum_k n_k \frac{1}{\tau_k} \approx \int d^2k \, \mathrm{th}\frac{\beta\omega_k}{2} \frac{1}{e^{\beta\omega_k} - 1} \beta_k \,. \tag{30}$$

Furthermore, assuming quadratic dispersion for ω_k, we find immediately the linear temperature dependence of the resistivity. This result is interesting, but, on the other hand, it is rather general. Indeed, such dependence would follow from any model with boson carriers scattered elastically on other quasiparticles or impurities. There should be more subtle features to differentiate various models.

5. SCENARIO FOR SUPERCONDUCTIVITY

All weak coupling models would provide a scenario for superconductivity more or less similar to that of the standard BCS theory/1/. On the other hand, for the strong coupling approach this question is still widely open. The superconductivity may appear as boson (holon) condensation, or pairing due to interlayer tunneling/20/, or due to a special gauge field induced by fractional statistics of quasiparticles/21/.

We would like to point out here that an alternative scenario may be possible within the present scheme of composite vs constituent particle picture.

We start from the zero doping limit when there is an AF long range order, or, at least, a two-dimensional quasi- long-range order, on the Cu sublattice. Upon doping the singlet states are formed which locally distort the AF background. At the further increase of doping, the AF quasi-long-range order becomes unstable, giving its way to an RVB state which is characterized by the appearance of the condensation of the singlet states $\langle d_{k\uparrow} d_{-k\downarrow} \rangle$ on the Cu sublattice. Now consider the scattering process shown in Fig.1, where the wavy line represents Cu, while the bold and dash lines represent composite particles and p electrons, respectively. The anomalous Cu pairing may induce the condensation of the composite particles $\langle ee \rangle$ and the Cooper-type pairing of O $\langle p_{-k\uparrow} p_{k\downarrow} \rangle$. This is consistent with the original proposal of Anderson/2/, namely, the RVB singlet pairing induces superconductivity upon doping. In our

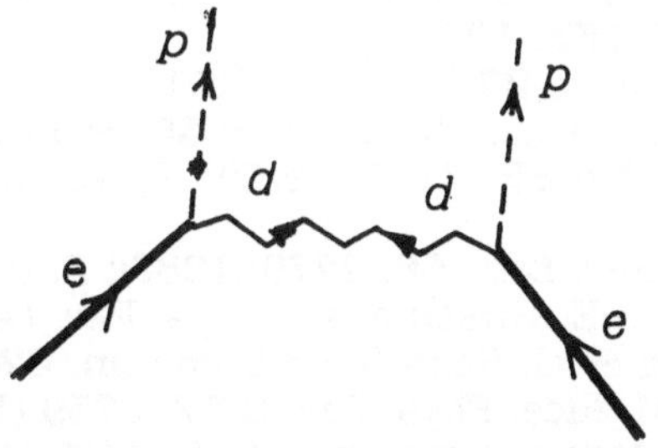

Fig.1 Possible scattering process responsible for superconductivity (See the text)

picture, the pairings <*ee*> and <*pp*> are the two aspects of essentially the same physics. Each of them will give rise to some observable effects. Roughly speaking, <*pp*> corresponds to the non-conservation of the number of O holes, while <*ee*> corresponds to the non-conservation of the number of the singlet states formed by them with Cu. This process is also similar to the one considered in Ref. 20, where the scattering of holons on spinons is discussed in connection with the tunneling of genuine holes between different layers. The process considered there is an interplane one, whereas here we discuss the intraplane process. The common feature of these two processes is the presence of an anomalous propagator due to RVB condensation. This is the crucial ingredient leading to the pairing of fermions and bosons.

We should mention that the recent experimental data on NMR relaxation/22/ and on the Knight shift/23/ seem to show the coexistence of isotropic and anisotropic pairings and a big difference in the relaxation rates of the nuclear spins on the Cu(2) sites, on the one hand, and on the Cu(1) and O sites, on the other hand. This is consistent with the picture of different types of pairings in the oxide superconductors described here. The interesting result on the dependence of critical currents upon the twinning angles/24/ at the grain boundaries can also be interpreted as due to the overlap of the isotropic and anisotropic order parameters/25/.

The above discussion on a plausible scenario of superconductivity is rather tentative, but it indicates on an interesting possibility. The details of the theory are being worked out explicitly now.

To summarize, we have proposed in this paper a scheme to consider the composite vs constituent particle aspects of the hole motion in the CuO_2 plane. Some implications of this scheme are consistent with the available experimental data, but there is still much work to be done to make more explicit predictions.

ACKNOWLEDGEMENTS

We would like to thank B. Alt'shuler, J. Birman, L. Bulaevskii, V.J. Emery, B.I. Halperin, D. Khomskii, D.H. Lee, P.A. Lee, E.J. Mele, T.M. Rice, B. Sakita and E. Tosatti for helpful discussions.

REFERENCES

1. Proceedings of the Interlaken Conference, ed. by J. Müller and J.L. Olsen, Physica **153-155**, 1988.
2. P.W. Anderson, Science **235**, 1196 (1987).
3. See, e.g., A. Fujimori et al., Phys. Rev. B **35**, 8814 (1987), J. M. Tranquada et al., ibid,**35**, 7187 (1987); N. Nücker et al., Z. Phys. B **67**, 9 (1987).
4. V.J. Emery, Phys. Rev. Lett. **58**, 2279 (1987); P. Prelovsek, Phys. Lett. A **126**, 287 (1988); J.E. Hirsch et al., Phys. Rev. Lett. **60**, 1668 (1988); C.M. Varma et al. Solid State Commun. **62**, 681 (1987).
5. F.C. Zhang and T.M. Rice, Phys. Rev. B **37**, 3759 (1988).
6. V.J. Emery and G. Reiter, Phys. rev. B **38**, 4547 (1988); *ibid*, Preprint.
7. V.J. Emery, Phys. Rev. B **14**, 2989 (1976).

8. M. Ogata and H. Shiba, J. of Phys. Soc. Jpn **57**, 3074 (1988).
9. P. Prelovsek, Ref. 4; E.Y. Loh et al., Phys. Rev. B **38**, 2494 (1988).
10. S. Maekawa et al., Physica C **152**, 133 (1988).
11. N. Andrei and P. Coleman, Preprint.
12. See, e.g., P. A. Lee et al., Comments on Solid State Phys.**128**, 99 (1986).
13. L. Yu would like to thank V.J. Emery for emphasizing this point.
14. Z.B. Su, L. Yu, J.M. Dong, and E. Tosatti, Z. Phys. B **70**, 131 (1988).
15. G. Kotliar, P.A. Lee, and N. Read, in Ref. 1, p.538.
16. See, e.g., B.I. Shraiman and E.D. Siggia, Phys. Rev. Lett. **60**, 740 (1988); S. Schmitt-Rink, C.M. Varma, and A.E. Ruckenstein, Phys. Rev. Lett. **60**, 2793 (1988); C.L. Kane, P.A. Lee,and N. Read, MIT Preprint.
17. Z.B. Su et al., unpublished.
18. P.W. Anderson and Z. Zou, Phys. Rev. Lett. **60**, 132 (1988); *ibid*, 2557 (1988).
19. C. Kallin and A.J. Berlinsky, Phys. Rev. Lett. **60**, 1556 (1988).
20. J. Wheatley, T. Hsu, and P.W. Anderson, Phys. Rev. B **37**, 5897 (1988).
21. R.B. Laughlin, Science **242**, 525 (1988).
22. Y. Kitaoka, K. Ishida, K. Asayama, H. Katayama-Yoshida Y. Okabe, and T. Takahashi, Preprint; M. Horvatic, P Segransan, C. Berthier, Y. Berthier, P. Butaud, J.Y. Henry, M. Couach, and J.P. Chaminade, Preprint.
23. M. Takigawa, P.C. Hammel, R.H. Heffner, and Z. Fisk, Los Alamos Preprint.
24. D. Dimos, P. Chaudhari, J. Mannhart, and F.K. LeGoues, Phys. Rev. Lett. **61**, 219 (1988).
25. L.Yu would like to thank D.H. Lee for this observation.

S PAIRING BY DOUBLE EXCITATION OF TRIPLET IN TWO BAND SYSTEM WITH INTERMEDIATE ELECTRON CORRELATIONS

Keiichiro Nasu

Institute for Molecular Science, 38, Nishigonaka, Myodaiji
Okazaki, 444, Japan
and
Institute for Science Interchange, Villa Gualino
Viale Settimio Severo, 65-10133, Torino, Italy

ABSTRACT

A two band system composed of a half-filled band and a almost fully filled band (hole band), with no charge transfer in between, is studied as a model system for high temperature superconductors. The intraband electron-electron repulsions are assumed to be intermediate strength, while the interband repulsion V is assumed to be weak. It is shown, however, that this V results in an attraction between holes through a virtual excitation of two triplet excitons (two magnons), which are low lying excitations in the correlation gap of the half filled band. Because of its retardation effect, this attraction can overcome the direct repulsion between holes, only at around the Fermi level, and can give Tc of about 100K. The large damping of this excitation is also obtained.

INTRODUCTION

Problems related with the high temperature superconductors of Cu-O type new ceramics are matters of considerable interest in these few years, and a large amount of experimental and theoretical studies have already been devoted to these problems. In spite of these efforts, the mechanism of this new phenomena are still left unclarified.

As far as the recent experimental studies are concerned, however, there have been important progresses. The results of the NMR and the NQR measurements[1], the photoemission spectra[2], the electron-energy-loss spectra, and the optical spectra; all show that there are two electronic energy bands near the Fermi level of this material. One is the d-like band of Cu, strongly hybridized with the p-like band of O, polarized pallarel to the Cu-O plane. The second is another p-like band of O, and the angle resolved photoemission spectra[2] shows that the Fermi level is in this band, although the polarization of this p orbital relative to the Cu-O plane is not well known yet.

On the other hand, the presence of the anomaly in the temperature dependence of the NMR relaxation time[1], and the absence of the so-called "T-linear" specific heat[3], suggest us that the pairing is an ordinary s wave, rather than the d wave. The ratio between the energy gap Δ and the transition temperature Tc; $2\Delta/k_B Tc$, is also found not to be so much different from the standard value. The isotope effect is also confirmed

Interacting Electrons in Reduced Dimensions
Edited by D. Baeriswyl and D.K. Campbell
Plenum Press, New York

to exist, although it is small. All these experimental results tell us that this superconductivity is an ordinary BCS type except the hight of Tc. The temperature dependence of the magnetic penetration depth also well agrees with the traditional s-wave BCS theory[4].

Thus, we can infer that the origin of the attraction in this material is a virtual excitation of some quasi-bosonic excitations, not so much different from phonons, but with electronic origins. In this connection, the recently obtained Raman scattering spectrum is of great importance, since it shows the presence of a low lying singlet excitation of an electronic origin[5]. It is the double excitation of triple excitons or magnons in the energy gap opened by the electron correlation. The presence of such a singlet excited state, being not observed by the one photon process, is the characteristic of a system with the electron correlation, as already well known in polyacetylene[6]. In the case of Cu-O type ceramics, it is a low lying singlet excitation with an energy about three times greater than the breathing mode phonon of O, and expected to cooperate with this mode in building up the superconductivity.

The purpose of the present paper is to clarify the effect of this excitation on the superconductivity. The essential difference between this problem and the magnetic polaron problem is that the conducting electron can create this excitation without changing its spin. It is triggered, not by the exchange interaction, but by the Coulomb interaction. In this paper, this effect will be studied taking a two-band system as a model for Cu-O type new ceramics.

TWO BAND MODEL BY CPA

Let us consider a two-band system with the following Hamiltonian H,

$$H = H_0 + \Delta H_c, \tag{1.1}$$

where H_0 is the Hamiltonian of the two bands and ΔH_c is the interband Coulomb interaction. H_0 and ΔH_c are given as

$$H_0 = \sum_{k,\sigma} \sum_{i=1,2} E_i(k)\, A^{+}_{ki\sigma} A_{ki\sigma} + \sum_{l} \sum_{i=1,2} U_i\, N_{li\alpha} N_{li\beta}, \tag{1.2}$$

$$N_{li\sigma} \equiv A^{+}_{li\sigma} A_{li\sigma},$$

$$\Delta H_c = V \sum_{l} \sum_{\sigma,\sigma'} N_{l1\sigma} N_{l2\sigma'}. \tag{1.3}$$

Here, $A^{+}_{li\sigma}$ is the creation operator of electron in band i (=1,2) with spin σ(=α,β) at a site l, and $A^{+}_{ki\sigma}$ is its Fourier component with a wave vector k. $E_i(k)$ is the energy of an electron in the band i with a wave vector k, and it is given as

$$E_1(k) = -4T(\cos(k_x) + \cos(k_y)),\ E_2(k) = -4T(\cos(k_x) + \cos(k_y))-\Delta E, \tag{1.4}$$

where, 8T is the band width assumed to be common to both bands. k_x and k_y are x and y components of k, and ΔE is the energy difference between the two bands. In eq.(1.2), U_i (i=1,2) is the intraband electron-electron repulsion, and V, in eq.(1.3), is the interband one. Since we are interested mainly in the case of intermediate strength of the electron correlation, we assume 8T, U_i and ΔE are of the same order, and V is smaller than them:

$$(8T,\ U_i,\ \Delta E) >> V. \tag{1.5}$$

In this model, band 1 corresponds to the d-like band, and we assume it is half filled: $\Sigma_l N_{1l\sigma} = N/2$, where N is the total number of the sites. The band 2 corresponds to the p band, assumed to be almost fully filled:

$$N_2 \equiv \sum_l N_{12\sigma},\ N_2/N = 0.8 \sim 0.9. \quad (1.6)$$

Our scenario for the superconductivity is such, that the electrons (holes) in band 2 make pairs through an attraction mediated by a singlet excitation of the electronic system in band 1, which are in an insulating state because of U_1. This excitation occurs through V. For simplicity, the hybridization between two bands are assumed to be absent. Even though such a hybridization is absent, the interband exchange interaction H_{ex}

$$H_{ex} \equiv -J\sum_l S_{l1}\cdot S_{l2}, \quad (1.8)$$

is expected to be present, where J is this exchange integral and S_{li} is the spin matrix of an electron in site l and band i. In general, however, V is always much biggar than J, and hence we will treat this H_{ex} subsidiary as compared with ΔH_c. In the magnetic polaron theory, only H_{ex} is taken into account, and it is based on the assumption that U_1 is infinite; $U_1 \gg (8T,V)$. In our system, however, U_1 is of the same order of 8T, or only a few times greater than 8T.

In order to describe the states in such an intermediate strength of the electron correlation, we use the single-site coherent potential approximation (CPA), and for this sake, we replace $N_{li\sigma}$ in eq.(1.2) by a random number Q_{li} as

$$H_0 = H_0(Q) + \Delta H_0(Q), \quad (1.9)$$

$$H_0(Q) \equiv \sum_{k,\sigma}\sum_{i=1,2} E_i(k)\, A^+_{ki\sigma} A_{ki\sigma} - \sum_l \sum_{i=1,2} U_i Q_{li}(N_{li\alpha}-N_{li\beta}) + \sum_l \sum_{i=1,2} U_i Q^2_{li}, \quad (1.10)$$

$$\Delta H_0(Q) \equiv \sum_l \sum_{i=1,2} U_i (N_{li\alpha}-Q_{li})(N_{li\beta}+Q_{li}). \quad (1.11)$$

Here, $H_0(Q)$ denotes the one-body Hamiltonian with only a statistically fluctuating random numbers, and Q symbolically denotes 2N dimensional vector $Q=(Q_{11},Q_{12},\ldots,Q_{N1},Q_{N2})$. $\Delta H_0(Q)$ denotes the residual part of H_0 which gives, not static, but dynamical fluctuations. We assume that this static random number is real and takes only two values;

$$Q_{li} = \pm Q_i, \quad (1.12)$$

with an equal weight. $+Q_i$ means that site l is occupied by an up spin electron, while $-Q_i$ means the down spin one. $|Q_i|$ is determined to minimize the total energy of electrons within the CPA. Thus we can get the one-electron green's functions, and in the case of band 1, the energy gap due to the correlation opens. In the case of band 2, the one electron density of states is also somewhat modulated from the original one. In these calculations, ΔE is so chosen that the Fermi level is always at the center of the band 1 irrespective of N_2/N.

S PAIRING BY DOUBLE EXCITATION OF TRIPLET

Let us now calculate the anomalous part of the green's function, using the Nambu formalism, wherein an up spin electron is represented by the electron picture, while the down spin one is represented by the hole picture. The anomalous part can be expanded with respect to V and $\Delta H_0(Q)$, and assuming V is small, we take the diagram shown in Fig.1.

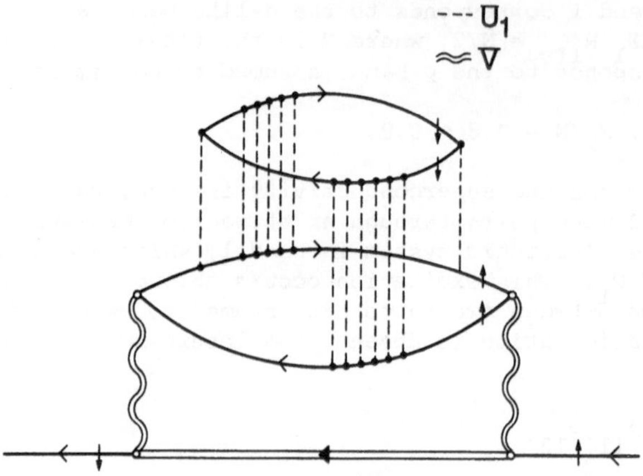

Fig.1 The pairing due to the double excitation of triplet. ═╪═ denotes the anomalous part of green's function in band 2. The two bubbles are in band 1.

In this diagram, an up-spin electron (or hole) in band 2 creates an electron-hole pair across the correlation gap in band 1. This virtual excitation occurs through V and the energy required for this excitation is of the order of U_1. The electron or hole, thus created in band 1, can also create another electron-hole pair in band 1 through $\Delta H_0(Q)$, and the spin of the second electron (or hole) is opposite to the first electron (or hole). At this stage, the partner should be exchanged within two pairs, since the electrons (or holes), with opposite spins with each other, can make a bound state. We can call it a triplet exciton or magnon in the correlation gap. Since its binding energy is of the order of U_1, it is a low lying excitation in the gap, and its double excitation has no total spin.

Using these results, we can now estimate the effective on-site interaction U_{eff} between two holes, which are near the Fermi level. It is given as

$$U_{eff} = (U_2 - M^2/2E_M)\, R_{vc}, \tag{2.1}$$

where M is the coupling constant between this singlet excitation and the electron (or hole) in band 2. It is proportional to V, but much reduced than this value, since it also encludes the matrix element of the second excitation of the electron-hole pair shown in Fig.1. In practical calculations of M, we also have to include other 7 diagrams that contribute to this virtual excitation. E_M is the averaged energy of this triplet excitation, obtained within the random phase approximation. R_{vc} is the vertex correction for this interaction in band 2, and reflects the metallic screening for both U_2 and the attraction mediated by this virtual excitation. R_{vc} decreases rapidly from 1 as U_2 and $(1-N_2/N)$ increases.

Practical calculations for Tc is performed for some typical cases, and one of these results is shown in Fig.2. In this case, we have used

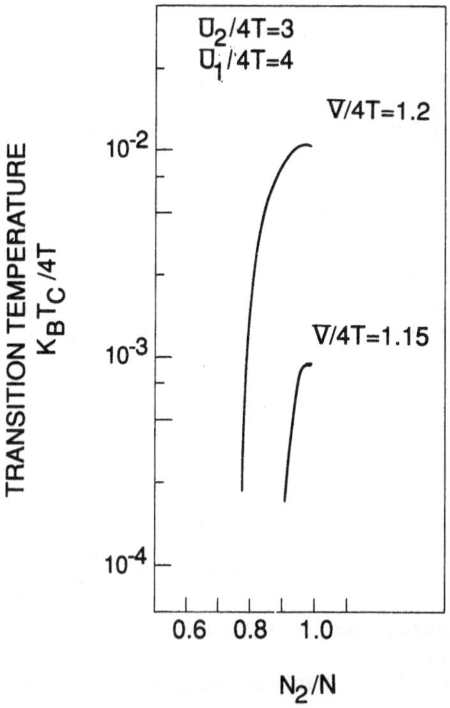

Fig.2 Tc as a function of N_2/N.

following values for the parameters: $U_2/4T=3$, $U_1/4T=4$, $V/4T\lesssim1.2$. In this case calculated values of E_M and M are as follows,

$$2E_M/4T \sim 0.2, \quad (M/V)^2 \sim 0.6.$$

Since 4T is about 1eV, $2E_M$ is about 0.2eV and M^2 is about $0.6(\mathrm{eV})^2$. Although M itself is smaller than U_2, the denominator $2E_M$ is also small, and it makes the effective attraction slightly overcome the repulsion, only around the Fermi level, resulting in Tc of the order of 100K. This is the so-called retardation effect. Such high-Tc's can be obtained only for small values of $(1-N_2/N)$, and as it increases, Tc decreases rapidly as shown in Fig.2. It means that both the attraction and the repulsion are well screened by the metallic nature of band 2.

We can also estimate the damping rate of this triplet excitation in band 1, due to the exchange interaction given by eq.(1.8). Through this interaction, the excitation in band 1 is resonantly transferred to the continuum in band 2, and it results in the damping Γ, which is given as

$$\Gamma = 2\pi(j/4)^2\int dE_1\int dE_2\rho(E_1)\rho(E_2)(1-n(E_1))n(E_2)\delta(E_M-E_1-E_2), \qquad (2.2)$$

where we have used the second order perturbation theory for J, and $\rho(E)$ is the one-electron density of state in band 2. $n(E_i)$ is the Fermi distribution function. Γ rapidly increases as $(1-N_2/N)$ increases. However, this damping gives no alteration of our conclusion for U_{eff} and Tc, since these quantities are concerned with only the virtual excitations and not real excitations. One can easily infer that this attraction can cooperate directly to the phonon mechanism to increase Tc.

CONCLUSION

We have thus shown that the attraction mediated by the double excitation of triplet can overcome the direct repulsion and results in high Tc's of about 100K. This mechanism can directly cooperate to the phonon mechanism.

ACKNOWLEDGEMENTS

This work is financially supported by the Ishida Foundation and also by Grant-in-Aid for general science research (No. 62540273) of the Japanese Ministry of Education, Science and Culture.

REFERENCES

1. Y. Kitaoka, Parity (Maruzen, Tokyo, in Japanese) Vo.03, No.8, 52 (1988).
2. T. Takahashi, Parity (Maruzen, Tokyo, in Japanese) Vo.03, No.9, 46 (1988).
3. M. Sera and M. Sato, Parity (Maruzen, Tokyo, in Japanese) Vo.03, No.06, 66 (1988).
4. A. Fiory, A. Hebard, P. Mankiewich and R. Howard, Phys. Rev. Letters, 61, 1419 (1988).
5. S. Sugai, S. Shamoto, and M. Sato, Phys. Rev. B 38, 6436 (1988).
6. K. Schulten, in this volume.

FERMI LIQUID THEORY FOR

HIGH TEMPERATURE SUPERCONDUCTORS

Kevin S. Bedell

Los Alamos National Laboratory
MS B262
Los Alamos, NM 87545

ABSTRACT

In this article the Fermi liquid theory of metals is discussed starting from Luttinger's theorem. The content of Luttinger's Theorem and its implications for microscopic theories of high temperature superconductors are discussed. A simple quasi-2d Fermi liquid theory is introduced and some of its properties are calculated. It is argued that a number of experiments on $YBa_2\,Cu_3\,O_{6+x}$, x>0.5, strongly suggest the existence of a Fermi surface and thereby a Fermi liquid normal state.

I. INTRODUCTION

The discovery of the new high temperature superconductors [1,2] (HTS) has resulted in an intense theoretical effort to understand the origin of the pairing mechanism responsible for the high transition temperatures, T_c. A crucial issue is whether the normal state, where $T>T_c$, can be characterized as a Fermi liquid. This is important for understanding the superconducting transition in these materials. In the standard picture it is the Cooper pairing of the Fermi liquid quasiparticles that results in a superconducting transition. In some of the more exotic theories [3, 4, 5] of HTS one is far removed from the standard theory. For example, in the RVB theories [3, 4] the superconducting transition is understood as a Bose condensation of pre-formed Bosons in the normal state. These exotic theories of HTS are most imaginative however, it is not at all clear if they are applicable to the high temperature superconductors.

In addition to the high transition temperatures the unusual normal state properties of the HTS have also been sources of inspiration for theory. Some of the unusual normal state properties of $YBa_2Cu_3O_{7-\delta}$ are, the apparent linear temperature dependence of the resistivity in the a-b (Cu-O) plane and the semiconducting like feature in the c-axis direction, [6] the linear temperature dependence of the Hall coefficient, [7] and the temperature independent thermopower, [7] to name a few. To date none of the theories of HTS has provided a consistent account of these properties. What I would like to address in this article is whether Fermi liquid theory can account for these unusual normal state properties. The type of Fermi liquid one can envisage for the HTS materials must include the strong anisotropy, i.e., quasi-2d character, and the strong electronic correlations. Very little is known about the properties of such a Fermi liquid so it is not at all obvious if this is applicable. For Fermi liquid theory to be applicable we must have a Fermi surface. In the absence of the usual direct probes of the Fermi surface, e.g., the de Hass-van Alphen effect, we can not be sure that a Fermi surface exists and if it does, what the volume might be. As we will see there are a number of experiments that provide indirect evidence for a Fermi surface. Given this we will consider some of the consequences of a Fermi liquid theory for the high temperature superconductor $YBa_2Cu_3O_{7-\delta}$.

II. LUTTINGER'S THEORUM AND FERMI LIQUID THEORY

Before we begin to study the Fermi liquid features of the high temperature superconductors we state what this means for a metal. To do this we outline below the exact statement of Luttinger's theorem.[8] The theorem depends on certain analytic properties of the single particle propagator matrix, $\hat{G}(\omega_n)$, where ω_n is the Matsubara frequency. This is the exact propagator and it satisfies Dyson's equation,

$$\hat{G}^{-1}(\omega_n) = \hat{\omega}_n - \hat{\varepsilon}^{\circ} - \hat{\Sigma}(\omega_n) \tag{1}$$

with $\hat{\varepsilon}^{\circ}$ the bare particle energy matrix and $\hat{\Sigma}(\omega_n)$ the self energy. For the band case $\hat{G}$, $\hat{\varepsilon}$, and $\hat{\Sigma}$ are diagonal in the momentum $\vec{p}$ thus we can express the matrix elements as follows,

$$\left(\hat{\varepsilon}\right)_{rr'} = \varepsilon^{\circ}_{\vec{p}\nu}\, \delta_{\vec{p}\vec{p}'}\, \delta_{\nu\nu'} = \left(\hat{\varepsilon}^{\circ}_{\vec{p}}\right)_{\nu\nu'} \delta_{\vec{p}\vec{p}'} \tag{2a}$$

and

$$\left(\hat{\Sigma}\right)_{\nu\nu'} = \delta_{\vec{p}\vec{p}'}\, \hat{\Sigma}_{\vec{p},\nu\nu'}(\omega_n)\, \delta_{\vec{p}\vec{p}'} = \left(\hat{\Sigma}_{\vec{p}}\right)_{\nu\nu'} \delta_{\vec{p}\vec{p}'}. \tag{2b}$$

Here ν and ν' represent both the spin and the band indicies.

For each index ν we can define a single particle distribution function $n_{\vec{p}\nu}$ where,[8]

$$n_{\vec{p}\nu} = \frac{1}{\beta} \sum_n \exp(\omega_n 0^+) (\hat{G}_{\vec{p}})_{\nu\nu} \tag{3a}$$

with

$$\hat{G}^{-1}_{\vec{p}} = \hat{\omega}_n - \hat{\varepsilon}^{\circ}_{\vec{p}} - \hat{\Sigma}_{\vec{p}}(\omega_n) \tag{3b}$$

and

$$\beta = 1/k_B T$$

From this we define the momentum distribution function $n_{\vec{p}}$,

$$n_{\vec{p}} = \sum_{\nu} n_{\vec{p}\nu} = \frac{1}{\beta} \sum_n \exp(\omega_n o^+)\, \mathrm{Tr}\,(\hat{G}_{\vec{p}}) \tag{4}$$

If we define the eigenvalues of the matrix $\hat{\varepsilon}^{\circ}_{\vec{p}} - \hat{\Sigma}_p(\omega_n)$ as $L_{\vec{p}\rho}(\omega_n)$, where ρ represents the "true" band index, we can write Eq. (4) as follows,

$$n_{\vec{p}} = \frac{1}{\beta} \sum_n \sum_\rho \exp(\omega_n 0^+) \frac{1}{\omega_n - L_{\vec{p}\rho}(\omega_n)} \tag{5a}$$

If we now let $T \to 0$ we can write the distribution function as,

$$n_{\vec{p}} = \frac{1}{2\pi i} \sum_\rho \int_{-\infty}^{\mu} d\varepsilon \left\{ \frac{1}{\varepsilon - i\eta - L_{\vec{p}\rho}(\varepsilon - i\eta)} - \frac{1}{\varepsilon + i\eta - L_{\vec{p}\rho}(\varepsilon + i\eta)} \right\} \tag{5b}$$

where μ is the chemical potential.

The function $L_{\vec{p}\rho}(\varepsilon \pm i\eta)$ has the following analytic properties, [8]

$$L_{\vec{p}\rho}(\varepsilon \pm i\eta) = Q_{\vec{p}\rho}(\varepsilon) \mp i\, J_{\vec{p}\rho}(\varepsilon)$$

and

$$J_{\vec{p}\rho}(\varepsilon) > 0. \tag{6}$$

If we now assume that,

$$J_{\vec{p}\rho}(\varepsilon) = \begin{cases} C^{(2)}_{\vec{p}\rho}(\varepsilon - \mu)^2 \ln|\varepsilon - \mu| , & 2d \\ C^{(3)}_{\vec{p}\rho}(\varepsilon - \mu)^2 , & 3d \end{cases} \tag{7}$$

a number of rigours consequences follow. To begin with there is a set of momentum values, $\vec{p}_F$, that satisfy the equation,

$$\mu - Q_{\vec{p}_F\rho}(\mu) = 0 \tag{8}$$

where the momentum distribution function is discontinuous. This set of momentum values define the Fermi surface of the metal. In general only a few of the bands in a metal will satisfy Eq. (8) and these will be referred to as the "unfilled bands". It also follows that the total number of electrons, N, is given by,

$$N = \sum_{\vec{p}\rho} \Theta(\mu - Q_{\vec{p}\rho}(\mu)) = N_c + N_f \quad , \tag{9}$$

where N_c is the number of electrons in the partially filled bands and N_f is the number of electrons in the full bands. The volume of the Fermi surface is determined by N_c and this is the same as for the non-interacting system.

The theorem of Luttinger [8] is exact, depending only on certain analytic properties of the single particle propagator. It is sometimes claimed that this is a theorem which depends on a perturbative relationship between the interacting and non-interacting system. This is clearly wrong. The only place a perturbative approach is necessary is in the actual calculation of the single particle propagator for the Hamiltonian of the system, this is the "If" part of the theorem, Eq. (7). This is the part of the theorem that microscopic theories must concentrate on. If it can be demonstrated that a particular microscopic theory yields a self-energy that satisfies Eq. (7) then Luttinger's theorem guarantees that we have a Fermi liquid whose volume is fixed by N_c. It has not yet been demonstrated in any of the models of HTS that claim the existence or non-existence of a Fermi liquid that Eq. (7) is or is not satisfied. Of course to unambiguously show that a given Hamiltonian does or does not have a Fermi liquid regime the exact propagator must be calculated. This is a task that is currently beyond most theoretical approaches, with the possible exception of Monte Carlo techniques, thus, microscopic models have not been helpful in deciding the issue of the existence or non-existence of a Fermi liquid regime in the HTS.

This issue is particularly complicated in the high temperature superconductors due to the strong electronic correlations as well as the quasi-2d nature of the CuO_2 planes. For example the compound $YBa_2 Cu_3 O_{6+x}$, with $x<0.4$ is an antiferromagnetic insultator, [9] which results from the strong electronic correlations, this is clearly not a Fermi liquid. As oxygen is added the antiferromagnetism is suppressed and for $x>0.4$ a metallic normal phase appears above T_c.[9] The exact nature of this metallic phase is not yet certain. I will argue that the normal phase is a Fermi liquid. However, due to the strong correlations, the quasi-2d nature of the CuO_2 planes, as well as the possibility of several bands crossing the Fermi level, this Fermi liquid will most likely have a number of unusual properties. Until detailed calculations are carried out for this Fermi liquid we cannot decide if some unusual property, e.g., the temperature dependence of the Hall coefficient, [7] is or is not a signature for a Fermi liquid.

III. A FERMI LIQUID THEORY FOR $YBa_2 Cu_3 O_{6+x}$, $x>0.5$

For Fermi liquid theory to be applicable to $YBa_2 Cu_3 O_{6+x}$, $x>0.5$, we must have a Fermi surface. The traditional experiments, e.g., the de Hass-van Alphen effect, are not available, however, there is some strong indirect evidence for a Fermi surface. Some of the most striking evidence for a Fermi surface, although this is somewhat indirect, is the photoemission experiment of List et al. [10] This experiment shows clear evidence for a sharp Fermi edge in $EuBa_2 Cu_3 O_{6+x}$, for $x>0.5$. This is certainly consistent with the existence of a Fermi surface. More recently List et al. have also seen evidence for band dispersion.[11] Another experiment that strongly suggests the existence of a Fermi surface is the position-annihilation experiment of Bansil et al. [12] on $YBa_2 Cu_3 O_7$. The structure of the Fermi surface suggested by these experiments [11, 12] is quite similar. I will come back to this point later on.

Given the possibility of a Fermi surface a question we can ask is, what is the volume enclosed by this surface? For a simple metal with a spherical Fermi surface the number density, n_c, of conduction electrons can be obtained from the Hall coefficient. In general the Hall coefficient can not be used to determine n_c since it depends on details of the band structure etc. (see section IV). We can envisage a number of Fermi liquid scenarios for the high temperature superconductors with rather different Fermi surface volumes, two of these I will outline below.

The first is the model proposed by Emery. [13] In this model the copper d-band is split by the correlations with the lower Hubbard band full and the upper one empty. When holes are added they go into the oxygen p-band, eventually suppressing the long range antiferromagnetic order. The Fermi liquid that forms has a Fermi surface that encloses a volume δn_c where $\delta < 1$ is the fraction of additional holes per unit cell and $n_c \approx 10^{22}$ cm^{-3} is the number density of copper holes. Another Fermi liquid we can have is one in which the copper d-state and oxygen p-state are strongly hybridized. The addition of holes here eventually suppresses the long range antiferromagnetic order. However, the Fermi surface in this case will enclose a volume of $n_c + \delta n_c$ holes.

Of these two Fermi liquids I believe that the second version, with some modifications, is appropriate to the HTS, in particular to $YBa_2 Cu_3 O_{6+x}$, x>0.5. This choice is supported by the photoemission [11] as well as the positron annihilation [12] experiments. From these experiments, in particular the photoemission [11], it is clear that there is a strong copper d-state character at the Fermi level as well as oxygen p-state. It is also clear from these experiments that more than one band crosses the Fermi surface as suggested by bandstructure calculations. [12,14] This need for more than one band at the Fermi surface is also evident in the Knight shift [15] and NMR [16] experiments on $YBa_2 Cu_3 O_{6+x}$.

Before starting the next section some additional comments about the Fermi liquids we have discussed are needed. It has been argued by Kim et al. [17] that the first Fermi liquid we discussed above does not satisfy Luttinger's theorem. This claim is without justification since Emery [13] has not calculated the exact propagator in this model. The claim by Kim et al. [17] is probably based on the assumption that the non-interacting reference system begins with the density nc, i.e., one hole per CuO_2 plane. Adding enough holes, δn_c, to suppress the antiferromagnetism and then turning on the interactions would yield a Fermi liquid state with $n_c + \delta n_c$ holes, if of course the self energy has the property given by Eg. (7). However, Luttinger's theorem does not tell us how to count the number of electrons in partially filled bands. The Fermi liquid scenario proposed by Emery [13] could also satisfy Luttinger's theorem if we make the right choice for the non-interacting reference system. The non-interacting system would be δn_c holes, enough to suppress the antiferromagnetism, in the oxygen p-band with the copper d-band split and the lower Hubbard band full. If we assume that the Hubbard bands remain rigid as we turn on the interactions between the holes in the oxygen p-band, this includes as well the interactions with the copper band, and the self energy satisfies Eq. (7) this would satisfy Luttinger's theorem, i.e., this would be a Fermi liquid. Of course experiment does not support this choice for the Fermi liquid state of $YBa_2 Cu_3 O_{6+x}$. However, it is possible that the model of Emery [13] does have a Fermi liquid of the type he proposed but this has not been demonstrated.

Another point should be made in regard to the dimensionality of the Fermi liquid. Luttinger's theorem is valid for a 2d system. It is straightforward to go through the analysis of Luttinger [8] for the 2d system using the 2d form for the imaginary part of the self energy, see Eq. (7). Thus, at the level of Luttinger's theorem there is no difference between a 2d or 3d Fermi liquid. However, the presence of the $\ln|\varepsilon-\mu|$ in the 2d quasiparticle lifetime suggest that this is a marginal dimension for a Fermi liquid. It is more difficult in 2d to show that Eq. (7) is satisfied for a given Hamiltonian since all of the diagrams that appear in a perturbative calculation of $J^{(2)}_{\vec{p}\vec{p}}(\varepsilon)$ are potentially important.

IV. A FERMI LIQUID MODEL FOR $YBa_2 Cu_3 O_{6+x}$

In the previous section we argued that some experiments 10, 11, 12 on YBa2 Cu3 O_{6+x} x>0.5, suggest the existence of a Fermi surface that encloses a volume of the order

$n_c \approx 10^{22}$ cm^{-3}. It is also evident from several experiments 10, 11, 12, 15, 16 that more than one band crosses the Fermi surface. For now we will consider only a single band Fermi liquid to stress more the role of correlations and the quasi-2d nature of the quasiparticle excitations. We will come back and discuss the need for a multiple band Fermi liquid when considering the Hall effect and the Knight shift.

The excitation spectra for a quasi-2d Fermi liquid is

$$\varepsilon_{\vec{p}} - \mu = v_F\,(p_{\|} - k_F) + 2t_{\perp} \cos p_{\perp} c\,. \tag{10}$$

Here $p_{\|}$ is the component of the momentum in the a-b plane and $p_{\perp}$ is the component along the c-axis and $c \approx 11.7$ Å is the unit cell length in this direction. Here v_F is the Fermi velocity in the a-b plane and it is given by $v_F = k_F/m^*$ with m^* the effective mass in the a-b plane. From k_F, c and m^* we have the density $n = k_F^2/\pi c$, the density of states at the Fermi level $N(0) = k_F/(\pi v_F c)$, and the Fermi energy $T_F = k_F v_F/2$. The quasi -2d nature of this spectrum requires the hopping energy between nearest neighbor planes, $t_{\perp}$, to satisfy the condition $t_{\perp} \ll T_F$.

With this form for the quasiparticle excitation spectrum what are the characteristic signatures of this Fermi liquid? To begin with the low temperature specific heat $C_v = \gamma T$ with $\gamma \propto N(o)$. The linear specific heat has not been clearly seen in the high T_c materials, thus, we have no independent determination of N(o). We can get an estimate of this from other experiments. The quasiparticle susceptibility, in the absence of spin orbit interactions is isotropic, and is given to order T^2 by,

$$\chi(T) = \chi(o)\,\left(1 - \beta \left(\frac{T}{T_F}\right)^2\right) \tag{11}$$

with $\chi(o) = \mu^2 N(o)/(1+F_0^a)$ and β is a dimensionless quantity of the order of $(1 + F_0^a)^{-2}$. The value of $\chi(o)$ is estimated to be 3×10^{-4} emu/mole after taking into account core diagmagnetism etc. [18] The weak temperature dependence of $\chi(T)$ [15, 18] suggests that β is of the order of one or that F_0^a is close to zero. This gives a large effective mass enhancement $m^*/m \approx 10$. If we set $m^*/m \approx 10$ then $F_0^a = -0.3$ and $T_F \approx 1600$ °K, here we have used $n = n_c = 10^{22}$ cm^{-3}. This choice of the effective mass is consistent with the measured value of the London penetration depth [19] and reflectivity data [20] on the $YBa_2Cu_3O_{6.9}$ material.

With such a large effective mass we might expect a large contribution to transport properties coming from electron-electron scattering. For example the resistivity of a quasi-2d metal would behave in the following way: For $T \ll t_{\perp}$ we have a three dimensional metal, thus

$$\rho_{\|}(T) = \rho_{\|}^{\circ} + A_3 T^2 \tag{12a}$$

and

$$\rho_{\perp}(T) = \rho_{\perp}^{\circ} + B_3 T^2 \tag{12b}$$

where $\|(\perp)$ is the resistivity in (perpendicular to) the a-b plane. For $T >> t_\perp$ we have that [21]

$$\rho_{\|}(T) = \overset{\circ}{\rho}_{\|} + A_2\, T^2 \ln \frac{T_F}{T} \tag{13a}$$

and

$$\rho_{\perp}(T) = \overset{\circ}{\rho}_{\perp} + B_2\, T^2 \ln \frac{T_F}{T} \tag{13b}$$

The residual resistivities are related if we assume that the impurity scattering is isotropic, thus

$$\overset{\circ}{\rho}_{\|} = \frac{m^*}{ne^2\tau_0} \tag{14a}$$

and

$$\overset{\circ}{\rho}_{\perp} = \frac{1}{2}\left(\frac{v_F}{ct_\perp}\right)^2 \overset{\circ}{\rho}_{\|} \tag{14b}$$

where τ_0 is the impurity scattering lifetime. If we further assume that the electron-electron scattering is also isotropic then,

$$B_2 = \frac{1}{2}\left(\frac{v_F}{ct_\perp}\right)^2 A_2 \tag{14c}$$

From the temperature dependence of $\rho_{\|}(T)$ for a quasi-2d metal it does not look as if this could be applicable to the HTS materials. In single crystal samples of $YBa_2Cu_3O_{6.9}$ the resistivity in the a-b plane can be fit approximately with a straight line. [7] In Fig. 1 we plot a straight line (solid curve) which is a fit to the resistivity data of Tozer et al. [7] This fit to the data can be given as $\rho_{\|}(T) = \overset{\circ}{\rho}_{\|} + CT$ with $\overset{\circ}{\rho}_{\|} = 70\ \mu\Omega-\text{cm}$, and $C = 1.3\ \mu\Omega-\text{cm}/^\circ K$. Also shown in Fig. 1 is a plot of Eq. (13a) (dashed curve) with $\overset{\circ}{\rho}_{\|} = 122\ \mu\Omega-\text{cm}$, $A = 30.3 \times 10^{-4}\ \mu\Omega-\text{cm}/^\circ K^2$ and $T_F = 10^{3\circ}K$. Note that this T_F is comparable to the value we obtained from $\chi(o)$. We also get a good fit to the data if we use $T_F = 1600\ ^\circ K$, $\overset{\circ}{\rho}_{\|} = 135\mu\Omega-\text{cm}$, and $A = 21.8 \times 10\ ^\circ K^{-4}\ \mu\Omega\ \text{-cm}/^\circ K^2$. The main point to emphasize by this data fitting is that the cutoff in the logarithm is of the order of T_F. If F_0^a was closer to -1 we would expect the low temperature Fermi liquid behavior to persist over a much smaller temperature range. The fact that this logarithm appears to persist over such a broad temperature is again suggestive that the quasiparticle interaction F_0^a is small. Another point we note is that the size of the coefficient A_2 is comparable to the coefficient, A_3, of the T^2 term in the strongly correlated A-15 compounds. [21] For example in Nb_3 Sn we have that $A_3 = 70 \times 10^{-4}\ \mu\Omega-\text{cm}/^\circ K^2$.

We can estimate the size of A_2 if we assume that the resistivity lifetime, $\tau_\rho(T)$, can be approximated by the quasiparticle lifetime, $\tau_{qp}(T)$. This at best only gives an order of magnitude estimate for simple metals [22] and so we should not expect this to be any better for the HTS materials. If we assume that only s-wave scattering contributes we have, [19]

$$\tau_{qp}^{-1}(T) \approx \frac{\pi}{2}\frac{\overline{W}}{T_F}\, T^2 \ln \frac{T_F}{T} \tag{15a}$$

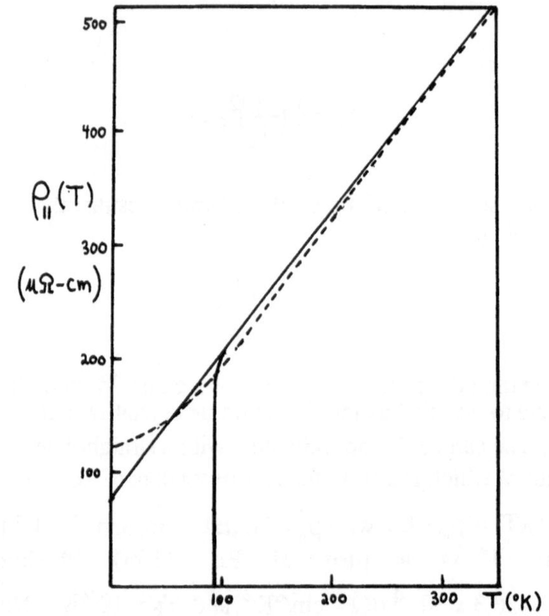

Fig. 1 A plot of the resistivity in the a-b plane $\rho_{||}(T)$. The solid surve is a straight line fit to the data of reference 7 for a single crystal of $YBa_2Cu_3O_{6.9}$. The dashed curve is a plot of Eq. (13a) with $T_F = 10^3$°K, $A_2 = 30.3 \times 10^{-4}\ \mu\Omega$ - cm/°K^2 and $\mathring{\rho}_{||} = 122\mu\Omega$ - cm .

and

$$\rho_{\parallel}(T) - \overset{\circ}{\rho}_{\parallel} \approx \frac{\pi^2 c}{4e^2} \overline{W} \frac{T^2}{T_F^2} \ln \frac{T_F}{T} \quad . \tag{15b}$$

For s-wave scattering we have that $\overline{W} = \overline{W}^{\uparrow\downarrow} = | A_0^s - A_0^a |^2$ where for a charged system $A_0^s = 1$. If we only have s-wave scattering then the triplet scattering amplitude must vanish, thus $A_0^s + A_0^a = 0$ or $F_0^a = -0.5$. Note that this value of F_0^a is rather close to our earlier estimate. If we now plug these values of the Landau parameters into Eq. (15b) with $T_F = 10^{3}$°K we get that $A_2 \approx 40 \times 10^{-4}\ \mu\Omega$ - cm/°K^2 and for $T_F = 1.6 \times 10^{3}$°K we get $A_2 \approx 16 \times 10^{-4}\ \mu\Omega$ - cm/°K^2. Clearly, with these small values of T_F electron-electron scattering can contribute a large fraction to the resistivity.

In the c-axis direction the resistivity, in most of the early experiments on single crystals,[7] showed a semiconducting like up turn just above T_c. This is clearly inconsistent with the prediction of Eq. (13b). More recently this upturn has been questioned by Iye et al.[23] In a number of their samples[23] $\rho_\perp(T)$ is roughly parallel to $\rho_\parallel(T)$ as a function of T. This is what we expect for a quasi-2d metal if Eqs. (14b) and (14c) are satisfied. As noted earlier for this quasi-2d band structure to be valid we must have $t_\perp << T_F$. We can estimate $t_\perp$ using Eq. (14b) and the experimental value of $\overset{\circ}{\rho}_\perp / \overset{\circ}{\rho}_\parallel$ which is of the order of 10^2.[7, 23] From this we have that $t_\perp \approx 20$°k ($T_F = 10^3$°K) or $t_\perp \approx 30$°K ($T_F = 1.6 \times 10^{3}$°K). Since T_c is larger than $t_\perp$ we are well into the regime where the resistivity is characterized by Eqs. (13a) and (13b). Given the likelihood of a Fermi surface and the potentially important electron-electron scattering contribution to the resistivity I believe the metalic value of $\rho_\perp(T)$ seen by Iye et al.[23] is the correct behavior.

The one band Fermi liquid model described above leads to two rather general conclusions: The first is that the band structure is quasi-2d and the other is that these are strongly correlated metals. These features will remain regardless of the details that might be necessary for a more realistic Fermi liquid. A one band Fermi liquid is almost certainly incomplete. The Knight shift[15] on $YBa_2Cu_3O_{6.9}$ shows clear evidence for at least two different bands at the Fermi level. The Hall effect also strongly suggests a more complicated band structure than a single band Fermi liquid.

If we consider a simple tight binding band structure with only nearest neighbor hopping then the Hall coefficient vanishes when the band is half full. Clearly, this is not the way to go about counting the number of conduction electrons! Including more realistic band structure effects,[17, 24] e.g., next nearest neighbor hopping, additional bands, etc. changes this in a rather non-trivial way. For example in the ab-plane it is found[12, 24] that R_{ab} is positive (hole-like) while in the c-axis direction R_c is negative (electron-like). Including a number of the details from the band structure is important for understanding some of the features of the Hall coefficient. However, the linear temperature dependence observed in the Hall coefficient is certainly beyond our current understanding of this effect. Most likely the strong electronic correlations and the multi-band character of the Fermi surface are responsible for this effect. These must be included in the calculation of the Hall coefficient before any conclusions regarding the implications of the temperature dependence of this coefficient can be made.

V. CONCLUSION

In this article I have discussed the notion of Fermi liquid theory for layered metals and the application to HTS. One of the early Fermi liquid treatments of $YBa_2Cu_3O_{6.9}$ was that proposed by Bedell and Pines.[25] This assumed an isotropic 3d Fermi liquid. This is clearly inconsistent with the resistivity anisotropy of the HTS. However, the strong electronic correlations leading to large effective masses was already evident from this analysis.[25] This is a feature that will survive as more realistic Fermi liquid descriptions emerge. It is also clear that the quasi-2d character of the quasiparticles of this Fermi liquid will be important. Many of the properties of a strongly correlated, quasi-2d, multi-band, Fermi liquid are not known and much research is still needed to elucidate these properties.

Acknowledgements: I want to thank D. Rainer and S. Trugman for many useful discussions. I would also like to thank the Institute for Scientific Interchange, Torino, Italy, for providing a stimulating environment where some of this work was completed.

References

1. J. G. Bednorz and K. A. Müller, Z. Phys. B 64, 189 (1986).

2. M. K. Wu, J. R. Ashburn, C. J. Tong, P. H. Hor, R. L. Meng, L. Gao, Z. J. Huang, Y. Q. Wang, and C. W. Chu, Phys. Rev. Lett. 58, 908 (1987); R. J. Cava, B. Batlogg, R. B. van Dover, D. W. Murphy, S. Sunshine, T. Siegrist, J. P. Remeika, E. A. Rietman, S. Zahurak, and G. P. Espinosa, ibid. 58, 1676 (1987).

3. P. W. Anderson, Science 235, 1196 (1987); G. Baskaran, Z. Zhou, and P. W. Anderson, Solid State Communications 63, 973 (1987).

4. S. A. Kivelson, D. S. Rokhsar, and J. P. Sethna, Phys. Rev. B 35, 8865 (1987); S. A. Kivelson, Phys. Rev. B. 36, 7237 (1987).

5. D. Emin, preprint 1988; P. Prelovesk, T. M. Rice, and F. C. Zhang, J. Phys. C 20 L (1987).

6. T. Penny, M. W. Shafer, B. L. Olsen, and T. S. Plaskett, Adv. Cer. Mater. 2, 577 (1987); N. P. Ong, Z. Z. Wang, J. Clayhold, J. M. Tarascon, L. H. Green, and W. R. McKinnon, in "Novel Superconductivity", Proceedings of the International Workshop on Novel Mechanisms of Superconductivity, Berkley, 1987, edited by S. A. Wolf and V. Z. Kresin (Plenum, NY 1987); S. W. Cheong, S. E. Brown, J. R. Cooper, Z. Fisk, R. S. Kwok, D. E. Peterson, J. D. Thompson, G. L. Wells, E. Zirngieble, and G. Gruner, Phys. Rev. B. 36, 3913 (1987).

7. S. W. Tozer, A. W. Kleinsasser, T. Penny, F. Holtzberg, Phys. Rev. Lett. 59, 1768 (1987).

8. J. M. Luttinger, Phys. Rev. 119, 1153 (1960); J. M. Luttinger and J. C. Ward, ibid. 118, 1417 (1960).

9. J. M. Tranquada, A. H. Moudden, A. I. Goldman, P. Zolliker, D. E. Cox, G. Shirane, S. K. Sinha, D. Vaknin, D. C. Johnston, M. S. Alverez, and A. J. Jacobson, Phys. Rev. B 38, 2477 (1988).

10. R. S. List, A. J. Arko, Z. Fisk, S. W. Cheong, J. D. Thompson, J. A. O'Rourke, C. G. Olson, A-B Wang, Tun-Wen Pi, J. E. Schirber, and N. D. Shinn, preprint 1988 A. J. Arko, et al., Jour. Mag. and Mag. Mat. 75, L1, (1988).

11. A. J. Arko and R. S. List private communications.

12. A. Bansil, R. Pankaluoto, R. S. Rao, P. E. Mijnarends, W. Dlugosz, R. Prasad, and L. C. Smerdskjaer, Phys. Rev. Lett. 61, 2480 (1988).

13. V. J. Emery, Phys. Rev. Lett. 58, 2794 (1987).

14. S. Massida, J. Yu, A. J. Freeman, and D. D. Koelling, Phys. Lett. A122, 198 (1987).

15. M. Takigawa, P. C. Hammel, R. H. Heffner, Z. Fisk, J. L. Smith, and R. Schwarz, Phys. Rev. B 39, 300 (1989); M. Takigawa, P. C. Hamil, R. H. Heffner, and Z. Fisk, Phys. Rev. B (Rapid Communications), to be published.

16. H. Yasuoka, T. Shimizu, T. Imai, S. Sasaki, Y. Ueda, and K. Kosuge, International Conference on Nuclear Methods in Magnetism, (Munich) 1988.

17. Ju H. Kim, K. Levin, and A. Auerbach, preprint 1988.

18. Cheong et al. ref. 7 and Takigawa, et al. Phys, Rev. B 39, 300 (1989), ref. 15.

19. D. R. Harshman, G. Aeppli, B. Batlogg, R. J. Cava, E. J. Ansaldo, J. H.Brewer, W. Hardy, S. R. Kreitzmann, G. M. Luke, D. R. Noakes, and M. Senba, Phys. Rev. B. 36, 2386 (1987); A. T. Fiory, A. F. Hebard, P. M. Mankiewich, and R. E. Howard, Phys. Rev. Lett. 61, 1419 (1988).

20. G. A. Thomas, J. Orenstein, D. H. Rapkine, M. Capizzi, A. J. Millis, R. N. Bhatt, L. F. Schneemeyer, and J. V. Waszczak, Phys. Rev. Lett. 61, 1313 (1988).

21. M. Kaveh and N. Wiser, Adv. in Phys. 33, 257, (1984); and references therein.

22. W. E. Lawrence and J. W. Wilkins, Phys. Rev. B 7, 2317 (1973).

23. Y. Iye, T. Tamegai, H. Takeya, and H. Takei, JJAP, Superconducting Materials, 1, 46 (1988).

24. P. B. Allen, W. E. Pickett, and H. Krakauer, Phys. Rev. B 36, 3926 (1987); ibid., 37, 7482 (1988).

25. K. S. Bedell and D. Pines, Phys. Rev. B, 37, 3730 (1988).

INDEX